中外物理学精品书系

本书出版得到“国家出版基金”资助

中 外 物 理 学 精 品 书 系

高 瞻 系 列 · 1 8

Math Physics Foundation of Advanced Remote Sensing Digital Image Processing

高级遥感数字图像处理数学物理基础

晏 磊 赵红颖 林 沂 孙岩标 著

图书在版编目 (CIP) 数据

高级遥感数字图像处理数学物理基础 = Math Physics Foundation of Advanced Remote Sensing Digital Image Processing：英文 / 晏磊等著. — 北京：北京大学出版社，2024. 1
（中外物理学精品书系）
ISBN 978-7-301-34732-4

Ⅰ. ①高… Ⅱ. ①晏… Ⅲ. ①遥感图像－数字图像处理－英文 Ⅳ. ① TP751.1

中国国家版本馆 CIP 数据核字 (2024) 第 004854 号

书　　名	Math Physics Foundation of Advanced Remote Sensing Digital Image Processing（高级遥感数字图像处理数学物理基础）
著作责任者	晏磊　等著
责 任 编 辑	王剑飞
标 准 书 号	ISBN 978-7-301-34732-4
出 版 发 行	北京大学出版社
地　　址	北京市海淀区成府路 205 号　100871
网　　址	http://www.pup.cn
电 子 邮 箱	zpup@pup.cn
新 浪 微 博	@ 北京大学出版社
电　　话	邮购部 010-62752015　发行部 010-62750672　编辑部 010-62754271
印 刷 者	北京中科印刷有限公司
经 销 者	新华书店
	787 毫米 ×960 毫米　16 开本　32. 5 印张　778 千字
	2024 年 1 月第 1 版　2024 年 1 月第 1 次印刷
定　　价	168. 00 元

“中外物理学精品书系”
（三期）
编　委　会

序　言

物理学是研究物质、能量以及它们之间相互作用的科学。她不仅是化学、生命、材料、信息、能源和环境等相关学科的基础，同时还与许多新兴学科和交叉学科的前沿紧密相关。在科技发展日新月异和国际竞争日趋激烈的今天，物理学不再囿于基础科学和技术应用研究的范畴，而是在国家发展与人类进步的历史进程中发挥着越来越关键的作用。

我们欣喜地看到，随着中国政治、经济、科技、教育等各项事业的蓬勃发展，我国物理学取得了跨越式的进步，成长出一批具有国际影响力的学者，做出了很多为世界所瞩目的研究成果。今日的中国物理，正在经历一个历史上少有的黄金时代。

为积极推动我国物理学研究、加快相关学科的建设与发展，特别是集中展现近年来中国物理学者的研究水平和成果，在知识传承、学术交流、人才培养等方面发挥积极作用，北京大学出版社在国家出版基金的支持下于2009年推出了“中外物理学精品书系”项目。书系编委会集结了数十位来自全国顶尖高校及科研院所的知名学者。他们都是目前各领域十分活跃的知名专家，从而确保了整套丛书的权威性和前瞻性。

这套书系内容丰富、涵盖面广、可读性强，其中既有对我国物理学发展的梳理和总结，也有对国际物理学前沿的全面展示。可以说，“中外物理学精品书系”力图完整呈现近现代世界和中国物理科学发展的全貌，是一套目前国内为数不多的兼具学术价值和阅读乐趣的经典物理丛书。

“中外物理学精品书系”的另一个突出特点是，在把西方物理的精华要义“请进来”的同时，也将我国近现代物理的优秀成果“送出去”。这套丛书首次成规模地将中国物理学者的优秀论著以英文版的形式直接推向国际相关研究的主流领域,使世界对中国物理学的过去和现状有更多、更深入的了解,不仅充分展示出中国物理学研究和积累的“硬实力”,也向世界主动传播我国科技文化领域不断创新发展的“软实力”,对全面提升中国科学教育领域的国际形象起到一定的促进作用。

习近平总书记2020年在科学家座谈会上的讲话强调:“希望广大科学家和科技工作者肩负起历史责任,坚持面向世界科技前沿、面向经济主战场、面向国家重大需求、面向人民生命健康,不断向科学技术广度和深度进军。”中国未来的发展在于创新,而基础研究正是一切创新的根本和源泉。我相信“中外物理学精品书系”会持续努力,不仅可以使所有热爱和研究物理学的人们从书中获取思想的启迪、智力的挑战和阅读的乐趣,也将进一步推动其他相关基础科学更好更快地发展,为我国的科技创新和社会进步做出应有的贡献。

“中外物理学精品书系”编委会主任

中国科学院院士,北京大学教授

王恩哥

2022年7月于燕园

内 容 简 介

本书是北京大学研究生必修课“高级遥感数字图像处理”教学和研究生专题研讨相结合历经十八年的结晶，力图为遥感数字图像处理提供一部尽可能详细的数学物理手册。

本书包括三个部分：第一部分系统与整体处理基础，是遥感数字图像处理的出发点，由第一至四章构成，包括遥感数字图像处理的系统概述、系统支撑条件，以及遥感数字图像整体处理分析的数学基础和物理学基础。第二部分像元处理理论与方法，是遥感数字图像处理的核心和细节所在，由第五至十章构成，包括遥感数字图像像元处理理论 I——时空域卷积线性系统、理论 II——时频域卷-乘傅里叶变换、理论 III——频域滤波、理论 IV——时域采样，以及遥感数字图像变换基础 I——时空等效正交基、变换基础 II——时频组合正交基。第三部分技术与应用，是遥感数字图像处理的目的和落脚点，由第十一至十六章构成，包括遥感数字图像处理技术 I——复原降噪声、技术 II——压缩减容量、技术 III——模式识别(图像分割)、技术 III——模式识别(特征提取及分类)、技术 IV——彩色变换与三维重建以及应用举例。

本书可用于空间信息、遥感等地球观测领域研究人员学习、教学，是理解其他遥感图像处理书籍中数学物理本质和相互关系的重要参考书。

Foreword 1

The need to craft a comprehensive remote sensing book that introduces physics-based mathematical modeling arises from the demand to bridge the gap between theoretical foundations and practical applications in this dynamic field. Remote sensing, with its ever-evolving technologies, plays a pivotal role in understanding and monitoring the Earth's processes. However, the intricate interplay of mathematical and physical principles underlying remote sensing techniques often poses a challenge for students and professionals alike. However, currently available books on remote sensing often focus more on teaching technical methods and algorithms, with less emphasis on analyzing and discussing the mathematical and physical foundations inherent in remote sensing image processing. This has led readers to not fully grasp the core of image processing technology.

The authors of this book have been engaged in optical physics modeling and remote sensing image processing technology for many years. Through continuous accumulation and reflection during 18 years of graduate teaching, the author has summarized the mathematical methods behind image processing technologies, resulting in the creation of this book. This forthcoming book aims to address this challenge by providing a thorough exploration of the mathematical physics concepts underpinning remote sensing methodologies. By elucidating these principles in a clear and accessible manner, the book seeks to empower students, researchers, and practitioners with the necessary knowledge to navigate the complexities of models used in remote sensing applications. The integration of mathematical modeling of physics into the educational framework will not only enhance the understanding of the subject, but also foster innovation and advancements in the field.

The significance of this book extends beyond individual learning pursuits. Its emergence is poised to make a valuable contribution to the international remote sensing community by cultivating a pool of well-equipped professionals. Through a holistic approach that combines theoretical understanding with practical insights, the book aims to serve as a cornerstone for talent development in the global remote sensing domain.

Alper Yilmaz

Alper Yilmaz, Ph.D.
Professor, The Ohio State University, USA
Editor-in-Chief of the Photogrammetry Engineering & Remote Sensing Journal

Foreword 2

Remote sensing (RS) is a technology that utilizes electromagnetic waves to observe objects of interest in a non-contact manner, enabling quantitative analysis of their presence and changes and providing insight into their nature. It represents a complex information transformation process, where signals obtained from RS sensors or detectors undergo conversion from data to information and eventually to knowledge, enriching our understanding of the world. Generally, RS primarily employs the images of the Earth's surface to showcase the environment. Through image processing techniques, it extracts, expresses, and displays the form and pattern of Earth objects, exploring and analyzing their features and studying their characteristics of change. Consequently, the image processing of remotely sensed data has become an indispensable means for acquiring knowledge, a crucial avenue for understanding the world, and an essential course for professionals in RS technology. This field has evolved along with the development of RS technology. At present, there are a variety of mature techniques and software in this field. However, at present, the field of RS image processing pays more attention to technology and procedures and lacks in-depth analysis of mathematical and physical foundations, resulting in a lack of systematic theoretical support for application-based RS digital image processing technology, which affects its deepening and improvement process.

This book systematically refines the basic technology of RS digital image processing, explains its physical nature through mathematical language, and improves the theory of RS image processing technology. The book validates theoretical methods through a diverse range of practical examples to ensure practical application support. With its comprehensive and in-depth content presented in an accessible manner, the book represents an innovative attempt to analyze RS digital image processing methods with mathematical fundamentals and physical essence. This book mainly discusses the dominant optical image processing in RS, which involves less microwave RS processing and needs to be improved. Nevertheless, its comprehensive and in-depth discussion of the mathematical foundations of optical RS image processing is still commendable in this field. This book is a valuable reference for researchers and educators engaged in RS digital image processing. It is also suitable as a desk manual for professionals in spatial information technology and applications.

I am delighted to see the publication of this book and sincerely hope that it will contribute significantly to the enhancement of RS digital image processing and the cultivation of talent in this field in China.

Above is the preface to the book.

Qingxi Tong
Academician, Chinese Academy of Sciences (CAS)
Professor, Peking University, China
Professor, Aerospace Information Research Institute, CAS, China

Foreword 3

The careful crafting of a literature resource that introduces mathematical and physics-based models of remote sensing, driven by a need to bridge the gap between theoretical foundations and practical applications, has been successfully delivered in this book by Prof Lei Yan and colleagues. Remote sensing, with its ever-evolving technologies and relentless computational dynamism, plays a pivotal role in understanding and monitoring the Earth's processes. However, the intricate interplay of mathematical and physical principles underlying remote sensing techniques often poses a challenge for learners and professionals alike. In many cases, the current selection of books on remote sensing image processing focus somewhat more on technical methods and algorithms, and somewhat less on analyzing and discussing the mathematical and physical aspects. This leads to readers not fully grasping the core of image processing technology.

Prof. Lei Yan, lead author, has been engaged in optical physics modeling and remote sensing image processing technology for many years. Together with his colleagues, Prof. Yan summarizes the mathematical methods behind various image processing technologies. They provide a thorough explanation of the mathematical and physical concepts that underpin remote sensing methodologies. By elucidating these principles in a clear and accessible manner, the book's content seeks to empower students, researchers, and practitioners with the necessary knowledge to navigate the complexities of remote sensing applications and assist in brokering their solutions. The formulas presented in the book are supported and elaborated with worked examples and challenges in remote sensing research. The integration of mathematical physics into the educational framework will not only enhance the understanding of the subject but also foster innovation and advancements in the field.

The significance of this book extends beyond individual learning pursuits. Its importance is to inspire and contribute to the training of the next generation of learned- and knowledgeable researchers and professionals. The book should serve as a cornerstone for talent development in the global remote sensing domain.

Dr. Kevin Tansey
Professor, University of Leicester, UK
Editor-in-Chief for International Journal of Remote Sensing

Introduction

This book focuses on remote sensing digital image processing in the field of spatial information and seeks to provide a detailed mathematical physics manual for remote sensing digital image processing. This book consists of three parts. Part I, from the first to fourth chapter, covers the system and the overall processing foundation. Part II, from the fifth to the tenth chapter, discusses the pixel processing theory and methods and covers the core and details of remote sensing mathematical image processing. Part III, from the eleventh to sixteenth chapter, presents a discussion on the technology and its application, explaining the purpose and application of remote sensing digital image processing.

Part I discusses the system and the overall processing foundation. This part introduces the amplitude processing of a remote sensing image. While improving the overall level of the remote sensing image, it does not change the interrelationship of remote sensing images, which is discussed in Part II of the book, which focuses on processing. The main contents are as follows: Chap. 1 is a systematic overview of remote sensing digital image processing, including the overall architecture and a detailed explanation and application extensions of remote sensing digital image processing to establish the global concept of remote sensing digital image processing systems. Chapter 2 presents the system support conditions for remote sensing digital image processing analysis. The first step in the analysis of remote sensing digital image processing is introduced, which is understanding the influence of system support conditions on image performance, including image input and acquisition and output display and processing software design. Chapter 3 discusses the mathematical basis of the overall analysis and the analysis of remote sensing digital images. The overall image performance analysis and improvement methods based on histograms, that is, image preprocessing, and convolution theory are used to eliminate the irrational appearance of the image as much as possible, that is, to focus on the surface regardless of its origin. Chapter 4 presents the physical basis of the overall analysis and the analysis of remote sensing digital images. The main factors that affect the overall quality of remote sensing images derive from the imaging process, which cannot be artificially changed in pretreatment and must be changed by the "root" analysis of the physical imaging process. The "tackling" of the issue of the overall treatment of the remote sensing image by image pixel processing is addressed in Part II of the book.

Part II of the book discusses pixel processing theory and methods. The crucial feature of digital images is that each discrete pixel can be processed and analyzed. Based on the image preprocessing described in Part I, Part II presents the method for analyzing and processing each pixel, i.e., the core of remote sensing digital image processing, which embodies the essence of remote sensing digital image pixel processing content. The content of Part II is as follows: Chap. 5, Remote Sensing Digital Image Pixel Processing Theory I: Linear System with Space–Time Domain Convolution, clarifies the remote sensing digital image linear system inertia delay caused by the convolution effect and elaborates the convolution theory. Chapter 6, Remote Sensing Digital Image Pixel Processing Theory II: Time–Frequency Fourier Transform from Convolution to Multiplication, introduces the time domain and its reciprocal frequency domain mutual duality and discusses the convolution–simple product transformation relationship that achieves time–frequency conversion using the Fourier transform theory. Chapter 7, Remote Sensing Digital Image Pixel Processing Theory III: Frequency Domain Filtering, presents the frequency domain filtering theory in the extraction and retention method for different frequency scale information. Chapter 8, Remote Sensing Digital Image Pixel Processing Theory IV: Time Domain Sampling, discusses the nature of the continuous spatial information captured by sampling that is input into the computer processing required for creating discrete information, the establishment of the Shannon sampling theorem, and related theory. Chapter 9, Basics of Remote Sensing Digital Image Pixel Transformation I: Space–Time Equivalent Orthogonal Basis, presents the basis function establishment, base vectors, base image linear representation theory, a description of the Fourier transform, remote sensing digital image processing transmission and storage, and the mathematical essence of compression transformation. In Chap. 10, Basics of Remote Sensing Digital Image Pixel Transformation II: Time–Frequency Orthogonal Basis, a discussion is presented on how the time–frequency combined orthogonal basis is used to extract the time–frequency characteristics of the image through the local window compression and expansion of the same image, that is, the wavelet transform, which is the time–frequency information unit.

Part III of the book contains sections on technology and application. In the technical process of remote sensing digital image processing, the principle, methods, and mathematical physics of the main steps are introduced. The detailed contents are as follows: Chap. 11, Remote Sensing Digital Image Processing: Noise Reduction and Image Reconstruction, discusses the mathematical physics in removing the convolution effect and equivalent noise in digital image processing or the input. Chapter 12, Digital Image Compression, discusses compression reduction capacity. To ensure the restoration of an image under the premise of removing redundant data and highlighting useful information, there is a need to compress image processing. There are two types of image compression: One is based on the information entropy under the redundant limit theory of lossless compression, and the second is based on the loss of value under the theory of loss; that is, under the premise of the minimum allowed engineering error, the amount of stored data retains as much useful information as possible. Chapter 13, Pattern Recognition (Image Segmentation), elaborates that image information recognition is the most important purpose of remote sensing digital image processing. To extract useful information in an image, it is necessary to conduct pattern recognition of the image. Pattern recognition is generally divided into three steps: feature "search" "extraction" "classification", that is, image segmentation (the means of remote sensing application, elaborated in this chapter); and feature extraction and classification judgment (the purpose of remote sensing application; elaborated in Chap.14). Chapter 14, Pattern Recognition (Feature Extraction and Classification), discusses the basic goal of remote sensing applications, that is, the identification of target objects. Chapter 15, Applications of Remote Sensing Digital Image Processing: Color Transform and 3D Reconstruction, discusses how the information can be properly expressed so that the image is more suitable for the human eye or how to make the image more closely reflect the real world after proper processing of the remote sensing image and extracting useful information. This requires proper color transformation and 3D reconstruction of the image processing results. Chapter 16 provides an application example of remote sensing digital image processing. It introduces four typical applications of remote sensing digital image processing to enhance and deepen the understanding of the essential methods of remote sensing digital image processing and to explain the overall image processing methods, image pixel processing theory, and the effectiveness of important technical means.

Contents

Part I
The System and the Overall Processing Foundation

Chapter 1
Overview of Remote Sensing Digital Image Processing

Chapter Guidance This chapter provides a general description of remote sensing digital image processing systems, including the overall framework, details and application extension of remote sensing digital image processing. This chapter not only establishes for readers the global concept of remote sensing digital image processing systems but also lays a foundation for the discussion and development of other chapters in this book.

Remote sensing is an important means of collecting earth data and its changing information and has been widely used worldwide. However, the cultivation and accumulation of remote sensing disciplines have gone through several hundred years of history. As a technical discipline, it is an emerging comprehensive and interdisciplinary subject that has gradually developed on the basis of modern physics, space science, computer technology, mathematical methods and earth theory science. At present, remote sensing platforms, sensors, remote sensing information processing, and remote sensing applications have developed by leaps and bounds, especially in the remote sensing information processing of digital, visualization, intelligence and networking data, leading to many improvements and innovations. However, this development has not met the needs of the majority of users. Increasingly rich remote sensing information has not been fully exploited. The processing of remote sensing information, especially the processing of remote sensing digital images, has become one of the core problems in remote sensing technology research.

The details of this chapter are as follows: A discussion on remote sensing systems (Sect. 1.1), which mainly include remote sensing data acquisition systems and remote sensing digital image processing systems, is presented first, illustrating the root and indispensable role of remote sensing digital images in remote sensing systems. Next, remote sensing digital images (Sect. 1.2), including the concept, characteristics, resolution and format of remote sensing digital images, are elaborated. A discussion on remote sensing digital image processing (Sect. 1.3) introduces remote sensing digital image processing methods and remote sensing image inversion. The framework of this book is then presented (Sect. 1.4), introducing the structure and content of the entire tutorial and framework.

1.1 Remote Sensing Systems

A remote sensing system is the prerequisite for and the basis of remote sensing digital image generation. This section includes two parts: remote sensing information acquisition systems and remote sensing digital image processing systems. In addition, this section systematically introduces the whole process of remote sensing data from acquisition to processing, explains the root and important position of remote sensing digital images in remote sensing systems, and introduces the principle, types, and characteristics of remote sensing information and the digital processing of remote sensing.

1.1.1 Remote Sensing Information Acquisition Systems

Remote sensing information acquisition systems acquire data through a variety of remote sensing technologies.

1. Remote Sensing Principle and Types of Remote Sensing Electromagnetic Spectra

(1) Remote sensing principle

Remote sensing is based on the principle of different objects having different spectrum responses. Remote sensors are used to detect the electromagnetic wave reflection of surface objects from the air and thus extract the information on these objects, ultimately completing the remote identification of the objects.

(2) Spectral characteristics of ground objects

Remote sensing images are divided into smaller units, which are called image elements or pixels. The luminance value of each pixel in a remote sensing image represents the average radiation value of all objects in a pixel, which varies with the composition, texture, state, surface feature and electromagnetic band use of the object. Characteristics that vary with the above factors are called the spectral characteristics of the objects [1]. Note that the brightness of the image is a quantified radiation value and is a relative measurement. The spectral information of ground objects is based on the principle that different objects have different brightness in the same spectral band and that the same objects also have different brightness in different wavelengths. Therefore, different electromagnetic bands can also reflect this difference and the different characteristics of ground objects [2].

(3) Types of remote sensing electromagnetic spectra

At present, the electromagnetic bands mainly applied in remote sensing are as follows [3, 4].

① Visible remote sensing

This is the most widely used remote sensing method. The remote sensing of visible light with a wavelength of 0.4–0.7 μm is typically used as a photosensitive element (image remote sensing) or as a photodetector sensing element. Visible light photography remote sensing has a high ground resolution but can only be used in the daytime.

② Infrared remote sensing

This category is divided into three types. One is near-infrared or photographic infrared remote sensing, which operates in a wavelength range of 0.7–1.5 μm using a directly sensed photosensitive film; the second type is mid-infrared remote sensing, which uses a wavelength range of 1.5–5.5 μm; and the third type is far-infrared remote sensing, which uses a wavelength range of 5.5–1000 μm. Mid- and far-infrared remote sensing are typically used for remote sensing radiation, with operating capacity in both the day and night. Some commonly used infrared remote sensors include optical mechanical scanners.

③ Multihyper spectral remote sensing

Simultaneously and remotely observing the same ground objects or regions with several or multiple different spectral bands enables various information corresponding to each spectrum band to be obtained. The combination of remote sensing information of different spectral bands can obtain more information about objects, which is beneficial for interpretation and recognition. The commonly used multispectral remote sensing devices include multispectral cameras and multispectral scanners. The term multispectral mainly refers to a spectrum width of a few to several hundred nanometers; hyperspectral primarily refers to a spectrum width of a few to dozens of nanometers; and ultrahigh spectrum mainly refers to a spectrum width of 1 nm or finer scales.

④ Ultraviolet remote sensing

The main remote sensing method for ultraviolet light with wavelengths of 0.3–0.4 μm is ultraviolet photography.

⑤ Microwave remote sensing

This type of sensing refers to remote sensing using microwaves at wavelengths of 1–1000 μm. Microwave remote sensing can work in both day and night conditions but has a low spatial resolution. Radar is a typical active microwave system, and synthetic aperture radar is often used as a microwave remote sensor.

The development trend of modern remote sensing technology has been gradually expanding from the UV spectrum to X-rays and γ-rays and has been converting from a single electromagnetic band to a synthesis of sound waves, gravitational waves, seismic waves and several other waves. The differences between visible and near-infrared images, thermal infrared images, and radar images in the context of imaging principles and image characteristics are summarized in Table 1.1.

2. Characteristics of Remote Sensing Information

The remote sensing information not only captures the spectral information of ground objects and different electromagnetic bands but also captures, through instantaneous imaging, the spatial information and morphological features of the ground objects,

Table 1.1 Summary of the characteristics of visible light, near-infrared, thermal infrared and radar images

Imaging method	Visible and near-infrared images	Thermal infrared images	Radar images
Band	0.4–0.5 μm	8–10 μm	1 mm–1 m
Acquisition condition	Day	Day, night	Day, night
Acquisition method	Passive	Passive	Active
Fog penetration capability	No	Yes	Yes
Sensor types	Photography, scanning	Scanning	Radar, nonimaging
Projection method	Central projection	Central projection	Oblique projection
Physical meaning	Solar radiation reflected by objects	Radiation emitted by ground objects	Intensity of backscatter
Spatial resolution	Instantaneous field width $S = \frac{H}{f}D$	Instantaneous field width $S = \frac{H}{f}D$	Range direction: $R_{\mathrm{r}} = \frac{\tau C}{2} \sec \beta$ Cross-range direction: $R_{\mathrm{a}} = d/2$
Radiometric resolution	The signal is greater than 2–6 times. $P_{\mathrm{EN}} = \frac{P}{S/N} = \frac{N}{R}$	The ground temperature is greater than 2–6 times. $\Delta T_{\mathrm{EN}} = \sqrt[4]{\frac{P_{EN}}{\varepsilon\sigma}}$	The signal is greater than 2–6 times. $P_{\mathrm{EN}} = \frac{P}{S/N} = \frac{N}{R}$
Spectral resolution	The band number of the imaging spectrometer is 386, and the interval of each band is less than 5 nm		Single band; no spectral resolution
Time resolution	Time resolution refers to the time interval needed to repeatedly acquire images in the same area		

namely, the temporal and spatial information. Therefore, remote sensing images at different phases have differences in both spectral and spatial information.

The differences in spatial information are mainly reflected in the spatial frequency information, edge and linear information, structure or texture, and geometric information. Spatial information is reflected in the spatial variation of the image brightness values. The spatial position, length, area, and distance of points, polylines, polygons or regions in the image are considered spatial information. Texture is an important type of spatial information in the remote sensing of images and can assist in the recognition of images and the extraction of object features. In addition, spatial structure information is also very useful in remote sensing. In remote sensing digital image processing, the enhancement and extraction of spatial structure information is an important part of the image information enhancement process [2].

However, remote sensing image information is the record of the present situation from instantaneous imaging. Many objects change with time. One reason for this is the occurrence, development and evolution of the natural change process. A second reason is rhythm; that is, the development of events demonstrates certain periodic repetition rules in a time series, and the spectral and spatial information also changes with time. Therefore, in remote sensing image digital image processing, transient information cannot cover an entire development process. The time information of the remote sensing image is related to the temporal resolution of the remote sensor and is also related to the entire imaging seasons.

3. Remote Sensing Information Collection and Remote Sensing Satellites

(1) Remote sensing information collection

Remote sensing information collection refers to data collection through various remote sensing techniques. We typically use aircraft or equipment in man-made resource satellites to explore, measure or reconnoiter various targets and their changes on Earth (including the atmosphere) over a long period of time and then identify, separate and collect the identified geographic entities and their attributes to obtain source data that can be processed.

(2) Platform for remote sensing information collection

Remote sensing information collection platforms: air remote sensing platforms and satellite remote sensing platforms.

Air remote sensing platforms generally refer to platforms with heights below 20 km, mainly including aircraft and balloons. Remote sensing platforms have the advantages of low altitude and flexibility and are not limited by ground conditions, short investigation periods and data recovery convenience. Therefore, they are widely used, and the remote sensing data are obtained primarily from air remote sensing platforms.

A satellite remote sensing platform is a platform using satellite technology. Usually, remote sensing satellites can orbit for several years. The satellite orbit can be determined as needed. The remote sensing satellite can cover the entire earth or any area specified for a specified period of time, and it can continuously map a designated area on the surface of the earth as it travels along the Earth's synchronous orbit.

(3) Remote sensing satellites

Remote sensing satellites are used as satellites for outer space remote sensing platforms. At present, the main parameters of remote sensing satellites commonly used at home and abroad are listed in Table 1.2.

(4) The orbit of a remote sensing satellite

If we assume that the earth is a homogeneous sphere surrounded by satellites, its gravitational field is the center force field, and the mass center is also the gravity center. Then, to make the artificial earth satellite (also called a satellite) move in a circle in the center force field, the orbit must be a complex curve that has a subtle difference from the Kepler elliptical orbit. Kepler's elliptical orbit is often used to describe the general motion of a satellite. In the case of a man-made satellite, its orbit is divided into low and high orbits by height and into forward and backward orbits by the direction of the Earth's rotation. There are specialized orbits, including the equatorial orbit, geosynchronous orbit, geostationary orbit, polar orbit and solar geosynchronous orbit.

The shape and size of the satellite orbit are determined by the long axis and short axis, while the intersection angle, near position amplitude angle and orbital inclination angle determine the orientation of the orbit in space. These five parameters are called satellite orbit elements (root numbers). Sometimes, the perigee moment is added, which is called the sixth element. With these six elements, the satellite's position in space at any time can be known. Figure 1.1 shows a schematic diagram of a satellite orbit.

1.1.2 Remote Sensing Digital Image Processing Systems

A remote sensing digital image processing system refers to the analysis, processing and disposal of remote sensing digital images through computers to meet the visual, psychological and other requirements of the technical system. Image processing is the application of signal processing in the image domain, is a subclass of signal processing, and is closely related to computer science, artificial intelligence and other fields [5].

The three basic components of the remote sensing digital image processing system are the image digitizer, the computer and the image display device used to process the image. In their natural form, objects cannot be directly processed by a computer. Because the computer can only handle numbers rather than pictures, an image must be in digital form before being processed with a computer. Figure 1.2 shows a remote sensing image that is represented in a digital matrix. The process of converting ground object information into digital values is called digitization. At each pixel

Table 1.2 List of major satellite parameters at home and abroad

Satellite name (Country/ Region)	Weight/kg	Lifetime/a	Height/km	Dip angle/°	Revisit period/d	Payload performance		
						Camera	Width/km	Resolution/m
SPOT-4,5 (France)	2755	5	822	98.7	26	CCD	60	2.5, 5, 10
GEOEYE-1 (USA)								0.5
EROS 1A (Israel)	250		480		4	CCD	12	1.8
EROS 2B (Israel)	350	2	600	97.3	4	CCD	16	0.8
IRS-P 5 (India)	1560	3	618	97.87	5	CCD	30	2.5
QUICKBIRD (USA)	825	7	450	98	2.4	CCD	16.5	0.61, 2.44
ORBVIEW-3 (USA)	356	5	470	98.3	2.5	CCD	8	1. 4
IKONOS-1 (USA)	817	5	674	98.2	3	CCD	11	1. 4
ROCSAT (Taiwan)	620	5	891	98.99		CCD	60	2. 15
ALOS (Japan)	4000	5	691	98.16	46	CCD	35, 70, 350	2.5, 10, 7–100
NEMO (USA)	574	5	605	97.81	7	hyperspectral	30	5, 30
RESOURCE-DK (Russia)						CCD		1,3
WORLDVIEW-1/2 (USA)	2500	7	770	98.0	1.7, 1.1	CCD	17.6,16.4	0.45–0.51
PLEIADES (France)	1000	5	694	98.64	26	CCD	20	0.7
RADARSAT-2 (Canada)	1650	7	798	98.6	24	SAR	500	3
TERRASAR-X (Germany)	1230	5	514	97.44	11	SAR	10, 30, 100	1, 3, 16
COMOS-SKYMED (Italy)	600	5	619	97.86	16	SAR	10, 30, 100	1–100

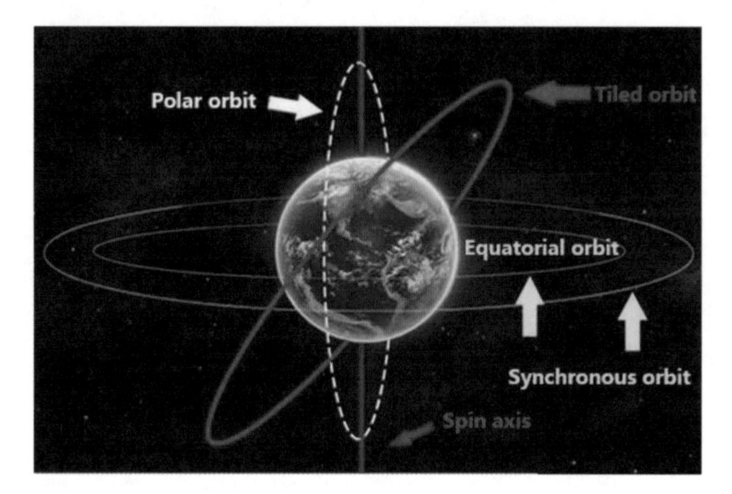

Fig. 1.1 Satellite orbits

position, the brightness of the image is sampled and quantized to obtain a value at the corresponding point in the image that represents its brightness level. After all the pixels have completed the above transformation, the image is represented as a digital matrix. Each pixel has two properties: position and grayscale. This matrix is the object that the computer can process. In the process of remote sensing digital images, the pixels in the image can be modified as required. The result of the processing is displayed by a process that is inverse to the digitized phase; that is, the gray value of each pixel is used to determine the brightness of the corresponding point on the display. This result is transformed into an image that people can interpret intuitively.

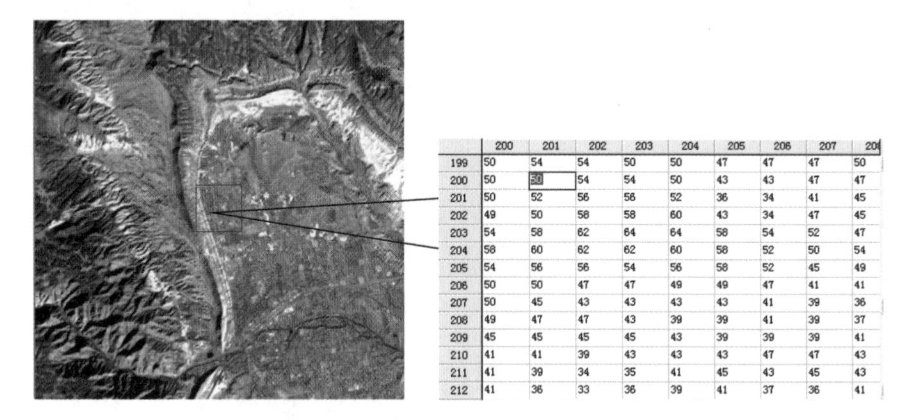

	200	201	202	203	204	205	206	207	20
199	50	54	54	50	50	47	47	47	50
200	50	50	54	54	50	43	43	47	47
201	50	52	56	56	52	36	34	41	45
202	49	50	58	58	60	43	34	47	45
203	54	58	62	64	64	58	54	52	47
204	58	60	62	62	60	58	52	50	54
205	54	56	56	54	56	58	52	45	49
206	50	50	47	47	49	49	47	41	41
207	50	45	43	43	43	43	41	39	36
208	49	47	47	43	39	39	41	39	37
209	45	45	45	45	43	39	39	39	41
210	41	41	39	43	43	43	47	47	43
211	41	39	34	35	41	45	43	45	43
212	41	36	33	36	39	41	37	36	41

Fig. 1.2 Luminance values of remote sensing images and their corresponding regions

1.2 Remote Sensing Digital Images

This section introduces the concept of remote sensing digital images and introduces the concepts, features, resolution and format of remote sensing digital images. With the increasing maturity of remote sensing image acquisition methods, remote sensing images have gradually become one of the important data sources of spatial geographic information. In particular, the widespread use of transmission-type remote sensing satellites provides a reliable guarantee for rapid and periodic access to ground-based observations.

1.2.1 Concept of Remote Sensing Digital Images

Remote sensing digital images are remote sensing images expressed in digital form, which are the product of information obtained by various sensors and are the information carrier of remote sensing targets. Similar to the pictures we take in our lives, remote sensing images can also "extract" useful information. A digital recording mainly refers to the scanning tape, disk, CD or other electronic record and uses a photodiode as a detection element to detect the reflection or emission energy of an object. It converts the light radiation energy difference to an analog voltage difference (analog electrical signal) through a photoelectric conversion process, and then through an analog-to-digital (A/D) conversion, the analog value is converted into a value (brightness value) and stored on digital tape, disk, CD-ROM or other media. The remote sensing image reflects the continuously changing physical field. Image acquisition imitates the visual principle, and the display and analysis of recorded images must take into account the characteristics of the visual system [1, 5].

A remote sensing digital image refers to a two-dimensional matrix composed of small areas. For monochrome grayscale images, the brightness of each pixel is represented by a numerical value, usually in the range of 0–255, which can be represented by a byte, where 0 is black, 255 is white and others indicate grayscale. A remote sensing digital image is a two-dimensional continuous optical function of equidistant rectangular grid sampling, followed by amplitude equalization of the two-dimensional data matrix. Sampling is the measurement of each pixel value, and quantization is the process of digitizing the value. Figure 1.2 shows digital images taken by a thematic mapper (TM), and the right figure shows the brightness of the corresponding box in the left figure.

As remote sensing digital images are essentially two-dimensional signals, the basic techniques in signal processing can be used in remote sensing digital image processing. However, because the remote sensing digital image is a very special two-dimensional signal reflecting the visual properties of the scene and represents a very sparse sampling of the two-dimensional continuous signal, that is, a meaningful description or characteristics obtained from a single sampling or a small number of samplings, we cannot copy one-dimensional signal processing methods and need specialized technology. In fact, remote sensing digital image processing is more dependent on specific application problems and is a collection of a series of special technologies that lack a consistent strict theoretical system.

1.2.2 Characteristics of Remote Sensing Digital Images

The application of remote sensing technology is very broad, and there are nearly 30 fields that can use remote sensing technology, such as land water resource surveys, land resource surveys, marine resources surveys, map surveys, archaeological surveys, environmental monitoring and planning management. Numerous applications depend on its unique sets of advantages.

(1) Large coverage, large amount of effective information, and a certain periodicity

Remote sensing is a macroscopic observation with a large coverage of remote sensing images; for example, the width of a LANDSAT image is 185 km $\times$ 185 km, and that of a SPOT image is 60 km $\times$ 60 km. Remote sensing satellites have a fixed revisit cycle, are periodic, and can form a time series, which helps in subsequent analysis and processing, the extraction of information and so on. Remote sensing digital images not only cover a large area but also include multiple bands that are different from ordinary color images; therefore, an image includes a large amount of data. In view of this feature, the improvement of algorithmic efficiency is important in remote sensing digital image processing.

(2) Good reproducibility and strong abstraction

Digital images are stored in a binary form and will not degrade or lose image quality through a series of transformation operations, such as image storage, transmission, or duplication. As long as the original image is accurately represented digitally, the digital image processing process always enables the reproduction of the image to be maintained [6, 7].

(3) High processing accuracy

According to the current technology, if the image's digital device capacity is strong enough, almost any analog image can be digitized into a two-dimensional array of any size. Modern scanners can quantify the grayscale of each pixel to 16 bits or higher, which means that the digital precision of the image can meet any application requirement.

(4) Wide application area

The images from different information sources can be converted into a digital code form and are composed of grayscale images represented by a two-dimensional array; thus, they can be processed by a computer; that is, the digital image processing method is applicable to any image as long as the corresponding image information acquisition measures are adopted for different image information sources.

(5) High flexibility

Digital image processing can complete not only linear operations but also nonlinear operations. In theory, all operations that can be expressed in mathematical or logical relationships can be implemented using digital image processing.

1.2.3 Resolution Characteristics of Remote Sensing Digital Images

Remote sensing digital images are the product of information obtained by various sensors and are the information carrier of remote sensing detection targets. Remote sensing digital images can obtain important information on the following three aspects: the size, shape and spatial distribution characteristics of target features; the shape and characteristics of the target features; and the dynamic characteristics of the target features. This in turn corresponds to the three aspects of a remote sensing image: geometric characteristics, physical characteristics and temporal characteristics. The performance parameters of these three aspects are the spatial resolution, spectral resolution, radiation resolution, and temporal resolution described below. With the continuous development of remote sensing technology, the bands available for remote sensing have expanded from the original visible and near-infrared bands to the thermal infrared bands and the microwave radar bands; the detectable targets and application areas have also been greatly expanded, but the sensors and working principles of the two are significantly different, which inevitably leads to inconsistencies in the two types of remote sensing images.

In visible and near-infrared remote sensing and thermal infrared remote sensing, the type of sensor used to detect radiant energy is the scan sensor. Visible and near-infrared remote sensing detects the reflection of solar radiation, reflecting the reflectivity of the objects, and can identify different objects according to the different characteristics of the reflection spectrum. Thermal infrared remote sensing detects the radiation energy radiated by the objects themselves, so the thermal infrared image reflects the radiation emission capability of the objects. In other words, the targets of the detection of visible and near-infrared remote sensing and thermal infrared remote sensing are ground radiation energy.

Radar, originally known as "radio detection and ranging", uses radio methods to find targets and determine their spatial locations. The principle is that the radar equipment transmitter transmits electromagnetic wave energy through space in the direction of the target using an antenna, and the objects in this direction reflect the electromagnetic wave; the radar antenna receives the reflected wave and sends it to the receiving device for processing, enabling the extraction of information, such as the distance from the target to the electromagnetic wave launch point, the distance change rate (radial velocity), the azimuth, and the height.

Although visible and near-infrared images and thermal infrared images have many similarities in the sensor acquisition phase, the radar image sensor and its working principle are very different from the former two, as well as the transmission signal and the received signal. Therefore, the image features and differences between the visible and near-infrared images, thermal infrared images and radar images are introduced in the following discussion.

1. Spatial Resolution

The spatial resolution of remote sensing images refers to the size of the ground range represented by the pixels, that is, the instantaneous field of view of the scanner or the smallest unit that can be distinguished by the ground object. It usually refers to the recognition of the critical geometric dimensions of the object, reflected as the scale of the remote mapping. For example, in Landsat's TM 1–5 and 7 bands, one pixel represents a ground range of 28.5 m × 28.5 m, or the spatial resolution is generally 30 m.

(1) Spatial resolution of visible and near-infrared images

For the visible and near-infrared images acquired by the photographic type sensor, the resolution of the imaging system can be evaluated using the Rayleigh criterion:

$$\theta = 1.22\lambda/D \tag{1.1}$$

where D is the aperture of the camera, λ is the wavelength, and θ is the observation angle.

The above formula shows that the larger the aperture is, the greater the resolution of the imaging system and the higher the image resolution.

For the visible and near-infrared images acquired by the scanning type sensor, it can be seen from the instantaneous field of view of the infrared scanner (Fig. 1.3) that the spatial resolution can be defined as:

$$S = \frac{H}{f}D \tag{1.2}$$

where H is the height of the satellite from the ground, D is the scale of the detector, and f is the focal length of the optical system. For the scanning-type sensor, the radiation energy of all objects within the instantaneous field of view is imaged in a pixel. Therefore, the spatial resolution of the scanning-type sensor is the width of the ground field observed in the instantaneous field of view of the sensor.

Although both thermal infrared images and visible near-infrared images can use scanning-type sensors, the geometric resolution of thermal infrared images is far smaller than that of visible and near-infrared images. Since the radiant energy is inversely proportional to the wavelength, the radiated intensity of the infrared band is much smaller than that of the visible and near-infrared bands in a certain field of view, and the radiated energy decreases as the wavelength increases. The principle

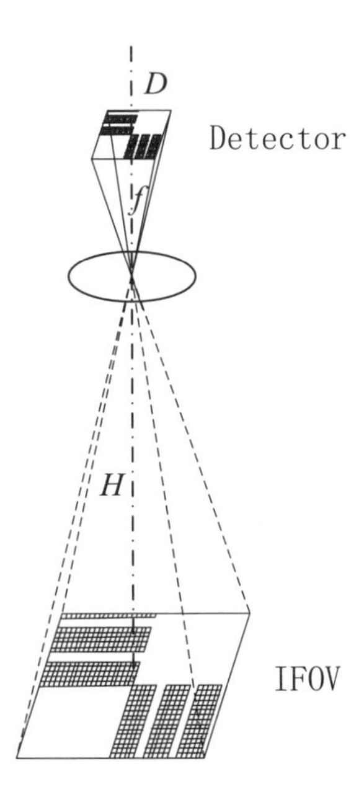

Fig. 1.3 Instantaneous field of view (IFOV) of an infrared scanner

of thermal infrared imaging is the same as that of the visible band, and the pixel brightness is an integration value of the radiation energy within the field of view. To obtain a higher radiation brightness, the geometric resolution must be sacrificed. For example, in TM sensors, at visible and near-infrared wavelengths, the resolution reaches up to 30 m, while in the thermal infrared band, it can only reach 60 m.

(2) The spatial resolution of the imaging radar

For radar imaging type sensors, the resolution of the cross-range and range direction to the image is different because it is different in the range and cross-range direction from the signal processing method.

The minimum distance at which the radar is able to distinguish two ground targets is called the distance resolution. The general radar distance resolution can be expressed as:

$$R_{\mathrm{r}} = \frac{\tau C}{2} \sec \beta \tag{1.3}$$

where τ is the pulse width, C is the electromagnetic wave propagation velocity, and β is the depression angle. In the range direction, different target objects need to reflect the different parts of the pulse and thus can be distinguished in the image, as shown in Fig. 1.4. The U part of the pulse is first reflected by point X, and then the V part is reflected by point Y. To be able to distinguish two points on the image, the two parts of the pulse reflected by the objects must arrive at the antenna at different times. It can be inferred that if the two targets are too close or the pulse is long, then the

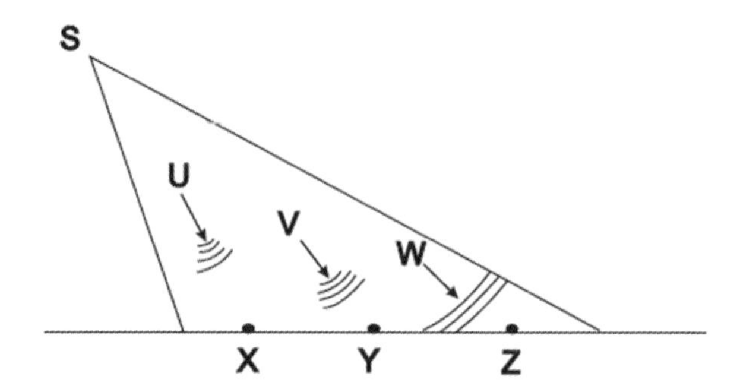

Fig. 1.4 Schematic diagram of pulse reflection

reflection pulse parts of the two targets may overlap and reach the antenna at the same time, which is difficult to distinguish in the image.

From the above analysis, we can see that the radar distance resolution is independent of the height and is correlated to the pulse width and depression angle. The wider the pulse bandwidth is, the lower the resolution of the distance; the greater the depression angle is, the higher the resolution of the distance. Therefore, unlike optical imaging, radar imaging usually adopts side-view imaging. Pulse compression technology is typically used in practical applications to improve the range resolution.

The cross-range resolution of the radar refers to the minimum distance between the two target objects that can be distinguished along the heading. To distinguish two targets along the heading, the two target objects must be in different beams. The cross-range resolution can be expressed as follows:

$$R_{\mathrm{a}} = \omega R \tag{1.4}$$

where ω is the lobe angle and R is the slope distance. The lobe angle is proportional to the wavelength λ and inversely proportional to the antenna aperture d; if the synthetic aperture radar synthetic aperture is L_{p} and equal to R_{a}, then the above equation can be expressed as

$$R_{\mathrm{a}} = \frac{\lambda}{d} R = \frac{\lambda}{\left(\frac{\lambda}{d}\right) R} R = d \tag{1.5}$$

Due to the two-way phase shift of the synthetic aperture antenna, its cross-range spatial resolution is doubled, that is, $d/2$.

2. Radiation Resolution

Even if the spatial resolution is high enough, whether the object can be distinguished also depends on the sensor's radiation resolution. The so-called radiation resolution refers to the ability of the sensor to distinguish the minimum difference between the two radiation intensities. In the remote sensing image, the radiation quantization value of each pixel is shown. It is reflected as the grayscale of the image, such as 64 grayscale, 128 grayscale, and 256 grayscale. In the visible, near-infrared band, it is described as the equivalent reflectance of the noise (radar images usually do not consider the radiation resolution), while in the thermal infrared band, the equivalent temperature difference, the minimum detectable temperature difference and the minimum resolution of the temperature difference are used as substitutes. The radiation resolution algorithm is commonly expressed as.

$$R_{\mathrm{L}} = (R_{\max} - R_{\min})/D \tag{1.6}$$

where $R_{\max}$ is the maximum radiation value, $R_{\min}$ is the minimum radiation value, and D is the quantization level. The smaller the R_{L} is, the more sensitive the sensor.

The output of the sensor includes the signal and the noise. If the signal is less than the noise, the output is noise. If the minus between the two signals is less than the noise, the two signals cannot be distinguished on the output records. Noise is a kind of random electrical fluctuation, and its arithmetic mean value is zero. The square root is used to calculate the noise voltage N, and the equivalent noise power is obtained:

$$P_{\mathrm{EN}} = \frac{P}{S/N} = \frac{N}{R} \tag{1.7}$$

where P is the input power, S is the output voltage, and R is the detection ratio. The signal can only be displayed when the signal power is greater than the equivalent noise power (see Fig. 1.5). When the actual input signal power is greater than or equal to 2–6 times the equivalent noise power, the signal can be distinguished.

For thermal infrared images, the equivalent noise power should be converted to the equivalent noise temperature ΔT_{EN}:

$$\Delta T_{\mathrm{EN}} = \sqrt[4]{\frac{P_{\mathrm{EN}}}{\varepsilon\sigma}} \tag{1.8}$$

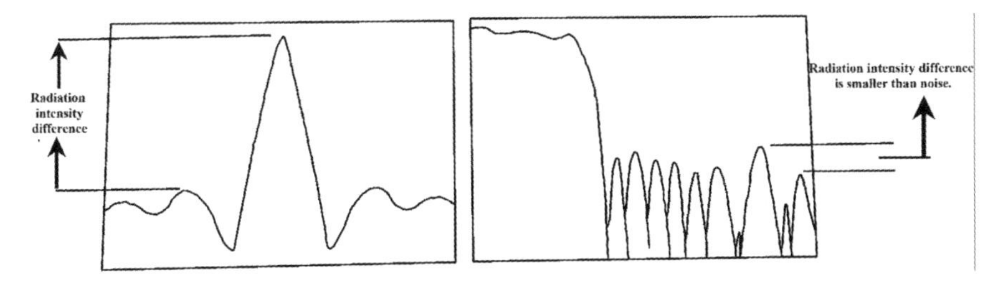

Fig. 1.5 Diagram of radiation resolution

Additionally, only when the ground temperature difference is greater than or equal to 2–6 times ΔT_{EN} can the signal be distinguished in the thermal infrared image.

3. Spectral Resolution

Unlike geometric resolution, the spectral resolution refers to the minimum wavelength interval for detecting the radiated energy of the spectrum and thus denotes a spectral detection capability. It includes the sensor's total detection band width, the number of bands, the wavelength range and the interval of each band. At present, the number of wavelength bands of the imaging spectrometer is 386, the interval of each band is as small as 5 nm, and the sensor with the highest spectral resolution has reached the nanometer level.

Figure 1.6 provides a good description of spectral resolution; using the color spectrum to represent the entire electromagnetic band, Fig. 1.6a can be understood as the infinitesimal spectral resolution. However, this is only an ideal state; when the image is obtained, all spectral information is aliased together, that is, displayed in a "full color image"; to obtain different spectral information, spectra must be divided into different sections. Figure 1.6b shows that the spectral spectrum increases as the band interval increases; Fig. 1.6c shows that as the range of each band becomes narrower, the number of bands increases and the spectral resolution increases. Obviously, the greater the number of bands is, the higher the spectral resolution.

Sensors with different spectral resolution exhibit great differences in detecting the same ground object. Multispectral imaging technology has 10–20 spectral channels, and the spectral resolution is $\lambda/\Delta\lambda \approx 10$ (λ is the wavelength; denoted similarly below); hyperspectral imaging technology has 100–400 spectral channels, and the general spectral resolution rate reaches $\lambda/\Delta\lambda \approx 100$. Ultrahigh spectral imaging has approximately 1000 spectral channels, and the spectral resolution is generally $\lambda/\Delta\lambda \geq 1000$. The latter two have high spectral resolution. The almost continuous spectral curve can distinguish small differences in the spectral characteristics of the object, which is helpful to identify more targets and plays a significant role in the identification of mineral composition. Take the MODIS loaded on the EOS-AM1 series satellite as an example: it belongs to a type of hyperspectral sensor with low

Fig. 1.6 Improvement of two forms of spectral resolution

discontinuity (spectral range of 0.4–14.5 μm), a small number of bands (36 bands) and a low ground resolution. The detectable spectral range of MODIS covers the visible to thermal infrared band. In the visible and near-infrared band, it has 19 channels, and the band interval reaches tens of nanometers. In the thermal infrared band, it has 10 channels, and the band interval reaches tens of nanometers.

Since the imaging radar antenna can only transmit or receive a single band of signals at the same time, there is no concept of spectral resolution for radar images. Imaging radar only works in a single band, such as the Terra SAR-X satellites equipped with SAR sensors working in the X-band and Envisat equipped with ASAR working in the C-band. The wavelength ranges for the different band names are shown in Table 1.3.

4. Time Resolution

The time resolution is the time interval required to repeatedly acquire images in the same area, that is, the time frequency of sampling. For aerial imagery, it is the interval between two adjacent images. The extreme situation is the "gaze", which currently is now on the level of seconds; it is also known as the revisiting period or covering period for satellite images. The temporal resolution of remote sensing images is limited by the spatial resolution. The time resolution of most of the satellites in orbit that do not perform stereoscopic observations is equal to their repetition periods. The time resolution of satellites that have intertrack stereoscopic observations is shorter than the repetition periods. For example, for the SPOT satellite, to take a stereo image between a track and another track at the equator, the time resolution is 2 days. The time resolution is directly related to the dynamic change of the detectable target.

The time resolution range of remote sensing is relatively wide. Time resolution is very important for dynamic monitoring. Different remote sensing objectives require the use of different temporal resolutions. For example, based on the natural historical evolution of the research object and the cycle of the social production process, satellite images are divided into five types: (1) ultrashort term, such as typhoons, cold waves, sea conditions, fish, and urban heat islands, which require resolutions in hours; (2) short term, such as floods, ice, drought and floods, forest fires or pests, crop growth, and green indices, requiring resolutions in days; (3) medium term, such as land use, crop yield, and biomass statistics, generally requiring resolution in months or quarters; (4) long term, such as soil and water conservation, nature conservation, glacial advances and retreats, lake growth and decline, coastal changes, desertification and greening, which require resolution in years; and (5) ultralong term, such as neotectonic movement, volcanic eruptions and other geological phenomena, can require up to several decades.

Table 1.3 Table for microwave wavelengths and their names

Band name	P	L	S	C	X	Ku	K	Ka
Wavelength (cm)	136–77	30–15	15–7.5	7.5–3.75	3.75–2.40	2.40–1.67	1.67–1.18	1.18–0.75

For typical remote sensing sensors, the temporal resolution has different forms of expression. The details are as follows.

(1) Visible and near-infrared images

For visible and near-infrared images and thermal infrared images, higher temporal resolution means more accurate dynamic monitoring of the observed target. This is very beneficial in many ways, such as for the dynamic monitoring of crops, the monitoring of ocean surfaces and tides, and the dynamic monitoring of water resources.

(2) Imaging radar

An important application using radar images is interferometric synthetic aperture radar (InSAR). The principle is to use SAR to obtain two or more monocular complex images of the same area in parallel orbit to form interference and obtain three-dimensional surface information of the area. In the application of InSAR to obtain terrain information, radar images are required to have good coherence. The factors that affect the coherence of interference are system thermal noise, registration error, spatial baseline, time baseline and Doppler frequency center deviation. Among them, system thermal noise is inherent to SAR, and multiview processing can partially inhibit its impact. SAR image registration can reach the subpixel level, and improving the registration accuracy can inhibit its impact. The spatial baseline refers to the difference between the position of the interferometric image on the satellite at the time of acquisition. The time baseline refers to the difference in the acquisition time of the interference image pair. The longer the time baseline is, the worse the coherence between the interference image pairs, the greater the interference noise, and the more difficult it is to obtain the relevant terrain information. Figure 1.7 shows a temporal and spatial distribution baseline of an InSAR analysis, from which the time and spatial distribution of the image can be clearly known. The time resolution is important for InSAR; if a high temporal resolution image can be obtained, then it can not only lead to high precision results for InSAR but also be helpful in polarization–interference (P-In) SAR technology, which demands a large amount of data.

1.2.4 Format of Remote Sensing Digital Image

In the digital processing of remote sensing images, in addition to the image data, there is still a need for other data that are related to remote sensing imaging conditions, such as lighting conditions and the time of the remote sensing imaging. Computer compatible tape (CCT) and computer disc–read only memory (CD-ROM) store and provide data mainly in the following formats.

(1) Band Sequential (BSQ) format

The BSQ format records remote sensing image data in band order. Each band of the image file forms a separate image file. The data file is stored one record at a time in

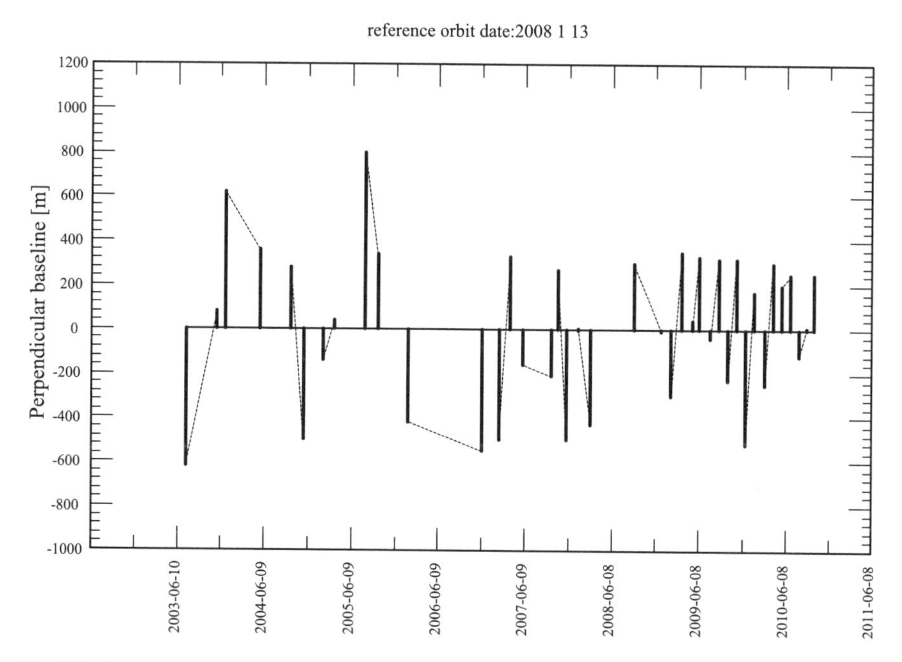

Fig. 1.7 Spatiotemporal baseline distribution of image sets

the order of being scanned; the first band is stored first, and then the second band is stored, etc., until all bands are stored. There are four types of CCT in BSQ format: tape catalog files, image properties files, impact data files, and tail files. The BSQ format is the most common way to record images, and it is also very easy to extract the bands separately.

(2) Band Interleaved by Line (BIL) format

The BIL format is a remote sensing data format that is cross arranged according to band order. The BIL format is similar to the BSQ format and is also composed of four types of files, but each type has only one file.

(3) Band Interleaved by Pixel (BIP) format

The BIP format is a remote sensing data format that records data per pixel in band order. It arranges the brightness values of n bands of each pixel in the dataset in order and then places an end-of-file (EOF) tag at the end of the dataset.

(4) Hierarchical Data Format (HDF) format

HDF is a self-describing and multiobject file format for storing and distributing scientific data. HDFs can represent many necessary conditions for scientific data storage and distribution. HDF supports readme files and provides versatility, flexibility, scalability and cross-platform functionality. HDF supports six basic data types: raster images, palettes, scientific datasets, annotations, virtual data, and virtual groups.

1.3 Remote Sensing Digital Image Processing

Remote sensing has broken through the bottleneck of data acquisition and has been or is moving toward a new stage of comprehensive application, and with the improvements in remote sensing image resolution (including spatial resolution, spectral resolution, radiation resolution and time resolution), remote sensing applications have evolved from a simple qualitative application stage to a new stage of a combination of qualitative and quantitative application. Additionally, remote sensing image processing methods have transitioned from optical processing, visual interpretation and manual drawing to the digital processing stage [8–13]. In this process, internationally, a number of high-level remote sensing image processing systems have been launched and gradually accepted by the majority of users. The promotion and usage of these systems has greatly accelerated the application of remote sensing images and promoted the transition from a coarse application to a fine application of remote sensing images. The processing of remote sensing digital images involves importing remote sensing images into computers in digital form and then transforming one image into another modified image using a certain mathematical method according to the law of digital images. It is a process extending from image to image.

1.3.1 Remote Sensing Digital Image Processing Method

To carry out the processing of the image data, it is necessary to create mathematical descriptions of these related images.

There are two methods for digital image processing: discrete methods and continuous methods.

The discrete method regards the image as a collection of discrete sampling points, in which each point has its own attributes and the processing of the image involves the operation of these discrete units. An image is stored and represented in a digital form, which comprises a discrete number; therefore, using discrete methods to deal with digital images is reasonable. A concept related to this method is the spatial domain. Spatial domain image processing takes the image plane itself as a reference and processes the pixels in the image directly.

The continuous method assumes that as the image is usually the actual reflection of the physical world, the image obeys the continuous mathematical description law. Image processing is an operation of a continuous function, and continuous mathematical methods are used for image processing. The main concept related to this method is the frequency domain (the relevant concepts of the frequency domain and the spatial domain are described in detail in the following sections). The image in the frequency domain is an image that is processed by a Fourier transformation and reflects frequency information. After the image processing is completed in the frequency domain, the image is usually transformed in the spatial domain for image display and contrast.

The theory behind multiple processing operations is actually based on the analysis of continuous functions, which can effectively solve the processing problem. However, other processes are more suitable for conceptualizing individual pixels by using logical operations, for which the discrete method is more suitable. Often, a process can be described by both methods, and we must choose one. In many cases, we find that using continuous analyses or discrete technology solutions can lead to the same answer. However, the understanding of the problem is very different for different concepts. If there is a one-sided emphasis on either a continuous or a discrete method, there could be significantly different results as a result of the sampling effect. During digital image processing, after the implementation of the level of digitization, the original continuous form of the image should be maintained: the processing method used should support the analog-to-digital conversion process and ensure that the original content is not lost—or at least not obviously lost. This process should be able to identify the occurrence of sampling effects and then take effective methods to eliminate or reduce the effects to an acceptable level.

1.3.2 Inversion of Remote Sensing Digital Images

Optical remote sensing imaging is a process involving energy transmission and transformation, as shown in Fig. 1.8. Optical remote sensing assumes the sun as a radiation source, and the energy of solar radiation spreads to the surface in the form of electromagnetic waves. The atmosphere between the sun and the surface is approximately 50 km thick. During the process of penetrating the atmosphere, solar radiation is scattered or absorbed by the atmosphere, aerosols and other particles. Different objects have different absorption and reflection effects on solar radiation, and solar radiation reflected by ground objects passes through the atmosphere and is then received by optical remote sensing imaging systems on aeronautical or aerospace remote sensing platforms. The solar radiation of different chemical compositions and spatial geometries is different in terms of the absorption and reflection intensity of solar radiation. Therefore, the solar radiation reflected by the ground objects carries the geometric, physical and chemical information of the objects, which is also the basis of the inversion of the remote sensing parameter.

In the early development stages of optical remote sensing, the treatment of remote sensing images was mainly based on visual interpretation, and remote sensing workers obtained object information by comparing the reflection intensity and geometric shape of the objects. With the development of remote sensing technology, there was a growing dissatisfaction with obtaining object information through the visual interpretation and computer-aided image processing methods; in many remote sensing applications, the physical and chemical parameters of remote sensing images needed to be quantitatively calculated [14–16]. The process of quantitatively calculating the reflectivity of the surface in remote sensing images is called remote sensing quantitative inversion. The process is shown in Fig. 1.9.

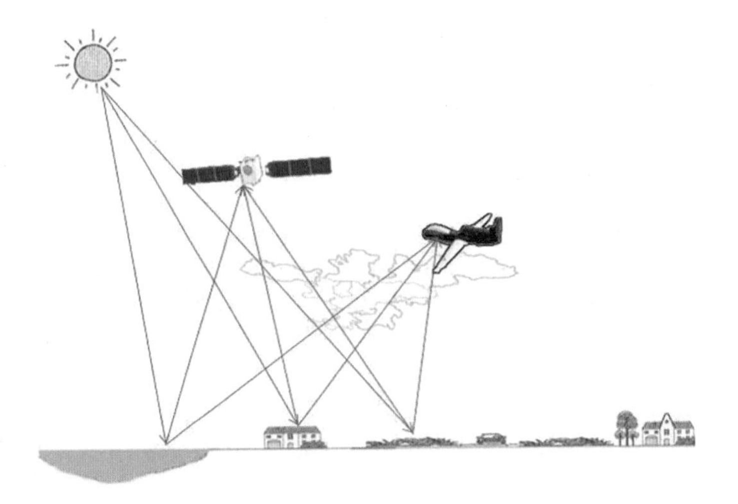

Fig. 1.8 Diagrammatic sketch of optical remote sensing

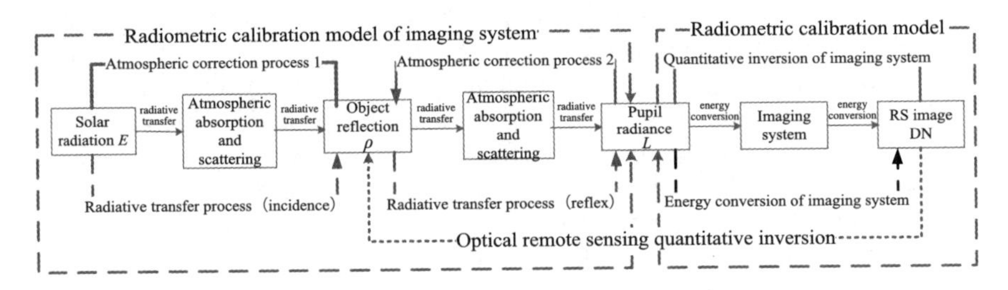

Fig. 1.9 Schematic diagram of the quantitative inversion process of optical remote sensing

The optical remote sensing quantitative inversion process is the inverse process of the optical imaging process. It consists of at least four steps:

The first is the quantitative inversion process of the imaging system. Based on the radiation calibration model of the imaging system, we can establish the mathematical relationship between the DN value of the imaging system and the entrance pupil radiation L. In the case in which the DN value is known, we can calculate the pupil radiance L. Second, atmospheric correction is applied to the entrance pupil radiation L to obtain the radiation exposure of the object's reflection; the atmospheric radiation is corrected again for the solar radiation to obtain the irradiance to the ground. Finally, the reflectivity of the object in the imaging direction can be obtained by dividing the radiance reflected by the radiance received by the ground object. The reflectivity is the basis for calculating the other physical parameters, and the user can calculate the required geomorphic parameters according to the specific remote sensing application.

1.4 Framework of This Book

This book is composed of three parts:

The first part (Chaps. 1–4) discusses remote sensing digital image processing systems, mainly introducing content on the remote sensing digital image overall processing, which is used to solve the image integrity question and determine the characteristics of the entire image; this is often regarded as preprocessing. Chapter 1 presents a system overview and establishes a holistic conceptual framework. The second chapter discusses the system support conditions, specifically for the digital image, the display and the processing software, which form the premise of the computer processing of remote sensing images. The third chapter discusses the mathematical basis of the overall processing of digital images, such as gray histograms, point operations, algebraic operations and geometric operations, enabling the understanding and mastery of the overall image. The fourth chapter presents the physical basis of the overall processing of digital images, namely, the analysis of the physical characteristics of the imaging process to ensure that the imaging process is correct and that the image quality is good.

The second part of the book (Chaps. 5–10) discusses the theory and methods of pixel processing and the theoretical part of advanced remote sensing digital images; it is also the main part of this book. It mainly introduces the mathematical processing inside and between pixels of the remote sensing image that solves the theoretical method problems of each pixel processing of the image, determines the characteristics of each image pixel, and is the basis of the remote sensing digital image processing analysis, which is often regarded as the deep processing of images. The fifth chapter discusses the space–time domain convolution linear system, which is the basic theory of remote sensing digital image processing. The image transformation system is an inertial system with linear system characteristics. Chapter 6 describes the time–frequency domain roll–product transformation, which refers to the dual transformation of the time domain (dimension is T) and its reciprocal frequency domain (the dimension is T^{-1}). This theory provides the possibility of a simplified operation of digital image processing. Chapter 7 presents frequency domain filtering, which retains the different frequency information in the frequency domain according to different production needs. Chapter 8 elaborates time domain sampling. The continuous spatial information in the natural world must be transformed into the discrete information needed for computer processing, that is, time series information. Therefore, the sampling process becomes the prerequisite guarantee of the computer expression analysis of the remote sensing information. Chapter 9 discusses the space–time equivalent orthogonal basis. The expression, storage, and computation of any image can be expressed through a two-dimensional matrix vector representation consisting of the basis function–basis vector–basis image. Chapter 10 covers the time domain combined orthogonal basis. Image processing can also be characterized by the time–frequency basis image method; that is, the image is compressed, enhanced and fused by wavelet transform.

The third part of the book (Chaps. 11–13) contains the technology and application sections. Based on the knowledge theories elaborated in the theoretical and methodological parts of the book, a wealth of digital image processing techniques and applications have been developed, such as restoration and noise reduction (Chap. 11), compression reduction (Chap. 12), pattern recognition for image segmentation (Chap. 13), pattern recognition for feature extraction and classification (Chap. 14), and color transformation and 3D reconstruction (Chap. 15). These image processing methods can be implemented by computer-related algorithms. Chapter 16 offers an application example, primarily introducing remote sensing digital image application processing methods and typical cases of remote sensing digital image processing. This chapter addresses the problem of image application and the purpose of remote sensing digital image processing, presenting some selected cases that concretely analyze several essential features of remote sensing digital images.

Presenting in general a progressive functional relationship, Chap. 16 discusses several typical cases that are used to feed back to the above three parts of the book to verify the validity of the content, thus forming a closed loop of the whole structure of this book. The specific content framework is shown in Fig. 1.10.

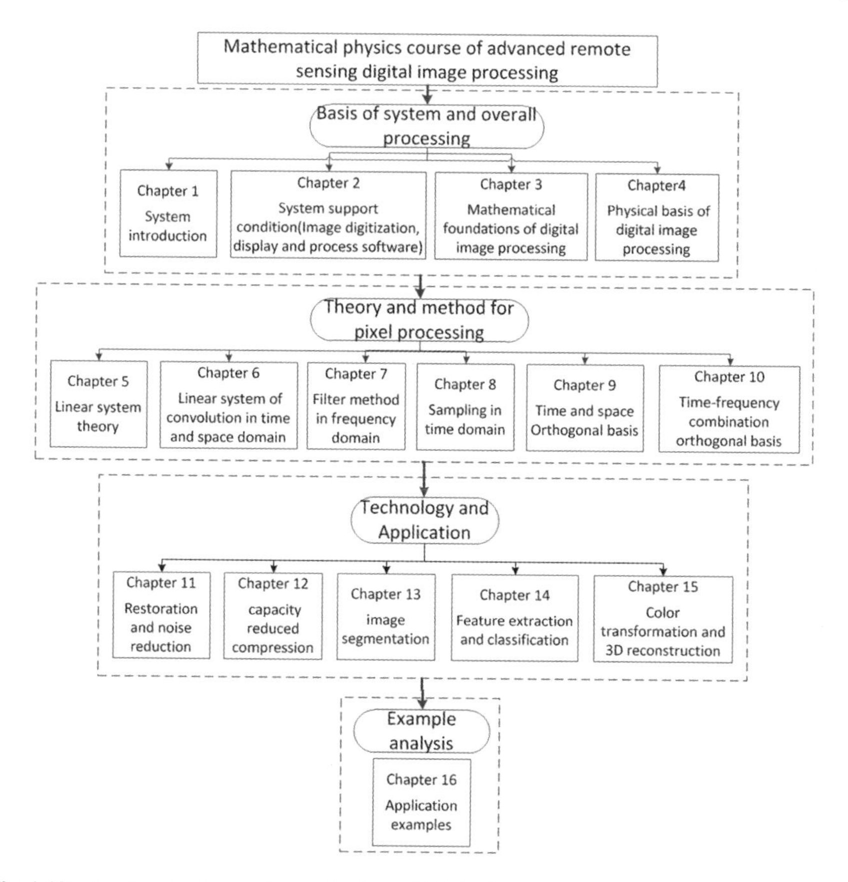

Fig. 1.10 Chapter structure and organization of this book

References

1. Mei A, Peng W, Qin Q, et al. Guidance of remote sensing. Beijing: Higher Education Press. 2001. (In Chinese).
2. Zhao Y. Principles and methods of remote sensing application analysis. Beijing: Science Press. 2003. (In Chinese).
3. Chen S. Dictionary of remote sensing. Beijing: Science Press. 1990. (In Chinese).
4. Xiru X. Remote sensing physics. Beijing: Peking University Press. 2005. (In Chinese).
5. Tang G, Zhang Y, Liu Y, et al. Remote sensing digital image processing. Beijing: Science Press. 2004. (In Chinese).
6. Gonzalez RC, Woods RE. Digital image processing. Nueva Jersey. 2008.
7. Szeliski R. Computer vision: algorithms and application. Springer Sci Bus Media. 2010.
8. Liu P, Kuang G. Methods of image mosaic in remote sensing images. Comput Knowl Technol (Acad Commun). 2007;01. (In Chinese).
9. Yang Z. Discussing the application of wavelet transform in remote sensing image processing. Sci Technol Trans. 2011;14. (In Chinese).
10. Ding Y. Research and progress on radiation correction and enhancement of remote sensing images. Sci Friends (Sect. B). 2009;10. (In Chinese).

11. Tang J, Nie Z. Geometric correction of remote sensing images. Mapp Spat Geogr Inf. 2007;02. (In Chinese).
12. Tian R, Peng L, Feng W. Research on remote sensing digital image mosaic based on genetic algorithm. Agricult Mechaniz Res. 2004;06. (In Chinese).
13. Liu Q, Zhang G. Research on acquisition methods of remote sensing images. J Harbin Inst Technol Univ. 2003;12. (In Chinese).
14. Li X, Shi B, Liu X et al. Remote sensing image technology of atmospheric path radiation. J Shanghai Univ (Nat Sci Sect). 2007;02. (In Chinese).
15. Fan N, Li Z, Fan W et al. PHIG-3 Hyperspectral data pretreatment. Soil Water Conserv Res. 2007;02. (In Chinese).
16. Li X, Xu L, Zeng Q et al. Remote sensing images of atmospheric path radiation and monitoring of urban air pollution. J Remote Sens. 2008;05. (In Chinese).

Chapter 2
System Support Conditions for Remote Sensing Digital Image Processing and Analysis

Chapter Guidance Based on the framework of remote sensing digital image processing systems, as described in the first chapter, this chapter highlights the important components of image processing systems, namely, the system support conditions, including the input, output and three support software modules. Before processing remote sensing digital images, we must determine whether the input and output interface state image processing system module has been optimized in terms of technology. If an image is repeatedly processed and cannot satisfy the requirements, it is necessary to evaluate if the input and output parameters for extracting and processing the software are unreasonable. In this context, it is necessary to comprehensively understand the basic support conditions. This chapter focuses on the input gamma correction curve, point spread and modulation of the function of the output display, which pertain to the first step of system design and analysis of the supporting conditions of remote sensing digital image processing. The chapter first describes the image processing task, which lays the foundation for the interface support.

Because computers can only process digital images, and images may appear in several natural forms, digital images, which require a digital input, are a prerequisite for image processing. Additionally, as the digital output of digital image processing, image displays must transform digital images into usable forms and analyze the image effect in the process of image monitoring and interactive control. Thus, the premise of the design and application of an efficient remote sensing image processing system is the suitability of the image processing and analysis software.

This chapter introduces and compares the two most commonly used digital image sensors and describes the physical concepts, basic digital image input, response characteristics and other key concepts related to the digital image output. Finally, this chapter outlines the software commonly used in the processing of remote sensing images and the design of the remote sensing image processing software. A UAV aerial remote sensing digital image processing software is considered as an example for the analysis. The logical framework of the chapter is shown in Fig. 2.1.

The chapter is organized as follows: 2.1 Digital Image Sensors: We describe CCD and CMOS based mainstream imaging sensors and their performance evaluation; 2.2

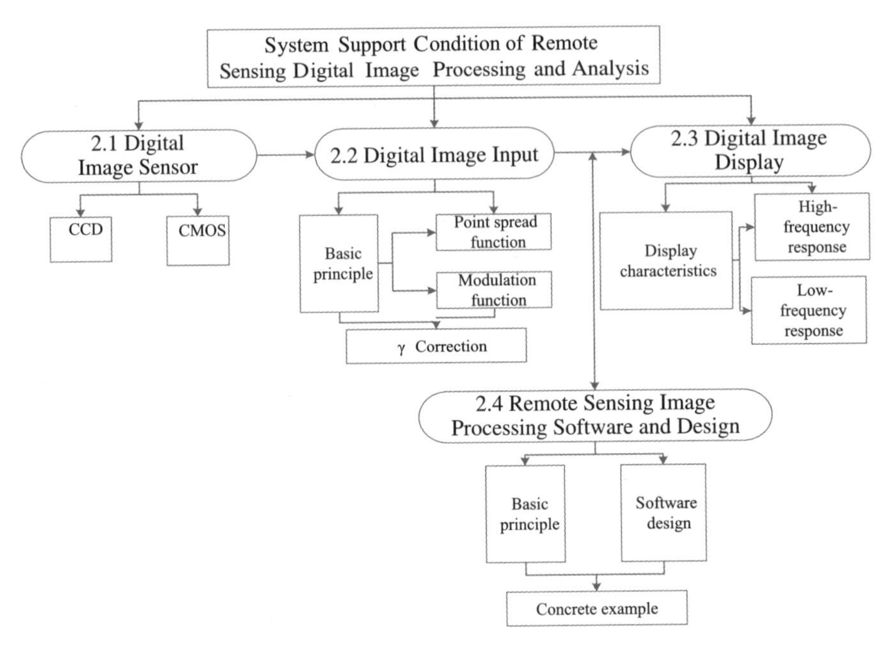

Fig. 2.1 Logical framework of this chapter

Digital Image Input: Instruments and a γ correction method are used to minimize the effect of nonlinear input image; 2.3 Digital Image Display: Responses and diffusion effect balance at high frequencies are used to alleviate the effect of fuzzy image output to an appropriate controllable level; 2.4 Remote Sensing Image Processing Software and Design: Software programs are reasonably selected or designed according to the top-to-bottom software design principle.

2.1 Digital Image Sensors

This section describes the development of imaging devices, i.e., image sensors, in digital imaging systems. Specifically, the chapter introduces the principles of current mainstream digital input CMOS and CCD sensors and compares the differences between the two techniques and structure.

The core component of the digital imaging system is the imaging device, i.e., the image sensor, whose function is to transform the optical image into an electrical signal, that is, to transform the light intensity information spatially distributed on the photosensitive surface of a sensor into a serial output according to timing. The electrical signal may be a video signal that can reproduce the incident optical image. Before the 1960s, the imaging task was mainly accomplished using various electron beam camera tubes (such as light guide tubes and flying spot tubes). In the late 1960s, with the maturity of semiconductor integrated circuit technology, particularly CMOS and CCD processes, various solid-state image sensors were rapidly developed.

The two main image sensors are based on CMOS and CCD frameworks. CCD devices exhibit a notable potential for development due to their inherent characteristics. The CMOS sensor, which appeared at the end of the 1960s, exhibited an inferior performance, which critically affected the image quality and limited the development and application of such sensors. In the 1970s and 1980s, CCD sensors, the technology of which was more mature than that of CMOS sensors, attracted considerable attention in the imaging field. Since the 1990s, with the realization of manufacturing technology advancements and the miniaturization and development of low power consumption and low-cost imaging systems and chip and signal processing technologies, CMOS sensors have been considerably enhanced and have emerged as the mainstream image sensors. The following sections describe the two kinds of sensors [1].

2.1.1 CCD Image Sensor

Since 1970, the development of the first CCD sensor by Bell Labs has relied on mature integrated circuit technology, and CCD sensor manufacturing technologies have rapidly developed. CCD sensors, as a new type of photoelectric converter, have been widely used in cameras and image acquisition, scanning and industrial measurement systems. Compared with other imaging devices and camera tubes, CCD image sensors exhibit a wide range of system responses, geometric accuracy, small size, low weight, low power consumption, high stability, long service life, high resolution, high sensitivity, and high spectral and photosensitive elements. According to the working characteristics of the utility model, CCD sensors can be characterized as linear matrix CCD sensors. According to the process characteristics, the sensors can be divided into CCD, 3CCD and Super CCD. In terms of the light, the sensors can be divided into positive light and back light CCD devices. According to the spectrum, the sensors can be divided into visible light CCD, infrared CCD, X-ray CCD and ultraviolet CCD. Moreover, such sensors can be categorized as linear array and face array CCD sensors.

The basic unit of a CCD includes many discrete photosensitive units known as pixels (Fig. 2.2). Each independent pixel contains a photosensitive element, such as a photodiode or light charge capacitance. The photosensitive cell outputs a charge proportional to the incident light. These charges are transferred to the CCD shift register and output by the shift register to the CCD device. These charges accumulate during exposure, and the amount of accumulated charge depends on the incident light intensity, integration time, and quantum efficiency of the photosensitive cell. In the absence of a light period, the amount of charge compensation accumulates.

As shown in Fig. 2.2, pixels can be arranged in a linear array or areal array. The clock signal transfers the charges from the pixel to the shift register after the high-frequency clock signal outputs discrete pixel charge transfer to the CCD. The frequency range of a typical shift register is approximately 1–10 MHz.

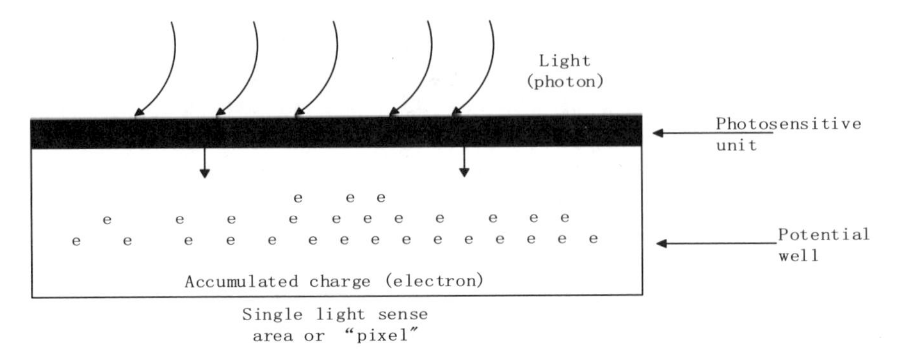

Fig. 2.2 CCD photosensitive unit

The typical CCD output and output voltage waveform are shown in Fig. 2.3. The CCD output is reset to the reference level due to the voltage across the sensing capacitor, which generates reset noise. The difference between the reference level and video signal level represents the amount of light. The number of charges of the CCD can be as low as 10, and the typical CCD output sensitivity is 0.6 μV/electron. The saturation output voltage of most electronic array CCDs ranges from 500 mV to 1 V. The range for linear CCDs is 2–4 V, and the DC level is 3–7 V. The film processing ability of CCDs is limited; therefore, the output signal of a CCD usually corresponds to an external circuit. Before processing the CCD digital output, the original output signal must be dislocated. In addition, the shift register must be biased and amplified. The CCD output voltage is extremely small and often submerged in noise. The noise source is the thermal noise in the reset switch, typically corresponding to approximately 100–300 electrons. The working principle of the CCD sensor is shown in Fig. 2.4.

2.1.2 CMOS Image Sensor

A CMOS sensor integrates the image sensing part, signal readout circuit, signal processing circuit, and control circuit on a single chip, along with a lens and other accessories, to form a complete camera system. CMOS image sensors can be divided into passive pixel sensors (PPSs), active pixel sensors (APSs) and digital pixel sensors (DPSs) according to the different methods of production of light charge PPSs, which can be divided into three types: photosensitive diode, grating and logarithm. A complete CMOS chip is mainly composed of a horizontal (vertical) control and timing circuit, photosensitive array, analog signal readout processing circuit, A/D conversion circuit, analog signal processor, interface and other components. The photosensitive array reads the current generated by light to the analog signal readout processing circuit under the effect of horizontal (vertical) control and sequential circuit, is

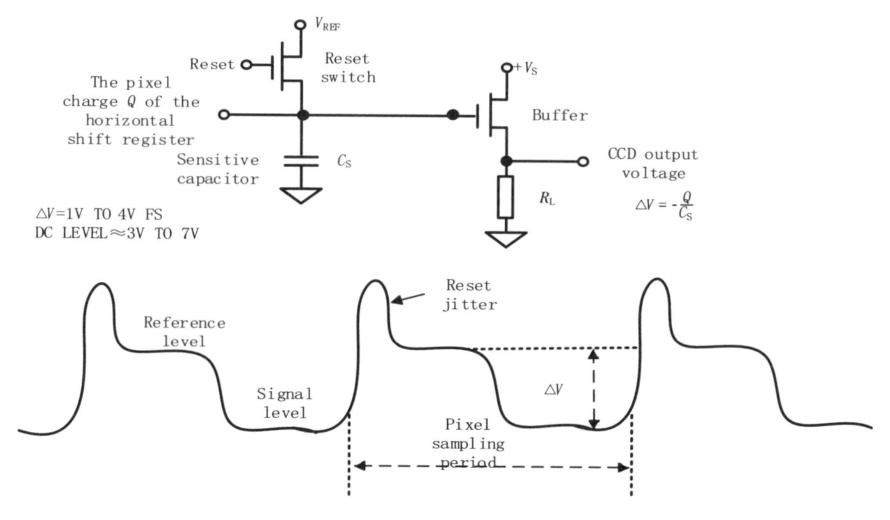

Fig. 2.3 CCD output and output waveform

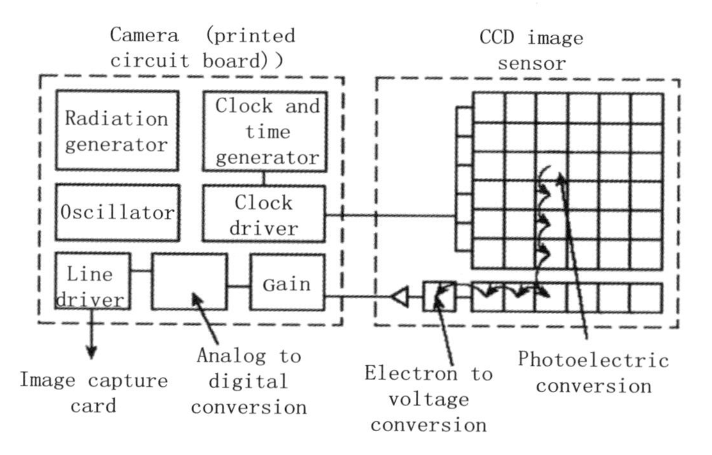

Fig. 2.4 CCD sensor principle

converted into a digital signal through A/D conversion, and is processed through digital signal processing and circuit processing. The result is output (Table 2.1).

The CMOS image sensor structure diagram is shown in Fig. 2.5. The CMOS image sensor structure can be divided into three layers: The upper layer includes a pixel array and row gating logic, the middle layer includes the analog signal processor and timing and control module, and the lower column is mainly composed of a parallel A/D converter [2]. The row strobe logic and column strobe logic may pertain to shift registers or decoders. The timing and control circuit limits the signal readout mode settings and integral time and controls the output rate of the data. The analog signal processor realizes signal integration, amplification, sampling and holding, and

Table 2.1 Internal structure of the CMOS imaging chip

Vertical control box timing circuit	Horizontal (vertical) control and sequential circuits
	Photosensitive array
	Analog signal readout processing circuit
	A/D conversion circuit
	Analog signal processor
	Interface

related double sampling. A parallel A/D converter is required in a digital imaging system. The working principle of the CMOS sensor is shown in Fig. 2.6.

A typical CMOS pixel array is a two-dimensional addressable sensor array. Each column of the sensor is connected to a bit line, and each sensitive unit output signal in the row that allows the selected line is sent to its corresponding bit line. The end of the bit line is a multiplexer, and each column is independently addressed. Another choice is to develop and test the phase of the DPS-CMOS image sensor. The traditional PPS and APS correspond to A/D conversion in pixels, while the DPS corresponds to the A/D conversion integrated in every pixel unit. Each pixel unit output is a digital signal, which leads to faster operation and lower power consumption.

As the technology has matured, CMOS and CCD image sensors have undergone rapid development, especially in terms of size: devices sized 5.6 μm × 5.6 μm unit-size CCD have been developed, and a color filter and microlens array were sequentially manufactured. The existing technology can use 0.25 μm feature size process technology to produce more integrated CMOS image sensors. However, a smaller sensor size is preferable. Under natural light conditions, to produce a gray level that the human eye can distinguish, it is necessary to stimulate 20–50

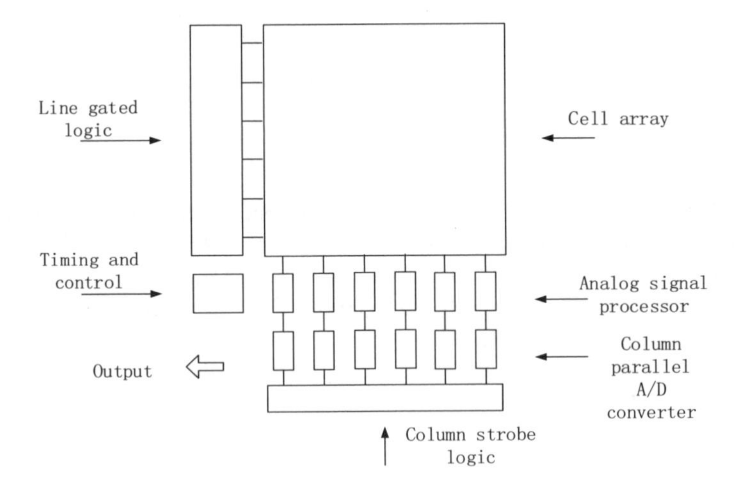

Fig. 2.5 CMOS main architecture

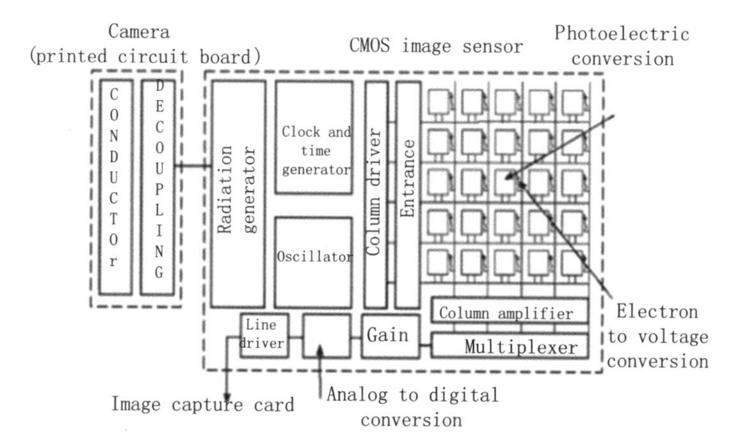

Fig. 2.6 CMOS sensor principle

electronic, high-precision sensors (which can display 1024 Gy levels) and excite approximately 20 thousand electrons, which requires a size of the order of 1 μm. To decrease the size and maintain precision, image sensor manufacturers must consider the abovementioned aspects.

2.1.3 Comparison of CMOS and CCD Image Sensors

As CCD and CMOS image sensors are considerably different in terms of the internal and external structures, the two sensors exhibit notable differences at the technical level, mainly in terms of in the following four aspects of the charge information:

(1) Information reading method. The charge information stored in the CCD must be read after the real-time transfer in the synchronization signal control. The charge information transfer and read output must include a clock control circuit and three sets of power supplies. The circuit is highly complex, leading to a lower stability. In the CMOS framework, signal reading can be simply realized after the photoelectric conversion of the direct current (or voltage) signal.
(2) Speed. The processing speed of the image sensor is related to the information reading method. CCD, under the control of the synchronous clock, is slower in terms of the output information. The CMOS signal collection method can more rapidly output signals and manage image information.
(3) Power consumption. CCDs require three sets of power supplies, and the power consumption is large. CMOS needs to use a single power supply, the power consumption is extremely low, 1/8 to 1/10 times that of the CCD.
(4) Image quality. CCD production technology was introduced early, and thus, the technology and noise isolation method are mature. The imaging quality is higher than that of CMOS. In particular, the imaging quality of CMOS with a high

degree of integration between the photoelectric sensor and circuit is slightly inferior, and the optical, electrical and magnetic interference associated with significant noise considerably influences in image quality. In recent years, with the progress of technology, the imaging quality of CMOS sensors has considerably enhanced; however, the effect remains inferior to that of CCD imaging sensors that achieve a high precision. Overall, the CCD sensors outperform the CMOS sensors. In recent years, with the continuous advancement in technology, the imaging quality of CMOS sensors has been enhanced; however, the imaging results are inferior to those of high-precision CCD sensors.

The two sensors exhibit the following differences in terms of structure:

(1) Internal structure. The imaging points of a CCD are arranged in an X–Y matrix, and each imaging point consists of a photodiode and a nearby charge storage area. Photodiodes convert light into electrons, and the number of focused electrons is proportional to the light intensity. As these charges are read, the data in each row are transferred to a buffer that is perpendicular to the charge transfer direction. Each row of charge information is read continuously and amplified by a charge/voltage converter and amplifier. This structure produces images with low noise and high performance. However, this complex structure of the CCD corresponds to a higher power consumption, lower processing speed, and higher cost. The electronics surrounding the CMOS sensor can be integrated in the same manufacturing process. The CMOS sensor is constructed similar to a memory. Each imaging pixel includes a photodiode, a charge/voltage conversion unit, a reset and selection transistor, and an amplifier. The sensor is covered with a metal interconnect (for timekeeping and reading signals). The vertical arrangement of the output signal interconnects are simpler in construction, which can decrease the power consumption while increasing the processing speed. However, the resulting image is not as accurate as that obtained using the CCD sensor.
(2) External structure. The electrical signals output by the CCD must be processed by subsequent address decoders, analog-to-digital converters, and image signal processors, and three sets of power supplies with different voltages and synchronous clock control circuits must be provided. The integration level is low. Digital cameras composed of CCDs usually include six or eight chips, with at least three chips required. Consequently, the cost of digital cameras based on CCDs is high.

In general, CMOS sensors exhibit unique advantages over CCD sensors. Although the current imaging quality is not satisfactory, as the technology continues to advance, the quality of CMOS imaging is expected to continually enhance. CMOS image sensors are expected to be applied in an increasing number of fields and promote the development of digital image technologies.

2.2 Digital Image Input

This section comprehensively describes digital images, including the principle of digital image input, evaluation indices for digital instruments, methods to enhance the linearity, and digital γ correction expression imaging systems that yield the point source response and point spread function, which can be used to quantitatively describe the spatial frequency of the image quality (resolution or contrast and the degree of the optical modulation transfer function).

2.2.1 Principle of Image Digitization

The original copies of traditional painting pictures, video tapes or printed printing products require the conversion results for analog images. Such images cannot be directly processed on a computer and must be transformed into digital images represented by a series of data. This transformation is referred to as image digitization. The operating technique corresponds to sampling.

Sampling is performed by the computer according to certain rules and characteristics of the simulated image site representation by considering the process of recording the data. This process involves two key considerations: determination of the number of sampling points in a certain area and calculation of the number of pixels. The resolution of the digital image is determined by the record of one factor for each data bit, for instance, for a certain depth. For example, the brightness of a certain point is represented by one byte, and this brightness can have 256 Gy levels. The 256 Gy level differences are evenly distributed between all black (0) and white (255). The full value of every pixel is expressed with a number between 0 and 255 and chroma hue. Hue and saturation factor are the same as the brightness factor record [3]. The record of color also involves the issue of patterns, as discussed in the following section. From the point or recorded data depth and plane, the digital image and simulated image sampling form must have a certain difference. A considerably small gap is preferable because the human eye cannot distinguish the images. The difference in the digital image and simulated image must be minimized to obtain the digital image.

2.2.2 Digital Image Instruments

The advantages and disadvantages of the measures of different digital image instruments can be summarized follows:

(1) Pixel size. The size of the sampling hole and distance between adjacent pixels determine the pixel size. The pixel size determines the resolution of the digitized image and quality of the digitized image.

(2) Image size. Different scanners can input images of different sizes.
(3) Measured image. The image features in the digital input that can be maintain in the simulated images are important indicators, and in different applications, the local characteristics of the simulated images may not be the same.
(4) Linearity. The degree of linearity is a key aspect in digitization. According to the digital light intensity and linearity, the actual accuracy of the gray value may be proportional to the brightness of the image. The effectiveness of nonlinear digitizers influences the subsequent processes.
(5) Grayscale number. The number of gray levels determines the delicateness of the digitized image. Most digital devices can generate 256 grayscale values, and with the development of technology, digital equipment can respond to higher grayscale values.
(6) Noise. The noise introduced by digital equipment can deteriorate the image quality. The noise must be less than the contrast of the analog image for the next step.

2.2.3 *γ Correction*

As discussed, the linearity of the digital equipment determines the image quality. In a digital image display and printing device, the gray value of the pixels in the image file and corresponding luminance value in the displayed image exhibit nonlinear projection relations. A digital projection imaging system in the image processing module implements γ correction to maintain digital equipment linearity. In an image acquisition system, the response of each color component in response to the light intensity values pertaining to the RGB values is nonlinear. By implementing γ correction, the response of each pixel of the image acquisition system value can be rendered linearly proportional to the light intensity. This conversion is known as the γ correction.

As shown in Fig. 2.7, the cathode ray tube (CRT) exhibits an exponential relationship between the image gray value Y and display brightness value E when displayed on the PAL image.

Because the light intensity produced by CRT is not directly proportional to the input voltage, it is proportional to the gamma exponentiation of the input voltage. Gamma correction is performed according to the following functional relationship:

$$R_{gc} = 255 \times \left(\frac{R}{255}\right)^{\frac{1}{\gamma_{red}}}, G_{gc} = 255 \times \left(\frac{G}{255}\right)^{\frac{1}{\gamma_{green}}}, B_{gc} = 255 \times \left(\frac{B}{255}\right)^{\frac{1}{\gamma_{blue}}} \tag{2.1}$$

γ_{red}, γ_{green}, and γ_{blue} denote the γ values of red, green and blue primaries, respectively. Figure 2.8 shows the inverse mirror index growth rate for the values shown in Fig. 2.7.

The values of the parameters in the abovementioned three formulas—γ_{red}, γ_{green}, and γ_{blue}—are determined experimentally. After γ correction, the gray values of

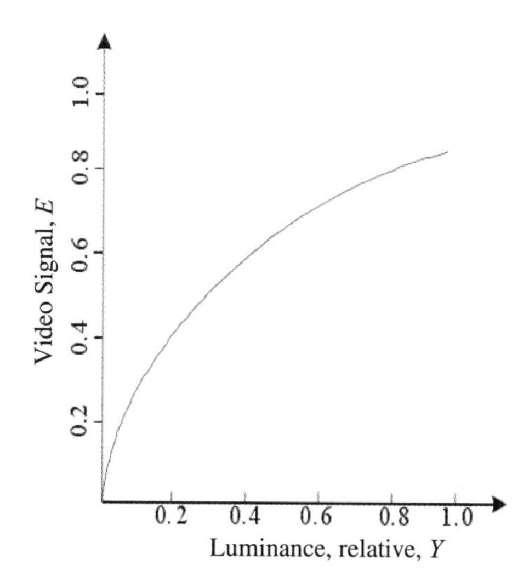

Fig. 2.7 Brightness curve of CRT

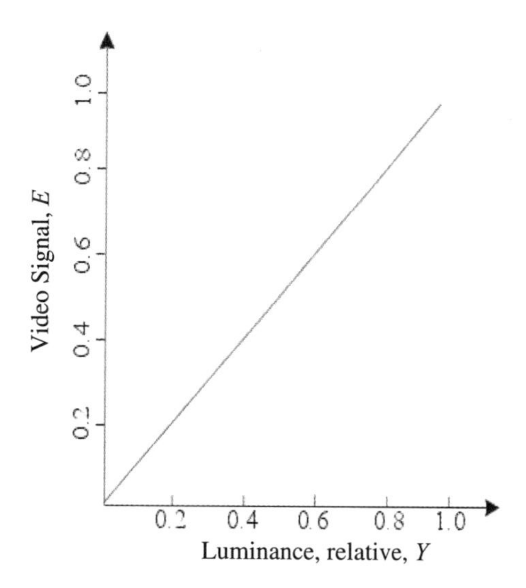

Fig. 2.8 γ correction curve

Y and E exhibit an exponential relationship for the brightness values, as shown in Fig. 2.9, and the linear relationship can be satisfied. Figure 2.9 shows the average amplitude of the findings shown in Figs. 2.7 and 2.8.

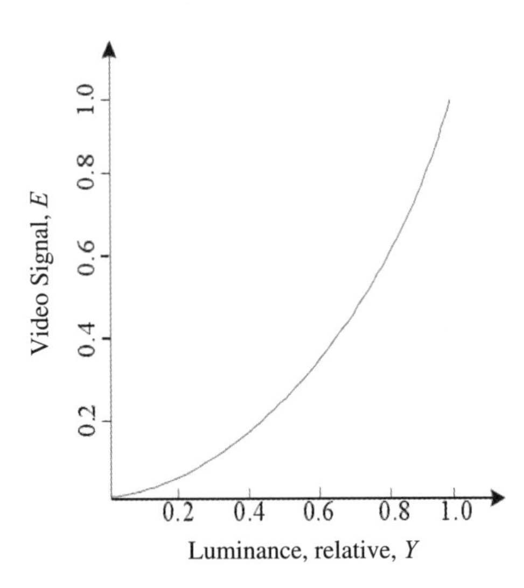

Fig. 2.9 Linear relationship between grayscale and brightness

2.2.4 *Point Spread Function*

Any object plane can be considered a combination of surface elements, and each face can be viewed as a weighted δ function. For an imaging system, if the distribution of the optical vibration caused by the vibration of any surface element on the object plane can be determined through the imaging system, the image can be obtained through linear superposition, that is, the convolution process. The distribution of the planar light field and intensity distribution of the image plane face-to-face light vibration yuan for the unit pulse. The surface element is known as the point source. The response imaging system for the point light source pertains to the point spread function (PSF), and as describes in the following section, the point spread function degeneration for most imaging system is Gaussian. Therefore, the point spread function can be expressed as $h(x, y) = \exp(-r^2)$. Because any light source can be seen as a collection of point light sources, the point spread function influences the imaging mechanism and imaging quality. When the point spread function is related to the spatial position of the image pixel, it is described in terms of the spatial change, and the point spread function is known as the space invariant.

The main properties of the point spread function can be described as follows:

(1) Deterministic. For time invariant systems, the point spread function is determined;
(2) Nonnegative. Determined by the physical causes of ambiguity;
(3) Finite support domain. In linear space invariant systems, the point spread function can be considered a finite impulse response (FIR) filter with limited response;

(4) Energy conservation. Because the imaging system is usually a passive process, image energy is not lost.

The Gaussian system function as the degradation function is the most common form for many optical imaging systems and measurement systems, such as cameras, CCD cameras and CMOS cameras [4]. Since the Fourier transform of the Gaussian function is still a Gaussian function and no zero crossing occurs, the Gaussian degenerate function cannot be identified using frequency-domain zero-crossings. In many cases, outlier observation images can provide the information necessary to identify the Gaussian boundary point spread function. The following section discusses the point spread function based on the Gaussian function.

2.2.5 Modulation Transfer Function

The modulation transfer function (MTF) can be used to describe the imaging quality of the spatial frequency, which is a key parameter to evaluate the imaging performance of the optical imaging system. The function can be determined in terms of the output/input image contrast.

According to the Fourier optics imaging theory, the relation between the pupil function, point spread function and modulation transfer function is shown in Fig. 2.10.

2.3 Digital Image Display

Image processing depends on the quality of the image display and exhibits several characteristics. This section focuses on the linear characteristics and low- and high-frequency responses. Specifically, this section presents the low-frequency response corresponding to the uniform brightness display formula and high-frequency response corresponding to the straight line and checkerboard formulas. The balance between high- and low-frequency responses and point spacing is analyzed.

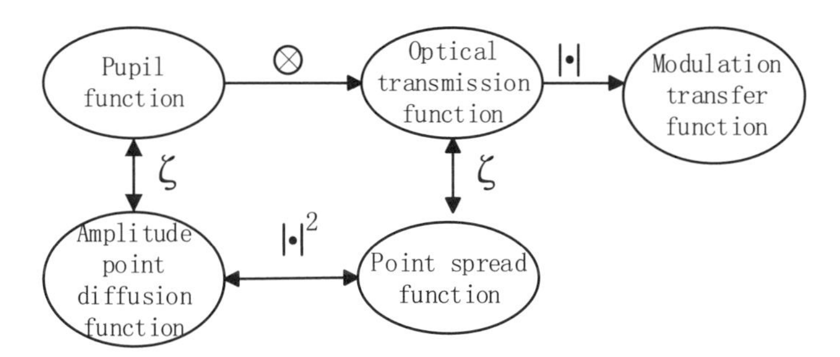

Fig. 2.10 Sketch map of the conversion relation of the optical function

2.3.1 Display Characteristics

Digital image display is an integrated process involving multiple display characteristics, including the image size, luminosity resolution, linearity, low-frequency response, high-frequency response noise influence, sampling, and display. The image size includes the display size and display resolution. Photometric resolution refers to the display's ability to display the correct brightness and optical density accuracy. As mentioned, using the γ correction curve, the linearity of the display system can be enhanced. Sampling problems and noise affect the accuracy of the image display. The final image display method includes soft and hard copies.

The key display characteristics are the low- and high-frequency responses of the display. The essence of the display is the mutual influence between the display points. The brightness of each display point changes with the distance to satisfy the following Gaussian distribution:

$$P(x, y) = \mathrm{e}^{-z(x^2+y^2)} = \mathrm{e}^{-zr^2} \tag{2.2}$$

Z is a parameter, and R is the radial distance of the center of the light. If R represents the radius at which the brightness is half the maximum value, the curve function of the point distribution is

$$P(x, y) = 2^{-\left(\frac{r}{R}\right)^2} \tag{2.3}$$

Each pixel gray value is based on the Gaussian curve and spread in the lower regions, which is a natural phenomenon in inertial space. Therefore, the effect of this diffusion on the gray values of other adjacent pixels is the most important feature of the output image.

2.3.2 Low-Frequency Response

The low-frequency response mainly refers to the display ability of the system to reproduce a large gray area. This ability mainly depends on the shape of the display points, pitch of the points, and amplitude noise and position noise of the display system. The Gaussian points corresponding to the brightness, separated by approximately two times the radius from the center to the outside region and peak, lie within 1% of the distance. Therefore, in most cases, the points overlap. Figure 2.11 shows the luminance distribution $P(r)$ along a straight line connecting two equal-amplitude Gaussian points separated by $d = 2R$. For the two center brightness middle positions of 12.5%, the display system of the low-frequency response of $D(r)$ is inferior.

To satisfy the requirements of the flat point spacing, the following special circumstances can be considered: center of pixel $D(0,0)$, midpoint (two pixels midpoint) $D(1/2,0)$ and diagonal midpoint (four pixel points) $D(1/2,1/2)$. The distribution of

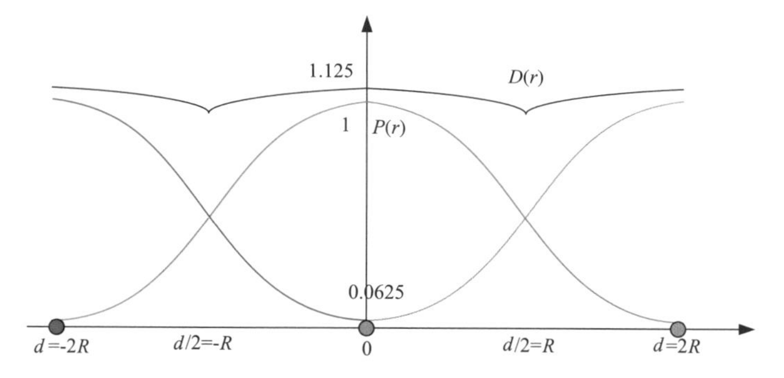

Fig. 2.11 Overlap between adjacent Gaussian points

several points is shown in Fig. 2.12. Ideally, the gray smooth regions are equal; however, this condition is not possible. Therefore, we select the appropriate pixel spacing of the three points of $D(x, y)$ to be equal.

The display brightness at the center of the pixel in the flat area consisting of bright points, with a unit amplitude brightness, can be expressed as

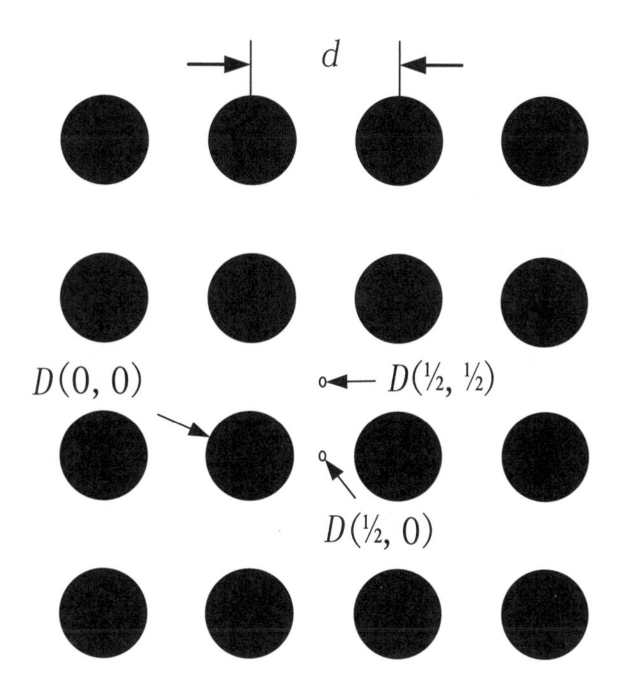

Fig. 2.12 Key positions of the flat region

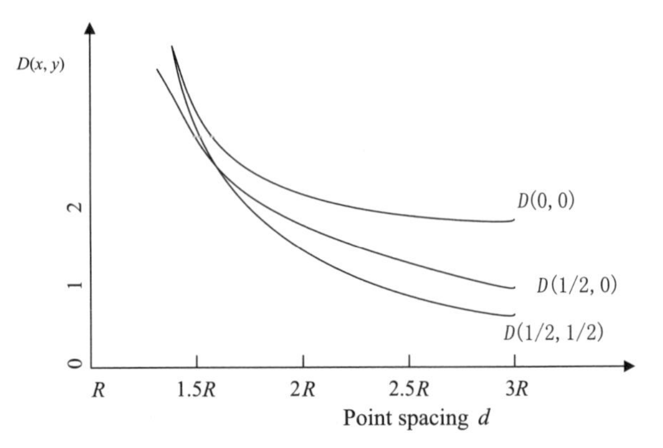

Fig. 2.13 Influence of luminance overlap on region flatness

$$\begin{aligned} D(0,0) &\approx 1 + 4p(d) + 4p\left(\sqrt{2}d\right) \\ D\left(\tfrac{1}{2},0\right) &\approx 2p\left(\tfrac{d}{2}\right) + 4p\left(\sqrt{5}\tfrac{d}{2}\right) \\ D\left(\tfrac{1}{2},\tfrac{1}{2}\right) &\approx 4p\left(\sqrt{2}\tfrac{d}{2}\right) + 8p\left(\sqrt{10}\tfrac{d}{2}\right) \end{aligned} \tag{2.4}$$

By substituting Formula (2.3) into Formula (2.4), we can obtain three expressions of the display brightness:

$$\begin{aligned} D(0,0) &\approx 1 + 2^{2-\left(\frac{d}{R}\right)^2}\left(1 + 2^{-\left(\frac{d}{R}\right)^2}\right) \\ D\left(\tfrac{1}{2},0\right) &\approx 2^{1-\left(\frac{d}{2R}\right)^2}\left(1 + 2^{1-\left(\frac{\sqrt{19}d}{2R}\right)^2}\right) \\ D\left(\tfrac{1}{2},\tfrac{1}{2}\right) &\approx 2^{2-\left(\frac{d}{\sqrt{2}R}\right)^2}\left(1 + 2^{1-\left(\frac{\sqrt{2}d}{2R}\right)^2}\right) \end{aligned} \tag{2.5}$$

Figure 2.13 shows the Gaussian point in the range $2R \le d \le 3R$. The three lines intersect at one point, and it is not possible to find three brightness values corresponding to d. In the range $1.55R \le d \le 1.65R$, the vertical distances between the three lines are small, and thus, this range is optimal. In the flat area of $d = 2R$, the brightness changes by 26%. At $d = 3R$, the brightness in the image area is visible, namely, the mosaic phenomenon, and a smooth gray value is observed instead of slowly varying values. In general, a smaller display point spacing corresponds to a superior flatness of the uniform area, i.e., a higher ability to maintain low-frequency information.

2.3.3 *High-Frequency Response*

In a display system, high-frequency responses appear as large contrast regions, which reflect the vertical display image details [5]. A commonly used high-frequency test

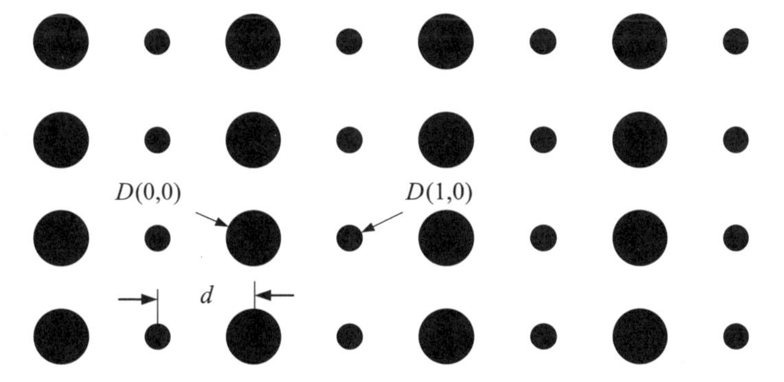

Fig. 2.14 Key positions in the line pattern

pattern consists of vertical (or horizontal) lines alternated by one pixel (alternating light and dark), referred to as a "pair of lines," in which each pair includes a dark line (consisting of zero intensity pixels) and an adjacent bright line (consisting of high-brightness pixels). Figure 2.14 shows a pattern in the high frequency of interest. The rough spots represent unit amplitude (maximum brightness of pixels) and black dots represent zero amplitude pixels (minimum brightness).

The brightness of two specific points can be expressed as

$$D(0,0) \approx 2p(d) + 2p(2d) \tag{2.6}$$

$$D(1,0) \approx 2p(d) + 4p\left(\sqrt{2}d\right) \tag{2.7}$$

Substituting Formula (2.5) into Formulas (2.8) and (2.9) yields the following expression:

$$M = \frac{D(0,0) - D(1,0)}{D(0,0)} \tag{2.8}$$

Similarly, Formula (2.5) is substituted for the expression form of M:

$$M = \frac{1 + 2^{\left(1-(d/R)^2\right)^2}\left(2^{-\left(\sqrt{2}d/R\right)^2-1}\right)}{1 + 2^{1-(d/R)^2}\left(1 + 2^{1-\left(\sqrt{3}d/R\right)^2}\right)} \tag{2.9}$$

According to Fig. 2.15, a larger M corresponds to a higher pattern contrast and superior high-frequency response. When the point spacing is less than $2R$, the modulation depth rapidly decreases. This finding shows that a larger distance between the display points corresponds to a superior reconstruction of the image details.

Another "worst case" high-frequency display pattern is a single-pixel checkerboard. The pixel brightness alternates horizontally and vertically. The key positions

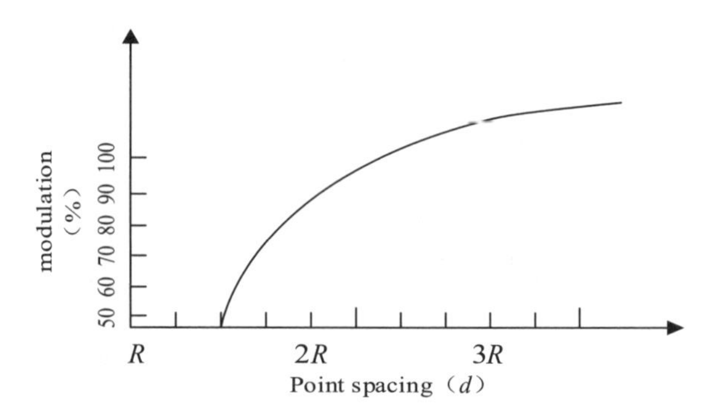

Fig. 2.15 Effect of distance on the vertical pattern

in the pattern are shown in Fig. 2.16. The maximum brightness is specified as

$$D(0,0) \approx 1 + 4p\left(\sqrt{2}d\right) \tag{2.10}$$

The minimum brightness is

$$D(1,0) \approx 4p(d) + 8p\left(\sqrt{5}d\right) \tag{2.11}$$

Similarly, Formula (2.5) is substituted into the abovementioned two formulas to obtain the following mathematical expressions:

$$D(0,0) \approx 1 + 2^{2-\left(\frac{\sqrt{2}d}{R}\right)} \tag{2.12}$$

$$D(1,0) \approx 2^{2-\left(\frac{d}{R}\right)^2}\left(1 + 2^{1-\left(\frac{2d}{R}\right)^2}\right) \tag{2.13}$$

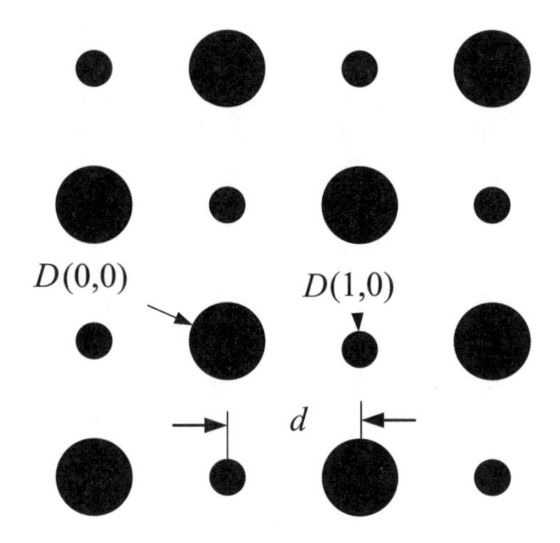

Fig. 2.16 Key position diagram for the case pattern

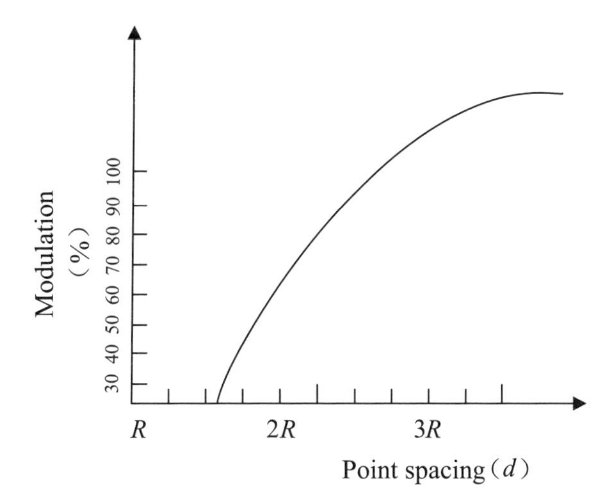

Fig. 2.17 Influence of pitch on the checkerboard pattern

To satisfy the high-frequency response, the selected point spacing must be as large as possible. At this point, the modulation factor M is expressed as

$$M \approx \frac{1 + 2^{2-\left(\frac{d}{R}\right)^2\left[2^{-(d/R)^2}\left(1-2^{1-\left(\sqrt{8}d/R\right)^2}\right)-1\right]}}{1 + 2^{2-\left(\frac{\sqrt{2}d}{R}\right)^2}} \tag{2.14}$$

According to Fig. 2.17, the loss of modulation depth is higher than that of the line pattern as the dot pitch decreases. Therefore, to satisfy the high-frequency response, the selected point spacing must be as large as possible.

This discussion indicates that the high-frequency response and low-frequency response characteristics for the point spacing exhibit opposite trends. Therefore, depending on the importance of the high-frequency and low-frequency information of the image and purpose of the study, the point spacing may be compromised to satisfy the balance requirements of low-frequency contours and high-frequency details of digital image processing [6].

Table 2.2 lists the main features of the high- and low-frequency responses of the output pixels. The distance between points can be set to ensure an effective balance between the high and low frequencies of related types of remotely sensed digital images. If the processing is not satisfactory, the image processing effects may deteriorate.

Table 2.2 List of the relevant characteristics parameters (for inkjet printers, as an example)

Correlation parameter	Lattice spacing	Material characteristics	Ink characteristics	Original image resolution
Image softness (low frequency)	Dense	Rough	High permeability	Low
Image brilliance (high frequency)	Rare	Smooth	Low permeability	High

2.4 Remote Sensing Image Processing Software and Design

This section describes the composition and function of remote sensing image processing software, such as pretreatment, extraction of region of interest, vegetation index calculation and image classification (considering the example of the commonly used ENVI software). The architecture for the remote sensing image processing software design, namely, the "top to bottom" principle of layered sequential structure and design software, is presented, and the design of drone aerial remote sensing digital image processing software is outlined [7].

2.4.1 Commonly Used Remote Sensing Image Processing Software

ENVI (environment for visualizing images) is the flagship product of ITT Visual Information Solutions. ENVI represents a set of powerful remote sensing image processing software programs developed by IDL scientists through remote sensing. The software is fast, convenient and accurate. The software uses geographical space images to extract information from leading software solutions and provides advanced, user-friendly tools to facilitate users to read, prepare, detect, analyze, and share information in images. At present, many imaging analysts and scientists are using ENVI to extract information from geospatial images. The programs have been widely used in scientific research, environmental protection, gas, oil and mineral exploration, agriculture, forestry, medicine, national defense security, earth science, public facilities management, remote sensing engineering, water conservancy, oceanography, surveying, and urban and regional planning industries [8].

ENVI provides automatic preprocessing tools that can rapidly and easily preprocess images. The software includes a comprehensive set of data analysis tools that include algorithms that can analyze images rapidly, easily, and accurately and advanced and easy-to-use spectral analysis tools that can facilitate scientific image analysis.

At present, the most popular image processing software programs in the remote sensing field, in addition to the remote sensing image software ENVI, include ERDAS, ER, Mapper, and PCI.

2.4.2 Design of Remote Sensing Digital Image Processing Software

Remote sensing digital image processing software represents a powerful tool to perform remote sensing image processing through computers. In the process of image processing, the software must be designed according to the specific processing requirements. Figure 2.18 illustrates the software architecture in terms of the demand function.

Remote sensing image processing software exhibits a layered sequential structure. The module includes the original image display layer, followed by the linear system

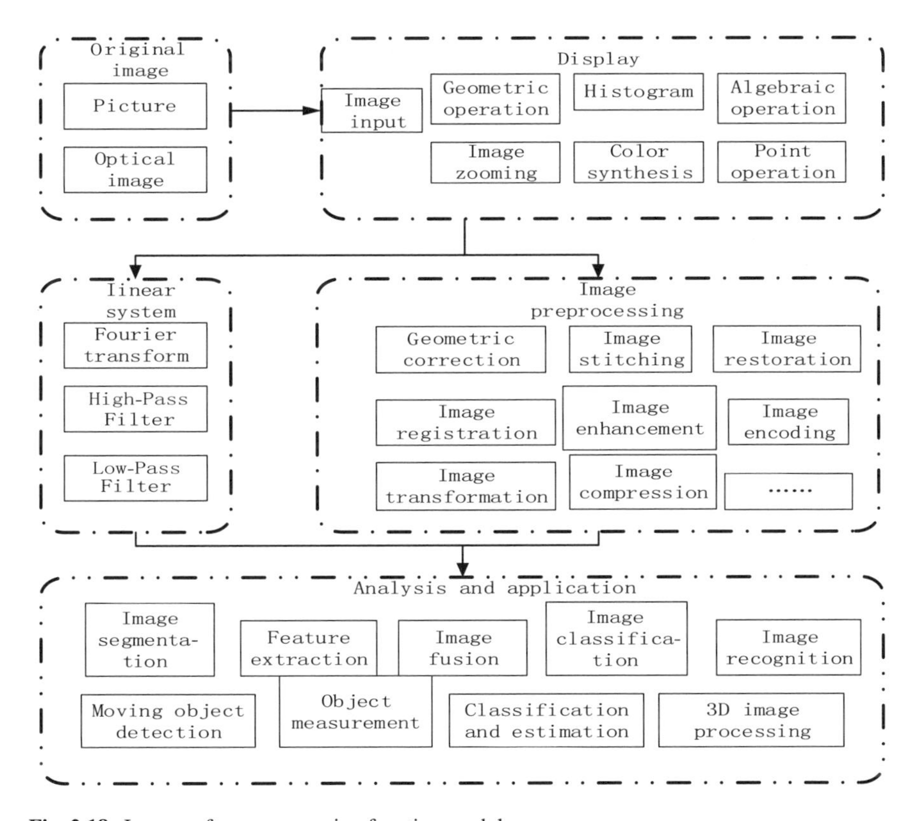

Fig. 2.18 Image software processing function module

pretreatment layer and the analysis and application layer. The three-layer interaction for the remote sensing images can help reflect fine details.

2.4.3 Example of Image Processing Software

A UAV system is considered as an example to describe the system processing flow of UAV remote sensing software, as shown in Figs. 2.19 and 2.20:

The system operates as follows: (1) Before the staff performs route planning, data are transferred to the control system; (2) UAV controlled flight is performed to implement the sensor; (3) the remote sensing control subsystem stores the captured data and transmits the fast view generated after sampling the original data to the drone platform through the remote sensing control subsystem. The drone platform uses the wireless transmission channel to transmit the fast view data to the ground receiving and distribution subsystem; (4) the ground receiving and distribution system unzips the data and performs storage and classification through the remote sensing data monitoring subsystem and remote sensing image processing and target recognition subsystems; (5) the ground staff can monitor the corresponding processing results on the ground through various processing subsystems.; (6) after the operation, the system is shut down, and data storage and recycling of equipment is performed.

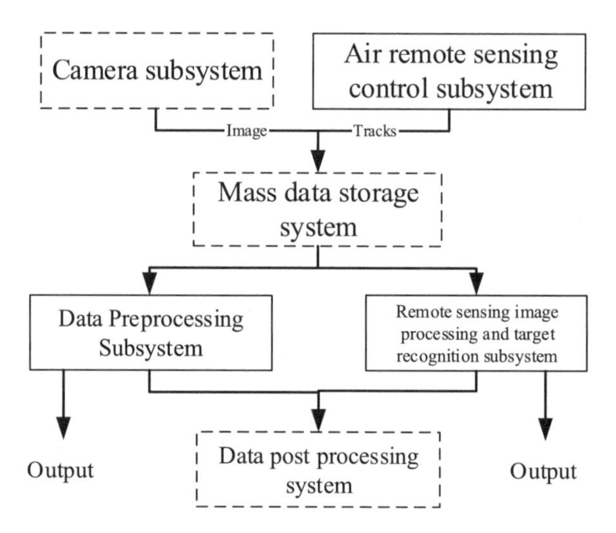

Fig. 2.19 Data acquisition process block diagram

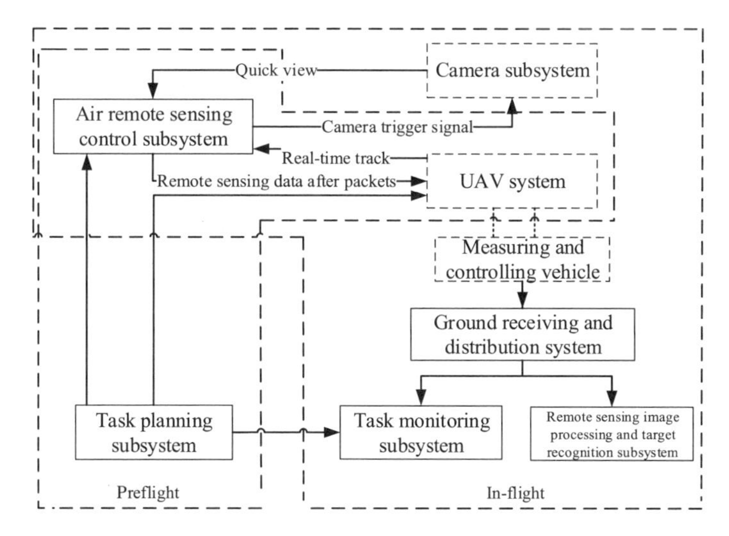

Fig. 2.20 Data processing block diagram after operation

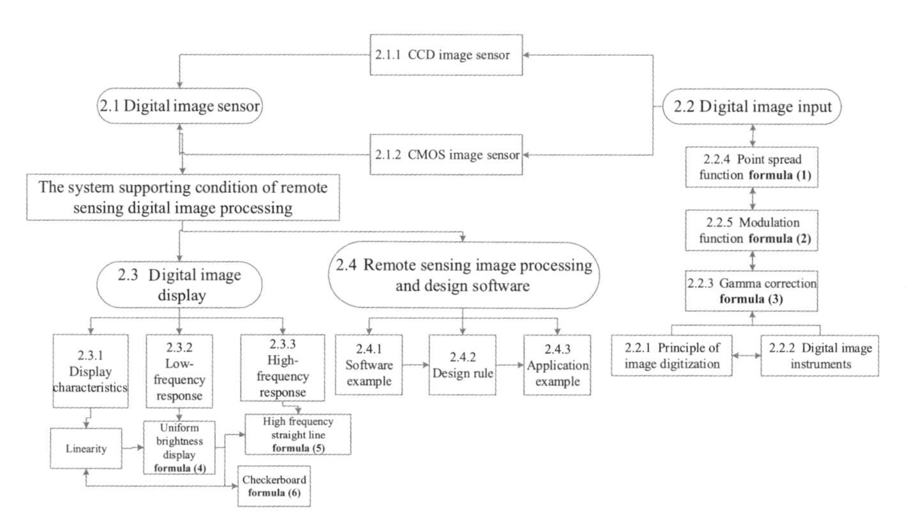

Fig. 2.21 Summary of the contents of this chapter

Table 2.3 Summary of mathematical formulas in this chapter

formula (1)	$h(x, y) = e^{-r^2}$	2.2.4
formula (2)	MTF = Contrast(Output Image)/Contrast(Input Image)	2.2.5
formula (3)	$R_{gc} = 255 \times \left(\frac{R}{255}\right)^{\frac{1}{\gamma_{red}}}, G_{gc} = 255 \times \left(\frac{G}{255}\right)^{\frac{1}{\gamma_{green}}}, B_{gc} = 255 \times \left(\frac{B}{255}\right)^{\frac{1}{\gamma_{blue}}}$	2.2.3
formula (4)	$D(0,0) \approx 1 + 4p(d) + 4p\left(\sqrt{2}d\right), D\left(\frac{1}{2}, 0\right) \approx 2p\left(\frac{d}{2}\right) + 4p\left(\sqrt{5}\frac{d}{2}\right),$ $D\left(\frac{1}{2}, \frac{1}{2}\right) \approx 4p\left(\sqrt{2}\frac{d}{2}\right) + 8p\left(\sqrt{10}\frac{d}{2}\right)$	2.3.2
formula (5)	$D(0,0) \approx 2p(d) + 2p(2d), D(1,0) \approx 2p(d) + 4p\left(\sqrt{2}d\right)$ $M = \frac{D(0,0) - D(1,0)}{D(0,0)}$	2.3.3
formula (6)	$D(0,0) \approx 1 + 4p\left(\sqrt{2}d\right)$ $D(1,0) \approx 2^{2-\left(\frac{d}{R}\right)^2}\left(1 + 2^{1-\left(\frac{2d}{R}\right)^2}\right)$	2.3.3

2.5 Summary

This chapter systematically introduces and compares the two most commonly used digital image sensors, CCD and CMOS, and describes the principles, indicators, and related diffusion functions, modulation transfer functions, and gamma corrections of digital imagers along with digital image displays. Features and responses are introduced. This chapter also describes the software and design of remote sensing image processing, and a UAV aerial remote sensing digital image processing software design is explained. The contents of this chapter are summarized in Fig. 2.21 and Table 2.3.

References

1. Song J. Research on Key Techniques in Remote Sensing Image Processing software. Institute of Remote Sensing Applications, CAS. 2005.
2. Yang B. CCD Subdivision Technology and Its Application Research. Institute of Geophysics, China Earthquake Administration. 2005.
3. Wang F, Wang J, Hu D. Comparison of three commercial remote sensing digital image processing software. Remote Sens Technol Appl. 1998;02. (In Chinese).
4. Xiong P. CCD and CMOS image sensor characteristic comparison. Semiconduct Photoelectr. 2004;01. (In Chinese).

5. Ni J, Huang Q. CMOS image sensor and its development trend. Opt Electromech Inf. 2008;05. (In Chinese).
6. Wang F, Wang J, Hu D. Three commercial remote sensing digital image processing software comparison. Remote Sens Technol Appl.1998;02. (In Chinese).
7. CCD. Encyclopaedia Britannica Online. Encyclopaedia Britannica Inc. 2013.
8. Balestra F. Nanoscale CMOS: innovative materials, modeling, and characterization. Hoboken, NJ:ISTE. 2010. Mohammad H, Mirshams M. Remote sensing satellites evaluation software (RSSE software). Aircraft Eng Aerospace Technol.2009;81. https://doi.org/10.1108/000226609 10967282.

Chapter 3
Mathematical Basis for the Overall Processing and Analysis of Remote Sensing Digital Images

Chapter Guidance After establishing the system support conditions for the processing and analysis of remote sensing digital imagery (Chap. 2), reasonable mathematical methods can be used to analyze the image and enhance the system performance. However, methods to analyze the overall performance of the image and the use of mathematical methods to enhance the overall image quality based on the analysis results must be clarified.These problems appear to be overcome through image preprocessing. Image preprocessing can eliminate the irrational appearance of the image, that is, "treat the symptoms", regardless of their origins. Notably, image preprocessing can only change the overall performance of the image and does not change the relationship of the image, thereby ensuring the performance (Part 2) for each unit in digital image processing analysis. The core of this chapter is the gray histogram theory and histogram convolution properties of the related computing.

Remote sensing digital image processing is aimed at the mathematical transformation of remote sensing images. The specific mathematical transformations include the following: 3.1 Grayscale Histograms and Probability to describe and analyze the image; 3.2 Point Operations and Function Transformation to realize the function transformation of the image; 3.3 Algebraic Operations and Convolution to realize the operations of two input images, including point-to-point add, subtract, multiply and divide calculations, which can be simplified to the two-image histogram of convolution (delay inertial space characteristics); 3.4 Geometric Operations and Matrices to realize the image translation, rotation, scaling, projection space transformation and grayscale numerical integral interpolation. The logical framework of this chapter is shown in Fig. 3.1.

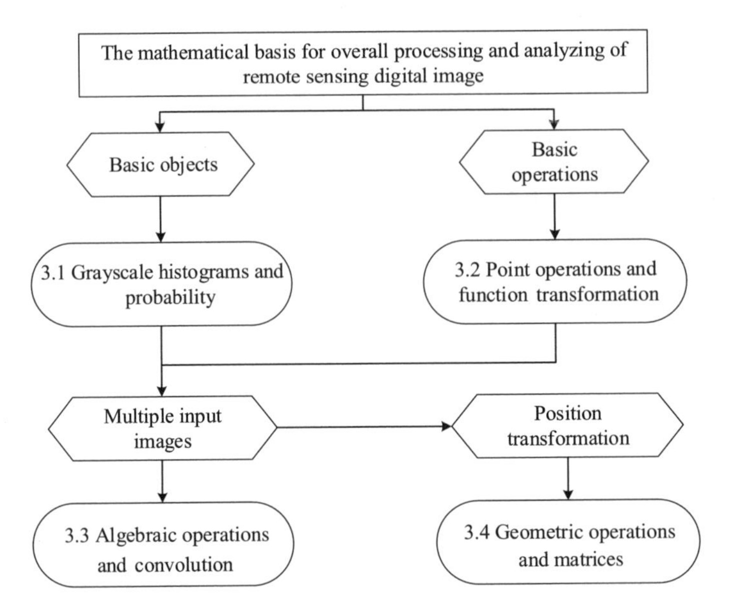

Fig. 3.1 Logical framework of this chapter

3.1 Grayscale Histograms and Probability

The grayscale histogram describes the proportion of pixels with different grayscale values in the entire image. This entity is a function of the grayscale level and does not consider the spatial position of each grayscale pixel. This function is used to analyze the characteristics of the whole image, as shown in Fig. 3.2.

For continuous images, the following steps can be used to define histograms: a continuous image is defined using function $D(x, y)$, and the grayscale value smoothly decreases. Considering grayscale D_1, all the image grayscales for D_1 points are

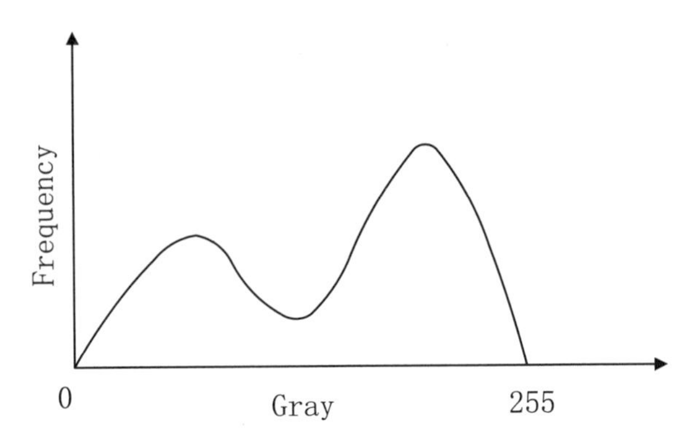

Fig. 3.2 Histogram of the digital image

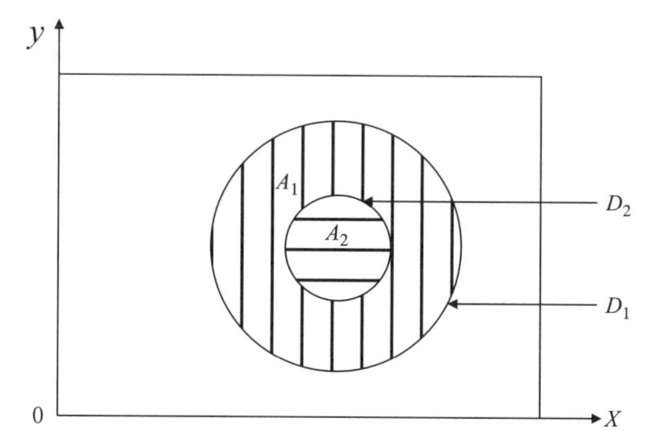

Fig. 3.3 Grayscale contour of the continuous image

connected to obtain a contour line surrounded by a regional gray level greater than D_1, as shown in Fig. 3.3.

The threshold area function $A(D)$ is the area enclosed by contour D, and the histogram of the continuous image is defined as

$$\begin{cases} H(D) = \lim\limits_{\Delta D \to 0} \frac{A(D) - A(D + \Delta D)}{\Delta D} = -\frac{\mathrm{d}}{\mathrm{d}D} A(D) \\ A(D) > A(D + \Delta D) \end{cases} \tag{3.1}$$

3.1.1 Probability Density of the Histogram

1. Threshold Area Function, $A(D)$

This value refers to the area enclosed by all the outlines with grayscale level D in the continuous image, defined as

$$A(D) = \int_D^{\infty} H(p) dp \tag{3.2}$$

The image area and histogram relations are shown in Fig. 3.4a and b.

2. Probability Density Function (PDF)

A histogram normalized to a unit area is defined as (Fig. 3.4c)

$$PDF = p(D) = \frac{1}{A_0} H(D) \tag{3.3}$$

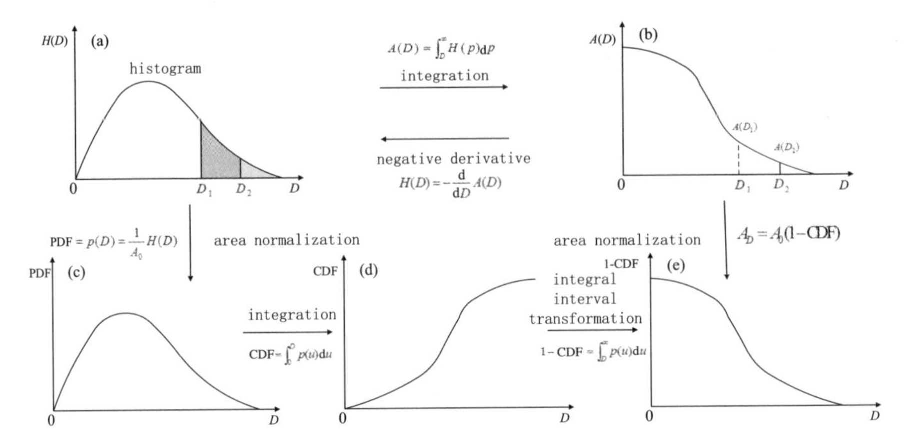

Fig. 3.4 Relationship between the histogram and its associated functions

A_0 is the area of the image.

3. Cumulative Distribution Function (CDF)

The area enclosed by the probability density function, which is also the normalized threshold area function [1], defined as (Fig. 3.4d).

$$\text{CDF} = P(D) = \int_0^D p(u)\mathrm{d}u = \frac{1}{A_0}\int_0^D H(u)\mathrm{d}u \tag{3.4}$$

From histogram $H(D)$ to image area function $A(D)$, the essence of mathematics is as follows: through the integration of the probability density function PDF, the integral interval sequence is changed. The explanation is as follows: the histogram is the number of different digital image grayscale bitmap statistical frequencies. When the samples (that is, the image bitmap) are adequately large and considerably more than the number of grayscale levels, the frequency is associated with different gray levels of the probability density function, and the sum of these probability densities is equal to 1. We multiply the area A_0, which is the area of the image, as shown in Fig. 3.4.

3.1.2 Histogram Applications

1. Rapid Image Detection

Gray histograms can be used to determine whether an image can enable the use of the full range of allowed grayscale and detect the problems existing in digital images.

2. Boundary Threshold Selection

Different grayscale contours can help effectively determine the boundaries of simple objects in an image. For example, if a dark object exists in a light background image, the threshold can be determined using the valley T between the two peaks of the histogram. The image is divided into two parts, as shown in Fig. 3.5, and expressed as the increase in the gray level and probability (light background).

3. Integrated Optical Density (IOD)

This metric reflects the "quality" of the image, defined as

$$\mathrm{IOD} = \int_0^a \int_0^b D(x, y)\mathrm{d}x\,\mathrm{d}y \tag{3.5}$$

where a and b are the boundaries of the delineated image region. For digital images,

$$\mathrm{IOD} = \sum_{i=1}^{N_L} \sum_{j=1}^{N_S} D(i, j) \tag{3.6}$$

$D(i, j)$ is a pixel gray value for (i, j), N_L is the number of rows in the image, and N_S is the number of columns in the image.

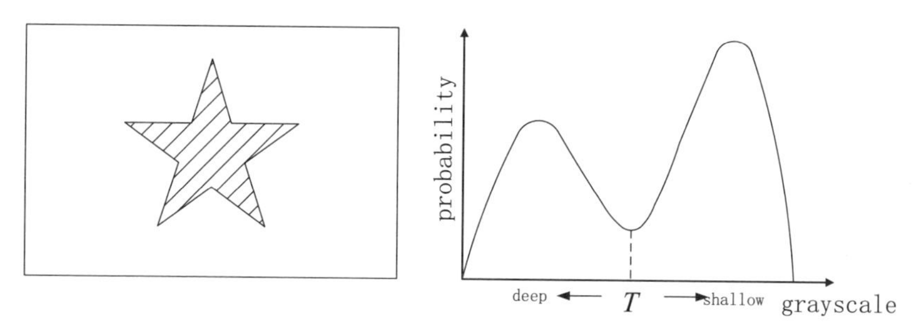

(a) Light background dark object digital image (b) Double peak histogram

Fig. 3.5 Image with notable differences between the background and object and histogram classification

3.2 Point Operations and Function Transformation

According to Sect. 3.1, the overall characteristics of the image can be represented by a grayscale histogram. Based on this aspect, point operations can realize the image conversion. Point operation pertains to contrast enhancement, contrast stretching or gray level transformation, and its essence is based on a certain gray level transformation function of the image gray value of the expected change, which can be regarded as a "transformation from pixel to pixel" operation and is an important part of remote sensing digital image processing. For the input image $A(x, y)$ and output image $B(x, y)$, the point operation can be expressed as

$$B(x, y) = f[A(x, y)] \tag{3.7}$$

where $f(A)$ is the grayscale function (grayscale transformation, GST).

3.2.1 *Types and Characteristics of Point Operations*

Point operations can be completely determined by the gray level transformation function, and the gray level transformation function describes the input and output of the mapping relationship between the gray levels. Notably, the point operation changes only the gray value of each point without changing their spatial relationship. Different types of point operations correspond to different gray level transformation functions, which can be divided into the following two types.

(1) Linear point operation, which is typically applied to grayscale distribution standardization

The linear point operation refers to the point operation in which the output gray level is linearly dependent on the input grayscale, and the grayscale transform function $f(D)$ is a linear function, i.e.,

$$D_B = f(D_A) = aD_A + b \tag{3.8}$$

If $a = 1$ and $b = 0$, $b(x, y)$ is a simple copy of $a(x, y)$. If $a > 1$, the contrast of the output image is enhanced. If $a < 1$, the contrast of the output image decreases. If $a = 1$ and $b \neq 0$, the output image becomes brighter or darker. If $a < 0$, the bright area darkens, and the dark area brightens.

The inverse function is

$$D_A = f^{-1}(D_B) = (D_B - b)/a \tag{3.9}$$

$H_A(D)$ and $H_B(D)$ represent the histogram of the input and output images, respectively.

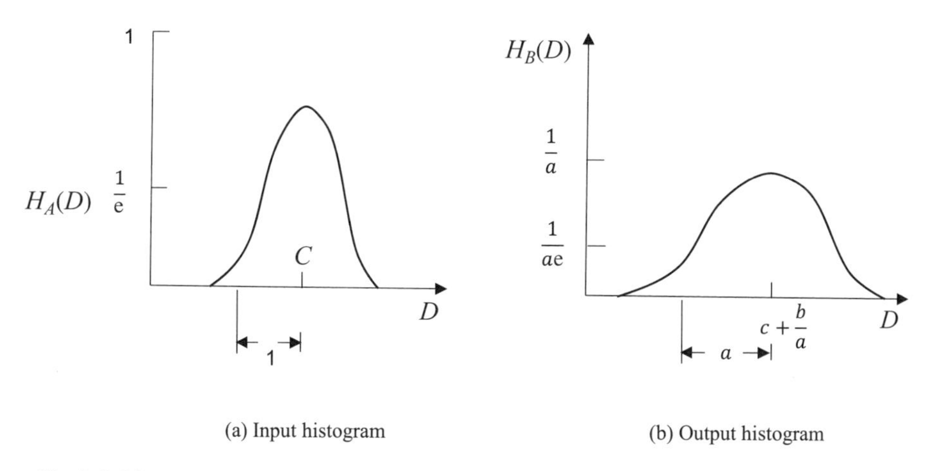

Fig. 3.6 Linear transformation of the Gaussian function histogram

$$H_B(D) = \frac{1}{a} H_A\left(\frac{D-b}{a}\right) \tag{3.10}$$

Assuming that the input histogram is a Gaussian function, $H_A(D) = \mathrm{e}^{-(D-c)^2}$ as shown in Fig. 3.6a, the output histogram is $H_B(D) = \frac{1}{a}\mathrm{e}^{-\left[\frac{D}{a}-\left(c+\frac{b}{a}\right)\right]^2}$ as shown in Fig. 3.6b. The output histogram is also a Gaussian function, and only the peak and width are changed.

(2) Nonlinear monotonic point operation, which is typically applied for threshold processing (image binarization)

The linear operation of an image involves many limitations, and the effect to be achieved often exceeds the ability of the linear operation. In this case, the nonlinear operation is of significance. This section describes nonlinear monotone point operations in which the drab gray level transformation function is satisfied. According to the operation of the middle range of grayscale, common nonlinear monotone point operations can be divided into the following three categories.

In the first type of operations, the gray level of the intermediate range of pixels is increased, while the brighter and darker pixels exhibit only a minor change, that is,

$$f(x) = x + cx(D_{\mathrm{m}} - x) \tag{3.11}$$

D_{m} is the maximum grayscale of the image.

In the second type of operations, the contrast of the brighter or darker parts of the object is decreased, and the contrast of the object in the middle range is enhanced, that is,

$$f(x) = \frac{D_{\mathrm{m}}}{2}\left\{1 + \frac{1}{\sin\left(a\frac{\pi}{2}\right)}\sin\left[a\pi\left(\frac{x}{D_{\mathrm{m}}} - \frac{1}{2}\right)\right]\right\} \tag{3.12}$$

This point operation has a slope greater than 1 in the middle and less than 1 at both ends, and the curve exhibits an S-shape.

In the third type of operations, the contrast of objects in the middle range is decreased, and the contrast between the brighter and darker parts is enhanced, that is,

$$f(x) = \frac{D_{\mathrm{m}}}{2}\left\{1 + \frac{1}{\tan\left(a\frac{\pi}{2}\right)}\tan\left[a\pi\left(\frac{x}{D_{\mathrm{m}}} - \frac{1}{2}\right)\right]\right\} \tag{3.13}$$

This point operation has a slope less than 1 in the middle and greater than 1 at both ends.

Thresholding represents a typical application of nonlinear point operations. The function is to select a threshold and binarize the image for image segmentation and edge tracking.

3.2.2 Function Transformation of Point Operations

Assume that one input image $A(x, y)$ subjected to the gray transformation function $f(D)$ defined by the point after processing produces output image $B(x, y)$. We derive the output image histogram expression for the known input image histogram $H_A(D)$. Assume that any output pixel gray value is defined as

$$D_B = f(D_A) \tag{3.14}$$

The function conversion relationship is shown in Fig. 3.7. Assuming that $f(D)$ is a nondecreasing function with a limited slope,

$$D_A = f^{-1}(D_B) \tag{3.15}$$

In Fig. 3.7, the grayscale D_A is converted to D_B, the grayscale $D_A + \Delta D_A$ is converted to $D_B + \Delta D_B$, and all the pixels in this interval are accordingly converted. The number of output pixels between gray levels D_B and $D_B + \Delta D_B$ is equal to the number of input pixels between gray levels D_A and $D_A + \Delta D_A$. The area of $H_B(D)$ between D_B and $D_B + \Delta D_B$ must be equal to the area of $H_A(D)$ between D_A and $D_A + \Delta D_A$:

$$\int_{D_B}^{D_B+\Delta D_B} H_B(D)\mathrm{d}D = \int_{D_A}^{D_A+\Delta D_A} H_A(D)\mathrm{d}D \tag{3.16}$$

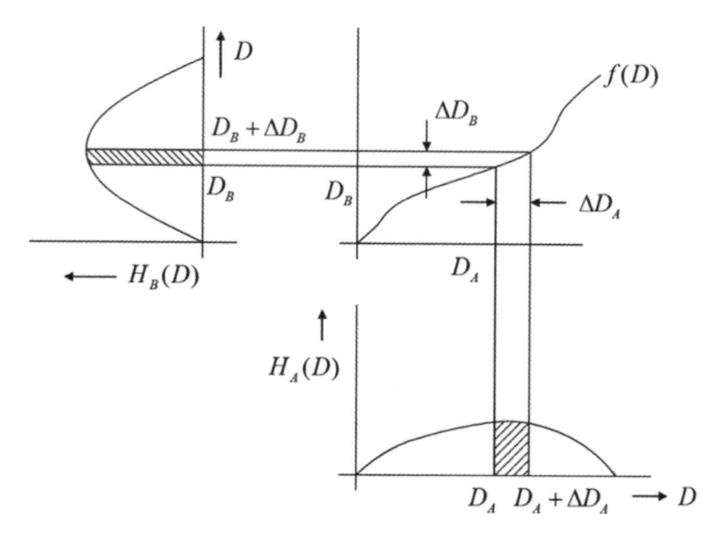

Fig. 3.7 Relationship between the linear point operation and histogram

When ΔD_A and ΔD_B are small, the area of integration can be approximated by a rectangle:

$$H_B(D_B)\Delta D_B = H_A(D_A)\Delta D_A \tag{3.17}$$

Therefore,

$$H_B(D_B) = \frac{H_A(D_A)\Delta D_A}{\Delta D_B} = \frac{H_A(D_A)}{\Delta D_B / \Delta D_A} \tag{3.18}$$

The limit yields

$$H_B(D_B) = \frac{H_A(D_A)}{\mathrm{d}D_B / \mathrm{d}D_A} \tag{3.19}$$

Substituting this expression into the front inverse function, we obtain the output histogram formula from the input histogram and transform function

$$H_B(D_B) = \frac{H_A(D_A)}{\mathrm{d}(f(D_A)) / \mathrm{d}D_A} = \frac{H_A[f^{-1}(D)]}{f'[f^{-1}(D)]} \tag{3.20}$$

where$f' = \mathrm{d}f/\mathrm{d}\ D$.

3.2.3 Applications of Point Operations

1. Histogram equalization

For a continuous image, the histogram equalization of the grayscale transform function is the maximum image gray level D_m multiplied by the histogram of the cumulative distribution function CDF, and the grayscale of the transformed image corresponds to a uniform density function [2], as shown in Fig. 3.8.

$$f(D) = D_m \times \text{CDF} = D_m P(D) = \frac{D_m}{A_0}\int_0^D H(u)\mathrm{d}u \tag{3.21}$$

For discrete images, the actual histogram may appear uneven. Figure 3.9 shows the image associated with a lack of exposure, with the image grayscale concentrated in the darker area. After deriving the transformation function through the principle of cumulative distribution function, the new image grayscale is uniformly distributed, and the image quality is enhanced, as shown in Fig. 3.10.

2. Histogram Matching

We transform an image such that its histogram matches that of another image or a particular function. This process is an extension of histogram equalization, known as histogram matching [3]:

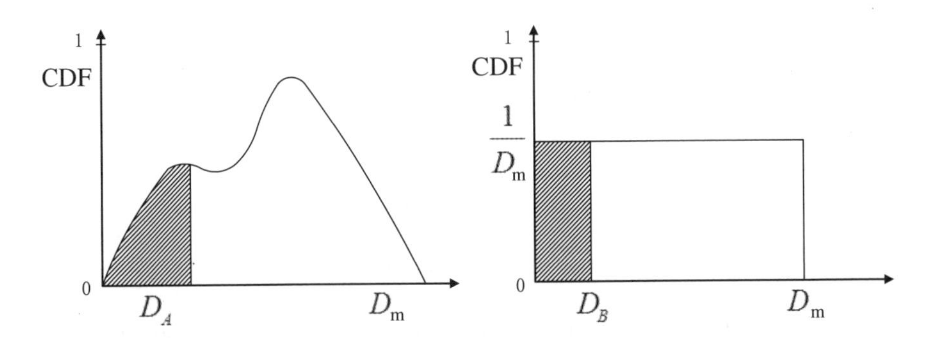

Fig. 3.8 Histogram equalization

Fig. 3.9 Before equalization of histograms

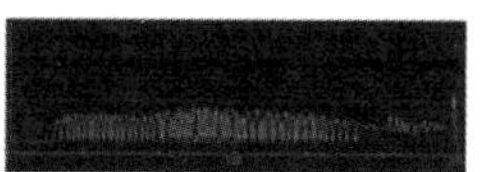

Fig. 3.10 After equalization of histograms

First, the histogram of image $A(x, y)$ is equalized by the point operation $f(D)$ to obtain image $B(x, y)$, that is,

$$B(x, y) = f[A(x, y)] = D_{\mathrm{m}} P[A(x, y)] \tag{3.22}$$

Next, image $B(x, y)$ is transformed into image $C(x, y)$ by the second point function $g(D)$

$$C(x, y) = g[B(x, y)] \tag{3.23}$$

According to (3.22) and (3.23),

$$C(x, y) = g\{D_{\mathrm{m}} P[A(x, y)]\} \tag{3.24}$$

$$g(D) = P^{-1}\left(\frac{D}{D_{\mathrm{m}}}\right) \tag{3.25}$$

Next, we obtain the histogram matching transformation formula:

$$C(x, y) = g\{f[A(x, y)]\} = P_3^{-1}\{P_1[A(x, y)]\} \tag{3.26}$$

where P_3 and P_1 represent the CDFs of images $C(x, y)$ and $A(x, y)$, respectively.

3.3 Algebraic Operations and Convolution

The point operation represents a function transformation of one image. For two or more images, algebraic operations can be performed. An algebraic operation refers the point-to-point addition, subtraction, and multiplication and division calculation of two or more input images to obtain the output image. The essence of algebraic operation for two images is the convolution of the histogram, and the grayscale histogram of the image is obtained after algebraic operation. The convolution expressions of the addition and subtraction algebraic operations (Eq. 3.27) and multiplication and division operations (Eq. 3.28) of the two images are as follows:

$$\left.\begin{array}{r}H_C(D_C) = H_A(D_A)*H_B(D_B)\\ H_C(D_C) = H_A(D_A) * H_B(\overrightarrow{D_B})\end{array}\right\} \tag{3.27}$$

$$\left.\begin{array}{r}H_C(D_C) = H_A(D_A)*H_B(D_B)\\ H_C(D_C) = H_A(D_A) * H_{B^{-1}}\left(D_B^{-1}\right)\end{array}\right\}\text{(Shift convolution)} \tag{3.28}$$

where the multiplication and division operations corresponding to Eq. (3.28) are shifted convolution operations. Algebraic operations correspond to only the addition and subtraction of the corresponding pixel gray value in the image, regardless of other points, or the position of the pixels in the image.

3.3.1 Addition Operations

The two-dimensional histogram of two unrelated histograms is the product of the respective histograms:

$$H_{AB}(D_A, D_B) = H_A(D_A)H_B(D_B) \tag{3.29}$$

The two-dimensional histogram can be reduced to a one-dimensional marginal histogram:

$$H(D_A) = \int_{-\infty}^{\infty} H_{AB}(D_A, D_B)\mathrm{d}D_B = \int_{-\infty}^{\infty} H_A(D_A)H_B(D_B)\mathrm{d}D_B \tag{3.30}$$

For every point in the image,

$$D_A = D_C - D_B \tag{3.31}$$

In this case,

$$H(D_A) = \int_{-\infty}^{\infty} H_A(D_c - D_B)H_B(D_B)\mathrm{d}D_B \tag{3.32}$$

The resulting histogram of the output can be expressed as

$$H_C(D_C) = H_A(D_A) * H_B(D_B) \tag{3.33}$$

where * represents convolution operations.

The image addition can be averaged over multiple images of the same scene to decrease the additive random noise, and one image can be added to another image to achieve the effect of double exposure.

3.3.2 Subtraction Operations

Subtraction can be performed to detect the differences between two images. The essence is the addition of nonoperations. Subtraction has the following three applications.

(1) Eliminating the background

The background effects can be removed to eliminate several system effects and highlight the object. After obtaining the object image, the object can be removed to obtain the image of the blank area, and the two images can be subtracted to obtain the image of the object. In this process, the grayscale range after subtraction must be focused on.

(2) Motion detection

For two images with the same area but different temporal instances, the image of the moving object, namely, the difference image, can be obtained by subtracting the two images.

(3) Gradient amplitude

In an image, the region with a large grayscale variation exhibits a large gradient value, which is generally the boundary of an object within the image; therefore, the gradient image of the image can help obtain the image object boundary [4].

$$\nabla f(x, y) = \boldsymbol{i}\frac{\partial f(x, y)}{\partial x} + \boldsymbol{j}\frac{\partial f(x, y)}{\partial y} \tag{3.34}$$

The gradient is

$$|\nabla f(x, y)| = \sqrt{\left(\frac{\partial f}{\partial x}\right)^2 + \left(\frac{\partial f}{\partial y}\right)^2} \tag{3.35}$$

3.3.3 Multiplication Operations

1. One-Dimensional Situation

Let the input functions be $D_A(x)$ and $D_B(x)$. The output function is $D_C(x) = D_A(x) * D_B(x)$. In the case of one dimension, the length is the "area", and the threshold area function can be expressed by the inverse function of $D_C(x)$:

$$x(D_C) = f^{-1}(D_A(x) * D_B(x)) \tag{3.36}$$

Therefore, the output function histogram can be expressed as

$$H_C(D_C) = \frac{\mathrm{d}f^{-1}(D_A(x) * D_B(x))}{\mathrm{d}(D_A(x) * D_B(x))} \tag{3.37}$$

2. Two-Dimensional Situation

Let the input images be D_A and D_B. The output image $D_C = D_A * D_B$. According to the histogram matching principle,

$$\int_0^D H_C(D_C)\mathrm{d}D_C = A_0 P_A[D_A(x, y)] = A_0 P_B(D_B(x, y)) \tag{3.38}$$

where P_A and P_B are the cumulative distribution functions of the input images A and B, respectively, and the left side of the formula represents the area of the output image C. The following expressions can be obtained:

$$H_C(D_C) = \frac{A_0\mathrm{d}\{P_A[D_A(x, y)]\}}{\mathrm{d}D_C} = \frac{A_0\mathrm{d}\{P_A[D_A(x, y)]\}}{\mathrm{d}\{D_A(x, y) * D_B(x, y)\}} = \frac{A_0 * \mathrm{d}\int_0^D H_A\{D_A(x, y)\mathrm{d}D_A(x, y)\}}{A_1 * \mathrm{d}\{D_A(x, y) * D_B(x, y)\}} \tag{3.39}$$

$$H_C(D_C) = \frac{A_0\mathrm{d}\{P_B[D_B(x, y)]\}}{\mathrm{d}D_C} = \frac{A_0\mathrm{d}\{P_B[D_B(x, y)]\}}{\mathrm{d}\{D_A(x, y) * D_B(x, y)\}} = \frac{A_0 * \mathrm{d}\int_0^D H_B\{D_B(x, y)\mathrm{d}D_B(x, y)\}}{A_2 * \mathrm{d}\{D_A(x, y) * D_B(x, y)\}} \tag{3.40}$$

A_1 and A_2 represent the areas of images A and B, respectively.

Multiplication operations can be used to mask certain parts of an image. When a mask binary image is set, the mask image value corresponding to the portion of the original image that must be retained is set as 1, and the portion to be masked is set as 0. The mask image multiplied by the original image is used to erase part of the area, as shown in Fig. 3.11.

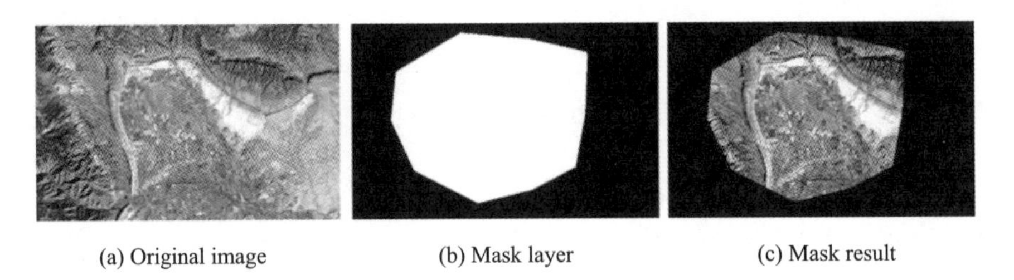

(a) Original image (b) Mask layer (c) Mask result

Fig. 3.11 Mask operation of digital image

Fig. 3.12 Remote sensing image and its RVI index

3.3.4 Division Operations

The essence of the division operation is the inverse operation of the multiplication operation, which produces the ratio image and is used for multispectral image analysis. For example, the relative vegetation index (RVI) obtained by dividing the near infrared band and visible red band can be used as a green vegetation index, as shown in Fig. 3.12. A brighter region corresponding to the remote sensing image vegetation area corresponds to a higher RVI index.

3.4 Geometric Operations and Matrices

Neither point nor algebraic operations involve changes in the position of the pixels in the image, whereas geometric operations can change the position of the pixel in the image. The essence of the geometric operation pertains to image translation, rotation, and scaling, the space projection matrix and the new pixel gray value of integer numerical value calculations. A geometric operation consists of two parts: space transformation and grayscale value determination. The former part describes the transformation of the spatial position of each pixel. The latter part determines the grayscale level of the transformed image pixels. A geometric operation can be implemented through the following two methods: forward and backward mapping. Mapping is the process of space transformation, and the mapping process after determining the mapped pixel values is a type of grayscale interpolation process.

3.4.1 Methods to Implement Geometric Operations

1. Forward Mapping

The forward mapping method transfers the grayscale of the input image pixel by pixel to the output image. If an input pixel is mapped to the position between four output pixels, the gray value is interpolated by the interpolation algorithm over the four output pixels; this process is known as pixel transfer. Since many input pixels may be mapped to the boundaries of the output image, the forward mapping algorithm is slightly wasteful. Moreover, the gray value of each output pixel may be determined by the gray value of many input pixels, corresponding to multiple calculations.

As shown in Fig. 3.13, the input pixel A_1 is transferred to the output point of B_1, and the A_1 gray value is assigned to pixels *a*, *b*, *c*, and *d*. Similarly, the gray value of A_2 is assigned to *d*, *e*, *f*, and *g*. The gray value of *d* from A_1, A_2 or additional input pixels must be calculated several times.

2. Backward Mapping

The backward mapping method sequentially maps the output pixels back to the input image. The backward mapping algorithm outputs the image pixel-by-pixel and line-by-line. The gray level of each pixel is uniquely determined by the interpolation of up to four pixels and calculated only once.

As shown in Fig. 3.14, the output pixel *A* is mapped to point *B* in the input image. The grayscale is determined by the grayscale value of four pixels *a*, *b*, *c*, and *d* around point *B*, and only one calculation is performed.

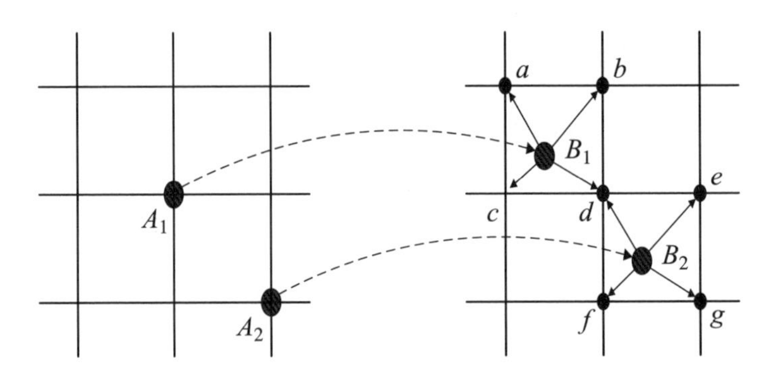

Fig. 3.13 Forward mapping

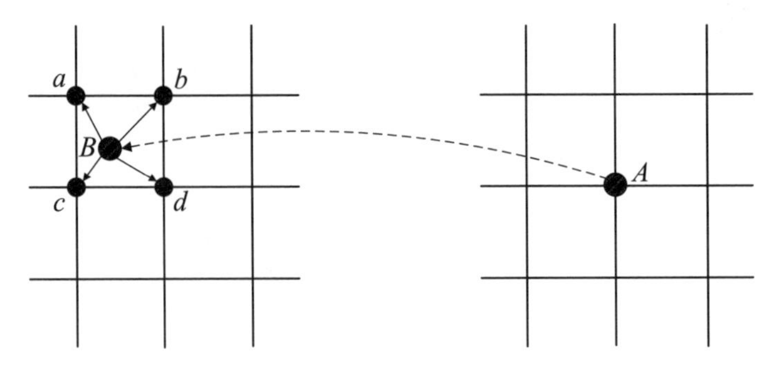

Fig. 3.14 Backward mapping

3.4.2 *Spatial Transformation*

In most applications, it is necessary to maintain the continuity of curves in the image and connectivity of objects. A less restrictive spatial transform algorithm is likely to disrupt straight lines and image, rendering the image content "fragmented". Although users can specify the movement of each pixel in the image point by point, this method may be tedious even for small images. A more convenient approach is to use mathematical methods to describe the spatial relationship between the input and output image points.

In geometric operations, the spatial relationship between the input and output image points can be described mathematically as

$$g(x, y) = f(x', y') = f[a(x, y), b(x, y)] \tag{3.41}$$

$f(x, y)$ and $g(x, y)$ represent the input and output images, respectively. The functions $a(x, y)$ and $b(x, y)$ uniquely describe the spatial transformation, and if the functions are continuous, the connectivity is expected to be maintained in the image.

3.4.3 *Grayscale Level Interpolation*

Commonly used grayscale-level interpolation algorithms include nearest neighbor interpolation, bilinear interpolation, and high-order interpolation.

1. Nearest neighbor interpolation

Nearest neighbor interpolation is the simplest interpolation method, also known as zero-order interpolation. In this framework, the gray value of the output pixel is equal to that of the input pixel closest to the position to which it is mapped.

2. Bilinear interpolation

The bilinear interpolation method uses the four points around the noninteger point to estimate the gray value, and the pixel value is interpolated according to a certain weight function. Figure 3.15 shows the simplified algorithm:

First, the upper two vertices are linearly interpolated:

$$f(x, y_0) = f(x_0, y_0) + [f(x_1, y_0) - f(x_0, y_0)]\frac{(x - x_0)}{(x_1 - x_0)} \tag{3.42}$$

Subsequently, the bottoms of the two vertices are interpolated:

$$f(x, y_1) = f(x_0, y_1) + [f(x_1, y_1) - f(x_0, y_1)]\frac{(x - x_0)}{(x_1 - x_0)} \tag{3.43}$$

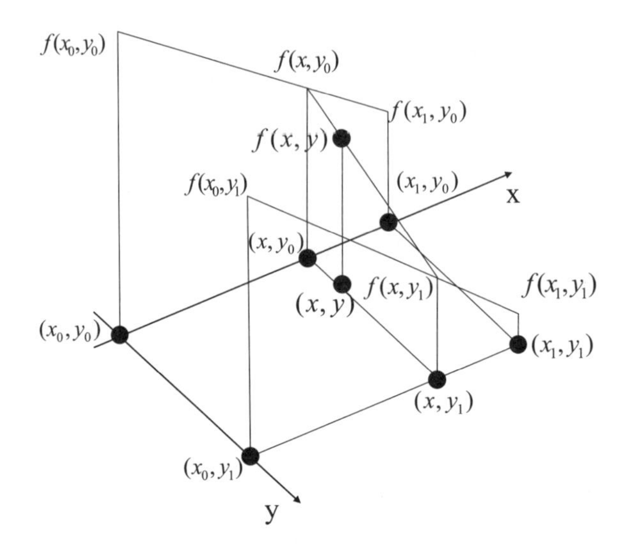

Fig. 3.15 Linear interpolation

Finally, we perform linear interpolation in the vertical direction to determine

$$f(x, y) = f(x, y_0) + [f(x, y_1) - f(x, y_0)]\frac{(y - y_0)}{(y_1 - y_0)} \tag{3.44}$$

Substituting (3.42) and (3.43) into Eq. (3.44) yields

$$\begin{aligned} f(x, y) =& f(x_0, y_0) + [f(x_1, y_0) - f(x_0, y_0)]\frac{(x - x_0)}{(x_1 - x_0)} \\ &+ [f(x_0, y_1) - f(x_0, y_0)]\frac{(y - y_0)}{(y_1 - y_0)} \\ &+ [f(x_1, y_1) + f(x_0, y_0) - f(x_1, y_0) - f(x_0, y_1)]\frac{(x - x_0)(y - y_0)}{(x_1 - x_0)(y_1 - y_0)} \end{aligned} \tag{3.45}$$

This expression represents the general formula for bilinear interpolation. If the derivation is performed in a unit square,

$$\begin{aligned} f(x, y) =& f(0, 0) + [f(1, 0) - f(0, 0)]x + [f(0, 1) - f(0, 0)]y \\ &+ [f(1, 1) + f(0, 0) - f(1, 0) - f(0, 1)]xy \end{aligned} \tag{3.46}$$

3. High-order interpolation

In geometric operations, the smoothing of the bilinear interpolation may degrade the details of the image, especially when magnification is performed. In other applications, bilinear interpolation slope discontinuities produce unexpected results. Both these cases can be corrected through high-order interpolation. Examples of high-order interpolation functions are the cubic spline, sin (ax) function, Legendre center function, and convolution [5].

3.4.4 Matrix of Geometric Operations

A geometric operation requires two separate algorithms. First, an algorithm must be used to define the spatial transform, which describes how each pixel moves from its initial position to the end position, that is, the motion of each pixel. Moreover, a gray-level interpolation algorithm is required because, in general, the position coordinates (x, y) of the input image are integers, and the position coordinates of the output image are nonintegers. The commonly used spatial transformations can be described as follows.

1. Simple transformation

   (1) Translation: Let $a(x, y) = x + x_0$ and $b(x, y) = y + y_0$. The translation operation can thus be performed.

   (2) Scaling: Let $a(x, y) = x/c$ and $b(x, y) = y/d$. The image is scaled $1/c$ and $1/d$ times on the x-axis and y-axis, respectively, and the image origin remains unchanged.

   (3) Rotation: $a(x, y) = x\cos\theta - y\sin\theta$ and $b(x, y) = x\sin\theta$-$y\cos\theta$ produce a rotation around the origin clockwise θ angle.

2. Composite transformation

The homogeneous coordinate system can be defined as follows.

$$\begin{bmatrix} a(x,y) \\ b(x,y) \\ 1 \end{bmatrix} = \begin{bmatrix} 1 & 0 & x_0 \\ 0 & 1 & y_0 \\ 0 & 0 & 1 \end{bmatrix} \begin{bmatrix} \cos\theta & -\sin\theta & 0 \\ \sin\theta & \cos\theta & 0 \\ 0 & 0 & 1 \end{bmatrix} \begin{bmatrix} \frac{1}{c} & 0 & -x_0 \\ 0 & \frac{1}{d} & -y_0 \\ 0 & 0 & 1 \end{bmatrix} \begin{bmatrix} x \\ y \\ 1 \end{bmatrix} \tag{3.47}$$

The image is first translated to ensure that position (x_0, y_0) is the origin. Subsequently, the angle is rotated and scaled. Other composite transformations can be similarly constructed.

3. General change

In many image processing applications, the required spatial transformations are quite complex and cannot be expressed in simple mathematical formulas. In addition, the required spatial transformation is often obtained from the measurement of the actual image, and thus, it is preferable to describe the geometric transformation with these measurement results rather than the functional form and create complex transform models by using the measurement results as parameters [6].

3.4.5 Applications of Geometric Operations

When a ground prototype is transformed to remote sensing information through remote sensing, the spatial characteristics of the ground target are partially distorted and deformed due to atmospheric transmission effects and characteristics of the remote sensor program. Therefore, to obtain a variety of scale map measurements of surface remote sensing images, it is necessary to perform geometric correction, map projection, and coordinate transformation, among other steps.

1. Geometric correction

Geometric correction eliminates the geometric distortion of the digital image caused by the sensor, that is, the original distortion of the remote sensing image, to determine the new image for the space transformation equation and implement the grayscale level interpolation process. The principle is to quantitatively determine the correspondence between the image coordinates and geographic coordinates, that is, the projection of the image data onto the plane to ensure that the data conform to the process associated with the map projection system. The process has two basic steps: pixel coordinate transformation and pixel brightness resampling.

The geometric correction of remote sensing images includes strict geometric correction (such as collinear equation correction) and approximate geometric correction (including polynomial correction, affine transformation, direct linear transformation and rational function modeling), which are geometric operations in the spatial transformation process. The two operations are described in the following text.

(1) Collinear equation correction

The collinear equation is strictly established for static sensors, and the dynamic sensor pertains to point-by-point or line-by-line multicenter projections. Each independent part of the image corresponds to its own sensor state parameters. The collinear equation is

$$x - x_0 = -f\frac{a_{11}(X - X_s) + a_{12}(Y - Y_s) + a_{13}(Z - Z_s)}{a_{31}(X - X_s) + a_{32}(Y - Y_s) + a_{33}(Z - Z_s)} \tag{3.48}$$

$$y - y_0 = -f\frac{a_{21}(X - X_s) + a_{22}(Y - Y_s) + a_{23}(Z - Z_s)}{a_{31}(X - X_s) + a_{32}(Y - Y_s) + a_{33}(Z - Z_s)} \tag{3.49}$$

where $a_{11}, a_{12}, a_{13}, a_{21}, a_{22}, a_{23}, a_{31}, a_{32}$, and a_{33} are the rotation matrix elements.

$$\begin{bmatrix} a_{11} & a_{12} & a_{13} \\ a_{21} & a_{22} & a_{23} \\ a_{31} & a_{32} & a_{33} \end{bmatrix}$$

For the corresponding position parameter, x and y are the image plane coordinates of the image point, and x_0, y_0 and f are the inner azimuth elements of the imaging camera. X_s, Y_s, and Z_s are the three-dimensional spatial coordinates of the imaging camera. X,

Y, Z are the three-dimensional space coordinates of the actual measurement control point on the ground corresponding to the image point.

(2) Polynomial correction

The basic concept of polynomial correction is to directly simulate the image without considering the spatial geometrical relationship in the image imaging process. The method considers the overall deformation of the image as a composite effect of translation, scaling, rotation, radiation, deflection, and bending, along with a higher level of basic deformation. The coordinates of the corresponding points before and after the image processing can be described by an appropriate polynomial. The basic equations for polynomial correction are as follows:

$$\begin{aligned} u &= \sum_{i=0}^{m}\sum_{j=0}^{n}\sum_{k=0}^{p} a_{ijk} X^i Y^j Z^k \\ v &= \sum_{i=0}^{m}\sum_{j=0}^{n}\sum_{k=0}^{p} b_{ijk} X^i Y^j Z^k \end{aligned} \quad (3.50)$$

where u and v are the plane coordinates of the image point (in pixels); X, Y, and Z are the three-dimensional spatial coordinates of the actual measurement control point corresponding to the image point; and a_{ijk} and b_{ijk} are the correction coefficients.

2. Map projection

The map projection (Fig. 3.16) calculates the geographic elements on the ellipsoid plane to a plane according to certain mathematical rules. The projection formula is the spatial transformation formula in the geometric operation. After the projection, gray level interpolation is performed to obtain the planar remote sensing image [7].

(1) Orthogonal projection (Fig. 3.17)

The features on the surface of the sphere are projected onto a plane tangent to the sphere at the "projection center." The features are projected in a direction parallel to the plane normal.

The mathematical transformation principle of the orthogonal projection method is

$$\begin{cases} x = \rho\cos\sigma = \mathrm{R}\sin Z\cos\theta \\ y = \rho\sin\sigma = R\sin Z\cos\theta \end{cases} \quad (3.51)$$

where x and y are the plane coordinates after projection; ρ and σ are the polar coordinate parameters of the projection plane; and R, Z, and θ represent the radius

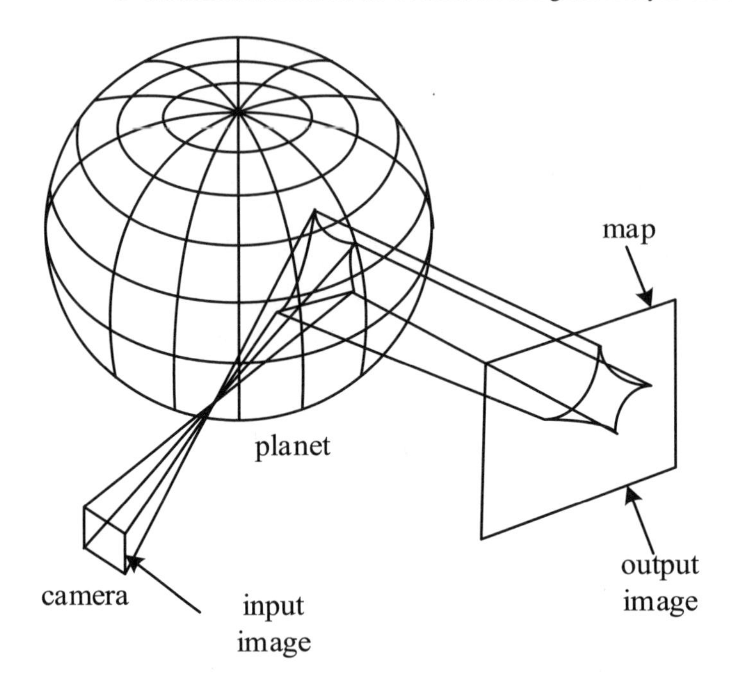

Fig. 3.16 Map projection

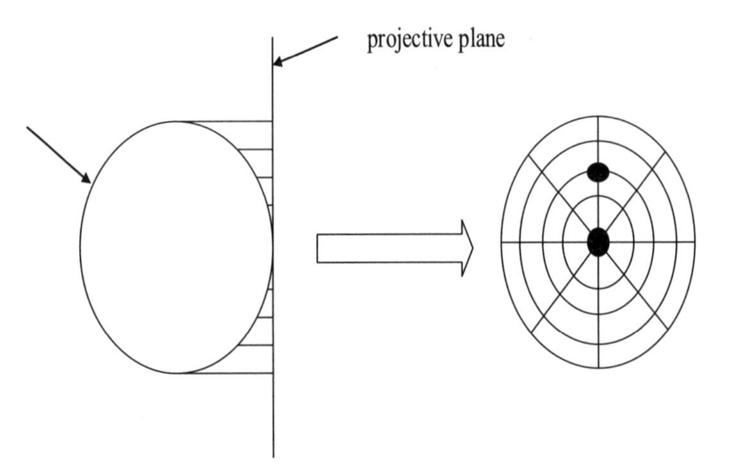

Fig. 3.17 Orthogonal projection method

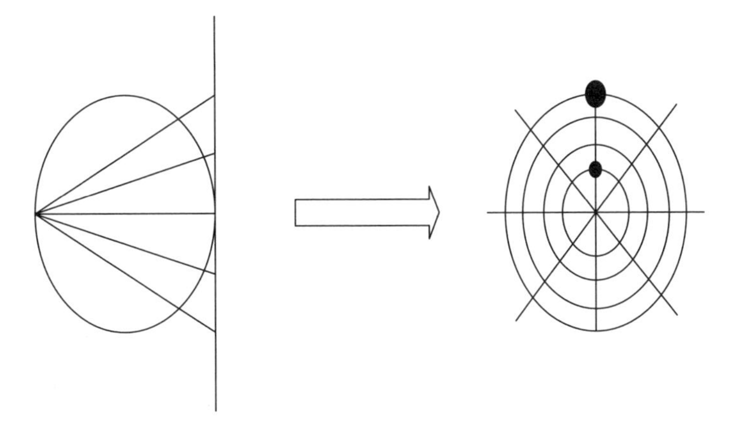

Fig. 3.18 Spherical plane projection

of the Earth, zenith distance and azimuth angle of any point on the spherical surface in the spherical coordinate system, respectively.

(2) Spherical plane projection method

The features on the surface of the ball are projected onto a plane tangent to the sphere at the "projection center" (Fig. 3.18).

The principle of mathematical transformation of the spherical plane projection method is

$$\begin{cases} x = \rho \cos \sigma = \frac{2R \sin Z \cos \theta}{\alpha + \cos Z} \\ y = \rho \sin \sigma = \frac{2R \sin Z \sin \theta}{\beta + \cos Z} \end{cases} \tag{3.52}$$

The parametric meaning is the same as that of the orthogonal projection method.

(3) Lambert equiangular conical projection method (Fig. 3.19)

In the Lambert equiangular conical projection method, the Earth's surface is projected onto a conical surface coaxial with the planets. The cone intersects the sphere on two wefts, corresponding to the standard weft. The meridian appears as a straight line on the map, and the weft appears as a circle on the cone of the map. When the cones are unfolded, the wefts become arcs, and the meridians intersect at the extreme. The distance between the wefts are adjusted to satisfy the conformal nature: two standard wefts are projected on a real scale, the scale between the standard wefts is decreased, and the scale outside the standard weft increases.

(4) Mercator projection method (Fig. 3.20)

The spherical surface features are mapped onto a cylindrical surface tangent to the sphere at the equator. The vertical position of the weft is

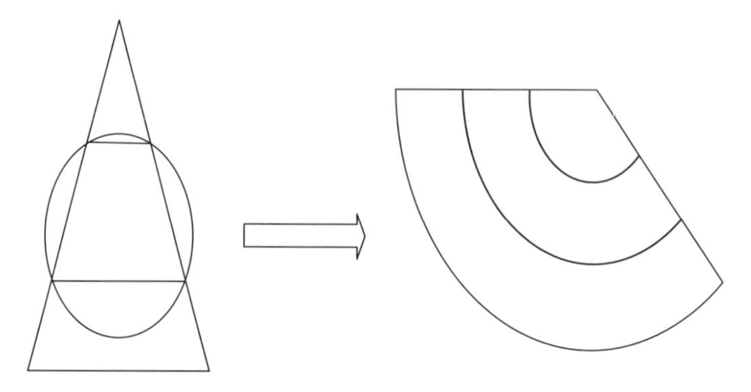

Fig. 3.19 Lambert equiangular conical projection method

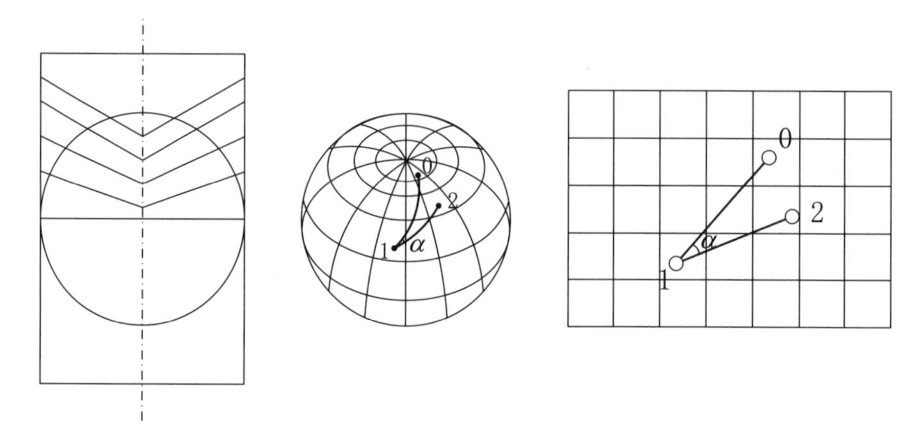

Fig. 3.20 Mercator projection

$$y = R\ln[\tan(45 + \alpha/2)] \tag{3.53}$$

where R is the radius of the earth, the Mercator projection is not angularly deformed, and the isocenter is presented as a straight line, as shown in Fig. 3.20.

3. Coordinate transformation

Coordinate transformation describes the position of the space entity to realize the transformation from one coordinate system to another system. Coordinate transformation is a necessary step in the measurement and compilation of maps at various scales. In plane geometry, it is necessary to perform a conversion between polar and rectangular coordinates, and map projection transformation or measurement system coordinate transformation must be implemented in the geographic information system concept, for instance, between the geodetic coordinate system to the map coordinate system.

Table 3.1 Ellipsoid parameters of the Beijing 54 and WGS84 coordinates

Coordinate	Ellipsoid	Semimajor axis *a*/m	Semiminor axis *b*/m	Flat rate *f*
Beijing 54	Krasovsky	6,378,245	6,356,863	1:298.30
WGS84	WGS84	6,378,137	6,356,752	1:298.26

Due to the differences between the global geocentric and regional coordinate systems, the ground position based on the WGS84 coordinate system obtained by GPS must be converted to China's current 1954 or 1980 national geodetic coordinate system or local independent coordinate system to be applied over the map, which is a key problem often encountered in the application of GPS.

(1) Beijing 54 coordinate system and WGS84 coordinate system

In 1954, the Beijing geodetic origin was established, named the Beijing 54 coordinate system. Table 3.1 lists the ellipsoid parameters of the Beijing 54 coordinate system.

The WGS84 coordinate system, which is used in GPS measurements, is a global geodetic coordinate system. Table 3.1 lists the ellipsoid parameters of the WGS84 coordinate system.

The origin of this coordinate system is the Earth's center of mass. As in the Earth space rectangular coordinate system, the Z axis pointing to the international time (BIH, Bureau International del 'Heure) 1984.0 protocol pertaining to the definition of the polar (CTP, but terrestrial pole) direction; the X axis pointing to the BIH 1984.0 zero meridian plane and intersection of the CTP equator, Y axis and Z axis; and the X axis vertical constitute a right-handed coordinate system.

(2) Coordinate conversion step

First, WGS84 coordinates are converted to Beijing 54 coordinates.

① The geodetic coordinates (B_{84}, L_{84}, h_{84}) of the GPS are converted from the WGS84 ellipsoid to the spatial rectangular coordinates (X_{84}, Y_{84}, Z_{84}). The formula is as follows:

$$\begin{cases} X_{84} = (N + h_{84}) \cos B_{84} \cos L_{84} \\ Y_{84} = (N + h_{84}) \cos B_{84} \sin L_{84} \\ Z_{84} = \left[N\left(1 - e^2\right) + h_{84}\right] \sin B_{84} \end{cases} \tag{3.54}$$

$N = \frac{a}{\sqrt{1-e^2 \sin^2 B_{84}}}$, a is the long radius of the Earth's ellipsoid, and e is the first eccentricity of the ellipsoid.

② The WGS84 space rectangular coordinates (X_{84}, Y_{84}, Z_{84}) are converted to the Beijing 54 space rectangular coordinates (X_{54}, Y_{54}, Z_{54}):

$$\begin{bmatrix} X_{54} \\ Y_{54} \\ Z_{54} \end{bmatrix} = \begin{bmatrix} X_0 \\ Y_0 \\ Z_0 \end{bmatrix} + (1+m)\begin{bmatrix} 1 & \varepsilon_Z & -\varepsilon_Y \\ -\varepsilon_Z & 1 & \varepsilon_X \\ \varepsilon_Y & \varepsilon_X & 1 \end{bmatrix}\begin{bmatrix} X_{84} \\ Y_{84} \\ Z_{84} \end{bmatrix} \tag{3.55}$$

X_0, Y_0, and Z_0 represent the translation parameters between two frames; ε_X, ε_Y, andε_Z represent the rotation parameters between the two frames; and m is the scaling parameter. The key to the solution is to solve seven parameters through a process known as the seven-parameter coordinate conversion. To calculate the seven parameters, more than three control points must be used. In practical applications, different areas usually share a set of seven parameters, and thus, the two coordinates can be converted using the known seven parameters.

The translation considers only three parameters, X_0, Y_0, and Z_0. The coordinate ratio is mostly consistent, and the scaling parameter m is 0 by default, which creates a three-parameter coordinate transformation. The three parameters are exceptions to the seven parameters:

$$\begin{bmatrix} X_{54} \\ Y_{54} \\ Z_{54} \end{bmatrix} = \begin{bmatrix} X_0 \\ Y_0 \\ Z_0 \end{bmatrix} + \begin{bmatrix} X_{84} \\ Y_{84} \\ Z_{84} \end{bmatrix} \tag{3.56}$$

③ The Beijing 54-space rectangular coordinates are converted to geodetic coordinates based on the Krasovsky ellipsoid:

$$\begin{cases} L_{54} = \arctan(Y_{54}/X_{54}) \\ B_{54} = \arctan\left[\left(Z_{54} + Ne^2 \sin B_{54}\right)/\sqrt{X_{54}^2 + Y_{54}^2}\right] \\ h_{54} = \sqrt{X_{54}^2 + Y_{54}^2} \sec B_{54} - N \end{cases} \tag{3.57}$$

In this formula, B_{54} must set the initial value for the iteration until the difference between the two B_{54} values is less than the limit, and the iteration is terminated.

④ The 54 earth coordinates of Beijing (B_{54}, L_{54}, h_{54}) are converted to Beijing 54 flat coordinates (X, Y).

$$\begin{cases} X = 6367558.4969\dfrac{\beta''}{\rho''} - \{a_0 - [0.5 + (a_4 + 0.6l^2)l^2]l^2 N\}\dfrac{\sin B}{\cos B} \\ Y = [1 + (a_3 + a_5 l^2)l^2]lN \cos B \end{cases} \tag{3.58}$$

In this case,

$$\begin{cases} l = \frac{(L-L_0)''}{\rho''} \\ N = 63699698.902 - [21562.267 - (108.973 - 0.612\cos^2 B)\cos^2 B]\cos^2 B \\ a_0 = 32140.404 - [135.3302 - (0.7092 - 0.0040\cos^2 B)\cos^2 B]\cos^2 B \\ a_4 = (0.25 + 0.00252\cos^2 B)\cos^2 B - 0.04166 \\ a_6 = (0.166\cos^2 B - 0.084)\cos^2 B \\ a_3 = (0.3333333 + 0.001123\cos^2 B)\cos^2 B - 0.1666667 \\ a_5 = 0.0083 - [0.1667 - (0.1968 + 0.0040\cos^2 B)\cos^2 B]\cos^2 B \end{cases} \tag{3.59}$$

where β is the undetermined coefficient of the corresponding variable, L is the longitude coordinate before conversion, L_0 is the central meridian coordinate of the projection zone, and $l = (L-L_0)/\rho$ is used to calculate the longitude difference of P and central meridian. If $L-L_0$ is in degrees, $\rho = 57.295779513$; if the value is in minutes or seconds, $\rho = 3437.7467708$ and $\rho = 206{,}264.80625$, respectively.

Subsequently, the WGS84 coordinates are converted to Beijing 54 coordinates. The conversion of the Beijing 54 coordinates to WGS coordinates is the inverse transformation of this transformation [8].

(3) Polar coordinates (generation of aerospace coordinate system)

The classic photogrammetry theory is based on the fact that two cameras with a certain distance on the ground shoot the same object at the same time through two angles of parallax images to calculate the plane and depth of field parameters. Space observation adopts this principle to obtain plane and elevation data; however, because the adjacent image overlaps the endpoint with the same objective, to achieve two different moment postures, the gesture position must be considered in terms of the speed and second order of acceleration. Thus, the equation of state space dimension increases linearly, and the calculation becomes complex. To maintain two cameras on the same platform, because the adjacent image overlaps the endpoint with the same objective, the parallax angle is extremely small, and a large elevation error is introduced. In this case, a considerable calculation error occurs, and the process becomes time-consuming. When the elevation increment approaches zero, the first derivative is infinite and does not converge. From the perspective of mathematical analysis, the rectangular coordinate system of the three axes has the same data "dimensionality" and same order of magnitude of each value increment. The relative increment in the Z axis is considerably smaller than that in the plane, and owing to the relative error, the Z axis parameters relative to the surface are extremely small and dimensionless. Notably, the three-dimensional relative error calculation matrix is full rank and not relevant, the Z tends to zero, and the calculation matrix becomes irrelevant. To ensure effective operation, the Z-axis parameter must "enlarged" to "adapt" to the plane parameters; however, this aspect amplifies the Z-axis error. Therefore, to fundamentally solve the problem of low speed and convergence, it is necessary to abandon the possibility of obtaining a sparse matrix and pathological solution of the traditional rectangular coordinate system. It is necessary to introduce a new aerospace coordinate system theory to decrease the number of weakly unrelated dimensions of sparse matrix by several orders of magnitude and achieve rapid and efficient convergence. Moreover, the polar coordinate data measured by the sensor can be used to extrapolate the track to make it more convenient and reliable to record and process the position data. Polar coordinates can be extended to multiple dimensions to solve the problem of complex data. Aerial triangulation is mainly used for measuring the location and digital product production and is the core of many measurement software modules. Bundle block adjustment is one of the most accurate and rigorous existing methods for air triangulation. This method is often used to obtain precise ground control point coordinates, and the result can be used for large digital cartography objects.

Polar coordinates are key components of the bundle block adjustment algorithm, and we express the coordinates of the image points of each feature point on each image as a function of the camera pose and feature point parameters, that is, the observation equation. Assuming that the main and secondary anchor points of the spatial three-dimensional feature point are t_m and t_a, respectively, the coordinate of the image point observed by feature point F_j in camera p_i can be expressed as

$$\begin{bmatrix} u_j^i \\ v_j^i \end{bmatrix} = \begin{bmatrix} x_j^i / t_j^i \\ v_j^i / t_j^i \end{bmatrix} \tag{3.60}$$

where

$$\begin{bmatrix} x_j^i \\ y_j^i \\ t_j^i \end{bmatrix} = \begin{cases} KR_m u_j^m, & \text{if } i = m \\ KR_i \overline{X}_j^m, & \text{if } i \neq m \end{cases} \tag{3.61}$$

K is the internal parameter matrix of the camera, and R_i is the rotation matrix of camera p_i, which is a function of the Euler angle parameters that represent the camera pose:

$$R_i = r(\alpha_i, \beta_i, \gamma_i) \tag{3.62}$$

The unit vector x_j^m extends from the main anchor point t_m to eigenpoint F_j and can be computed as

$$x_j^m = v(\psi_j, \theta_j) = \begin{bmatrix} \sin\psi_j \cos\theta_j \\ \sin\theta_j \\ \cos\psi_j \cos\theta_j \end{bmatrix} \tag{3.63}$$

The traditional Euclidean space *XYZ* parametric representation of three-dimensional feature points is more effective for closer feature points. However, if the feature points in the environment are far, and only a small parallax corresponds to the image, the feature points may involve significant errors and uncertainties in the depth direction, leading to divergence or nonconvergence of the algorithm.

Polar coordinates can express different types of feature points, such as those located near and far, but no depth information in three-dimensional space. The Gauss–Newton method can be used to solve polar optimization equations. The convergence speed of the Gauss–Newton method is related to the magnitude of the remaining quantity and degree of linearity of $r(x)$. A smaller residual quantity or higher linearity of $r(x)$ corresponds to a higher convergence rate. Under a strong nonlinearity of the remaining quantity, the residual function does not converge [9].

3.5 Summary

Focusing on the essence of mathematics, this chapter introduces four basic methods of image processing: the probability of grayscale histograms, the function transformation of point operations, the convolution of algebraic operations and matrices of geometric operations. These image processing methods are based on the remote sensing image and not the transformation of the pixels within the image. The contents of this chapter and mathematical formulas are summarized in Fig. 3.21 and Table 3.2.

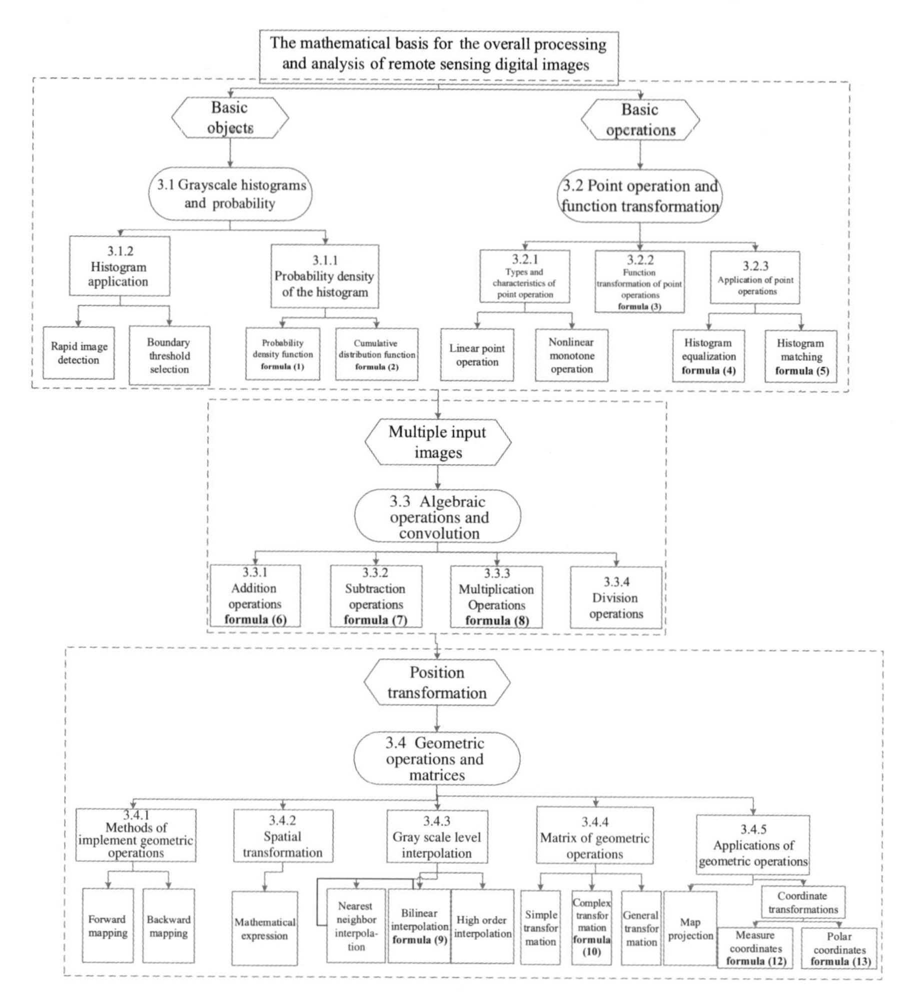

Fig. 3.21 Summary of the contents of this chapter

Table 3.2 Summary of the mathematical equations in this chapter

formula (1)	$PDF = p(D) = \frac{1}{A_0} H(D)$ (3.3)	3.1.1
formula (2)	$CDF = P(D) = \int_0^D p(u)\mathrm{d}u = \frac{1}{A_0}\int_0^D H(u)\mathrm{d}u$ (3.4)	3.1.1
formula (3)	$H_B(D_B) = \frac{H_A(D_A)}{\mathrm{d}(f(D_A))/\mathrm{d}D_A} = \frac{H_A[f^{-1}(D)]}{f'[f^{-1}(D)]}$ (3.20)	3.2.2
formula (4)	$f(D) = D_m \times \mathrm{CDF} = D_m P(D) = \frac{D_m}{A0}\int_0^D H(u)\mathrm{d}u$ (3.21)	3.2.3
formula (5)	$g\{f[A(x,y)]\} = P_3^{-1}\{P_1[A(x,y)]\}$ (3.26)	3.2.3
formula (6)	$H_C(D_C) = H_A(D_A) * H_B(D_B)$ (3.33)	3.3.1
formula (7)	$\nabla f(x,y) = \boldsymbol{i}\frac{\partial f(x,y)}{\partial x} + \boldsymbol{j}\frac{\partial f(x,y)}{\partial y}$ (3.34)	3.3.2
formula (8)	$H_C(D_C) = \frac{A_0\mathrm{d}\{P_B[D_B(x,y)]\}}{\mathrm{d}D_C} = \frac{A_0\mathrm{d}\{P_B[D_B(x,y)]\}}{\mathrm{d}\{D_A(x,y) * D_B(x,y)\}} = \frac{A_0 * \mathrm{d}\int_0^D H_B\{D_B(x,y)\mathrm{d}D_B(x,y)\}}{A_1 * \mathrm{d}\{D_A(x,y) * D_B(x,y)\}}$ (3.39)	3.3.3
formula (9)	$x - x_0 = -f\frac{a_{11}(X - X_s) + a_{12}(Y - Y_s) + a_{13}(Z - Z_s)}{a_{31}(X - X_s) + a_{32}(Y - Y_s) + a_{33}(Z - Z_s)}$ (3.48) $y - y_0 = -f\frac{a_{21}(X - X_s) + a_{22}(Y - Y_s) + a_{23}(Z - Z_s)}{a_{31}(X - X_s) + a_{32}(Y - Y_s) + a_{33}(Z - Z_s)}$ (3.49)	3.4.5
formula (10)	$\begin{bmatrix} a(x,y) \\ b(x,y) \\ 1 \end{bmatrix} = \begin{bmatrix} 1 & 0 & x_0 \\ 0 & 1 & y_0 \\ 0 & 0 & 1 \end{bmatrix}\begin{bmatrix} \cos\theta & -\sin\theta & 0 \\ \sin\theta & \cos\theta & 0 \\ 0 & 0 & 1 \end{bmatrix}\begin{bmatrix} 1 & 0 & -x_0 \\ 0 & 1 & -y_0 \\ 0 & 0 & 1 \end{bmatrix}\begin{bmatrix} x \\ y \\ 1 \end{bmatrix}$ (3. 47)	3.4.4
formula (11)	$f(x,y) = f(x_0,y_0) + [f(x_1,y_0) - f(x_0,y_0)]\frac{(x - x_0)}{(x_1 - x_0)}$ $+[f(x_0,y_1) - f(x_0,y_0)]\frac{(y - y_0)}{(y_1 - y_0)}$ $+[f(x_1,y_1) + f(x_0,y_0) - f(x_1,y_0) - f(x_0,y_1)]\frac{(x - x_0)(y - y_0)}{(x_1 - x_0)(y_1 - y_0)}$ (3.45)	3.4.3
formula (12)	$\begin{cases} X_{84} = (N + h_{84})\cos B_{84}\cos L_{84} \\ Y_{84} = (N + h_{84})\cos B_{84}\sin L_{84} \\ Z_{84} = \left[N\left(1 - e^2\right) + h_{84}\right]\sin B_{84} \end{cases}$ (3.54)	3.4.5

(continued)

Table 3.2 (continued)

formula (13)	$$\begin{bmatrix} x_j^i \\ y_j^i \\ t_j^i \end{bmatrix} = \begin{cases} KR_m u_j^m, & \text{if } i = m \\ KR_i \overline{X}_j^m, & \text{if } i \neq m \end{cases}$$	(3.61)	3.4.5
	$$x_j^m = v(\psi_j, \theta_j) = \begin{bmatrix} \sin\psi_j \cos\theta_j \\ \sin\theta_j \\ \cos\psi_j \cos\theta_j \end{bmatrix}$$	(3.63)	

References

1. Wegener M. Destriping multiple sensor imagery by improved histogram matching. Int J Remote Sens. 1990;11.
2. Pizer S M, Amburn EP, Austin JD et al. Adaptive histogram equalization and its variations. Comput Vis, Graphics, Image Process. 1987;39.
3. Shen D. Image registration by local histogram matching. Patt Recogn. 2007;40.
4. Sangwine SJ. Colour image edge detector based on quaternion convolution. Electr Lett.1998;34
5. Keys R. Cubic convolution interpolation for digital image processing. Acoust, Speech Signal Process, IEEE Trans. 1981;29
6. Civera J, Davison AJ, Montiel J. Inverse depth parametrization for monocular SLAM. Rob IEEE Trans. 2008;24
7. Pearson F. Map projections: theory and applications. Crc Press. 1990.
8. Wang J, Lu CP. Problem of coordinate transformation between WGS-84 and BEIJING 54. Crustal Deform Earthq. 2003.
9. Lei Y, Liang Z, Hongying Z et al. Method for aero photogrammetric bundle adjustment based on parameterization of polar coordinates, CN 102778224 B. 2014. (In Chinese).

Chapter 4
Physical Foundation and Global Analysis of Remote Sensing Digital Images

Chapter Guidance The results of the mathematical preprocessing of the overall performance of the image described in Chap. 3 indicate that the main factors that affect the overall quality of remote sensing images, such as geometric radiation distortion and environment of image collection, cannot be artificially changed through image preprocessing. These aspects must be changed through the "treat the root" analysis of physical imaging processing. In other words, the correlation distortion is directly related to the physical origin of the image processing system, which reflects the process of digital image collection, physical parameters and performance indices of the image, which can be solved by combining mathematical and physics concepts. Notably, the final effect of image processing is directly related not only to the physical process and parameters of image collection but also to the image processing process. If the former problem is solved only through follow-up image processing, the principle of "solving the problem where it occurs" is violated, and the follow-up processing becomes tedious. In this scenario, unsatisfactory digital imaging processing results may be obtained. It is necessary to understand and solve the problems of imaging process before implementing remote sensing digital image processing, and this task is an integral part of the overall processing of remote sensing images. The core of this chapter is the PSF, MTF, OTF, and image resolution of remote sensing digital imaging systems. Therefore, through the combination of the third chapter pertaining to mathematics and the physics described in this chapter, it is possible to simultaneously treat the entire remote sensing image, thereby laying the foundation for the system and overall processing of the image pixel processing described in the second part of the book.

Although remote sensing digital image processing focuses on only the process and method of image processing, clarification of the physical formation process of remote sensing images can help understand the systematic and methodological aspects of remote sensing image processing. The target object of remote sensing image processing pertains to images generated by various remote sensors. The characteristics of remote sensing images generated by remote sensing image systems

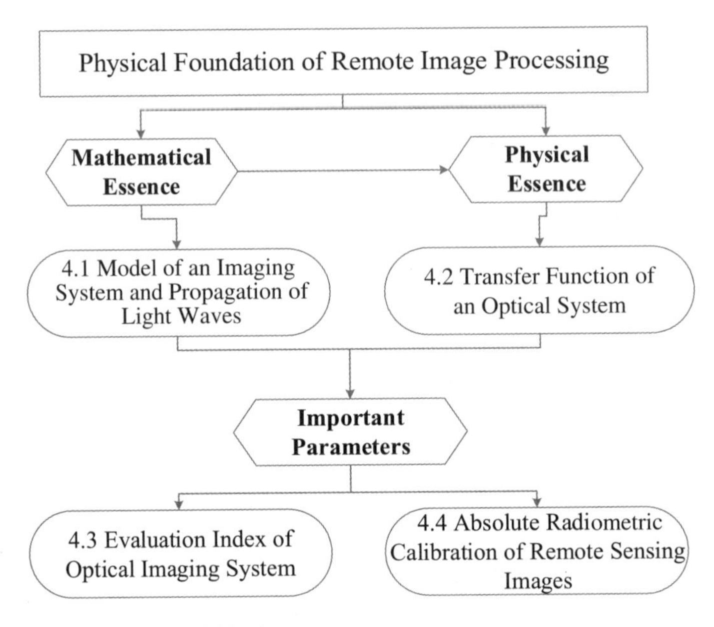

Fig. 4.1 Logical framework of this chapter

directly affect the usability of remote sensing images and the difficulty and precision of remote sensing image processing. Therefore, remote sensing digital image processing is based on not only strict mathematics but physics knowledge [1].

Optical imaging systems play a key role in the collection of remote sensing digital images because these systems are almost always present at the front end of the image processing system and often present at both ends. If the camera is involved before scanning, the lens system for this part must be considered in the analysis. A detailed analysis of all aspects of a digital imaging system is expected to be extremely complicated and is beyond the scope of this book. Therefore, this chapter analyzes the optical imaging system from a physical viewpoint to derive the physical essence of the remote sensing digital imaging process. As shown in Fig. 4.1, this chapter describes the imaging system model and propagation of light (Sect. 4.1) and introduces the physical process of imaging and error source. The optical system transfer function (Sect. 4.2) is presented to highlight the transmission process of imaging error. The evaluation index of the optical imaging system (Sect. 4.3) is described to clarify the relationship between the accuracy and error of the remote sensing image and quantification scale of the image. The absolute radiometric calibration of the remote sensing image (Sect. 4.4) allows the reader to ascertain the nature of the radiant energy intensity of the grayscale measurement of the image and use calibration to quantitatively quantify remote sensing.

4.1 Model of an Imaging System and Propagation of Light Waves

This section analyzes the distribution of the light waves emitted by the object points on the image plane after passing through a black-box model of the imaging system. The section describes the black-box model of the optical imaging system, Huygens principle of light propagation, Fresnel approximation and Abbe imaging principle and spatial filtering.

4.1.1 *Black-Box Model of Optical Imaging System*

Although several types of optical imaging systems exist, they can be abstracted into the following model: the light field distribution of the object plane is transformed by the optical system, the light field distribution of the image plane is obtained, and the process is shown in Fig. 4.2.

The optical aperture of the imaging system has a certain size. The light of the object plane passes through the aperture when entering the optical imaging system, and the spatial distribution of the plane transmittance of the aperture is defined by a pupil function. The object can be considered a collection of point light sources, and the imaging process of an object can be regarded as a superposition of these light sources. The distribution of the light field on the image plane after the point light source passes through the imaging system is known as the point spread function (PSF) [2]. The complex amplitude distribution of the light field on the plane is $U_o(x_o, y_o)$, and the complex amplitude distribution of the light field on the image plane is $U_i(x_i, y_i)$. The imaging process of the object can be expressed as follows:

$$U_o(x_o, y_o) = \iint_{\Omega} U_o(\xi, \eta)\delta(x_o - \xi, y_o - \eta)\,\mathrm{d}\xi\,\mathrm{d}\eta \tag{4.1}$$

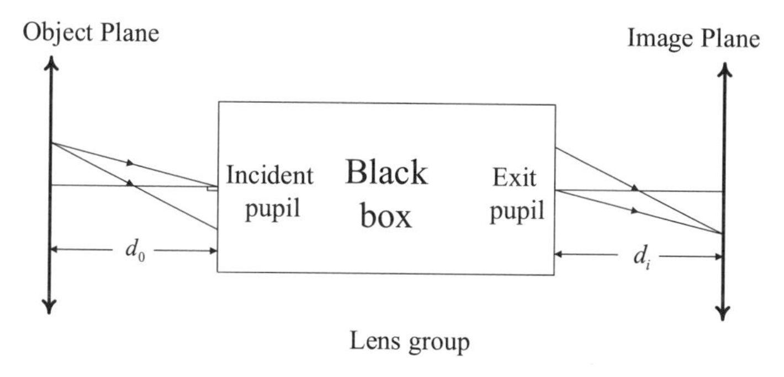

Fig. 4.2 Black-box model of imaging system

$$U_i(x_i, y_i) = F\left\{\iint\limits_{\Omega} U_o(\xi, \eta)\delta(x_o - \xi, y_o - \eta)\mathrm{d}\xi\,\mathrm{d}\eta\right\} \tag{4.2}$$

4.1.2 The Huygens Principle of Light Wave Propagation

An interesting and useful feature of light wave propagation is the Huygens–Fresnel principle. According to the Huygens–Fresnel principle, every point on the wavefront in the medium can be regarded as a point wave source of a new wavelet, and the propagation effect of a certain wave in space is the superposition of these waves. As the wave propagates through the aperture, the field at any point behind the hole is equal to the field generated when the holes are filled with a wavelet source of appropriate amplitude and phase. Mathematically, the Huygens principle assumes that the field at point (x_i,y_i) can be expressed as

$$u_i(x_i, y_i) = \frac{1}{\mathrm{j}\lambda}\iint\limits_{A} u_A(x_A, y_A)\frac{1}{r}\mathrm{e}^{\mathrm{j}kr}\cos(\theta)\mathrm{d}x_A\mathrm{d}y_A \tag{4.3}$$

As shown in Fig. 4.3, $U_A(x_A,y_A)$ is the field at the aperture, and the integral is performed over the entire aperture A. The distance between (x_i,y_i) and (x_A,y_A) is represented as r, and θ represents the angle between the connection of the two points and normal of the aperture plane.

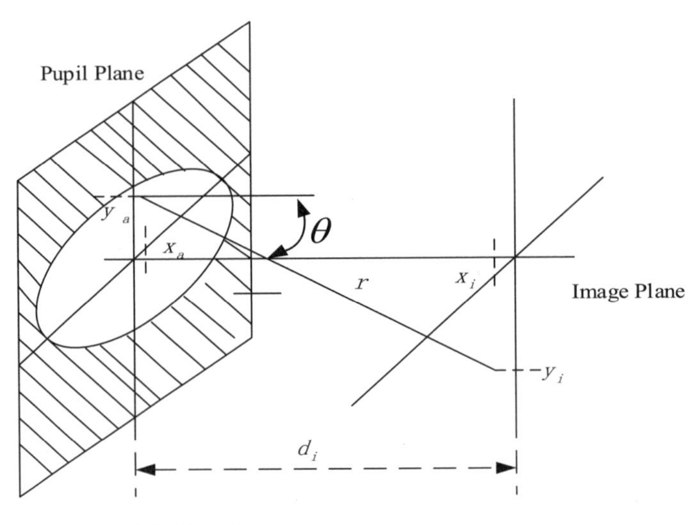

Fig. 4.3 Geometric model of imaging system

When θ is small, $\cos(\theta)$ may equal 1. If the pupil function is multiplied by the convergent light wave, the effect is equivalent to the field at all points on the hole plane except that the points inside the hole are 0. In this case, (4.3) can be written as.

$$u_i(x_i, y_i) = \frac{1}{\mathrm{j}\lambda}\int_{-\infty}^{\infty}\int_{-\infty}^{\infty} p(x_a, y_a)\frac{1}{R}\mathrm{e}^{-\mathrm{j}kR}\frac{1}{r}\mathrm{e}^{\mathrm{j}kr}\mathrm{d}x_a\mathrm{d}y_a \tag{4.4}$$

R is the distance from the origin of the plane of the image to (x_a, y_a):

$$R = \sqrt{x_a^2 + y_a^2 + d_i^2} \tag{4.5}$$

r is the distance between (x_a, y_a) and (x_i, y_i):

$$r = \sqrt{(x_i - x_a)^2 + (y_i - y_a)^2 + d_i^2} \tag{4.6}$$

In Formula (4.4), $1/R$ and $1/r$ can be approximated to $1/d_i$. However, in the exponential position, R and r have a large coefficient k, and thus, a better approximation is required.

4.1.3 Fresnel Approximation

The factor d_i in Formulas (4.5) and (4.6) is defined as

$$R = d_i\sqrt{1 + \left(\frac{x_a}{d_i}\right)^2 + \left(\frac{y_a}{d_i}\right)^2} \tag{4.7}$$

and

$$r = d_i\sqrt{1 + \left(\frac{x_i - x_a}{d_i}\right)^2 + \left(\frac{y_i - y_a}{d_i}\right)^2} \tag{4.8}$$

The binomial expansion of the square root is

$$\sqrt{1+q} = 1 + \frac{q}{2} - \frac{q^2}{8} + \cdots, \quad |q| < 1 \tag{4.9}$$

If we consider only the first two terms of the expansion, we can obtain the Fresnel approximations of the distances in Formulas (4.8) and (4.9):

$$R \approx d_i\left[1+\frac{1}{2}\left(\frac{x_a}{d_i}\right)^2+\frac{1}{2}\left(\frac{y_a}{d_i}\right)^2\right] \tag{4.10}$$

$$r \approx d_i\left[1+\frac{1}{2}\left(\frac{x_i-x_a}{d_i}\right)^2+\frac{1}{2}\left(\frac{y_i-y_a}{d_i}\right)^2\right] \tag{4.11}$$

Substituting Formulas (4.10) and (4.11) into Eq. (4.4) yields

$$u_i(x_i, y_i) = \frac{1}{\mathrm{j}\lambda d_i^2}\int_{-\infty}^{\infty}\int_{-\infty}^{\infty}\left\{p(x_a, y_a)\mathrm{e}^{-\mathrm{j}kd_i\left[1+\frac{1}{2}\left(\frac{x_a}{d_i}\right)^2+\frac{1}{2}\left(\frac{y_a}{d_i}\right)^2\right]}\right.$$
$$\left.\times\ \mathrm{e}^{\mathrm{j}kd_i\left[1+\frac{1}{2}\left(\frac{x_i-x_a}{d_i}\right)^2+\frac{1}{2}\left(\frac{y_i-y_a}{d_i}\right)^2\right]}\right\}\mathrm{d}x_a\,\mathrm{d}y_a \tag{4.12}$$

By expanding the index and merging the same item, the following formula can be obtained:

$$u_i(x_i, y_i) = \frac{\mathrm{e}^{(\mathrm{j}k/2d_i)(x_i^2+y_i^2)}}{\mathrm{j}\lambda d_i^2}\int_{-\infty}^{\infty}\int_{-\infty}^{\infty} p(x_a, y_a)\mathrm{e}^{(-\mathrm{j}2\pi/\lambda d_i)(x_i x_a+y_i y_a)}\mathrm{d}x_a\mathrm{d}y_a \tag{4.13}$$

In this case, $k\lambda = 2\pi$. Substituting the variables yields

$$x_a' = \frac{x_a}{\lambda d_i},\ y_a' = \frac{y_a}{\lambda d_i} \tag{4.14}$$

Formula (4.13) can be expressed as

$$u_i(x_i, y_i) = \frac{\lambda}{\mathrm{j}}\mathrm{e}^{(\mathrm{j}k/2d_i)(x_i^2+y_i^2)}\int_{-\infty}^{\infty}\int_{-\infty}^{\infty} p(\lambda d_i x_a', \lambda d_i y_a')\mathrm{e}^{-\mathrm{j}2\pi(x_i x_a'+y_i y_a')}\mathrm{d}x_a'\mathrm{d}y_a' \tag{4.15}$$

The complex coefficient in Eq. (4.15) affects only the phase on the image plane, which is negligible in commonly used image sensors. Thus, the item preceding the integral number is a complex constant. In addition to the difference in the complex coefficients, the PSF of coherent light is a two-dimensional Fourier transform of the pupil function.

In Fig. 4.2, the point source is located on the z-axis. If the point source deviates from the *z*-axis, the above derivation can still be performed except for the case of translation, as indicated in Eq. (4.2), and the same conclusion can be derived. In other word, under assumptions, the system is translationally invariant. However, as the image points gradually deviate from the optical axis, the conditions change. Thus, the point spread function of the imaging system deteriorates at the edge of the field.

However, it is customary to use the PSF located on the optical axis to evaluate the performance of the imaging system.

Equation (4.15) specifies the amplitude distribution of a point light source at the origin on the plane of the object in the image plane. The complex base in front of the integral number associates the brightness of the image with the brightness of the point light source and describes the phase change on the image plane. Because the general image sensor ignores the phase information, it is not discussed here. In addition, the overall brightness of the image can be simply determined by another method of analysis, that is, through the flux of light intercepted by the lens. Therefore, we are interested only in the parameters that can affect the image quality, that is, the shape of the PSF.

If the absolute calibration of the amplitude is not considered, and the terms before the integral number are omitted, Eq. (4.15) can be simplified. Thus, the convolution relationship between the object (subscript o) and image (subscript i) can be expressed as

$$u_i(x_i, y_i) = \int_{-\infty}^{\infty}\int_{-\infty}^{\infty} h(x_i - x_o, y_i - y_o)u_o(Mx_o, My_o)dx_o dy_o \tag{4.16}$$

where the impulse response is specified as

$$h(x, y) = \Im\{p(\lambda d_i x_a, \lambda d_i y_a)\} \tag{4.17}$$

In Eq. (4.16), $U_o(x_o, y_o)$ is the amplitude distribution of the object, and $u_o(Mx_o, My_o)$ is the projection of the object in the image plane (no image degradation). Thus, imaging can be considered a two-step imaging process: geometric projection, followed by convolution with the PSF on the image plane. The magnification factor M is always negative unless the axes of the image plane and object plane are rotated by 180°, and it is usually most convenient to perform the analysis on the plane of the object. In this case, it can be assumed that the convolution of the PSF is performed on the object plane. We use d_f to replace d_i in (4.15) to obtain the convolution of the PSF and $u_o(x_o, y_o)$ before imaging.

4.1.4 Abbe Principle of Image Formation and Spatial Filtering

According to the traditional imaging principle, when a beam of parallel light is used to illuminate objects, the object at any point is a wave source pertaining to the radiation spherical wave associated with the lens convergence. Each divergent spherical wave is converted into a converging spherical wave, and the center of the spherical wave is an image of a point on an object. A complex object can be considered to be composed of numerous points of different brightness, all of which form image points on the

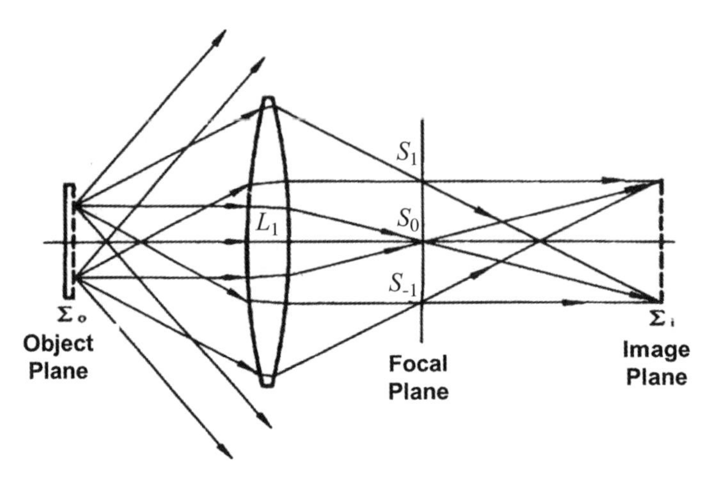

Fig. 4.4 Abbe imaging principle

image plane through the action of the lens. The image points are resuperimposed to form an image of the object. This traditional imaging principle focuses on the correspondence of points, and a point-to-point correspondence exists between the image and object [3].

According to Abbe's imaging principle, the imaging process of the lens can be divided into two steps. The first step is to form a spatial spectrum on the focal plane (i.e., the spectrum surface) of the lens by the diffracted light of the objective lens, which is the "frequency division" effect caused by diffraction. The second step is to form the image of the object by coherent superposition of beams of different spatial frequencies on the image plane, which is the "synthetic" effect caused by the interference. As shown in Fig. 4.4, the two steps of the imaging process are essentially two Fourier transformations.

Mathematically, a complex function can be Fourier expanded. According to this viewpoint, a complex picture is composed of many single frequency information points of different spatial frequencies. The ideal Fraunhofer diffraction system is a Fourier spectrum analyzer. When monochromatic light is incident on the image to be analyzed, the information of the spatial frequency is transmitted by a pair of plane-diffracted waves in a specific direction through Fraunhofer diffraction. These diffraction waves are intertwined in the near field and separated from each other in the far field to achieve "frequency division". The diffracted waves of different directions converge to different positions of the back focal plane x' by using a lens to form individual diffraction spots. Each pair of diffraction spots on x' represents a single frequency component in the original image. A higher frequency corresponds to a larger diffraction angle that is farther from the center. This task corresponds to the first step of Abbe's imaging principle for the "frequency division" process. In particular, the light emitted from the object is generated by Fraunhofer diffraction, and the Fourier spectrum is formed on the focal plane of the lens. For the second step of the imaging process, the wavefront on the object plane is set as

$$\tilde{U}_o(x, y) = A_1(t_0 + t_1 \cos 2\pi f x) \tag{4.18}$$

If three rows of planar diffraction waves are generated by the lens, three diffraction spots S_{+1}, S_0, and S_{-1} are formed on the back focal plane x'. If the three diffraction spots are regarded as three point sources, by considering the interference fields generated on the image plane $\tilde{U}_I(x', y')$, the complex amplitude of the three point light sources can be expressed as

$$\begin{cases} \tilde{A}_{+1} \propto \frac{1}{2} A_1 t_1 \exp\left[ik(BS_{+1})\right] \\ \tilde{A}_0 \propto A_1 t_0 \exp[ik(BS_0)] \\ \tilde{A}_{-1} \propto A_1 t_1 \exp\left[ik(BS_{-1})\right] \end{cases} \tag{4.19}$$

Due to the equal optical path between the object and image, $BS_0B' = BS_{+1}B' = BS_{-1}B' = BB'$. Therefore, the first phase factor is the same, and the second phase factor $e^{ik\frac{x'^2+y'^2}{z}}$ is identical. The two common phase factors are merged and the three waves are superimposed to describe the interference field as

$$\begin{aligned} \tilde{U}_I(x', y') &= \tilde{U}_0(x', y') + \tilde{U}_{+1}(x', y') + \tilde{U}_{-1}(x', y') \\ &= A_1 e^{i\varphi(x', y')} \left\{ t_0 + \frac{t_1}{2}\left[\exp(-ik \sin\theta'_{+1} x') + \exp(-ik \sin\theta'_{-1} x')\right] \right\} \end{aligned} \tag{4.20}$$

According to the Abbe sinus condition,

$$\frac{\sin\theta'_{\pm 1}}{\sin\theta_{\pm 1}} = \frac{y}{y'} = \frac{1}{V} \tag{4.21}$$

V is the lateral magnification of the imaging system, and thus,

$$k \sin\theta'_{\pm 1} x' = k \sin\theta_{\pm 1} x'/V \tag{4.22}$$

for which $k = 2\pi/\lambda$ and $\sin\theta_{\pm 1} = \pm f\lambda$. Substituting these terms into the expression $k \sin\theta_{+1} = \pm 2\pi\lambda$ yields

$$\tilde{U}_I(x', y') \propto A_1 e^{i\varphi(x', y')}\left[t_0 + t_1 \cos(2\pi f x')/V\right] \tag{4.23}$$

According to the comparison of the wavefront $\tilde{U}_o(x, y)$ on the object plane with $\tilde{U}_I(x', y')$ on the image plane, the two expressions are similar. The "synthesis" process sends a spherical wave forward to the primary diffraction pattern on the back focal plane of the objective lens. The interference is superimposed on the focal plane of the eyepiece and is the inverse Fourier transform of the spectral function. Specifically, the lens implements the Fourier transform.

4.2 Transfer Function of an Optical System

This section describes the mathematical and physical analysis of the transfer function of the optical system to clarify the imaging error transfer process, that is, the mathematical derivation of the coherent transfer function and noncoherent transfer function of the diffraction-limited optical imaging system is performed, and the mathematical relationship between the coherent transfer function and optical transfer function is derived.

4.2.1 Optical Imaging System Under the Diffraction Limitation

When light encounters obstacles or small holes in the process of propagation, the phenomenon in which the light deviates from the straight-line propagation path and propagates behind the obstacle is known as the diffraction of light. According to the characteristics of the incident light, diffraction can be divided into parallel light diffraction (Fraunhofer diffraction) and nonparallel light diffraction (Fresnel diffraction).

When an object is imaged through an optical system, it is often affected by diffraction. As shown in Fig. 4.5, the diffraction pattern formed by the point light source when passing through a small hole is a circular bright spot, which reflects that the optical system under diffraction constraints is no longer a rigid image point but a vague bright spot. This aspect can help understand the spatial resolution of sensors.

Fig. 4.5 Circular hole diffraction pattern

4.2.2 Coherent Transfer Function of a Diffraction-Limited Optical Imaging System

The imaging system is linear to the complex amplitude of the light field under coherent light illumination conditions. Therefore, under coherent light illumination conditions, (4.11) can be rewritten as

$$\begin{aligned} U_i(x_i, y_i) &= F\left\{\iint_{\Omega} U_o(\xi,\eta)\delta(x_o-\xi, y_o-\eta)\mathrm{d}\xi\mathrm{d}\eta\right\} \\ &= \iint_{\Omega} U_o(\xi,\eta)F\{\delta(x_o-\xi, y_o-\eta)\}\mathrm{d}\xi\mathrm{d}\eta \\ &= \iint_{\Omega} U_o(\xi,\eta)h(x_i, y_i, x_o-\xi, y_o-\eta)\mathrm{d}\xi\mathrm{d}\eta \\ &= U_o(x_o, y_o) * h(x_i, y_i, x_o, y_o) \end{aligned} \tag{4.24}$$

$h(x_i, y_i, x_o, y_o)$ is the transform *of* $\delta(x_o-\xi, y_o-\eta)$, which is the point spread function of the system. The process of visible optical imaging is the convolution of the complex amplitude distribution of the light field on the object plane with the point spread function of the system. The convolutional integral regards the object point as a primitive, and the image point is the coherent superposition of the diffraction pattern produced by the object point. The intensity distribution of the image under coherent light illumination conditions can be expressed as

$$I_i(x_i, y_i) = a|U_i(x_i, y_i)|^2 = a|U_0(x_0, y_0) * h(x_i, y_i, x_o, y_o)|^2 \tag{4.25}$$

where a is a real constant. The intensity distribution of the image under coherent light illumination is not linear.

This analysis uses a spatial distribution of the light field to describe the characteristics of an optical imaging system, which emphasizes the correspondence between object and image points. According to the theory of information optics, the input object plane information can be decomposed into components of various spatial frequencies. The information of the output image plane can also be decomposed into different spatial frequency components, and thus, it is possible to examine the components of the different spatial frequencies, loss, attenuation, phase shift and other characteristics in the transmission process of the optical system, that is, the transmission of the spatial frequency.

According to the theory of information optics, any two-dimensional object $f(x,y)$ can be decomposed into a linear superposition of a series of simple harmonic functions (physically representing a sinusoidal grating) with different spatial frequencies v_x *and* v_y in the x and y directions, respectively.

$$f(x, y) = \iint \psi_o(v_x, v_y) \exp\{j2\pi(v_x x + v_y y)\} dv_x dv_y \quad (4.26)$$

$\Psi_o(v_x, v_y)$ is the Fourier spectrum of $f(x, y)$, which represents the content of the spatial frequencies v_x and v_y contained in the object. The low-frequency component represents a slowly changing background and a large object contour, and the high-frequency component characterizes the details of the object.

The imaging process is analyzed considering the frequency domain, and the complex exponential function is used as the eigenfunction of the system to evaluate the transfer characteristics of the system against various frequency components. The input spectrum $G_{oc}(f_x, f_y)$ and output spectrum $G_{ic}(f_x, f_y)$ of the system are

$$G_{oc}(f_x, f_y) = \mathrm{FT}\{U_o(x_o, y_o)\} \quad (4.27)$$

$$G_{ic}(f_x, f_y) = \mathrm{FT}\{U_i(x_i, y_i)\} \quad (4.28)$$

The coherent transfer function $H_c(f_x, f_y)$ is expressed as

$$H_c(f_x, f_y) = \frac{G_{ic}(f_x, f_y)}{G_{oc}(f_x, f_y)} = \mathrm{FT}\{h(x_i, y_i, x_o, y_o)\} \quad (4.29)$$

According to the Kirchhoff diffraction formula, the complex amplitude distribution of the on-axis image point O can be expressed as

$$\begin{aligned} h(x_i y_i) &= \frac{1}{(\lambda d_i)^2} \iint P(\xi, \eta) \exp\left[-j\frac{2\pi}{\lambda d_i}(x_i \xi + y_i \eta)\right] d\xi d\eta \\ &= \iint P(\lambda d_i f_x, \lambda d_i f_y) \exp[-j2\pi(f_x x_i + f_y y_i)] df_x df_y \\ &= \Im\{P(\lambda d_i f_x, \lambda d_i f_y)\} \end{aligned} \quad (4.30)$$

and thus

$$H_c(f_x, f_y) = \mathrm{FT}\{\Im\{P(\lambda d_i f_x, \lambda d_i f_y)\}\} = P(-\lambda d_i f_x, -\lambda d_i f_y) \quad (4.31)$$

In other words, $H_c(f_x, f_y)$ equals the pupil function. There exists a certain coordinate scaling relationship between the spatial coordinates x *and* y and frequency domain coordinates f_x and f_y.

Generally, the pupil function always takes values of 1 and 0, and thus, the coherence transfer function takes values of 1 and 0. If the value of $x = -\lambda d_i f_x, y = -\lambda d_i f_y$ determined by x,y lies within the pupil, the exponential primitive of this frequency appears in the image distribution as an identical value, with neither amplitude nor phase change. In other words, the transfer function has a value of 1 for this frequency. If the value x,y lies outside the pupil, the system cannot pass the exponential primitive of this frequency; that is, the transfer function has a value of 0 for this frequency. In other words, the diffraction-limited system is a low-pass filter. There exists a finite frequency band in the frequency domain, which allows the highest frequency to pass through the system as the cutoff frequency, represented by f_0.

If P is defined in reflection coordinates, the burden of the negative sign can be eliminated, and Formula (4.31) can be rewritten as

$$H_c(f_x, f_y) = P(\lambda d_i f_x, \lambda d_i f_y) \tag{4.32}$$

In particular, the general pupil function is symmetrical about the optical axis, and thus, the abovementioned definition does not have any real impact.

4.2.3 Noncoherent Transfer Function of a Diffraction-Limited Optical Imaging System

In the case of noncoherent illumination, the amplitude and phase of the points on the object plane are independent of each other and statistically independent. In this context, although each point of the object exhibits a complex amplitude distribution after passing through the system, the illumination on the object plane is noncoherent. Thus, the complex amplitude distribution of the image cannot be obtained by the coherent superposition of these complex amplitude distributions. The corresponding intensity distributions must be obtained from the complex amplitude distributions, and the intensity distributions are superimposed (noncoherently superimposed) to obtain the image intensity distribution. The noncoherent superposition of light at the time of propagation time is linear with respect to intensity, and thus, the incoherent imaging system is a linear system of intensity. The imaging of the optical system is invariant, and thus, the noncoherent imaging system is a linear, space-invariant system of intensity. In an isohalo optical system, imaging is invariant in the spatial domain, and the incoherent optical imaging system is a linear invariant system of intensity. In incoherent linear space invariant imaging systems, the relationships between the image and object satisfy the following convolution integrals:

$$I_i(x_i, y_i) = \iint_{-\infty}^{\infty} I_o(x_o, y_o) h_I(x_i - x_o, y_i - y_o) \mathrm{d}x_o \mathrm{d}y_o \tag{4.33}$$

$I_o(x_o, y_o)$ is ideal for the intensity distribution of geometric optics, $I_i(x_i, y_i)$ is the intensity distribution, and $h_I(x_i - x_o, y_i - y_o)$ is the intensity impulse response (or noncoherent impulse response or intensity point diffusion function). This aspect represents the intensity distribution of the patch generated by the point object, which is the square of the complex amplitude spread function:

$$h_I(x_i - x_o, y_i - y_o) = |h(x_i - x_o, y_i - y_o)|^2 \tag{4.34}$$

Equations (4.33) and (4.34) show that under incoherent illumination, the image intensity distribution of the linear space invariant imaging system is the convolution of the intensity distribution of the ideal image and intensity point spread function. The imaging characteristics of the system are represented by $h_I(x_i, y_i)$, and $h_I(x_i, y_i)$ is represented by $h(x_i, y_i)$.

In terms of the intensities of a linear invariant system under incoherent illumination, it is convenient to describe the relationship between the image and object in the frequency domain. Fourier transform is performed on both sides of Eq. (4.34), and the irrelevant constants are omitted:

$$A_i(f_x, f_y) = A_o(f_x, f_y)H_I(f_x, f_y) \tag{4.35}$$

where

$$A_i(f_x, f_y) = \mathrm{FT}\{I_i(x_i, y_i)\} \tag{4.36}$$

$$A_o(f_x, f_y) = \mathrm{FT}\{I_o(x_i, y_i)\} \tag{4.37}$$

$$H_I(f_x, f_y) = \mathrm{FT}\{h_I(x_i, y_i)\} \tag{4.38}$$

$I_i(x_i, y_i)$ and $I_o(x_i, y_i)$ and $h_I(x_i, y_i)$ represent the intensity distributions and nonnegative real function, respectively. The Fourier transform must have a constant component that is a zero-frequency component, and the amplitude is greater than that of any nonzero component. Regardless of the image clarity, the total light intensity includes the zero-frequency component, and the ratio of the light intensity to the zero-frequency component carries the information. Therefore, the ratios of $A_i(f_x, f_y)$, $A_o(f_x, f_y)$, and $H_I(f_x, f_y)$ to the respective zero-frequency components are more significant. This aspect suggests that we normalize these entities with zero-frequency components and obtain the normalized spectrum as

$$\tilde{A}_i(f_x, f_y) = \frac{A_i(f_x, f_y)}{A_i(0, 0)} = \frac{\iint I_i(x_i, y_i)\exp[-\mathrm{j}2\pi(f_x x_i + f_y y_i)]\mathrm{d}x_i\mathrm{d}y_i}{\iint I_i(x_i, y_i)\mathrm{d}x_i\mathrm{d}y_i} \tag{4.39}$$

$$\tilde{A}_o(f_x, f_y) = \frac{A_o(f_x, f_y)}{A_o(0, 0)} = \frac{\iint I_o(x_o, y_o)\exp[-\mathrm{j}2\pi(f_x x_o + f_y y_o)]\mathrm{d}x_o\mathrm{d}y_o}{\iint I_o(x_o, y_o)\mathrm{d}x_o\mathrm{d}y_o} \tag{4.40}$$

$$\tilde{H}_I(f_x, f_y) = \frac{H_I(f_x, f_y)}{H_I(0, 0)} = \frac{\iint_{-\infty}^{\infty} h_I(x_i, y_i)\exp[-\mathrm{j}2\pi(f_x x_i + f_y y_i)]\mathrm{d}x_i\mathrm{d}y_i}{\iint_{-\infty}^{\infty} h_I(x_i, y_i)\mathrm{d}x_i\mathrm{d}y_i} \tag{4.41}$$

Since $A_i(f_x,f_y)=A_o(f_x,f_y)\,H_I(f_x,f_y)$ and $A_i(0,0)=A_o(0,0)\,H_I(0,0)$, the resulting normalized spectrum satisfies

$$\tilde{A}_i(f_x, f_y) = \tilde{A}_o(f_x, f_y)\tilde{H}_I(f_x, f_y) \tag{4.42}$$

where $\tilde{H}_I(f_x,f_y)$ is the optical transfer function of the noncoherent imaging system, which describes the effect of the noncoherent imaging system in the frequency domain.

$\tilde{A}_i(f_x,f_y)$, $\tilde{A}_o(f_x,f_y)$, and $\tilde{H}_I(f_x,f_y)$ are generally complex functions. We can use the corresponding modulus and argument for the replacement, such that

$$\tilde{A}_o(f_x, f_y) = \left|\tilde{A}_o(f_x, f_y)\right| \exp[\mathrm{j}\phi_o] \tag{4.43}$$

$$\tilde{A}_i(f_x, f_y) = \left|\tilde{A}_i(f_x, f_y)\right| \exp[\mathrm{j}\phi_i] \tag{4.44}$$

$$\tilde{H}_I(f_x, f_y) = m((f_x, f_y)) \exp[\mathrm{j}\phi] \tag{4.45}$$

According to the relationship between Eqs. (4.43) and (4.44),

$$m(f_x, f_y) = \frac{H_I(f_x, f_y)}{H_I(0,0)} = \frac{\tilde{A}_i(f_x, f_y)}{\tilde{A}_o(f_x, f_y)} \tag{4.46}$$

$$\phi(f_x, f_y) = \phi_i(f_x, f_y) - \phi_o(f_x, f_y) \tag{4.47}$$

Usually, $m(f_x,f_y)$ is known as the modulation transfer function (MTF), and $\varphi(f_x, f_y)$ is the phase transfer function. The former term describes the system's transfer characteristics for each frequency component contrast, and the latter term describes the system's phase shift for each frequency component.

4.2.4 Relationship Between CTF and OTF

The optical transfer function $\tilde{H}_I(f_x, f_y)$ and coherent transfer function $H_c(f_x, f_y)$ describe the transfer function when the same system adopts incoherent and coherent illumination, respectively, and these terms are determined by the physical properties of the system. Application of the autocorrelation theorem and Pashevall theorem yields

$$\tilde{H}_I(f_x, f_y) = \frac{H_I(f_x, f_y)}{H_I(0,0)} = \frac{FT\{h_I(x_i, y_i)\}}{\iint h_I(x_i, y_i)\mathrm{d}x_i\mathrm{d}y_i} = \frac{FT\{|h(x_i, y_i)|^2\}}{\iint |h(x_i, y_i)|^2\mathrm{d}x_i\mathrm{d}y_i}$$

$$= \frac{\iint H_c^*(\alpha, \beta) H_c(f_x + \alpha, f_y + \beta) \mathrm{d}\alpha \mathrm{d}\beta}{\iint |H_c(\alpha, \beta)|^2 \mathrm{d}\alpha \mathrm{d}\beta} \tag{4.48}$$

Thus, for the same system, the optical transfer function is equal to the autocorrelation normalization function of the coherent transfer function H_c.

For the diffraction-limited system under coherent illumination conditions, $H_c(f_x, f_y) = P(\lambda d_i f_x, \lambda d_i f_y)$, and this term is replaced in Formula (4.48) to obtain

$$\tilde{H}_I(f_x, f_y) = \frac{\iint\limits_{-\infty}^{\infty} P(\lambda d_i \alpha, \lambda d_i \beta) P\left[\lambda d_i (f_x + \alpha), \lambda d_i \left(f_y + \beta\right)\right] \mathrm{d}\alpha \mathrm{d}\beta}{\iint\limits_{-\infty}^{\infty} P(\lambda d_i \alpha, \lambda d_i \beta) \mathrm{d}\alpha \mathrm{d}\beta} \tag{4.49}$$

Assume that $x = \lambda d_i f_x$, $y = \lambda d_i f_y$, and the replacement of the integral variables does not affect the integration results. Therefore, the relationship between $\tilde{H}_I(f_x, f_y)$ and $P(\lambda d_i \alpha, \lambda d_i \beta)$ is

$$\begin{aligned} \tilde{H}_I(f_x, f_y) &= \frac{\iint\limits_{-\infty}^{\infty} P(x, y) P\left(x + \lambda d_i f_x, y + \lambda d_i f_y\right) \mathrm{d}x \mathrm{d}y}{\iint\limits_{-\infty}^{\infty} P^2(x, y) \mathrm{d}x \mathrm{d}y} \\ &= \frac{\iint\limits_{-\infty}^{\infty} P(x, y) P\left(x + \lambda d_i f_x, y + \lambda d_i f_y\right) \mathrm{d}x \mathrm{d}y}{\iint\limits_{-\infty}^{\infty} P(x, y) \mathrm{d}x \mathrm{d}y} \end{aligned} \tag{4.50}$$

For the pupil function with only two values of 1 and zero, P^2 in the denominator can be expressed as P. The formula shows that the OTF of the diffraction constrained system is the autocorrelation normalization function of the pupil function.

4.3 Evaluation Index of Optical Imaging System

This section discusses the resolution of optical imaging systems to understand the relationship between the accuracy and error of remote sensing images and the image quantization scale, including the spatial resolution, spectral resolution, and radiation resolution. Each resolution and the corresponding mathematical formulas are presented.

4.3.1 Spatial Resolution of the Remote Sensing Imaging System

Spatial resolution refers to the minimum ground distance or minimum target size that can be identified. The latter term corresponds to the remote sensor or image, which refers to the size of the smallest element that can be distinguished on the image, or the measurement of the minimum angle or linear distance for which the remote sensor can distinguish two objects. Three general expressions exist:

① Pixels. A pixel is the basic unit of the scanned image and the basic sampling point in the imaging process or computer processing, expressed by the brightness. The unit is square meters or square kilometers. For example, one pixel of the United States QuickBird commercial satellite corresponds to a surface area of 0.61 m × 0.61 m and spatial resolution of 0.61 m.

② Line pairs. For photographic systems, the minimum unit of the image is often determined by pairs of lines contained within a 1 mm interval, in units of/mm^{-1}. The line pair refers to a pair of equisized light and dark stripes or regularly spaced light and dark bar pairs "Aerial photography specification for terrain map" (GB/T 15,661–1995) specifies the lens resolution to be no less than 25 "pairs" per millimeter within the effective use area of the aerial camera. According to the objective lens resolution and photography scale, we can estimate the corresponding ground resolution on the aerial photography image.

$$D = M/R \tag{4.51}$$

where M is the denominator of the photographic scale and R is the lens resolution.

③ Instantaneous field of view (IFOV). The instantaneous field of view refers to the angle of view or field of view of a single detector element within the remote sensor, in milliradians (mrad). A smaller IFOV corresponds to a smaller smallest resolvable unit and higher spatial resolution. The IFOV depends on the size of the remote sensor optical system and detector. The information in an instant field of view represents a cell that records a composite signal response.

In an imaging optical system, the resolution represents the ability to measure the separation of two adjacent object points. Owing to diffraction, the system is no longer an ideal geometric point image but a spot (Airy spot) when the two points are extremely close and overlap. Therefore, the system may not distinguish between the images of two objects. In other cases, there exists a resolution limit of the optical system. This resolution limit is usually based on the Rayleigh criterion when the center of an Airy spot coincides with the first dark circle of another Airy spot. In this state, it is possible to distinguish two images. According to the Rayleigh criterion of resolution, the opening angle of the two objects (or the center of the corresponding two Airy spots) to the optical system is the minimum resolution angle of the optical system. The minimum resolution angle of the optical system is half the width of the Airy spot in the Fraunhofer diffraction pattern, that is, the diffraction angle of the first dark ring. The reciprocal of the minimum resolution angle is the resolution of the optical instrument.

Rayleigh distance (i.e., the resolution of the unit diameter): for a lens with a circular aperture, when the PSF of the image plane takes the first zero value, the radius is

$$r_0 = 1.22\frac{\lambda d_i}{a} \tag{4.52}$$

When imaging a nearly planar object with a camera, such as in high-altitude aerial photography and satellite imaging, it is more convenient to calculate the dimensions on the object plane rather than on the image plane, since the object plane is of interest. This process involves a 180° rotation and a scaling factor to ensure that the pixel pitch can be determined in meters on the ground. The spatial resolution can be marked by multiple meters on the object plane. When using the camera lens, $d_f \geq d_i \approx f$, and $M = d_i/d_f \leq 1$. In this case, the use of $f^{\#}$ to mark the diameter of the aperture is more convenient, that is,

$$f^{\#} = {}^{f}/_{a} \tag{4.53}$$

$f^{\#}$ is usually represented as *f/5.6*, that is, $a = f/5.6$. The camera lens $f^{\#}$ aperture settings correspond to $\sqrt{2}$ grading. In this case, changing the aperture by one step corresponds to doubling the aperture area or halving it, and the light intensity received by the negative film, i.e., the exposure, also increases or decreases. The cutoff frequency in the image plane coordinate system of noncorrelated light is

$$f_c = {}^{a}/_{\lambda d_i} \approx {}^{1}/_{(\lambda f^{\#})} \tag{4.54}$$

and the Abbe distance on the plane is

$$r_0 = {}^{\lambda d_i}/_{a} \approx \lambda f^{\#} \tag{4.55}$$

The Rayleigh distance is

$$r_0 = {}^{1.22\lambda d_i}/_{a} \approx 1.22\lambda f^{\#} \tag{4.56}$$

According to Fourier optics analysis, the object is composed of different spatial frequencies of the spectrum. The optical system is a low-pass filter, the cutoff frequency is determined by $1/1.22\lambda f^{\#}$ (λ represents the optical average wavelength, and $f^{\#}$ is the optical system f number), and the sampling frequency is the reciprocal of the detector pixel pitch (sampling pitch) P. According to the sampling theorem, the sampled signal can only be restored when the sampling frequency f_s is greater than or equal to twice the maximum frequency f_h of the sampled signal. In actual imaging systems, most systems do not meet the requirements of the sampling theorem, and spectral aliasing of the sampled signals is unavoidable. Therefore, to ensure that the signal-to-noise ratio does not decrease, the sampling frequency can be increased to decrease the frequency spectrum aliasing of the acquired image and enhance the spatial resolution of the image. This aspect is the essence of increasing the spatial resolution of the image. The Rayleigh criterion is a common method of describing the resolution of an imaging system. At smaller wavelengths and larger apertures, the resolution is enhanced (r_0 decreases) but the magnification of the instrument cannot be increased to enhance its resolution. Because of the increase in magnification, although the distance between the points is enlarged, the diffraction spots are also amplified. Regardless of the object enlargement, the things that cannot be distinguished by optical instruments cannot be distinguished by our eye or photography films. If the magnification of the optical instrument is insufficient, the object can be resolved by an optical instrument may not be able to be resolved because the image is extremely small, and the resolution of the instrument is not fully utilized. Thus, the magnification must be enhanced.

4.3.2 Spectral Resolution of Remote Sensing Imaging Systems

The interaction between electromagnetic waves of different wavelengths and materials is different. Therefore, the spectral characteristics of objects in different wavelength bands vary widely. Several types of detectors and different spectral channels have been designed to collect information. The multiband characteristics of remote sensing information are often described using spectral resolution. The spectral resolution is determined by the number of bands selected by the remote sensor, wavelength position of each band, and size of the wavelength interval. In other words, the number of selected channels, center wavelength of each channel, and bandwidth of each channel determine the spectral resolution. In the design of an imaging spectrometer, the metanumber of the detector is determined based on the spectral sampling interval and spectral range to be covered. If the range of the band is $(\lambda_1 - \lambda_2)$ and number of detectors is N, the spectral sampling interval is $\Delta\lambda = (\lambda_2 - \lambda_1)/N$. One detector generates one band covering the wavelength range of $\lambda_0 \pm 0.5\Delta\lambda$. The ideal spectral response curve of a band must be a rectangular gate function symmetrical to λ_0. When the line of sight of the system is a, the instantaneous field of view $\beta = a/f$, and the system's response to monochromatic light is proportional to the area of overlap of the image of the field of view in the focal plane and photodetector of the detector. According to the Rayleigh criterion, when the maximum and minimum values of the two lines with the same intensity profile overlap, the theoretical maximum spectral resolution is

$$R = \frac{\overline{\lambda}}{\Delta\lambda} = \frac{(\lambda_1 + \lambda_2)/2}{\lambda_2 - \lambda_1} \tag{4.57}$$

The actual spectral resolution is directly related to the system bandwidth. A smaller $\Delta\lambda$ corresponds to a higher resolution capability. Although the bandwidth is $\Delta\lambda$, it is possible to distinguish the absorption peak with an absorption bandwidth less than $\Delta\lambda$. When the center wavelength of the system consists of the center of the absorption peak, the spectral resolution of the system is the highest. The width of the spectral slit controls the luminous flux. A wider slit corresponds to a larger input energy. However, a wider slit corresponds to a larger width of each line in the

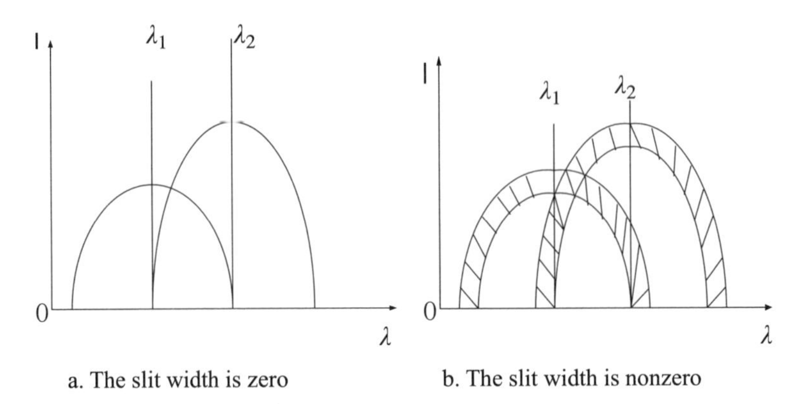

Fig. 4.6 Schematic of different spectral slit widths

spectral line, resulting in the overlapping of two adjacent spectral lines and inability of distinguishing each spectral line. Therefore, the resolution is lower. The width of the slit must not be extremely small because the detector sensitivity has a lower limit. If the input energy is extremely low, the detector does not effectively perform. Due to noise, the energy is extremely low, and the signal-to-noise ratio is not satisfactory. Figure 4.6 shows a schematic of the different spectral slit widths.

A more narrow band of the imaging corresponds to a greater band and higher spectral resolution. The current technology can reach a magnitude of 5–6 nm, with more than 400 bands. The subdivision spectrum can enhance the ability to automatically distinguish and identify target properties and constituent components. In terms of the spectral range of the sensor, a narrower range of a spectrum corresponds to a higher spectral resolution.

4.3.3 Radiation Resolution of Remote Sensing Imaging Systems

The identification of any image object depends on the difference in brightness between the detection target and feature. The radiation resolution refers to the sensitivity and distinguishing ability of the remote sensor to the strength of the spectral signal. In other words, the sensitivity of the detector is the minimum radiometric difference that can be resolved when the sensing element receives the spectral signal or resolving ability of two radiation sources. This value is generally expressed by the number of gray levels, that is, the number of gradations between the darkest and brightest gray values. In the visible and near infrared bands, the noise equivalent reflectivity is used to express the resolution. In the thermal infrared band, the noise equivalent temperature difference, minimum detectable temperature difference and minimum resolution temperature difference are used. The radiation resolution algorithm is defined as

$$\mathrm{RL} = {(R_{\max} - R_{\min})}/{D} \tag{4.58}$$

where $R_{\max}$ is the maximum radiation value, $R_{\min}$ is the minimum radiation value, and D is the quantization of the series. For the spatial and radiation resolutions, the instantaneous field of view (IFOV) is larger, the minimum pixel is larger, and the spatial resolution is lower. However, a larger IFOV corresponds to a greater incident energy that is instantaneously obtained by the flux, more sensitive radiation measurement, and higher ability to detect weak energy differences with a higher radiation resolution. Consequently, an increase in the spatial resolution is accompanied by a decrease in the radiation resolution.

4.3.4 Modulation Transfer Function of the Remote Sensing Imaging System

If the optical system is assumed to be a linear-invariant system, the process of objects passing through optical system imaging can be regarded as a process in which the transfer efficiency is unchanged, but the contrast decreases. The phase shifts, and the process is terminated at a certain frequency. This decrease in the contrast and phase shift varies with the frequency, and the corresponding function is known as the optical transfer function (OTF).

The modulation transfer function (MTF) is the modulus (amplitude) of the OTF. For a high-quality lens, the OTF can be converted to a real-valued MTF. This function only reflects the loss of the object's contrast during the imaging process, without reflecting the position (or phase) movement. A digital imaging system typically contains a series of devices that the image passes through. The MTF of each subsystem is multiplied to form the MTF of the system. Thus, if the MTF of a single device is known, the MTF of the entire imaging system can be determined by multiplying these MTFs.

Figure 4.7 shows a sinusoidal grating with a frequency of v, with the intensity distribution of $I(x) = I_0 + I_a\cos 2\pi vx$. After passing through the photographic system, for various reasons (temperature, pressure, imaging system and other factors), the image intensity distribution becomes $I'(x) = I'_0 + I'_a \cos 2\pi vx$. To express the degree of light and dark contrast of an image, the current common quantitative representation method is known as modulation (or contrast), which is defined as

$$M = \frac{I_{\max} - I_{\min}}{I_{\max} + I_{\min}} \tag{4.59}$$

In this case, $0 \le M \le 1$, $M_{\text{image}} = I_a/I_0$, $M_{\text{object}} = I_a/I_0$, and $M_{\text{image}} \le M_{\text{object}}$.

The degree of sinusoidal grating modulation M_{image} is decreased in relation to the frequency of the sinusoidal grating but is also related to the various media of the

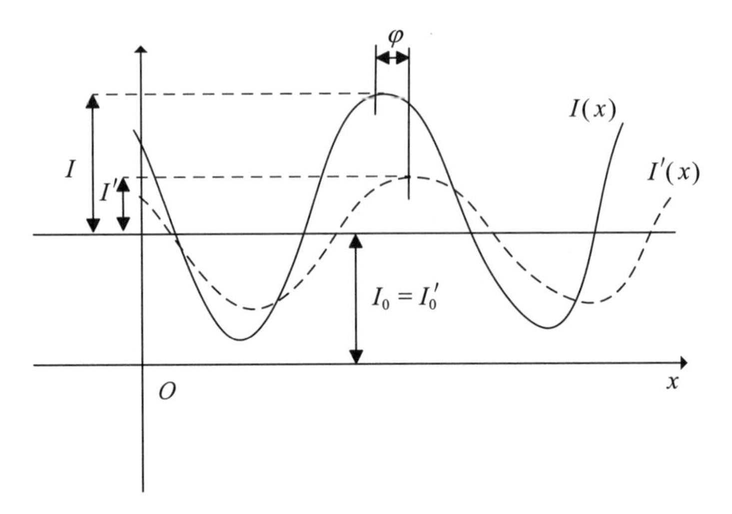

Fig. 4.7 Distribution of the intensity of the sinusoidal

photographic system:

$$T(v) = \frac{M_{\text{image}}(v)}{M_{\text{object}}(v)} \tag{4.60}$$

$T(v)$ is the modulation transfer function MTF. As shown in Fig. 4.8, the MTF is always equal to 1 at zero frequency, and the MTF decreases to almost zero at a certain frequency.

The radial target method is used to calculate the MTF of the imaging system. Figure 4.9 (f in the figure indicates the flight direction, and cf indicates the vertical flight direction) shows that the gray value corresponding to a radius of the target is determined to extract the gray values and arrange the grayscale values. Calculating

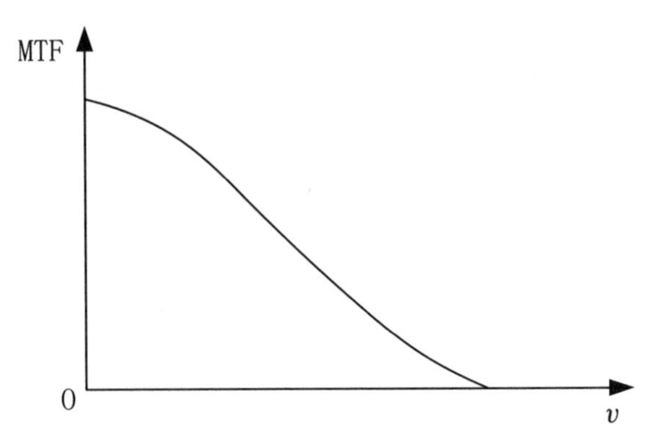

Fig. 4.8 Modulation transfer function curve grating before and after the photographic system

the radius of the radial target corresponding to the modulation on the arc yields the modulation degree at different frequencies, and the MTF at this frequency can be calculated. The MTF value at the unconnected frequency can be obtained to obtain the MTF function curve, as shown in Fig. 4.10.

In the MTF evaluation of the imaging system, because it is difficult to lay a sine target that satisfies the requirement, a rectangular target of a certain width is usually laid to calculate the contrast modulation transfer function at the spatial frequency. Subsequently, the MTF of the imaging system at the spatial frequency is obtained. The imaging quality of optical systems is often evaluated using the MTF.

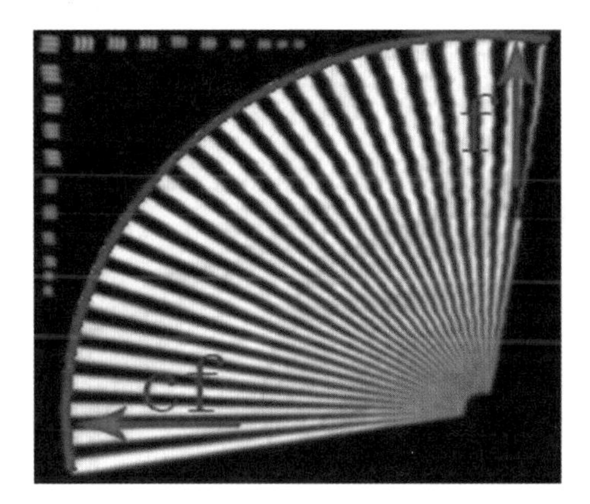

Fig. 4.9 Radial target

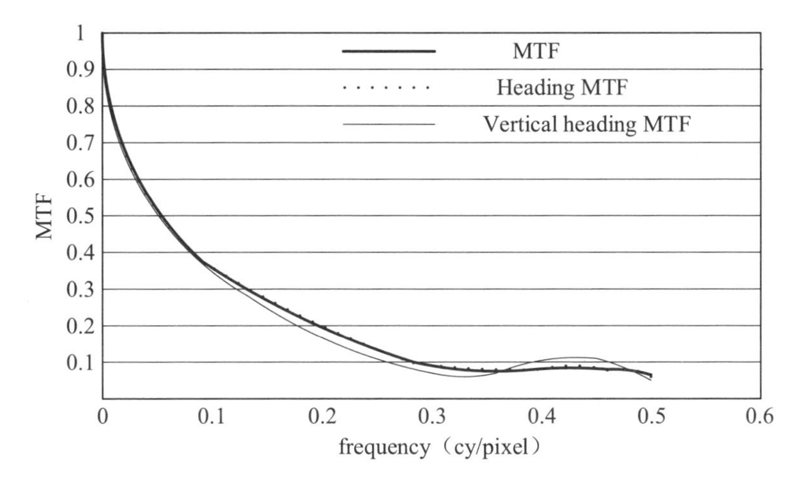

Fig. 4.10 MTF curves

4.3.5 Aberration

To ensure that the optical system produces a perfect image, the system must satisfy several conditions: the point and surface correspond between the object and image. The image of the magnification (off-axis distance) is constant, and the color is the same between the two entities. In general, the study of the ideal optical system requires paraxial small objects, and the incident light for paraxial light satisfies these conditions. The actual optical system is complex, and the main difference from the ideal system is that the paraxial conditions are difficult to fully satisfy. The object cannot be a material strictly converted to a point. The image plane on the various points of the magnification is not identical; and the medium for different wavelengths of different refractive indices exhibits dispersion, resulting in aberration.

According to the monochromatic character of incident light, aberrations can be divided into monochromatic aberrations and color differences. When the color difference pertains to nonmonochromatic light, the aberration of the different color images is different, and aberration occurs due to the different refractive indices of the differently colored light. Monochromatic aberration pertains to the optical system of monochromatic light that cannot be strictly imaged, and spherical aberration, coma, astigmatism, field song, and distortion may occur.

Figure 4.11a and b show the spherical aberration and coma diagram. Spherical aberration occurs because the large aperture of light emitted by the axis cannot be strictly gathered at a point, and a coma is formed close to the optical axis of large aperture light and cannot be gathered at a point. The astigmatism and field are formed by the fact that the vertical and horizontal directions of the out-of-axis rays do not converge on the same focal plane, and the distortion is the combined result of these effects.

In general, corrective aberrations correspond to the following steps: (1) add iris and limit the incidence angle; (2) use a nonspherical lens to enhance the convergence effect; and (3) use a lens or variable refractive index lens. These means are used in combination to eliminate aberrations to a certain extent.

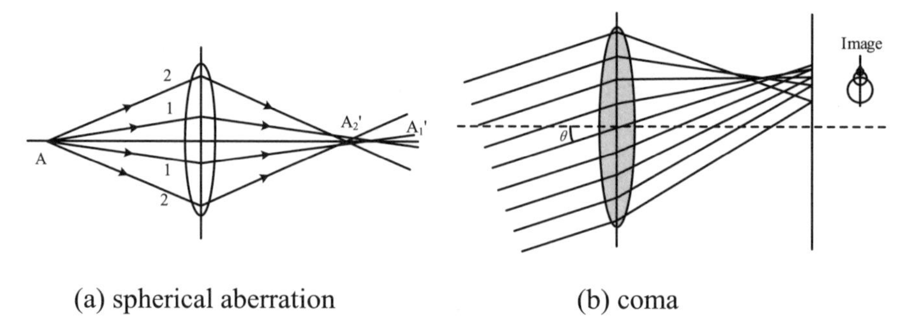

Fig. 4.11 Spherical aberration and coma

4.4 Absolute Radiometric Calibration of Remote Sensing Images

With the in-depth development and application of remote sensing technology in various fields, remote sensing quantification has emerged as a trend in remote sensing development. The basis of remote sensing quantification is the radiation calibration of remote sensing data. In this section, we introduce the nature of the radiation energy intensity measured by the grayscale of the image and describe the calibration method as the physical method of quantifying the scale of remote sensing. Radiation calibration is the process of establishing the connection between the amount of radiation at the entrance pupil of the space camera and output of the detector, including the relative radiation calibration and absolute radiation calibration. Absolute radiation calibration refers to the conversion of the digital number (DN) value that characterizes the sensor response into a physical quantity that represents the geophysical parameters [4]. In the visible and near infrared bands, absolute radiation calibration can be used to convert the DN value of the sensor response to the apparent radiance at the entrance pupil of the sensor. The absolute radiation calibration of the sensor is the basis of remote sensing quantification and is a prerequisite for the assimilation of multisource data. This method is a link for the application of different temporal remote sensing data and is of significance for remote sensing applications. Although the development of the space-borne absolute radiation calibration field in China has been developing rapidly, a specialized aerial remote calibration field to calibrate the airborne sensors on the track is lacking, which limits the quantification of the airborne remote sensing data. The UAV remote sensing load comprehensive verification field is China's first aerial remote calibration field, with multispectral optical camera site radiation calibration capabilities. Using the low-spectral radiation characteristic of the verification field, the wide-field multispectral camera of the test flight can be calibrated on orbit in the field, and the quantitative inversion of the ground information can be realized by using airborne remote sensing data in the future. It is necessary to enhance the level of quantitative remote sensing research in China.

4.4.1 Absolute Radiometric Calibration for Remote Sensing Images

The main methods for the absolute radiometric calibration of the sensor are the reflectivity method, radiance method (also known as the improved reflectivity method) and radiant basis method. The reflectivity base method is used to synchronize the ground reflectivity measurement, atmospheric aerosol characteristic observation and meteorological observation when the sensor is transected. Parameters such as the ground equivalent reflectance and aerosol optical thickness are obtained by processing the data. The parameters are input to the 6S radiation model for calculation. The irradiance-based method must be used to measure the ratio of diffuse to total radiation to avoid the assumption of the aerosol model and alleviate the associated errors and uncertainties compared to the reflectance-based method.

At present, 6S and MODTRAN are used for 0.3–0.4 μm and 0.3–100 μm, respectively. In this book, the radiation transmission model is 6S, and the basic principle is to assume that the sensor entrance pupil apparent reflectivity can be expressed as

$$\rho^* = \frac{\pi L * d_s}{E_0 \cos \theta_s} \tag{4.61}$$

$\rho*$ denotes the apparent reflectivity, d_s represents the distance correction factor, and E_0 is the apparent irradiance of the external environment. The basic correlation between the apparent reflectivity and apparent radiance can be deduced from Eq. (4.61). For the radiation characteristics of the target Lambertian and test sensor for the zenith observation, the apparent reflectivity and surface reflectivity exhibit the following relationship:

$$\rho^* = \left\{ \left[\rho_a + \frac{\rho_t}{1-\rho_t S} T(\theta_s) T(\theta_v) \right] \right\} T_g \tag{4.62}$$

where $\rho*$ is the apparent reflectance of the sensor, ρ_a is the atmospheric upward reflectance, ρ_t is the surface reflectance, S is the atmospheric hemisphere reflectivity, $T(\theta_s)$ is the solar-surface atmospheric transmittance, $T(\theta_v)$ is the surface-to-sensor atmospheric transmittance, and T_g is the absorbed gas transmittance.

4.4.2 Absolute Radiometric Calibration Based on Measurement of the Radiance on the Air Vehicle for Remote Sensing Images

In the radiance base method, an accurately calibrated spectral radiometer is placed at the same height as the aircraft to measure radiation to use the measured value as the actual received radiance and calibrate the sensor. For the UAV wide-field multispectral camera calibration, the basic process is shown in Fig. 4.12.

On November 14, 2010, the first scientific experiment of UAV remote sensing load comprehensive verification field technology was performed in Wulateqian, Inner Mongolia. Experiments were performed on all types of targets, including geometric characteristic targets (fan-shaped targets), radiation-specific targets, and multispectral performance targets. The layout is shown in Fig. 4.13 (false color composition).

The middle part of the figure is a fan-shaped target with a fan-shaped center angle of 114° and a radius of 50 m to detect the multispectral camera geometric resolution. The west side and north side of the fan-shaped target are three-line array targets

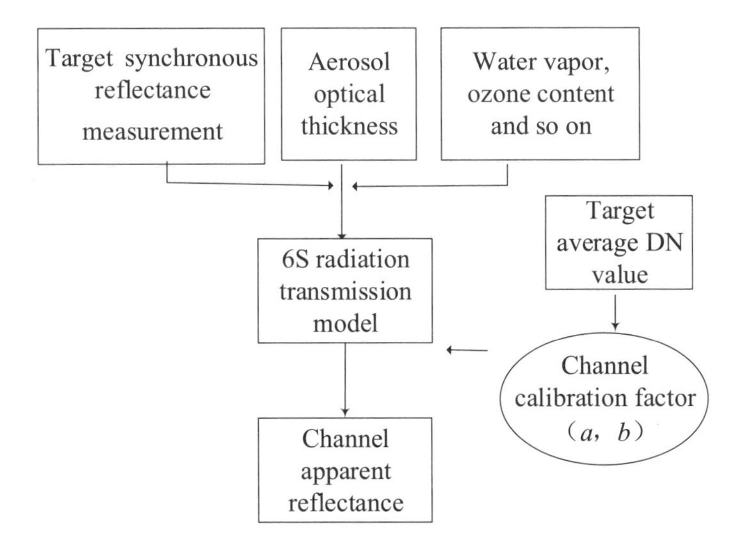

Fig. 4.12 UAV wide-field multispectral camera calibration flow chart

Fig. 4.13 Target distribution map

divided into different resolutions, and the layout requires three-line arrays. The target direction is perpendicular and parallel to the track, and the detection resolution is from 0.05 to 1.1 m. The target is the same as the sector target to detect the geometric resolution. The small square target of the fan-shaped target and three-line array target is a multispectral performance target. The spectral variation is not flat; there are peaks in several bands, and the size is 7 m × 7 m to evaluate the performance of multispectral cameras. In the northeastern part of the graph, four hyperspectral radiation target targets are selected, with nominal reflectance values of 50, 40, 30 and 20% (from left to right). The south side is a high-spectral target with nominal reflectivities of 60 and 4%, sized 20 m × 20 m for absolute radiometric calibration and MTF detection. The wide-field multispectral absolute radiation calibration target is used to identify the spectral flatness of the six hyperspectral performance targets.

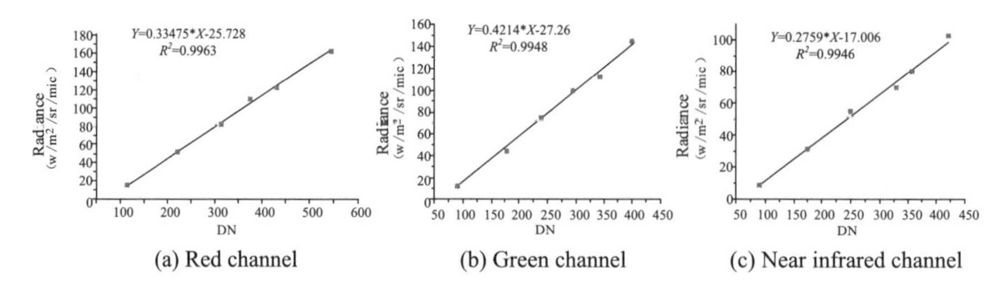

(a) Red channel (b) Green channel (c) Near infrared channel

Fig. 4.14 Radiation calibration results for three channels

Experiments are performed for the absolute radiation calibration of the green band, red band and near infrared band, and the channel response function of the three bands is interpolated to 2.5 nm and input to the 6S model. Next, the 6 reflectance parameters of the radiation target are measured by the spectrometer, and the 6S calculation mode is calculated. The brightness and radiation characteristics of each target in each channel of apparent radiance L_i can be calculated. The average DN of each target is calculated by averaging approximately 10 pixels by 10 pixels in the center of the target. Using the calculated apparent radiance and average DN value to obtain the minimum two-line linear fitting, we can obtain the calibration coefficient, and the scaling formula is as follows:

$$L_i = a * \mathrm{DN} + b \tag{4.63}$$

where a and b are the gain and offset in the scaling factor, respectively.

The radiation calibration results for the three channels are shown in Fig. 4.14. The results show that the linearity of the three bands is high, and the correlation coefficient is more than 99%.

As shown in Fig. 4.14, the three channels of the image are successfully obtained. It can be initially determined that the linearity of the sensor is within the range of 4%–60% of the reflectivity of the ground-laid target in the three channels for successful image acquisition. With a high degree of responsivity, the DN value output by the sensor can be linearly correlated with the radiance at the entrance pupil. Therefore, the evaluation of the effectiveness and availability of the calculated radiation calibration coefficient determines the uncertainty of the calculated radiation calibration coefficient, that is, the dispersion of the calibration coefficient.

Many factors affect the calibration, such as the ground reflectivity measurement error, atmospheric aerosol parameter measurement error and radiation transmission model error. At present, the calculation method for the uncertainty of radiation calibration is represented by the mean square root of each error component. The ground reflectance error has the highest contribution to the overall error calculation, and it can be considered that the error contribution is almost equal to the transmission.

The 6S inherent accuracy contribution is secondary. The change in the solar radiation intensity and atmospheric optical thickness measurement, type of atmospheric aerosols, and surface Lambertian properties also exert a certain influence, which contribute to the uncertainty of absolute radiation calibration.

4.5 Summary

This chapter clarifies the mathematical aspect, explains the transformation of light waves in imaging systems, and analyzes their mathematical characteristics and physical properties. A summary of the specific content is shown in Fig. 4.15. The corresponding formulas are listed in Table 4.1.

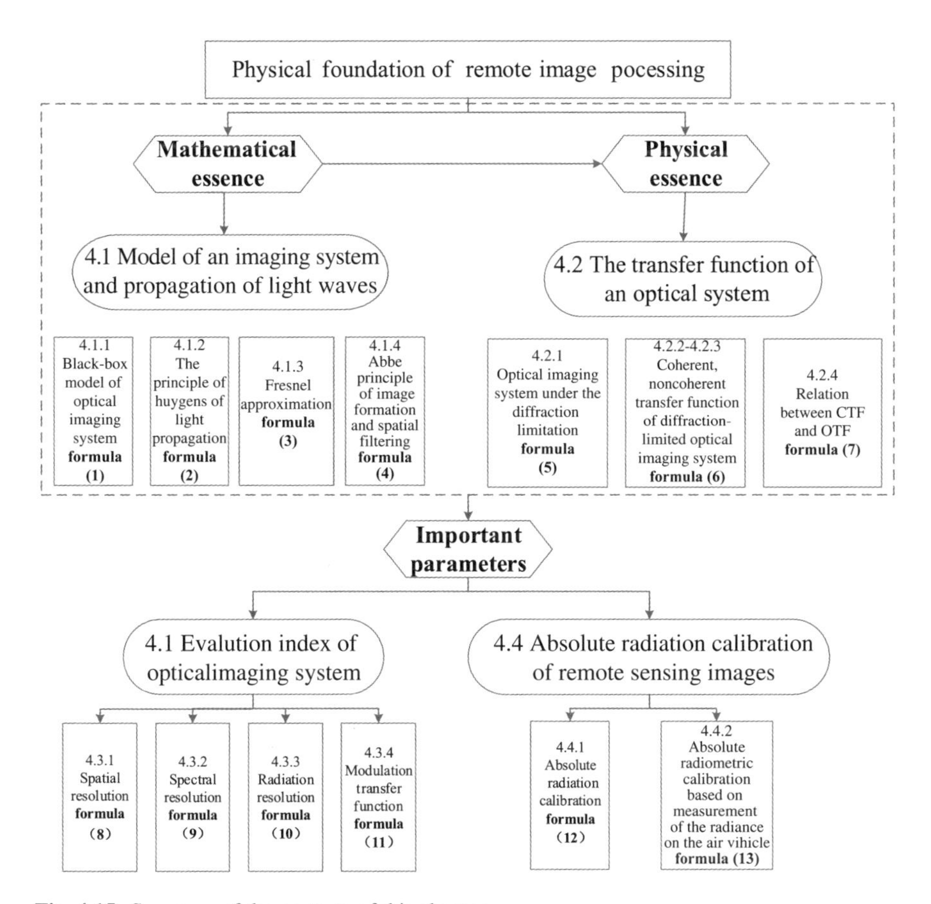

Fig. 4.15 Summary of the contents of this chapter

Table 4.1 Summary of the mathematical formulas in this chapter

formula (1)	$U_o(x_o, y_o) = \iint\limits_{\Omega} U_o(\xi, \eta)\delta(x_o - \xi, y_o - \eta)\mathrm{d}\xi\mathrm{d}\eta$ $U_i(x_i, y_i) = F\{\iint\limits_{\Omega} U_o(\xi, \eta)\delta(x_o - \xi, y_o - \eta)\mathrm{d}\xi\mathrm{d}\eta\}$	4.1.1
formula (2)	$u_i(x_i, y_i) = \frac{1}{\mathrm{j}\lambda}\int\int\limits_{A} u_a(x_a, y_a)\frac{1}{r}\mathrm{e}^{\mathrm{j}kr}\cos(\theta)\mathrm{d}x_a\mathrm{d}y_a$	4.1.2
formula (3)	$R \approx d_i\left[1 + \frac{1}{2}\left(\frac{x_a}{d_i}\right)^2 + \frac{1}{2}\left(\frac{y_a}{d_i}\right)^2\right]$ $r \approx d_i\left[1 + \frac{1}{2}\left(\frac{x_i - x_a}{d_i}\right)^2 + \frac{1}{2}\left(\frac{y_i - y_a}{d_i}\right)^2\right]$ $u_i(x_i, y_i) = \frac{\lambda}{\mathrm{j}}\mathrm{e}^{(\mathrm{j}k/2d_i)(x_i^2+y_i^2)}\int_{-\infty}^{\infty}\int_{-\infty}^{\infty} p(\lambda d_i x_a', \lambda d_i y_a')\mathrm{e}^{-\mathrm{j}2\pi(x_i x_a' + y_i y_a')}\mathrm{d}x_a'\mathrm{d}y_a'$	4.1.3
formula (4)	$\tilde{U}_I(x', y') = \tilde{U}_0(x', y') + \tilde{U}_{+1}(x', y') + \tilde{U}_{-1}(x', y')$ $= A_1\mathrm{e}^{\mathrm{i}\phi(x', y')}\left\{t_0 + \frac{t_1}{2}[\exp(-\mathrm{i}k\sin\theta_{+1}' x') + \exp(-\mathrm{i}k\sin\theta_{-1}' x')]\right\}$	4.1.4
formula (5)	$H_c(f_x, f_y) = \mathrm{FT}\{\Im\{P(\lambda d_i f_x, \lambda d_i f_y)\}\}$ $= P(-\lambda d_i f_x, -\lambda d_i f_y)$	4.2.1
formula (6)	$\tilde{H}_I(f_x, f_y) = \frac{H_I(f_x, f_y)}{H_I(0,0)} = \frac{\int_{-\infty}^{\infty}\int h_I(x_i, y_i)\exp[-\mathrm{j}2\pi(f_x x_i + f_y y_i)]\mathrm{d}x_i\mathrm{d}y_i}{\int_{-\infty}^{\infty}\int h_I(x_i, y_i)\mathrm{d}x_i\mathrm{d}y_i}$	4.2.2
formula (7)	$\tilde{H}_I(f_x, f_y) = \frac{\int_{-\infty}^{\infty}\int P(x, y)P(x + \lambda d_i f_x, y + \lambda d_i f_y)\mathrm{d}x\mathrm{d}y}{\int_{-\infty}^{\infty}\int P^2(x, y)\mathrm{d}x\mathrm{d}y}$ $= \frac{\int_{-\infty}^{\infty}\int P(x, y)P(x + \lambda d_i f_x, y + \lambda d_i f_y)\mathrm{d}x\mathrm{d}y}{\int_{-\infty}^{\infty}\int P(x, y)\mathrm{d}x\mathrm{d}y}$	4.2.3
formula (8)	$r_0 = 1.22\frac{\lambda d_i}{a} \approx 1.22\lambda f\#$	4.3.1
formula (9)	$R = \frac{\overline{\lambda}}{\Delta\lambda} = \frac{(\lambda_1+\lambda_2)/2}{\lambda_2-\lambda_1}$	4.3.2
formula (10)	$\mathrm{RL} = {(R_{\max} - R_{\min})}/{D}$	4.3.3
formula (11)	$T(v) = \frac{M_{\text{image}}(v)}{M_{\text{object}}(v)}$	4.3.4
formula (12)	$\rho^* = \left\{\left[\rho_a + \frac{\rho_t}{1-\rho_t S}T(\theta_s)T(\theta_v)\right]\right\}T_g$	4.4.1
formula (13)	$L_i = a * \mathrm{DN} + b$	4.4.2

References

1. Xihua Z. Modern optical foundation. Peking University Press. 2003.
2. Pal SK, King RA. Image enhancement using fuzzy sets. Electr Lett. 1980;16.
3. Cheng HD et al. A novel fuzzy logic app roach to mammogram contrast enhancement. Inf Sci. 2002;148.
4. Chang DC, Wu WR. Image contrast enhancement based on a histogram transformation of local standard deviation. Med Image, IEEE Trans. 1998;17.

Part II
Pixel Processing Theory and Methods

Chapter 5
Remote Sensing Digital Image Pixel Processing Theory I: Linear System with Space–Time Domain Convolution

Chapter Guidance Image pixel relations satisfy linear system relations. Linear system theory is the link between the signal input and output, and the basic theory to address time or frequency domain digital image information. The effectiveness of theory and method are described in the following chapters. The essence of a linear system is to use simple mathematical models to process and describe complex digital image processes. Because the computer exhibits a delay characteristic, the linear system exhibits a certain inertia, and the reflection of inertia is the convolution effect. Convolution is a prerequisite for continuous and discrete information processing and the core of linear system theory. In this chapter, we first introduce the definition and theory of linear systems and the origin, related properties and significance of the convolution effect. After introducing the convolution solution method, we explain the dynamic and static properties of the correlation function and relative effects in the convolution process. We introduce several basic functions to smooth, sample, and realize the deconvolution of digital images.

In the previous sections, we examined the effects of several image processing operations on the image. These applications can be explained by simple mathematical principles. In this chapter and following sections, we introduce an analytical tool for solving the relevant problems using the linear system theory.

The linear system theory is a mature theory that provides a solid mathematical basis for the study of sampling, filtering and spatial resolution. The structure of this chapter is shown in Fig. 5.1. The contents are as follows: The Definition of a Linear System (Sect. 5.1) introduces the linearly shift invariant delay characteristic and the use of a simple mathematical model to manage the complex digital image process. The Definition and Properties of Convolution (Sect. 5.2) is a key aspect of the spatial delay of linear systems. Convolution is the starting point of remote sensing digital image pixel processing. Calculation and Expression of Convolution (Sect. 5.3) describes the image representation of the convolution process. The matrix method is used to transform the remote sensing image cells into a matrix expression to solve the convolution. Five Basic Functions and their Relationships Related to Convolution (Sect. 5.4) are the link between the input and output and

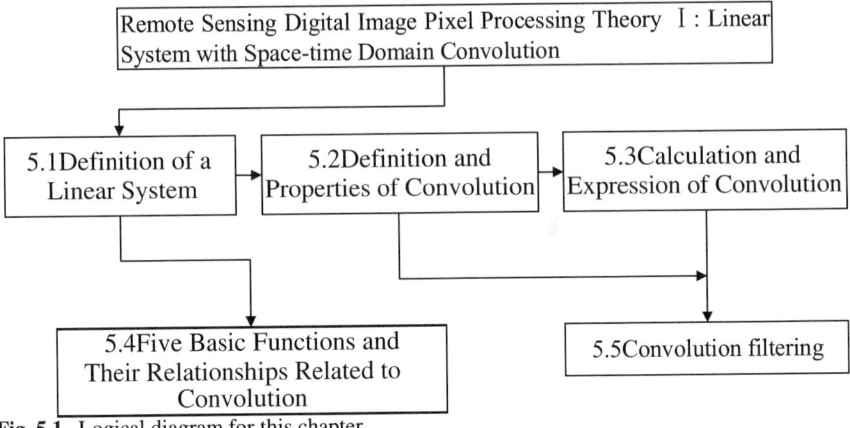

Fig. 5.1 Logical diagram for this chapter

the bridge between the matrix operation and ideal model of information processing. Convolution Filtering (Sect. 5.5) describes the filter characteristics of the convolution and the relevant advantages and disadvantages for different processing needs.

5.1 Definition of a Linear System

This section presents the definition of linear systems and describes the nature of the correlation properties by using a simple mathematical model to address and describe complex digital image processing.

A linear system is any entity that accepts an input and produces a corresponding output. The input and output may have one, two, or more than two dimensions. One-dimensional and two-dimensional linear systems are considered as examples. Figure 5.2 shows a schematic of one-dimensional and two-dimensional linear systems. In each case, the input of the system is a function of one or two variables, and the output produced by the system is another function of the same variable. For a two-dimensional case, each point on the image has a corresponding output after a system action.

1. Linear

A linear system exhibits the following characteristics. For a particular system, the input produces an output, i.e.,

$$x_1(t) \to y_1(t) \tag{5.1}$$

where "→" is intended to be generated, and the output of the other input is

$$x_2(t) \to y_2(t) \tag{5.2}$$

Fig. 5.2 Conceptual expression of linear systems with different dimensions

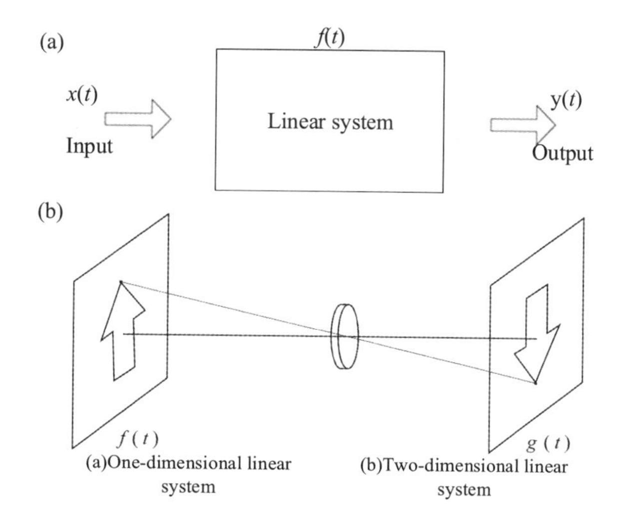

The system is linear if and only if it exhibits the following properties [1]:

$$x_1(t) + x_2(t) \to y_1(t) + y_2(t) \tag{5.3}$$

The output produced by the sum of the previous two signals as inputs is equal to the sum of the previous two outputs. Any system that does not meet this condition is nonlinear. Nonlinear system analysis is widely performed in many fields; however, the complexity of such analyses is considerably higher than linear system analysis, and for the considered application, this additional complexity is not necessary. Therefore, we limit the discussion to linear system analyses.

According to the definition of a linear system, the output produced by the sum of two input signals is equal to the sum of the outputs produced by the two signals acting alone on the system. Therefore, if the input signal is multiplied by a rational number, the output signal increases by the same multiple, i.e.,

$$ax_1(t) \to ay_1(t) \tag{5.4}$$

This formula also holds for irrational numbers and is used as an axiom. A linear system is defined by (5.1), (5.2), and (5.3) and its inference (5.4), as shown in Fig. 5.3.

The linear system theory is used to analyze a process, provided that the process is modeled (or at least approximated) by a linear system. If the system under consideration does not meet the linearity requirements and is nonlinear, the use of the linear system theory may lead to inaccuracies and errors. If the system is only slightly nonlinear, it can be assumed to be a linear system for analysis; however, the results of the analysis are valid only within the assumptions.

Weakly nonlinear systems are often examined using the linear system theory because this theory can be easily addressed and solves. However, caution must be exercised when addressing nonlinear systems since the linear system theory also

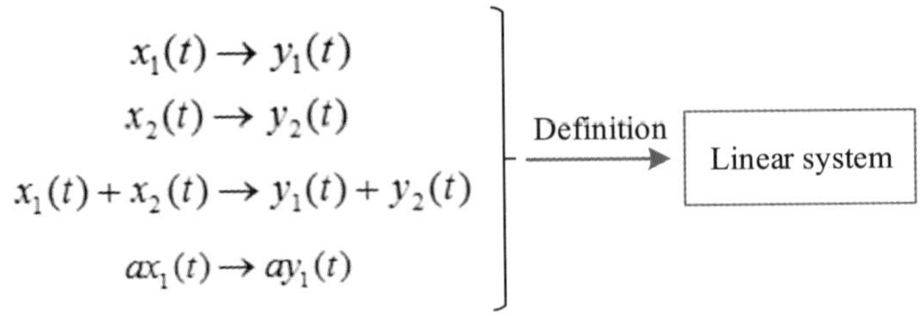

$$\left.\begin{array}{c} x_1(t) \rightarrow y_1(t) \\ x_2(t) \rightarrow y_2(t) \\ x_1(t) + x_2(t) \rightarrow y_1(t) + y_2(t) \\ ax_1(t) \rightarrow ay_1(t) \end{array}\right\} \xrightarrow{\text{Definition}} \boxed{\text{Linear system}}$$

Fig. 5.3 Definition of a linear system

fails when the linear assumptions cannot be met. In the analysis, in addition to mathematical methods, the validity of the premise hypothesis must be considered.

2. Shift Invariance

Another useful property of linear systems is the shift invariance. Assume that for a linear system,

$$x(t) \rightarrow y(t) \tag{5.5}$$

We shift the input signal T along the time axis t,

$$x(t-T) \rightarrow y(t-T) \tag{5.6}$$

Specifically, the output signal and length of translation are changed. In this case, the system exhibits shift invariance. For a shifted system, the translation of the input signal only shifts the output signal by the same length, and the nature of the output signal does not change. Spatial shift invariance is a two-dimensional generalization of the time translation invariance: if the input image has a translation relative to its origin, the output image is invariant except for the same translation, independent of the position of the starting point.

In the space-time domain, the linear system exhibits inertial characteristics, and the inertia reflects the convolution effect.

The system that simultaneously meets both linear and shift invariant properties is known as a linear shift invariant system, and the analysis described in the following chapters corresponds to linear shift invariant systems.

5.2 Definition and Properties of Convolution

5.2.1 Relation Between the Linear Shift Invariant System and Convolution

Consider the linear system shown in Fig. 5.2a. To perform linear system analyses, it is preferable to derive a general expression describing the relationship between input $x(t)$ and output $y(t)$. The expression of the linear function (superposition integral),

$$y(t) = \int_{\infty}^{+\infty} f(t, \tau)x(\tau)\mathrm{d}\tau \tag{5.7}$$

can generally express the relationship between $x(t)$ and $y(t)$ for any linear system. Any linear system can be established based on Function (5.7). To simplify the formula, we add the shift invariant constraint, described in (5.6), to Eq. (5.7),

$$y(t - T) = \int_{-\infty}^{+\infty} f(t, \tau)x(\tau - T)\mathrm{d}\tau \tag{5.8}$$

The abovementioned variables are replaced, and T is simultaneously added to t and τ to obtain

$$y(t) = \int_{-\infty}^{+\infty} f(t + T, \tau + T)x(\tau)\mathrm{d}\tau \tag{5.9}$$

When t and τ increase by the same amount, the value of $f(t, \tau)$ does not change, and we can define a function $g(t-\tau) = f(t, \tau)$ for the difference between t and τ. Equation (5.9) can be simplified as

$$y(t) = \int_{-\infty}^{+\infty} g(t - \tau)x(\tau)\mathrm{d}\tau \tag{5.10}$$

This expression corresponds to the famous convolution integral [2]. The output of a linearly shifted invariant system can be obtained through the convolution of the input signal with the function $g(t)$ of the characterization of the system and can be simply expressed as

$$y(t) = g * x \tag{5.11}$$

For a system S and input signal $x(t)$, the output signal is $y(t)$ and $y(t) = S\{x(t)\}$. Specifically, system S converts the input signal $x(t)$ into an output signal $y(t)$. S is linear. For any two signals x_1, x_2 and arbitrary constants c_1, c_2,

$$S\{c_1x_1(t) + c_2x_2(t)\} = c_1S\{x_1(t)\} + c_2S\{x_2(t)\} \tag{5.12}$$

In general, for any n input signals x_n,

$$S\left\{\sum_{i=1}^{n} c_i x_i(t)\right\} = \sum_{i=1}^{n} c_i S\{x_i(t)\} \tag{5.13}$$

C_i is an arbitrary constant. Similar to a continuous system, if S is shift invariant, such systems are known as linear shift invariant (LTI), which satisfy

$$\mathrm{LTI}\left\{\sum_{i=1}^{n} c_i x_i(t - t_i)\right\} = \sum_{i=1}^{n} c_i \mathrm{LTI}\{x_i(t - t_i)\} \tag{5.14}$$

where x_i is any input, and c_i, t_i are any constants (if delay $t_i < 0$, that is, the future signal affects the system in advance, we do not require the system to be causal, i.e., future signals can affect the current system output.) To simplify the problem, the signal is limited to the simplest pulse input of unit intensity, and the input is the discrete impulse function, i.e.,

$$\delta(t) = \begin{cases} 1, t = 0 \\ 0, t \neq 0 \end{cases} \tag{5.15}$$

where t is an integer. The integer has a nonzero value only at the origin, and the value is zero at the remaining points:

$$\delta(t - t_0) = \begin{cases} 1, t = t_0 \\ 0, t \neq t_0 \end{cases} \tag{5.16}$$

where t_0 is an arbitrary integer constant. This function has a filter feature, i.e.,

$$x(t) = \sum_{t_0=-\infty}^{+\infty} x(t_0)\delta(t - t_0) \tag{5.17}$$

This formula indicates that $x(t)$ must be considered a series of constants c_i that are no longer dependent on t_0. Therefore, we express the input signal $x(t)$ as the superposition of a series of shock signals with different coefficients.

The response of the LTI system to the impulse function (i.e., output) is known as the impulse response: $h(t) = \mathrm{LTI}\{\delta(t)\}$. For an LTI system, the response to any input can be calculated through only the impulse response function $h(t)$. In other words, if we can analyze the response of an LTI system to a unit-intensity pulse signal, we can describe the behavior of the system, at least from the perspective of the input and output [3].

The response of an LTI system to any input is.

$$y(t) = \mathrm{LTI}\,\{x(t)\} = \mathrm{LTI}\left\{\sum_{t_0=-\infty}^{+\infty} x(t_0)\delta(t - t_0)\right\} \tag{5.18}$$

Since the system is linear, the above equation can be changed to

$$y(t) = \sum_{t_0=-\infty}^{+\infty} x(t_0)\,\mathrm{LTI}\,\{\delta(t - t_0)\} \tag{5.19}$$

where $h(t) = \mathrm{LTI}\,\{\delta(t - t_0)\}$ is the system response to the impulse function for only delay t_0. Since the system is time invariant, $\mathrm{LTI}\,\{\delta(t - t_0)\} = h(t - t_0)$. This expression is substituted into the final relationship formula

$$y(t) = \sum_{t_0=-\infty}^{+\infty} x(t_0)h(t - t_0) \tag{5.20}$$

This expression is the convolution definition formula of discrete systems.

For both discrete and continuous LTI systems, the filtering properties are completely described by $h(t)$, and different LTI systems have different $h(t)$. For example, for a function $x(t) = \delta(t)$, the calculation $y(t) = h(t)$ is the impulse response function, which verifies that the previous derivation is self-consistent. If $h(t) = 3\delta(t)$ is known, $y(t) = 3x(t)$, i.e., the system is an ideal amplifier, the signal gain is 3 times, and the system of $h(t) = \delta(t)$ is equivalent to the wire. If

$$y(t) = \int_{-\infty}^{t} x(t_0)\,\mathrm{d}t_0, \tag{5.21}$$

which is an integral system, the output is a slight accumulation of the previous input signal's increment, and the impulse response function of this system can be calculated as [4]

$$h(t) = \int_{-\infty}^{t} \delta(t_0)\,\mathrm{d}t_0 = u(t) \tag{5.22}$$

This expression corresponds to a unit ladder function at the origin.

The input and output relations of linear shift invariant systems can be derived. After determining the impulse response function, the response of the system to any input can be obtained, and this output function is equal to the convolution of the impulse response function and input function. The relationship between the input and output in a linear shift invariant system can be expressed in two ways: in the first approach, the transfer plural function of the system is multiplied by the tuning input to obtain the corresponding tuning output. In the second approach, the real value impulse response of the system is considered, and the convolution between the system and input signal yields the corresponding output.

Since the transfer function and impulse response of a linear shift invariant system are unique and can completely characterize the system, it is reasonable to conclude that there exists a certain relationship between the two functions. The next chapter discusses this relationship.

5.2.2 Superposition and Expansibility of Convolution

A system is defined as a rectangular pulse $g(t)$ that begins at time $t = 0$. The input function is the impulse response $x_1(t)$ at time $t = 0$, and the convolution result of the two entities is the rectangular system impulse response $y_1(t)$ at $t = 0$, which is the same as the original system. The input function I_2 is the impulse response $x_2(t)$ at time $t = 1$. The convolution result of the two entities is the system's rectangular impulse response $y_2(t)$ at $t = 1$. $y_2(t)$ is the result of system $g(t)$ panning a unit of time to the right. The input function I_3 is the impulse response $x_3(t)$ at time $t = 2$, and the convolution result of the two is the rectangular impulse response $y_3(t)$ of the system at $t = 2$. Notably, $y_3(t)$ is the result of translation of system $g(t)$ to the right by 2 units. If the input function is a linear superposition $x(t)$ of functions I_1, I_2 and I_3, the convolution result of the system $g(t)$ is $y(t) = y_1(t) + y_2(t) + y_3(t)$. The specific process is shown in Fig. 5.4.

An arbitrarily shaped input function can be seen as the linear superposition of numerous pulse functions, and the response value at t is the system performance in the pulse function response delay t. If the pulse value is greater than 1 at time t, the system performance is amplified and enhanced. If the pulse value at time t is less than 1, the system performance is deteriorated. The final convolution result is a linear superposition of all impulse responses (Fig. 5.5). The range of t after convolution is larger than that before convolution, which reflects the expansion of the inertial system [5].

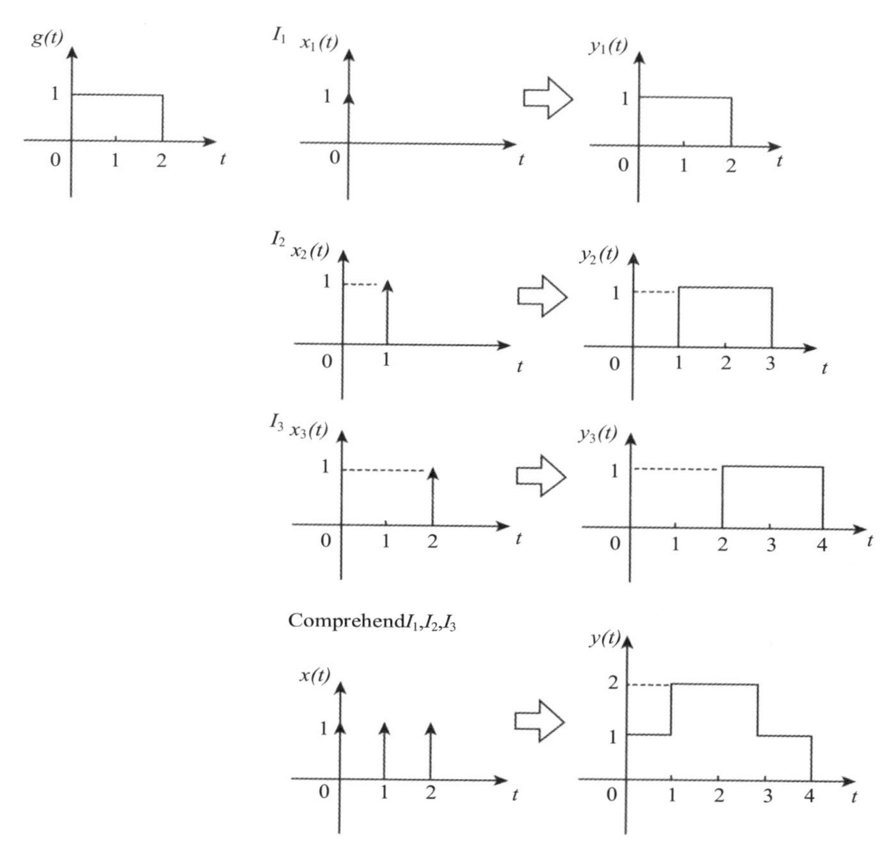

Fig. 5.4 Overlay and extensibility of convolution. ($y(t) = x(t) * g(t)$)

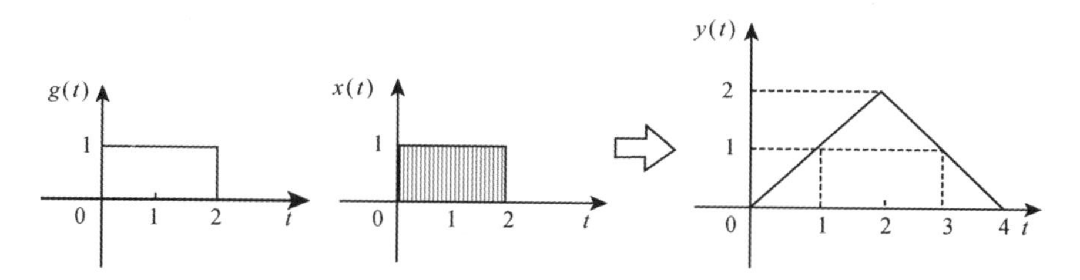

Fig. 5.5 The convolution of two rectangular pulse ($y(t) = x(t) * g(t)$)

5.2.3 Properties and Proof of Convolution

1. Commutative Law of Convolution

The commutative law of convolution can be expressed as

$$f_1(t) * f_2(t) = f_2(t) * f_1(t)$$

The proof can be presented as follows:

$$f_1(t) * f_2(t) = \int_{-\infty}^{+\infty} f_1(\tau) \cdot f_2(t - \tau)\,\mathrm{d}\tau \tag{5.23}$$

Let $t - \tau = \lambda$ such that $\mathrm{d}\tau = -\mathrm{d}\lambda$. In this case,

$$f_1(t) * f_2(t) = \int_{-\infty}^{+\infty} f_2(\lambda) \cdot f_1(t - \lambda)\,\mathrm{d}\lambda = f_2(t) * f_1(t) \tag{5.24}$$

The convolution result is independent of the order of the two functions. Since the integral area of the product is not affected by the inversion of $f_1(\tau)$ and $f_2(\tau)$, it is independent of t.

2. Distributive Law of Convolution

The distributive law of convolution can be expressed as

$$f(t) * [h_1(t) + h_2(t)] = f(t) * h_1(t) + f(t) * h_2(t) \tag{5.25}$$

The following analysis is performed from the viewpoint of the system. A block diagram of the system parallel representation is shown in Fig. 5.6.

Conclusion: When the subsystems are connected in parallel, the impulse response of the total system is equal to the sum of the impulse responses of the subsystems.

3. Associative Law of Convolution

The associative law of convolution can be expressed as

$$[f(x) * h_1(t)] * h_2(t) = f(x) * [h_1(t) * h_2(t)] = f(x) * h(t) \tag{5.26}$$

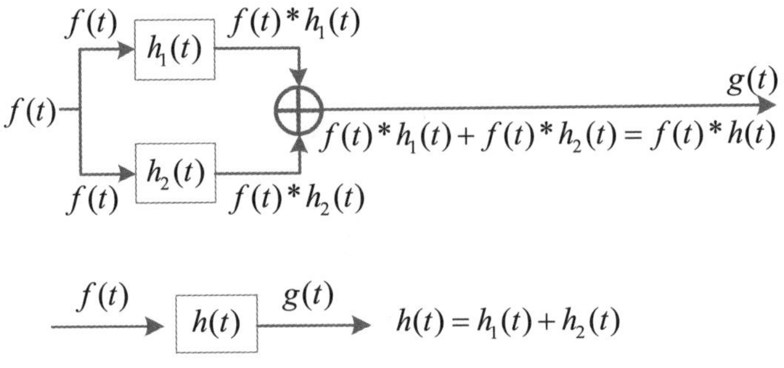

Fig. 5.6 Proof of distributive law

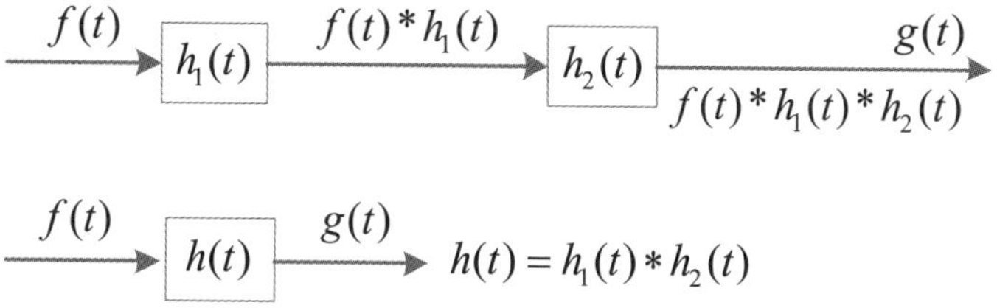

Fig. 5.7 Proof of the associative law of convolution

We consider the associative law of convolution from a systematic perspective. A block diagram of the system cascade is shown in Fig. 5.7.

Conclusion: When the subsystem of the time domain is cascaded, the total impulse response is equal to the convolution of the impulse response of the subsystems.

5.2.4 Physical Meaning of Convolution

The response of the system is related to not only the input of the current time system but also the input of the previous moments, which can be understood as the fact that the input signal of the previous time influences (this effect can be diminishing or weakening, among other influences) the system output in the current moment. When calculating the system output, it is necessary to consider the superimposition of the response of the signal input at the present time and "residual" effect of the response of the signal input several past instances.

Assuming that the system responds to $y(0)$ at time 0, if the response does not change at moment 1, the response at moment 1 is $y(0) + y(1)$, which is known as the cumulative sum of the sequence (which is different than the sum of the sequence). However, this phenomenon does not usually occur in the system because the response at time 0 is unlikely to remain unchanged at time 1. This change is thus expressed through the response function $h(t)$ and $x(0)$ as $x(n) \times h(m-n)$, and the cumulative sum operation to obtain a real system response.

The system response at a time is determined by not only the current time t and previous time $t-1$, and it may be coupled with times $t-2$, $t-3$ and $t-4$, among others. The range of influence is constrained by the range of m in $h(m-n)$ after the change in $h(n)$. In other words, the system response at the current time is related to the "residual effects" of the response at a previous moment [5].

5.3 Calculation and Expression of Convolution

5.3.1 One-Dimensional Convolution

Geometrically, convolution refers to a curve in the $(-\infty, +\infty)$ interval, and the value of each point on the curve is the integral value from the integral starting point to the point whose integrand is the product of the dynamic function and stationary function at that point. If the dynamic function is defined only in the finite interval (a, b) or semi-infinite interval $(0, +\infty)$, the function can expressed as a product with a step function or rectangular function. The convolution of a function on an infinite interval can be defined as a definite integral on a finite interval and can be solved using the fractional integral method.

Consider an example of solving the convolution $x(t) * g(t) = \int_{-\infty}^{+\infty} x(\tau)g(t-\tau)\mathrm{d}\tau$:

(1) change the coordinates of the graph $x(t) \to x(\tau)$, $g(t) \to g(\tau)$;
(2) obtain the converse of the signal (movable function) $g(-\tau) \to g(-\tau)$;
(3) shift the conversed signal by t units, $g(-\tau) \to g(t-\tau)$, as shown in Fig. 5.8;
(4) multiply the overlapping portion of $g(-t)$ and $x(t)$, $x(\tau)g(t-\tau)$;
(5) the integral of the graphics after multiplication is $\int_{-\infty}^{+\infty} x(\tau)g(t-\tau)\mathrm{d}\tau$.

Table 5.1 shows the solution for function $g(t)$.

The ranges of the independent variable t of the input function $x(t)$ and system function $g(t)$ are $[-f_1, f_2]$ and $[-g_1, g_1]$, respectively. The range of independent variable

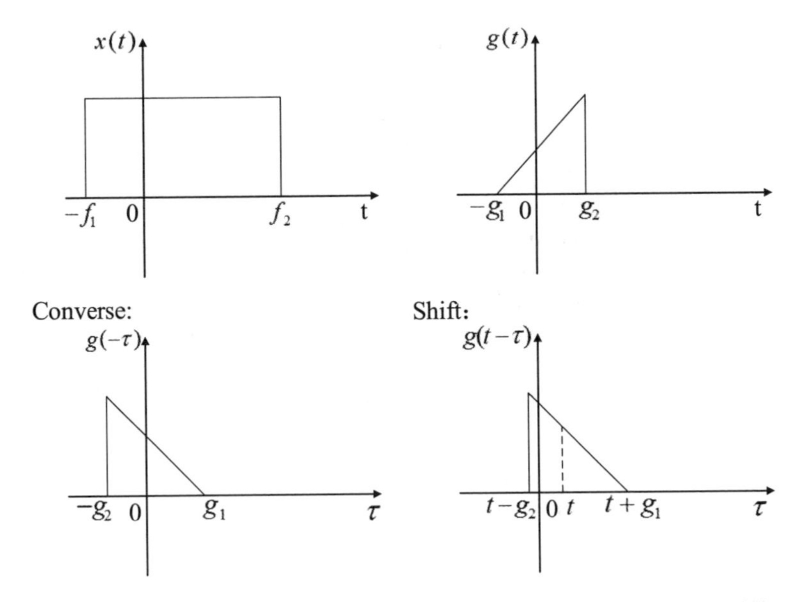

Fig. 5.8 Input function $x(t)$ and system function $g(t)$ and their convergence and time shift

Table 5.1 Convolution solver graphic process, where (a)–(e) correspond to the different moments of t, and (f) shows the convoluted pattern

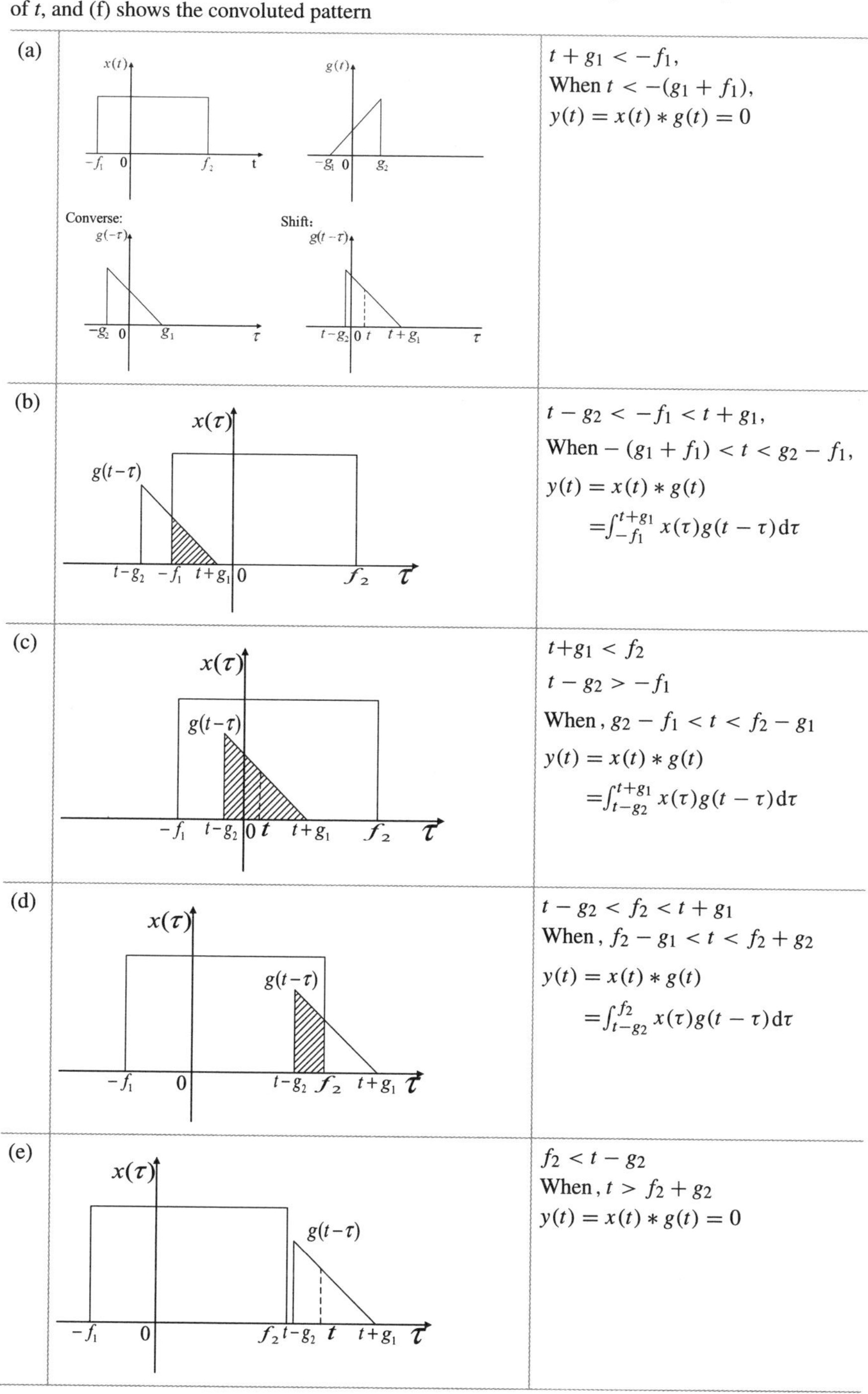

(a)	$t + g_1 < -f_1$, When $t < -(g_1 + f_1)$, $y(t) = x(t) * g(t) = 0$
(b)	$t - g_2 < -f_1 < t + g_1$, When $-(g_1 + f_1) < t < g_2 - f_1$, $y(t) = x(t) * g(t)$ $=\int_{-f_1}^{t+g_1} x(\tau)g(t-\tau)\mathrm{d}\tau$
(c)	$t+g_1 < f_2$ $t - g_2 > -f_1$ When, $g_2 - f_1 < t < f_2 - g_1$ $y(t) = x(t) * g(t)$ $=\int_{t-g_2}^{t+g_1} x(\tau)g(t-\tau)\mathrm{d}\tau$
(d)	$t - g_2 < f_2 < t + g_1$ When, $f_2 - g_1 < t < f_2 + g_2$ $y(t) = x(t) * g(t)$ $=\int_{t-g_2}^{f_2} x(\tau)g(t-\tau)\mathrm{d}\tau$
(e)	$f_2 < t - g_2$ When, $t > f_2 + g_2$ $y(t) = x(t) * g(t) = 0$

(continued)

Table 5.1 (continued)

(f)

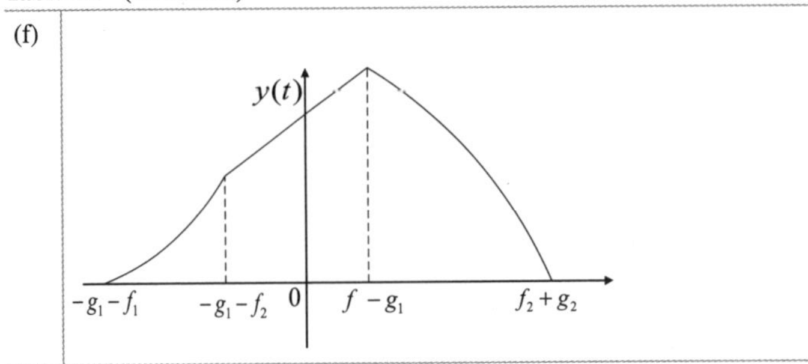

t of the convolution result $y(t)$ is $[-T_1, T_2]$.

$$y(t) = x(t) * g(t) = \int_{-\infty}^{+\infty} x(\tau)g(t-\tau)\mathrm{d}\tau$$

τ must satisfy the domain of f and g, and thus, $\begin{cases} -f_1 < \tau < f_2 \\ -g_1 < t-\tau < g_2 \end{cases} \Rightarrow$ $-(f_1 + g_1) < t < f_2 + g_2$, and $T_1 = f_1 + g_1$ with$T_2 = f_2 + g_2$.

5.3.2 *Matrix Solution of One-Dimensional Discrete Convolution*

To discretize functions,

$$f_p(i) = \begin{cases} f(i), 1 \le i \le m \\ 0, m < i < N \end{cases}, g_p(i) = \begin{cases} g(i), 1 \le i \le m \\ 0, m < i < N \end{cases}, N = n + m - 1$$

The expression of discrete convolution is

$$h = g * f = \begin{bmatrix} g_p(1) & g_p(N) & \dots & g_p(2) \\ g_p(2) & g_p(1) & \dots & g_p(3) \\ \dots & \dots & \dots & \dots \\ g_p(N) & g_p(N-1) & \dots & g_p(1) \end{bmatrix} \begin{bmatrix} f_p(1) \\ f_p(2) \\ \dots \\ f_p(N) \end{bmatrix} \tag{5.27}$$

For this formula, we can assume that $h = \left[h_p(1)\, h_p(2) \cdots h_p(N)\right]$. According to the rules of the linear matrix operation,

$$h_p(i) = g_p(i) f_p(1) + g_p(i-1) f_p(2) + \dots + g_p(1) f_p(i)$$

$$+ g_p(N)f_p(i+1) + g_p(N-1)f_p(i+2) + \dots$$
$$+ g_p(i+1)f_p(N),\ 1 \le i \le N \quad (5.28)$$

Considering that $f_p(i)$ is defined only within $1 \le i \le m$, its value is zero within $m \le i \le N$, $g_p(i)$ is defined only in $1 \le i \le n$, and it is zero within $n \le i \le N$. This formula can be transformed to

$$\begin{aligned} h_p(i) &= g_p(i)f_p(1) + g_p(i-1)f_p(2) + \dots + g_p(1)f_p(i) \\ &= \sum_{j=1}^{i} g_p(i+1-j)f_p(j),\ 1 \le i \le N \end{aligned} \quad (5.29)$$

Example: Solve the convolution of the two discrete vectors $\boldsymbol{g}(t) = \{1,2,3\}$ and $\boldsymbol{x}(t) = \{a, b, c, d, \mathrm{e}\}$.

The solution is as follows: $m = 3, n = 5, N = m + n - 1 = 7$,

The convolution diagram is as follows:

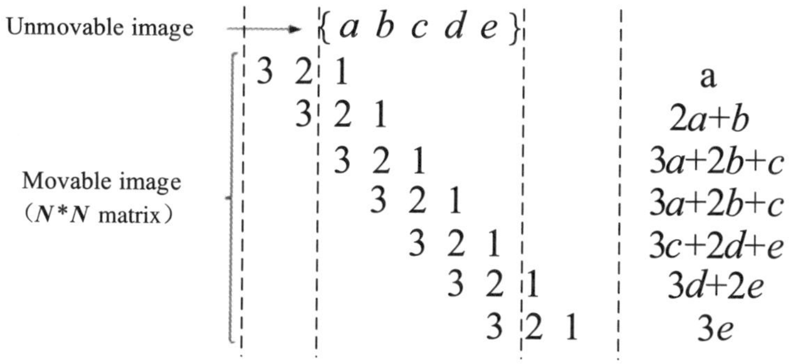

Unmovable image →			{a	b	c	d	e}			
Movable image (N*N matrix)	3	2	1							a
		3	2	1						2a+b
			3	2	1					3a+2b+c
				3	2	1				3a+2b+c
					3	2	1			3c+2d+e
						3	2	1		3d+2e
							3	2	1	3e

The matrix calculation form and result are

$$\boldsymbol{g}(t) * \boldsymbol{x}(t) = \begin{bmatrix} 1\,0\,0\,0\,0\,3\,2 \\ 2\,1\,0\,0\,0\,0\,3 \\ 3\,2\,1\,0\,0\,0\,0 \\ 0\,3\,2\,1\,0\,0\,0 \\ 2\,0\,3\,2\,1\,0\,0 \\ 0\,0\,0\,3\,2\,1\,0 \\ 0\,0\,0\,0\,3\,2\,1 \end{bmatrix} \cdot \begin{bmatrix} a \\ b \\ c \\ d \\ e \\ 0 \\ 0 \end{bmatrix} = \begin{bmatrix} a \\ 2a+b \\ 3a+2b+c \\ 3a+2b+c \\ 3b+2c+d \\ 3c+2d+e \\ 3e \end{bmatrix}$$

Four methods are available to calculate the discrete one-dimensional convolution:

(1) Table method

Given that $\boldsymbol{x}(n) = \{4, 3, 2, 1\}$, $\boldsymbol{h}(n) = \{3, 2, 1\}$, $\boldsymbol{y}(n) = \boldsymbol{x}(n) * \boldsymbol{h}(n)$ can be obtained as follows:

Fill the table with rows and columns, multiply the row and column elements, and add diagonal elements.

	4	3	2	1
3	12	9	6	3
2	8	6	4	2
1	4	3	2	1

The result is $y(n) = \{12, 17, 16, 10, 4, 1\}$.

(2) Alignment method

Given that $x(n) = \{4, 3, 2, 1\}$, $h(n) = \{3, 2, 1\}$, $y(n) = x(n) * h(n)$ can be obtained as follows:

Align the right side, multiply the corresponding value, and add the product value of the same column.

$$
\begin{array}{llcccccc}
\text{Align } x_1(n) & : & \underset{n=0}{\underset{\uparrow}{4}} & 3 & 2 & 1 & & \\
\text{Multiply} \times x_2(n) & : & & & \underset{n=0}{\underset{\uparrow}{3}} & 2 & 1 & \\
\hline
 & & & & 4 & 3 & 2 & 1 \\
 & & & 8 & 6 & 4 & 2 & \\
 & & 12 & 9 & 6 & 3 & & \\
\hline
\text{Add} \quad y(n) & : & \underset{n=0}{\underset{\uparrow}{12}} & 17 & 16 & 10 & 4 & 1
\end{array}
$$

The result is $y(n) = \{12, 17, 16, 10, 4, 1\}$.

(3) Sliding method

Given that $x(n) = \{4, 3, 2, 1\}$, $h(n) = \{3, 2, 1\}$, $y(n) = x(n) * h(n)$ can be obtained as follows:

Fold: $h(m) \rightarrow h(-m)$.

Shift: $h(-m) \rightarrow h(n - m)$.

Multiply: $\boldsymbol{x}(m) \cdot \boldsymbol{h}(n-m)$.

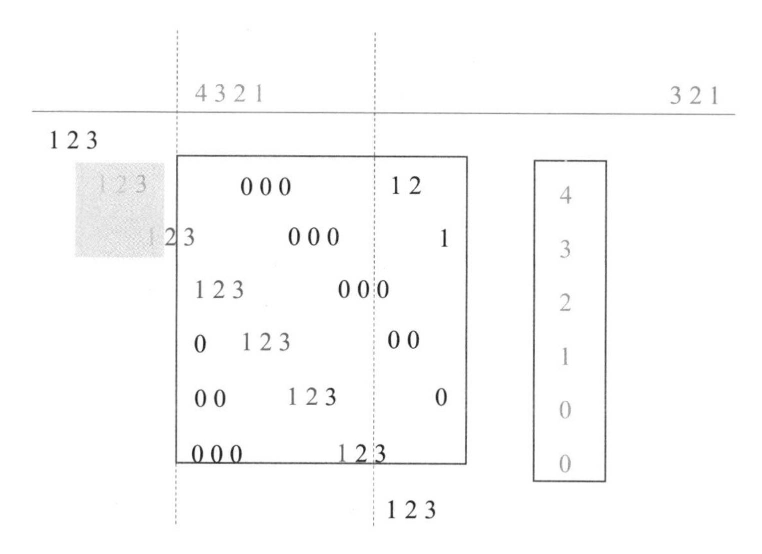

Add: $\sum \boldsymbol{x}(m) \cdot \boldsymbol{h}(n-m)$

$$4 \times 3 = 12$$
$$4 \times 2 + 3 \times 3 = 17$$
$$4 \times 1 + 3 \times 2 + 2 \times 3 = 16$$
$$3 \times 1 + 2 \times 2 + 1 \times 3 = 10$$
$$2 \times 1 + 1 \times 2 = 4$$
$$1 \times 1 = 1$$

The result is $\boldsymbol{y}(n) = \{12, 17, 16, 10, 4, 1\}$

4. Matrix Method

Given that $\boldsymbol{x}(n) = \{4, 3, 2, 1\}$, $\boldsymbol{h}(n) = \{3, 2, 1\}$, $\boldsymbol{y}(n) = \boldsymbol{x}(n) * \boldsymbol{h}(n)$ can be obtained as follows:

Calculate the number of elements in the sequence; fill in the sequence and shift it to form a matrix; and multiply the matrix to obtain the result.

The number of elements is determined as follows:

If the sequence length of $\boldsymbol{x}(n)$ is m, and the sequence length of $\boldsymbol{h}(n)$ is n, the sequence length of $\boldsymbol{y}(n)$ is $m+n-1$. In the considered example, $4+3-1=6$.

Form the following matrix: $\boldsymbol{h}(n)$ $\boldsymbol{x}(n)$.

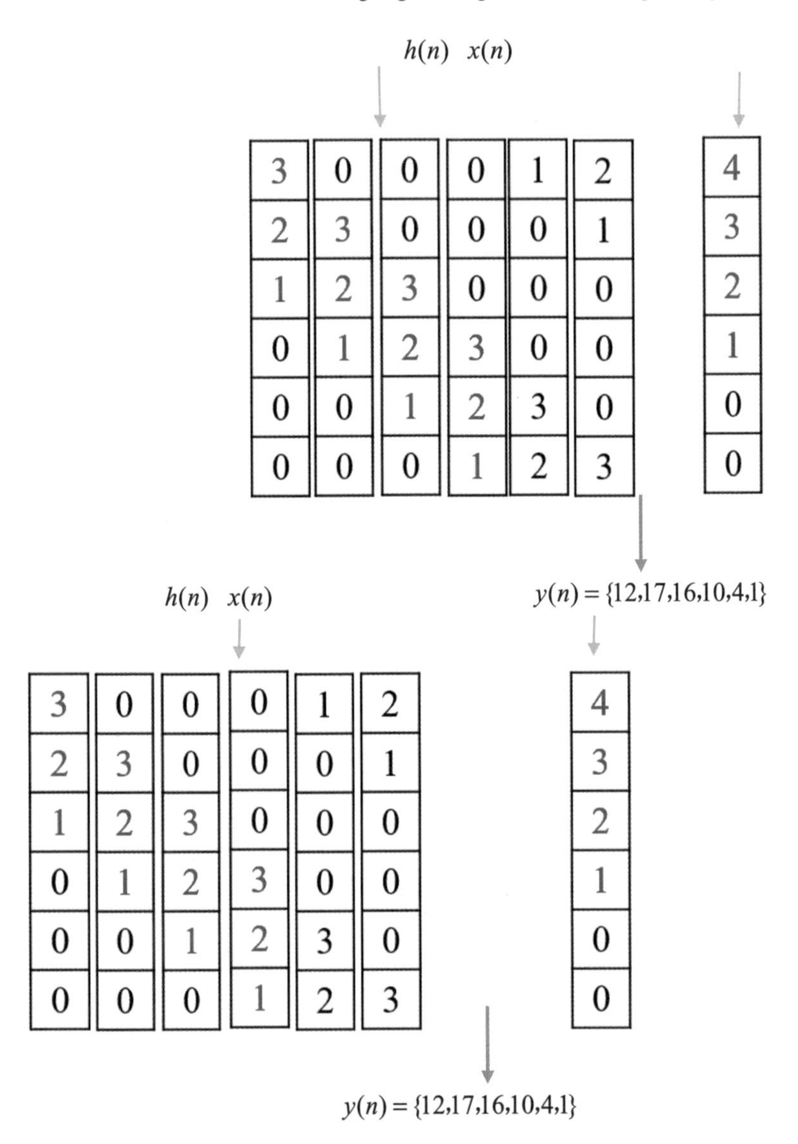

5.3.3 "Rolling" Effect of Convolution Operation

This section considers the convolution of two rectangular pulses. To illustrate the rolling effect of convolution operations in a graphical manner, graphs of the system function and input function are shown in Fig. 5.9. Figure 5.9 shows that the system function $g(t)$ and input function $x(t)$ are rectangular pulses with length n and m after discretization, respectively.

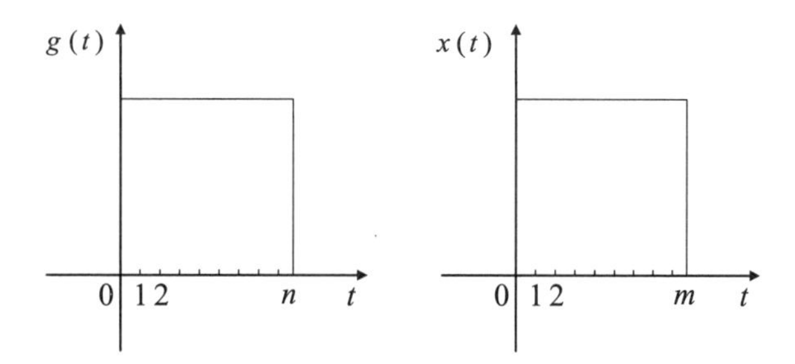

Fig. 5.9 Schematic of the system function $g(t)$ and input function $x(t)$

Section 5.3.1 explains the calculation of the convolution. In particular, the actual length of the convolution calculation process is only the sum of the actual length of the two functions minus 1, denoted by $N = n + m-1$. Objectively, the one-dimensional convolution process is a roll-wound process. As shown in Fig. 5.10, the discrete function $x(t)$ takes a normal value from time 1 to m. Between instances $m + 1$ to N, no value is extracted, and thus, zeros are set. When the 0 time of the winding body coincides with N, $f(0) = f(N) = 0$, and period N is formed. This function is a periodic (winding) function whose mode is N, recorded as mod (N). This "winding" computing feature is a vivid expression of the name "convolution", and thus, "convolution" has a vivid physical meaning. Similarly, a winding function $g(t)$ exists, as shown in the Fig. 5.10.

After the system function $g(t)$ is reversed, the intersection with $x(t)$ starts at $t = 0$, resulting in a convolution value. The $g(t)$ function can also be viewed as a rolling function, and at $t = 0$, only one edge intersects with $x(t)$, as shown in Fig. 5.11.

$g(t)$ moves clockwise around $x(t)$ over time, and more parts intersect to produce convolution values. Figure 5.12 shows the rolling situation at t_1, t_2, and t_3.

At t_1, the dynamic function $g(t)$ intersects partly with the definite function $x(t)$, the convolution of $0-t_1$ has values, and the remaining values are zero. At t_2, the dynamic function $g(t)$ intersects completely with the definite function $x(t)$, and the convolution of the intersecting part has a certain value. At t_3, part of the dynamic function has been rolled to $t > m$. Because the fixed function of this part has no value, the convolution has no value as well, although the remaining intersecting parts have convolution values.

When the dynamic function wraps around the fixed function, it returns to the spatial initial state shown in Fig. 5.10, and the convolution process ends. From the rolling process, we conclude that the space occupied by the convolution process of two functions is the sum of the length of the independent variable of the two discrete functions minus 1, which is neither more nor less.

The basic principle of image processing requires that the convolution operation must have an inverse operation that can uniquely return to the original state before image processing. Therefore, in a discrete one-dimensional convolution, the matrix must be full rank, owing to which, $N \leq m + n-1$. Otherwise, N is greater than $n + m-1$, the matrix is not positive, the inverse of the matrix does not exist, and we cannot return to the preconvolution state (the original image before convolution)

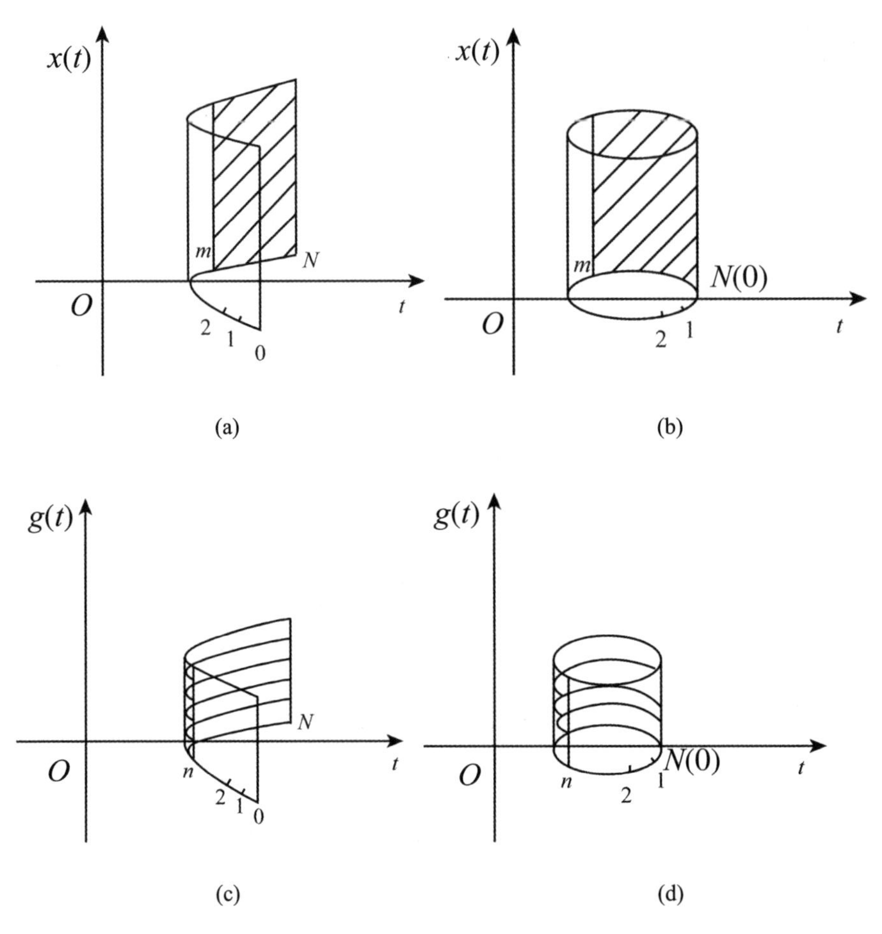

Fig. 5.10 "Rolling" of the input function $x(t)$ and system function $g(t)$ (the shaded portion represents no function value)

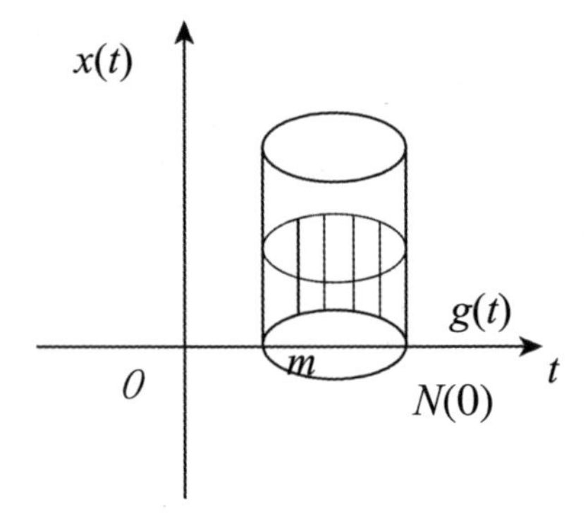

Fig. 5.11 Rolling situation at $t = 0$. $x(t)$ has values $0{-}m$, $g(t)$ has values $m{-}N$ and $n = N{-}(m{-}1)$

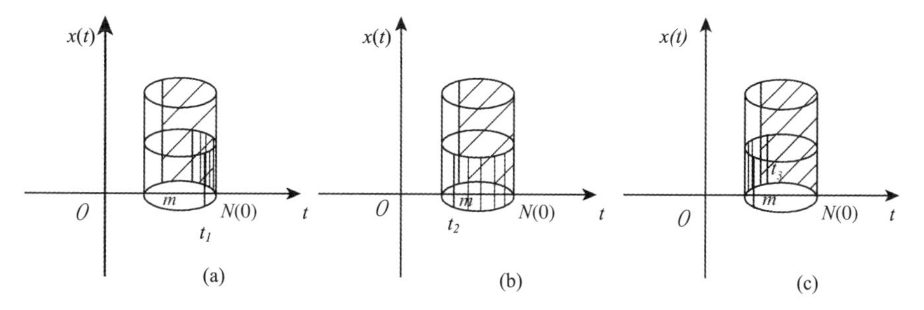

Fig. 5.12 Situation (a) at t_1, (b) at t_2, (c) at t_3

through the inverse operation after the convolution calculation. However, if N is less than $n+m-1$, several nonzero values of the coiling operation may overlap and may not be equal to the true values. Alias errors occur. Although the matrix is full and an inverse matrix exists, it is not possible to return to the matrix value before matrix calculation. Therefore, $N=n+m-1$ is a necessary and sufficient condition for the matrix operation to be reversible and the original image value to be invariant.

5.3.4 Two-Dimensional Convolution

The convolution of two-dimensional continuous functions is similar to that in the one-dimensional case. When the discussion is extended to two dimensions, x and y represent two independent variables. The two-dimensional convolution expression is

$$h(x,y)=f*g=\int_{-\infty}^{\infty}\int_{-\infty}^{\infty}f(u,v)g(x-u,y-v)\,\mathrm{d}u\mathrm{d}v \tag{5.30}$$

The relevant curve is shown in Fig. 5.13. Notably, $g(0-u,0-v)$ is only $g(u, v)$ rotated 180° around its origin, and $g(x-u, y-v)$ moves the origin of g after rotation to point (x, y). The two functions are multiplied by point, and the product function is two-dimensionally integrated, i.e.,

$$f(x,y)=A\mathrm{e}^{-(x^2+y^2)/2\sigma^2} \tag{5.31}$$

$$g(x,y)=\begin{cases}1, -1\le x\le 1, -1\le y\le 1\\ 0,\ \ \text{Other}\end{cases} \tag{5.32}$$

In this case, a two-dimensional rectangular pulse is convoluted with a larger two-dimensional Gaussian function. Since $g(x, y)$ is symmetrical about the origin, it

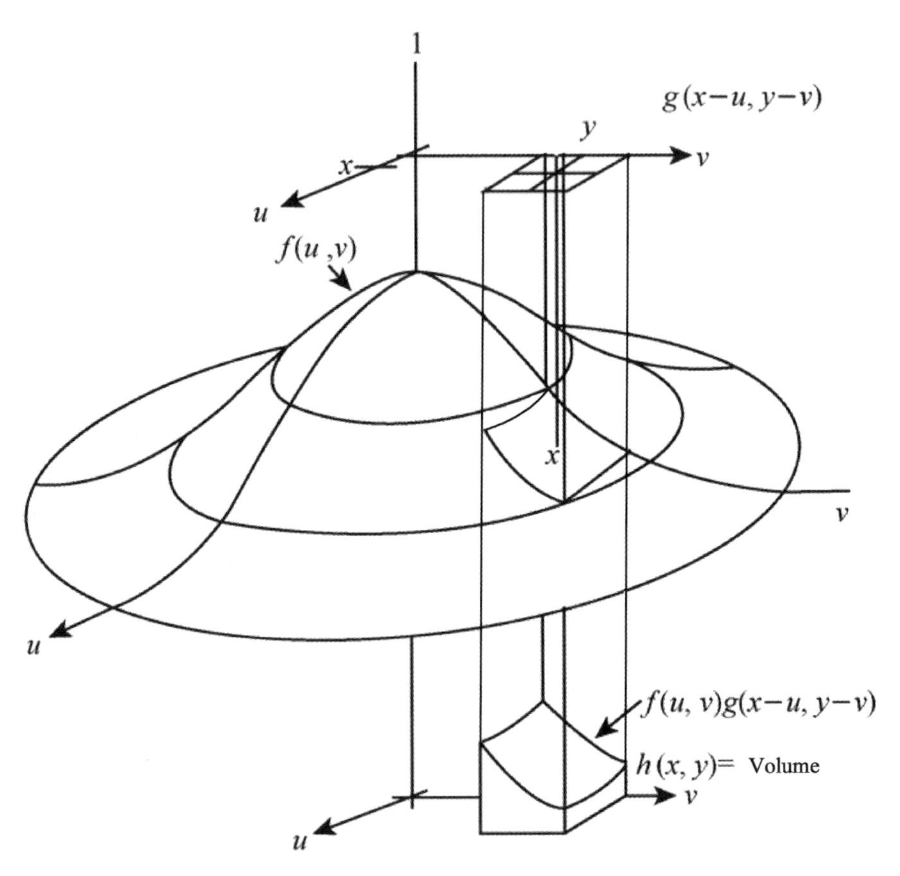

Fig. 5.13 Two-dimensional convolution

remains unchanged after a 180° rotation. The value of $h(x, y)$ is the volume of the product function when the square pulse is moved to position (x, y).

1. Finite Sampling

Consider the use of an image digitizer (such as a CCD sensor) to sample the image with a square sampling point. At each pixel position, the quantized gray value is the local average in a small square on the image. As shown in Fig. 5.13, $f(x, y)$ represents the quantized image, and $g(x, y)$ represents the spatial sensitivity function of the sampling point. The convolution $h(x, y)$ of $f(x,y)$ and $g(x, y)$ is the local mean that the digitizer "sees". In this way, convolution is an effective model that can describe the image sampling process. The function $g(x, y)$ can be any suitable function for describing the spatial sensitivity of the sampled aperture.

2. Discrete Two-Dimensional Convolution

The convolution of a digital image is similar to that of a continuous function, except that the independent variable is an integer value and the double integral is changed

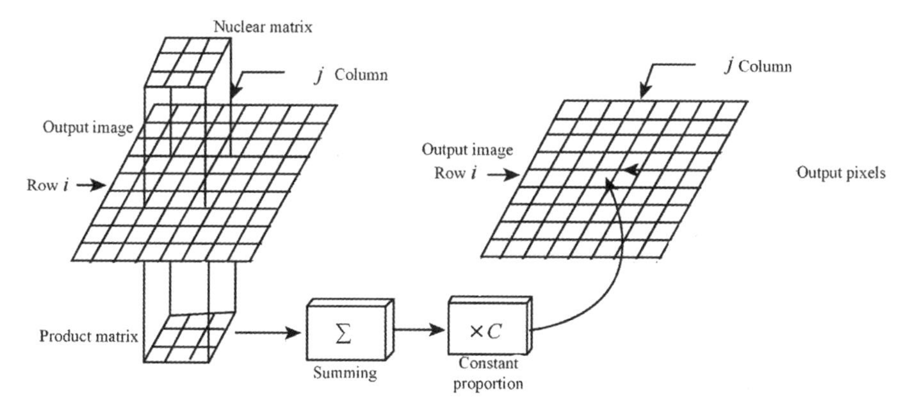

Fig. 5.14 Digital convolution

to a double sum (Fig. 5.14). Therefore, for a digital image,

$$H = F * G$$

$$H(i, j) = \sum_{m} \sum_{n} F(m, n)G(i - m, j - n) \tag{5.33}$$

Since F and G are nonzero only within a finite range, the summation calculation occurs only in areas in which the nonzero portion overlaps. The calculation of the discrete convolution is shown in Fig. 5.14. Array G is rotated by 180°, and its origin is moved to coordinates (i, j). The two arrays are multiplied by the elements, and the sum of the product is obtained.

As shown in Fig. 5.14, a 3×3 array G (convolution kernel) convolutes with a larger digital image F. The required multiplication and addition operands are equal to the product of the number of pixels in G and F (ignoring the effect at the edge of the image), respectively. The convolution is expected to consume considerable computational time unless the convolution kernel is small (or only a relatively small nonzero region is considered).

The pixels at the edge of the image can be attributed to the lack of a complete set of adjacent pixels. Therefore, in convolution operations, these areas must be categorically processed. In digital convolution calculations, four kinds of processing methods can be used for the pixels at the edges: In this first approach, the input image is complemented by repeating rows and columns on the edges of the image such that the convolution at the edge can be calculated. In the second approach, the input image is wound (similar to a cycle), assuming that the first column is followed by the last column. In the third approach, constant values (such as 0) are filled outside the input image. In the fourth approach, the rows that cannot be calculated are eliminated, and only the convolution of the pixels that can be calculated is considered [5].

The first and third methods are the commonly used. When quantizing an image, it is preferable to prevent important information from falling within an area that is less than half the width of the convolution kernel, and thus, the choice of convolution does not have critical consequences.

5.3.5 Matrix Form of Two-Dimensional Convolution

Digital images can be easily represented by matrices, and thus, linear algebra can be exploited. Matrix operations cannot be used directly for convolution operations but can be implemented through appropriate constructs [6].

Let arrays $\boldsymbol{F}$ and $\boldsymbol{G}$ be periodic in the x direction. The period length is at least equal to the sum of the horizontal lengths of the two arrays, and the y direction is assumed. Thus, if $\boldsymbol{F}$ is sized $m_1 \times n_1$ and $\boldsymbol{G}$ is sized $m_2 \times n_2$, we extend the analysis to $M \times N$ (where $M \geq m_1 + m_2 - 1$, $N \geq n_1 + n_2 - 1$) by padding 0. We name the new matrix F_p, G_p after the expansion. The following analysis assumes that $M = N$.

Second, we construct a $N^2 \times 1$ dimensional vector $\boldsymbol{f_p}$ from matrix $\boldsymbol{F_p}$ through row stacking. We transpose the first row of $\boldsymbol{F_p}$ such that it corresponds to the top N elements of $\boldsymbol{f_p}$, and place the transpose of the other lines in turn under it.

Subsequently, each row of matrix $\boldsymbol{G_p}$ generates a $N \times N$ circulant matrix, which produces a total of N matrices $(1 \leq i \leq N)$. The following text introduces the concept of a block matrix.

The block matrix is a matrix whose elements are matrices. That is, the block matrix is a relatively large matrix, which is actually an array constituted by matrices. The block circular matrix is a block matrix with the circulant matrix as the element. Using the block circulant matrix, we can generalize to a two-dimensional case.

According to the following method to generate a $N^2 \times N^2$ block circulant matrix $\boldsymbol{G_b}$, $\boldsymbol{G_b}$ is composed of $N \times N$ blocks, and each piece $(\boldsymbol{G_i}, i = 1, 2, \cdots N)$ is constructed by the ith line of $\boldsymbol{G_p}$:

$$\boldsymbol{G_b} = \begin{bmatrix} G_1 & G_N & \cdots & G_2 \\ G_2 & G_1 & \cdots & G_3 \\ \vdots & \vdots & \ddots & \vdots \\ G_N & G_{N-1} & \cdots & G_1 \end{bmatrix} \tag{5.34}$$

For example, in the upper left corner of block matrix $\boldsymbol{G_1}$, the first row of elements is formed by the first-row elements of $\boldsymbol{G_p}$, which are subjected to reverse order and shift transform. The other lines of $\boldsymbol{G_1}$ are rotated right with respect to their previous line. The other blocks of $\boldsymbol{G_b}$ are obtained by the other lines of $\boldsymbol{G_p}$ through a similar method. Thus, $\boldsymbol{G_b}$ is a block circulant matrix sized $N_2 \times N_2$ whose elements are $N \times N$ circulant matrices, and each element of $\boldsymbol{G_b}$ corresponds to one pixel of the output image.

The simple matrix form of convolution can be expressed as

$$h_p = G_b * f_p \tag{5.35}$$

where h_p is an output image that is padded and expressed in a column vector stacked row by row. G_b has N^4 elements. For example, when $N = 1000$, G_b has 10^{12} elements. Therefore, the matrix form is not computationally efficient but has other benefits. This form allows us to use the simple representations associated with linear algebra to design the image restoration filter. In addition, by using the symmetry properties of these matrices, the calculation can be considerably simplified.

Example $F = \begin{bmatrix} 1 & 2 \\ 3 & -1 \end{bmatrix}$, $G = \begin{bmatrix} -1 & 1 & 0 \\ -2 & 2 & 0 \\ 0 & 0 & 0 \end{bmatrix}$. We use the standard method of the matrix to calculate the two-dimensional convolution $H = F * G$.

Solution

First, N is set as $2 + 2 - 1 = 3$ to fill

$$F_p = \begin{bmatrix} 1 & 2 & 0 \\ 3 & -1 & 0 \\ 0 & 0 & 0 \end{bmatrix}, \quad G = \begin{bmatrix} -1 & 1 & 0 \\ -2 & 2 & 0 \\ 0 & 0 & 0 \end{bmatrix}$$

Next, we construct column vector $f_p = \begin{pmatrix} 1 & 2 & 0 & 3 & -1 & 0 & 0 & 0 & 0 \end{pmatrix}^{\mathrm{T}}$ with F_p. Establish the circulant matrix G_i by G_p:

$$G_1 = \begin{bmatrix} -1 & 0 & 1 \\ 1 & -1 & 0 \\ 0 & 1 & -1 \end{bmatrix}, \quad G_2 = \begin{bmatrix} -2 & 0 & 2 \\ 2 & -2 & 0 \\ 0 & 2 & -2 \end{bmatrix}, \quad G_1 = \begin{bmatrix} 0 & 0 & 0 \\ 0 & 0 & 0 \\ 0 & 0 & 0 \end{bmatrix}$$

Next, construct the block circulant matrix from the circulant matrix G_i:

$$G_b = \begin{bmatrix} -1 & 0 & 0 & 0 & 0 & 0 & -2 & 0 & 2 \\ 1 & -1 & 0 & 0 & 0 & 0 & 2 & -2 & 0 \\ 0 & 1 & -1 & 0 & 0 & 0 & 0 & 2 & -2 \\ -2 & 0 & 2 & -1 & 0 & 0 & 0 & 0 & 0 \\ 2 & -2 & 0 & 1 & -1 & 0 & 0 & 0 & 0 \\ 0 & 2 & -2 & 0 & 1 & -1 & 0 & 0 & 0 \\ 0 & 0 & 0 & -2 & 0 & 2 & -1 & 0 & 0 \\ 0 & 0 & 0 & 2 & -2 & 0 & 1 & -1 & 0 \\ 0 & 0 & 0 & 0 & 2 & -2 & 0 & 1 & -1 \end{bmatrix}$$

Calculate the convolution:

$$
\boldsymbol{h}_p = \boldsymbol{G}_b * \boldsymbol{f}_p \begin{bmatrix} -1 & 0 & 0 & 0 & 0 & 0 & -2 & 0 & 2 \\ 1 & -1 & 0 & 0 & 0 & 0 & 2 & -2 & 0 \\ 0 & 1 & -1 & 0 & 0 & 0 & 0 & 2 & -2 \\ -2 & 0 & 2 & -1 & 0 & 0 & 0 & 0 & 0 \\ 2 & -2 & 0 & 1 & -1 & 0 & 0 & 0 & 0 \\ 0 & 2 & -2 & 0 & 1 & -1 & 0 & 0 & 0 \\ 0 & 0 & 0 & -2 & 0 & 2 & -1 & 0 & 0 \\ 0 & 0 & 0 & 2 & -2 & 0 & 1 & -1 & 0 \\ 0 & 0 & 0 & 0 & 2 & -2 & 0 & 1 & -1 \end{bmatrix} \cdot \begin{bmatrix} 1 \\ 2 \\ 0 \\ 3 \\ -1 \\ 0 \\ 0 \\ 0 \\ 0 \end{bmatrix} = \begin{bmatrix} -1 \\ -1 \\ 2 \\ -5 \\ 2 \\ 3 \\ -6 \\ 8 \\ -2 \end{bmatrix}
$$

The calculation result is

$$
\boldsymbol{H} = \boldsymbol{F} * \boldsymbol{G} = \begin{bmatrix} -1 & -1 & 2 \\ -5 & 2 & 3 \\ -6 & 8 & -2 \end{bmatrix}
$$

The use of different kernel functions is expected to produce different outputs to meet the processing requirements. Next, we focus on several basic types of functions.

5.4 Five Basic Functions and Their Relationships Related to Convolution

The five basic functions and their relationships associated with convolution are the link between the input and output and connecting bridges of matrix operation and ideal model of information processing. This section focuses on five basic functions and their related properties.

5.4.1 Demand for Five Basic Functions in Remote Digital Image Processing and Their Effects

The five basic functions are the rectangular pulse function, triangular pulse function, Gaussian function, impulse function and step function. The functions are applied in remote sensing digital image processing:

(1) A rectangular pulse is often used as a rectangular sampling window and smooth function model.
(2) The application of triangular pulses is similar to that of rectangular pulses. A triangular pulse is used to model the triangular sampling window and smooth function.

(3) The Gaussian function has five important properties: ① rotational symmetry; ② single-valued function; ③ single Fourier transform spectrum; ④ Gaussian filter width characterized by parameter σ. The relationship between this parameter and smoothness is simple: a larger σ corresponds to a wider band; ⑤ separable. Owing to these five properties, the Gaussian smoothing filter is a highly efficient low-pass filter in both the spatial and frequency domains.
(4) The impulse function can discretize the actual continuous images to process the image.
(5) The step function is a discretization of the continuous image, but it is different from the impulse function in that it can only retain the region of interest of the image for image processing.

5.4.2 *Mathematical Expression and Relations of Five Basic Functions*

1. Rectangular Pulse Function

The rectangular pulse function (Fig. 5.15) is defined as in Formula (5.36)

$$\prod(x) = \begin{cases} 1, -\frac{1}{2} < x < \frac{1}{2} \\ \frac{1}{2}, x = \pm\frac{1}{2} \\ 0, \text{ Other} \end{cases} \tag{5.36}$$

2. Triangle Pulse Function

The triangle pulse function (Fig. 5.16) is defined as

$$\Lambda(x) = \begin{cases} 1 - |x|, |x| \le 1 \\ 0, |x| > 1 \end{cases} \tag{5.37}$$

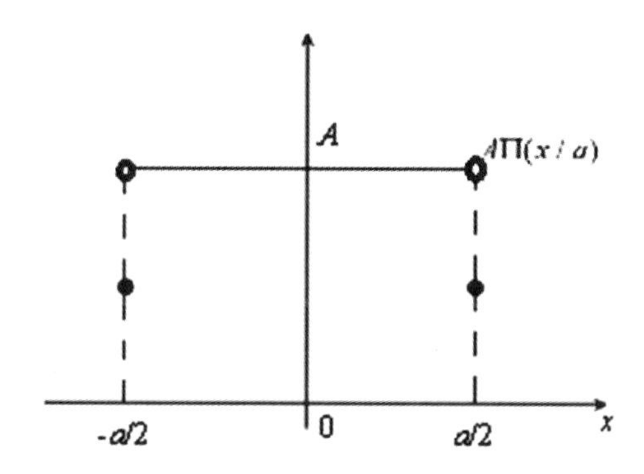

Fig. 5.15 Rectangular pulse function waveform

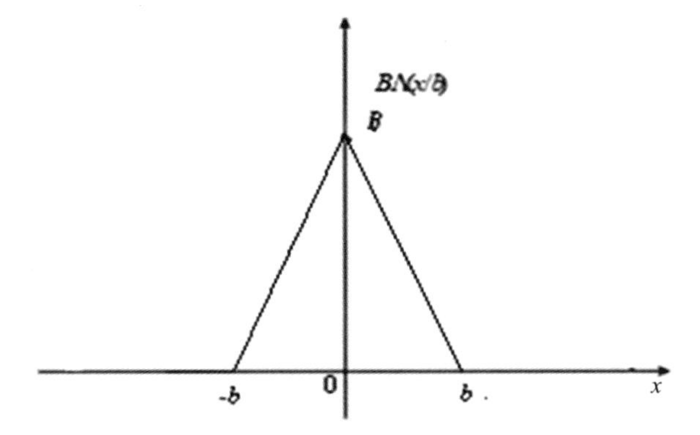

Fig. 5.16 Triangular pulse function waveform

The relationship between the triangular pulse and rectangular pulse is that the convolution of two identical rectangular pulses yields a triangular pulse.

3. Gaussian Function

The Gaussian function (Fig. 5.17) has the form

$$e^{-x^2/2\sigma^2} \tag{5.38}$$

4. Unit Impulse Function or Dirac Delta Function $\delta(x)$

Unlike a traditional function, this function is a symbolic function $\int_{-\infty}^{+\infty} \delta(x)\mathrm{d}x = \int_{-\varepsilon}^{+\varepsilon} \delta(x)\mathrm{d}x = 1$ defined by its integral property:

$$\int_{-\infty}^{+\infty} \delta(x)\mathrm{d}x = \int_{-\varepsilon}^{+\varepsilon} \delta(x)\mathrm{d}x = 1$$

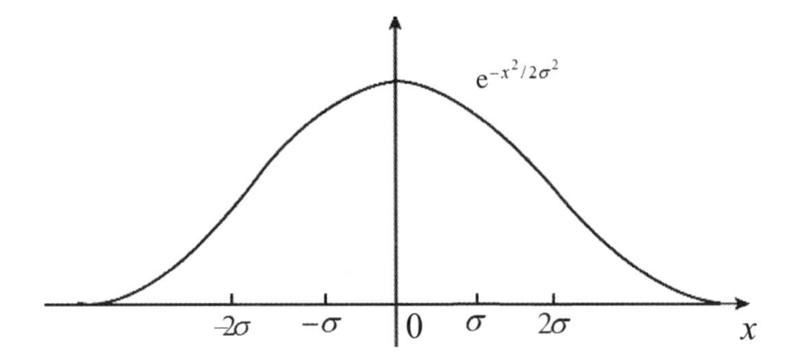

Fig. 5.17 Gaussian function

Fig. 5.18 Impulse function

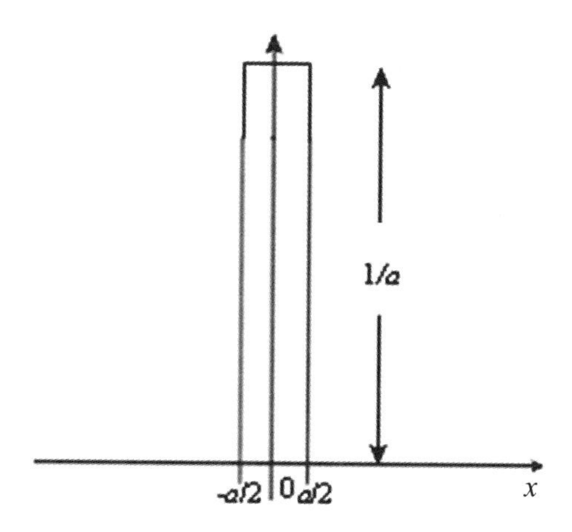

where ε is an arbitrarily small number greater than zero. When $x \neq 0$, $\delta(x) = 0$, and the unit impulse function is not defined at the zero point.

The relationship between the unit impulse function and rectangular pulse is as follows: The unit impulse function can be described as the limit of a narrow rectangular pulse (Fig. 5.18), i.e., $\delta(x) = \lim_{a \to 0} \frac{1}{a} \prod\left(\frac{x}{a}\right)$. When a decreases, the pulse becomes increasingly narrower, but it increases such that the unit area remains constant. In the extreme case, the pulse becomes infinitely high and infinitely narrow.

5. Step Function

The step function (Fig. 5.19) is a discontinuous symbolic function at $x = 0$ (Formula (5.39)), defined as

$$u(x) = \begin{cases} 1, x > 0 \\ \frac{1}{2}, x = 0 \\ 0, x < 0 \end{cases} \tag{5.39}$$

The integral property of the step function is

$$\int_{-\infty}^{+\infty} u(x) f(x) \mathrm{d}x = \int_{0}^{+\infty} f(x) \mathrm{d}x \tag{5.40}$$

where $f(x)$ is an arbitrary function. Figure 5.19 shows the waveform of the translation step function $u(x\text{-}x_0)$.

$$u(x - x_0) = \int_{-\infty}^{+\infty} \delta(\tau - x_0) \mathrm{d}\tau = \begin{cases} 1, x > x_0 \\ 0, x < x_0 \end{cases} \tag{5.41}$$

Fig. 5.19 Step function

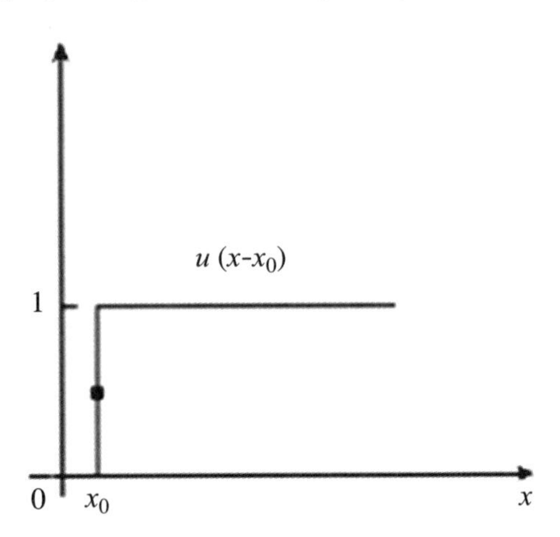

In turn, it can be expected that the unit impulse function is the derivative of the step function

$$u^{'}(x) = \frac{\mathrm{d}u(x)}{\mathrm{d}x} = \delta(x) \tag{5.42}$$

Equation (5.42) can be proved as follows: According to the integral by parts,

$$\int_{-\infty}^{+\infty} u^{'}(x) f(x) \mathrm{d}x = u(x) f(x)|_{-\infty}^{+\infty} - \int_{-\infty}^{+\infty} u(x) f^{'}(x) \mathrm{d}x \tag{5.43}$$

According to the definition of the step function (5.41),

$$\int_{-\infty}^{+\infty} u(x) f^{'}(x) \mathrm{d}x = -\int_{0}^{+\infty} f^{'}(x) \mathrm{d}x = -[f(+\infty) - f(0)] = f(0) \tag{5.44}$$

This phenomenon occurs because $f(+\infty) = 0$. According to the definition of the unit impulse function $\int_{-\infty}^{+\infty} f(x)\delta(x)\mathrm{d}x = f(0)$,

$$\int_{-\infty}^{+\infty} u^{'}(x) f(x) \mathrm{d}x = f(0) = \int_{-\infty}^{+\infty} \delta(x) f(x) \mathrm{d}x \tag{5.45}$$

This expression must hold for any $f(x)$, which is only possible if Formula (5.44) is established. Therefore, the relationship between the step function and impulse function is that the step function is the integral of the unit impulse function and the unit impulse function is the derivative of the step function.

These basic functions can be used alone or in combination to realize smoothing, edge enhancement, deconvolution and other processing, in a process known as convolution filtering.

5.5 Convolution Filtering

Convolution is often performed to achieve a linear operation of a signal or image, as illustrated by several examples in this section.

5.5.1 Smoothing

Figure 5.20 shows the case in which convolution is used to smooth the noise-interfered function $f(x)$. The rectangular pulse function $g(x)$ is the impulse response for the smoothing filter. As convolution progresses, the rectangular pulse moves from left to right, producing the function $h(x)$. The value of each point of $h(x)$ is the local average value of $f(x)$ over the unit length. This local average suppresses the high-frequency fluctuation while preserving the basic waveform of the input function. This application is a typical example of smoothing noise-polluted signals by using a filter with a nonnegative impulse response. We can use a triangular pulse or a Gaussian pulse as a smoothing function.

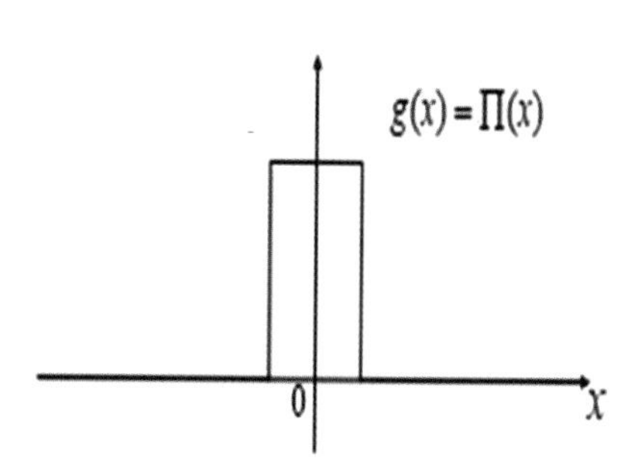

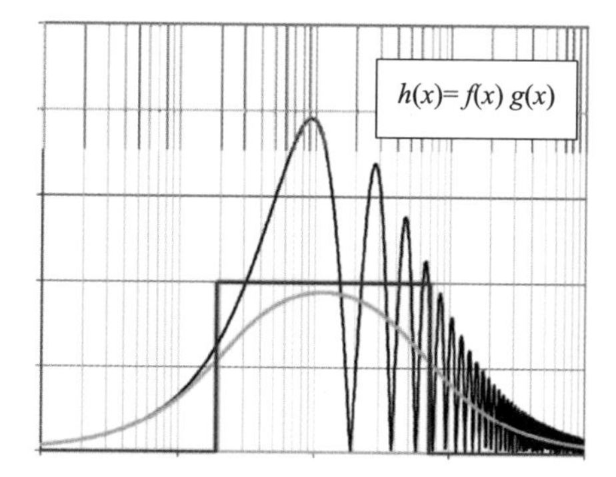

Fig. 5.20 Noise function for smoothness

5.5.2 Edge Enhancement

As another filter edge enhancement, the edge function $f(x)$ gradually increases, and the impulse response $g(x)$ is a positive peak function with a negative side lobe. With convolution, $g(x)$ moves from left to right, the side lobe and the main peak meet the edge in turn, obtaining output $h(x)$.

The illustrated edge enhancement filter has two adverse effects. First, this filter increases the slope of the edge gradient section. Second, at the two ends of the edge gradient section, an "overshoot" or "ringing" effect is produced. Commonly used edge enhancement filters exhibit this phenomenon.

As a second example of edge enhancement, consider the following impulse response:

$$g(x) = 2\delta(x) - \mathrm{e}^{-x^2/2\sigma^2} \tag{5.46}$$

In this case,

$$h = f * g = f(x) * 2\delta(x) - f(x) * \mathrm{e}^{-x^2/2\sigma^2} = 2f(x) - f(x) * \mathrm{e}^{-x^2/2\sigma^2} \tag{5.47}$$

The output function is twice the difference of the input function and the convolution of the input function and Gaussian function. The convolution with the Gaussian function blurs the edge; however, the output function obtains the enhanced edge in the graph. Overshoot occurs in this case as well, as shown in the figure.

This example illustrates that an edge enhancement effect can achieved when the blurred image is subtracted from the original image. This operation is similar to a camera darkroom technology known as unsharp masking.

5.5.3 Deconvolution

In many cases, the obtained images are affected by one or more linear processes that cannot be controlled. Image degradation caused by optical imperfections, low-quality recording and inferior displays can be modeled using convolution operations. The technique of using one convolution to remove another convolution effect is known as deconvolution. This aspect is further discussed in Chap. 11.

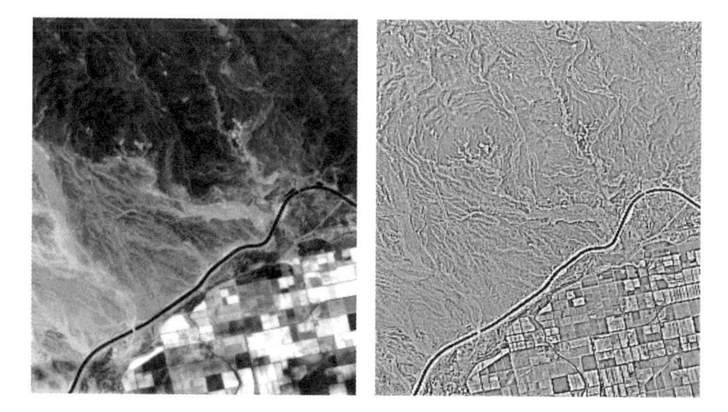

Fig. 5.21 Example of edge enhancement of remote sensing image

5.5.4 Examples of Remote Sensing Images

The system is a 3×3 template operator with a central value of 8 and other values of -1. The effect of convolution with the original remote sensing image is to enhance the slope of the edge gradient section and achieve the enhancement effect of the edge of the image extraction, as shown in Fig. 5.21.

5.6 Summary

This chapter establishes the basis for the analysis of optical systems, image sensors, electronic circuits and digital filters, which are components used in image processing systems, specifically, linear systems. The components and mathematical expressions of this chapter are shown in Fig. 5.22 and Table 5.2.

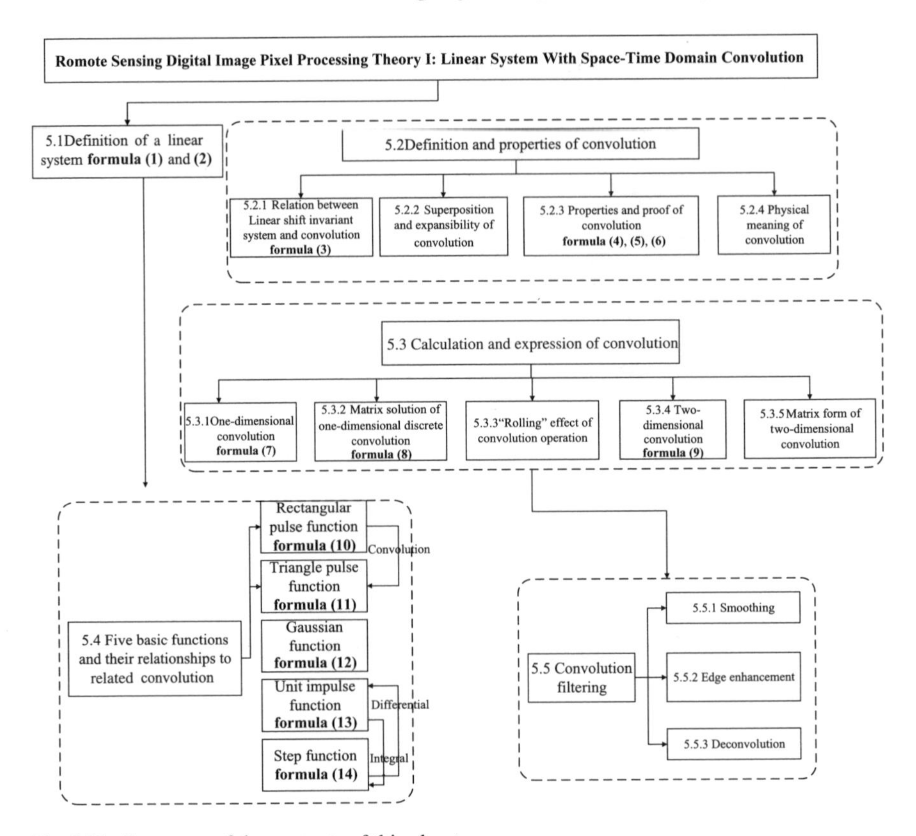

Fig. 5.22 Summary of the contents of this chapter

Table 5.2 Summary of the formulas in this chapter

formula (1)	$x_1(t) + x_2(t) \to y_1(t) + y_2(t)$ formula (5.3)	5.1.1
formula (2)	$x(t-T) \to y(t-T)$ formula (5.6)	5.1.2
formula (3)	$y(t) = \sum\limits_{t_0=-\infty}^{+\infty} x(t_0)h(t-t_0)$ formula (5.20)	5.2.1
formula (4)	$f_1(t) * f_2(t) = \int_{-\infty}^{+\infty} f_2(\lambda) \cdot f_1(t-\lambda)\mathrm{d}\lambda = f_2(t) * f_1(t)$ formula (5.24)	5.2.3
formula (5)	$f(t) * [h_1(t) + h_2(t)] = f(t) * h_1(t) + f(t) * h_2(t)$ formula (5.25)	5.2.3
formula (6)	$[f(x) * h_1(t)] * h_2(t) = f(x) * [h_1(t) * h_2(t)] = f(x) * h(t)$ formula (5.26)	5.2.3
formula (7)	$f(t) * h(t) = \int_{-\infty}^{+\infty} f(\tau)h(t-\tau)\,\mathrm{d}\tau$	5.3.1

(continued)

Table 5.2 (continued)

formula (8)	$h = g * f = \begin{bmatrix} g_p(1) & g_p(N) & \dots & g_p(2) \\ g_p(2) & g_p(1) & \dots & g_p(3) \\ \dots & \dots & \dots & \dots \\ g_p(N) & g_p(N-1) & \dots & g_p(1) \end{bmatrix} \begin{bmatrix} f_p(1) \\ f_p(2) \\ \dots \\ f_p(N) \end{bmatrix}$ formula (5.27)	5.3.2
formula (9)	$h(x, y) = f * g = \int_{-\infty}^{\infty} \int_{-\infty}^{\infty} f(u, v) g(x-u, y-v) \mathrm{d}u \mathrm{d}v$ formula (5.30)	5.3.4
formula (10)	$\prod(x) = \begin{cases} 1, & -\frac{1}{2} < x < \frac{1}{2} \\ \frac{1}{2}, & x = \pm\frac{1}{2} \\ 0, & \text{Others} \end{cases}$ formula (5.36)	5.4.2
formula (11)	$\Lambda(x) = \begin{cases} 1-\lvert x\rvert, \lvert x\rvert \le 1 \\ 0, \lvert x\rvert > 1 \end{cases}$ formula (5.37)	5.4.2
formula (12)	$\mathrm{e}^{-x^2/2\sigma^2}$ formula (5.38)	5.4.2
formula (13)	$\delta(x) = \lim_{a \to 0} \frac{1}{a} \prod\left(\frac{x}{a}\right)$	5.4.2
formula (14)	$u(x) = \begin{cases} 1, x > 0 \\ \frac{1}{2}, x = 0 \\ 0, x < 0 \end{cases}$ formula (5.39)	5.4.2

References

1. Kailath T. Linear Systems. Englewood Cliffs, N J: Prentice-Hall. 1980.
2. Wang X. Time-domain algorithm of LTI system response based on convolution technology. Foret Electron Measure Technol. 2005
3. Gonzalez RC, Woods RE. Digital Image Processing. New Jersey: Prentice Hall. 2007.
4. Gupta SC. Transform and State Variable Methods in Linear Systems. New York: John Wiley and Sons. 1966.
5. Kenneth R. Castleman Digital Image Processing. New Jersey: Prentice-Hall. 2000.
6. Hunt BR. A matrix theory proof of the discrete convolution theorem. IEEE Trans. 1971;(19).

Chapter 6
Remote Sensing Digital Image Pixel Processing Theory II: Time–Frequency Fourier Transform from Convolution to Multiplication

Chapter Guidance A remote sensing image is presented in the spatial domain and input to the computer as a time series, that is, as time domain (dimension T) information. The computer processes the remote sensing digital image by convolution. In remote sensing digital image processing, the time domain is the most commonly used dimension. In remote sensing feature analysis, this entity is equivalent to the spatial domain. The domain transform in this chapter refers to the dual transform between the time domain (dimension T) and its reciprocal frequency domain (dimension T^{-1}), which are mutually reversible and mapped sequentially. Because the spatial domain information is in inertial space, its operation pertains to convolution. When the processing information is excessively complex, it is converted to its reciprocal domain. The original convolution operation pertains to algebraic multiplication, which considerably simplifies the image processing. Fourier transform can transform the time and frequency domains by converting the infinite long time series operation into a unit circle operation with a period 2π (with 2π defined as mod $[2\pi]$).

Fourier transform can be used to quantitatively analyze effects, such as in digital systems, sampling points, electronic amplifiers, convolution filters, noise, and display points, and is a powerful tool for linear system analyses. The Fourier transform theory can be combined with its physical interpretation to solve most image processing problems. Image processing can be divided into spatial and frequency domain processing. The time domain is an accessible field, and the frequency domain is a mathematical field. The duality of the spatial and frequency domains renders the Fourier transform an important analytical tool in a large amount of present scientific research. The Fourier transform is a direct link between the spatial and time domains, and the convolution and multiplication transformation of the spatial and frequency domains can be realized by transformation. The multiplication is computationally simpler than the convolution calculation, which can help increase the speed of digital image processing. The Fourier transform is used to convert the difficult to process time domain signal into the easy to analyze frequency domain signal (signal spectrum) to ensure that several frequency domain signals can be processed by certain tools. Finally, the Fourier inverse transform converts these frequency domain signals into

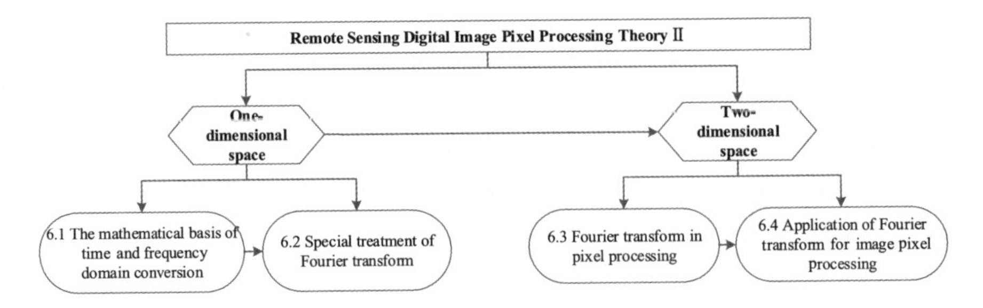

Fig. 6.1 Logical framework of this chapter

time domain signals. Fourier transform is essentially a universal method of solving problems, and from a basic viewpoint, it can facilitate the analysis of traditional areas from a different perspective.

In this chapter, we first consider the one-dimensional Fourier transform, extend the continuous Fourier transform to the discrete Fourier transform, and extend the properties of the Fourier transform to two-dimensional space. In this way, the Fourier transform can be used to study the correlation properties of remote sensing digital image processing. The structure of this chapter is shown in Fig. 6.1. We present the mathematical basis for time and frequency domain conversion (Sect. 6.1) by describing the Fourier transform method to realize time domain and frequency domain reversible transformation and physical nature of the dimensions of convolution and multiplication conversion being reciprocal to each other. Next, we describe the special treatment of the Fourier transform (Sect. 6.2), which can convert infinitely long time series into unit cycles and decrease the exponential level and computational complexity. Next, we describe the use of the Fourier transform in the image pixel processing (Sect. 6.3) by presenting the mathematical features of the Fourier transform matrix of images. The application of Fourier transform for image pixel processing (Sect. 6.4) is described to establish a direct relationship with the application of remote sensing image processing and demonstrate its effectiveness.

6.1 Mathematical Basis of Time and Frequency Domain Conversion

This section expands on the Fourier series and introduces the mathematical formula and existence condition of the Fourier transform. The function of the Fourier transform is described from the definition level, that is, the conversion of the spatial domain and frequency domain. The characteristics of the Fourier transform are presented, including the periodic and Fourier transforms of the random function. Subsequently, the properties of the Fourier transform are presented. The convolution theorem is the core and foundation of the widely used Fourier transform. The convolution of the spatial domain corresponds to the multiplication of the frequency domain, and the multiplication of the spatial domain is convoluted in the frequency domain.

6.1.1 Fourier Series

According to the Fourier series, to satisfy the Dirichlet condition,

(1) The number of discontinuities in the same cycle T_1 is limited
(2) Limited number of maxima and minima exist in the same period T_1
(3) The signal is "absolutely integrable" within the same period $T_1 \int_{T_1} |f(t)|\mathrm{d}t < \infty$.

Any periodic function can be expanded into an orthogonal function as a linear combination of the infinite series and Fourier series. The orthogonal function set can be a trigonometric set $\{1, \cos(\omega_1 nt), \sin(\omega_1 nt) : n \in N\}$ or complex exponential function set $\{\exp(\mathrm{j}n\omega_1 \mathrm{t}): n \in z\}$, and the function period is T_1, with angular frequency $\omega_1 = 2\pi f_1 = 2\pi/T_1$.

The Fourier series of function $f(t)$ is expanded as

$$f(t) = \frac{a_0}{2} + \sum_{n=0}^{\infty} a_n \cos\left(2\pi\frac{n}{T}t\right) + \sum_{n=0}^{\infty} b_n \sin\left(2\pi\frac{n}{T}t\right) \tag{6.1}$$

and

$$a_n = \frac{2}{T}\int_{-T/2}^{T/2} f(x)\cos\left(2\pi\frac{n}{T}x\right)\mathrm{d}x \tag{6.2}$$

$$b_n = \frac{2}{T}\int_{-T/2}^{T/2} f(x)\sin(2\pi\frac{n}{T}x)\mathrm{d}x. \tag{6.3}$$

This formula represents a function of period T by two infinite real coefficient sequences [1].

Formula (6.1) is the basis of the image Fourier transform, which indicates that a graph can be split into the sum of innumerable two-dimensional sine or cosine functions.

6.1.2 One-Dimensional Fourier Transform and Its Inverse Transform

The Fourier transform of the one-dimensional function $f(t)$ is defined as

$$F(\omega) = \Im[f(t)] = \int_{-\infty}^{\infty} f(t)\mathrm{e}^{-\mathrm{j}\omega t}\mathrm{d}t \tag{6.4}$$

The inverse of the function $F(\omega)$ is

$$f(t) = \Im^{-1}[F(\omega)] = \frac{1}{2\pi}\int_{-\infty}^{\infty} F(\omega)\mathrm{e}^{\mathrm{j}\omega t}\mathrm{d}\omega \tag{6.5}$$

The two transformations are reciprocal, i.e.,

$$\Im[f(t)] = F(\omega) \Leftrightarrow \Im^{-1}[F(\omega)] = f(t) \tag{6.6}$$

$F(\omega)$ is known as the image function of the Fourier transform of $f(t)$. $f(t)$ is the original function of $F(\omega)$, and the image function and original function constitute a pair of Fourier transform pairs, i.e.,

$$f(t) \leftrightarrow F(\omega) \tag{6.7}$$

The Fourier transform $F(\omega)$ for $f(t)$ is unique, and vice versa.

The frequency variable ω can be replaced by

$$s = \frac{\omega}{2\pi} \tag{6.8}$$

$$F(s) = \Im[f(t)] = \int_{-\infty}^{\infty} f(t)\mathrm{e}^{-\mathrm{j}2\pi st}\mathrm{d}t \tag{6.9}$$

The inverse is transformed into

$$f(t) = \Im^{-1}[F(s)] = \int_{-\infty}^{\infty} F(s)\mathrm{e}^{\mathrm{j}2\pi st}\mathrm{d}s \tag{6.10}$$

The functions $F(s)$ and $f(t)$ also constitute a pair of Fourier transform pairs, i.e.,

$$F(s) \leftrightarrow f(t) \tag{6.11}$$

6.1.3 Existence Conditions of Fourier Transform

1. Instantaneous Function

Several function values may rapidly approach zero when the argument reaches a large positive or negative value, and thus, the integrals of Eqs. (6.9) and (6.10) exist. In the considered case, the integral of the absolute value of a function exists if.

If the function is continuous or has only a finite number of discontinuities, the Fourier transform of the function exists for any value of *f*. We term these functions transient functions because the function value disappears when *t* is large.

2. Periodic Function and Constant Value Function

Consider the inverse function of a pair of impulse functions

$$f(t) = \Im^{-1}[\delta(s - f_0) + \delta(s + f_0)] = \int_{-\infty}^{\infty} [\delta(s - f_0) + \delta(s + f_0)]e^{j2\pi st}ds \tag{6.12}$$

According to only the impulsive property,

$$\begin{aligned} f(t) &= \int_{-\infty}^{\infty} \delta(s - f_0)e^{j2\pi st}ds + \int_{-\infty}^{\infty} \delta(s + f_0)e^{j2\pi st}ds \\ &= \int_{-\infty}^{\infty} \delta(s)e^{j2\pi (s+f_0)t}ds + \int_{-\infty}^{\infty} \delta(s)e^{j2\pi (s-f_0)t}ds \\ &= e^{j2\pi f_0 t} + e^{-j2\pi f_0 t} = 2\cos(2\pi f_0 t) \end{aligned} \tag{6.13}$$

We use Euler relations and divide both sides by 2 simultaneously:

$$\Im[\cos(2\pi f_0 t)] = \frac{1}{2}[\delta(s - f_0) + \delta(s + f_0)] \tag{6.14}$$

The Fourier transform of the cosine function with frequency f_0 is a pair of pulses, which are located in the frequency domain $f = \pm f_0$:

$$\begin{aligned} \Im[\sin(2\pi f_0 t)] &= \int_{-\infty}^{\infty} \frac{e^{j2\pi f_0 t} - e^{-j2\pi f_0 t}}{2j} e^{-j2\pi st}dt \\ &= \frac{j}{2}\int_{-\infty}^{\infty} [e^{-j2\pi (s+f_0)t} - e^{-j2\pi (s-f_0)t}]dt \end{aligned}$$

$$= \frac{\mathrm{j}}{2}[\delta(s + f_0) - \delta(s - f_0)] \tag{6.15}$$

We set $f_0 = 0$.

$$\Im[1] = \delta(s) \tag{6.16}$$

The Fourier transform of a constant is a pulse at the origin.

In other words, through the Fourier series theory, any periodic function with frequency f can be expressed as the accumulation of sinusoidal functions with frequency nf, where n is the integer value. According to the addition theorem, the Fourier transform of the periodic function is a series of isometric impulse functions in the frequency domain.

3. Time–Frequency Reciprocal Domain Duality

The time domain is an accessible field, and the frequency domain is a mathematical field. The correspondence between these two domains renders a complex problem simple. Although they belong to different domain, the two entities maintain the conservation of energy, and the compression of a function is equivalent to widening the transformation range but decreasing the amplitude to ensure that the energy in the two domains is equal.

Table 6.1 lists $f(t)$ from the "wide" function to the "narrow" function and its Fourier transform from the "narrow" to "wide" functions. Table 6.1 shows that the Fourier transform exhibits similarity and duality. The relationship between the spatial and frequency domains is expressed as "T" and "T^{-1}", respectively, in the dimension. The function that is a constant in the spatial domain is an impulse response in the frequency domain. The function with wide coverage in the spatial domain exhibits a small coverage in the frequency domain. In contrast, if the coverage in the spatial domain is small, the frequency domain is wide. The impulse response in the spatial domain is a constant in the frequency domain. The similarity and duality can also be explained by analysis and induction. When the spatial domain is extremely broad and it is difficult to summarize the function, the frequency domain is expected to be extremely thin and easy to summarize. The function that is thin and easy to analyze in the spatial domain is expected to be broad and easy to analyze in the frequency domain.

4. Random Function

Infinitely extended, nonperiodic, nonperiodic, and absolute integrals that do not exist can be collectively referred to as random functions. In most cases, only the autocorrelation function of the random function is used, i.e.,

$$R_f(\tau) = \lim_{T \to \infty} \frac{1}{2T} \int_{-T}^{T} f(t) f(t + \tau) \mathrm{d}t \tag{6.17}$$

Table 6.1 Similarity of fourier-time–frequency reciprocal domain transformation

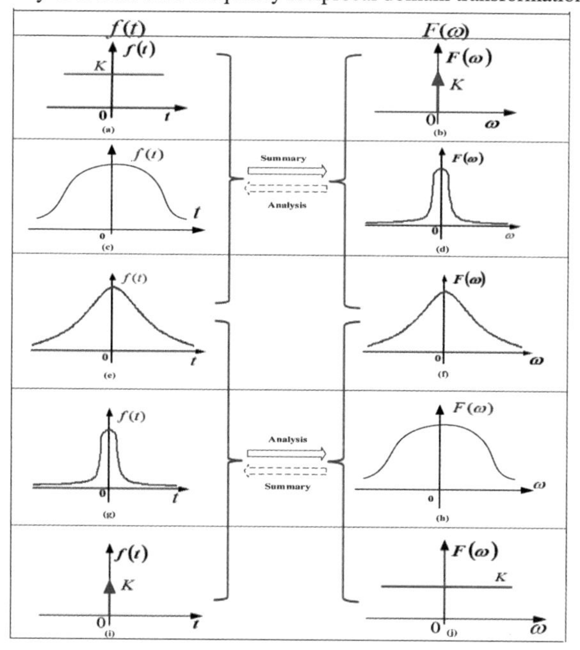

For many functions, this function exists. The autocorrelation function is a real even function, and its Fourier transform is the energy spectrum of $f(t)$. To transform a random function, it is necessary to redefine the Fourier transform of Eq. (6.9).

For inverse transformation, a similar treatment can be performed to manage the redefined transformation function. Only Eqs. (6.9) and (6.10) are used in this book because they are suitable for a finite extension bounded signal. The results obtained by this convention can be deduced using Eq. (6.6), and the results are generalized to all random functions of $R_f(\tau)$.

6.1.4 Property of Fourier Transform

1. Linear

The Fourier transform is a linear operation. If the Fourier transform corresponding to $f_1(t), f_2(t), \ldots, f_n(t)$ is $F_1(\omega), F_2(\omega), \ldots, F_n(\omega)_1$, the following transformation pair holds:

$$\sum_{i=1}^{n} a_i f_i(t) \leftrightarrow \sum_{i=1}^{n} a_i F_i(\omega) \tag{6.18}$$

where n is a finite positive integer, and a_i is a constant coefficient.

2. Parity

Function $f_e(t)$ is an even function only if

$$f_e(t) = f_e(-t) \tag{6.19}$$

Function $f_o(t)$ is an odd function only if

$$f_o(t) = -f_o(-t) \tag{6.20}$$

The non-odd/non-even function $f(t)$ can be divided into odd and even parts, i.e.,

$$f_e(t) = \frac{1}{2}[f(t) + f(-t)] \tag{6.21}$$

and

$$f_o(t) = \frac{1}{2}[f(t) - f(-t)] \tag{6.22}$$

In this case,

$$f(t) = f_e(t) + f_o(t) \tag{6.23}$$

The parity has a certain effect on the Fourier transform, proven as follows: according to the Euler formula,

$$\mathrm{e}^{\mathrm{j}x} = \cos(x) + \mathrm{j}\,\sin(x) \tag{6.24}$$

The Fourier transform of Eq. (6.10) can be modified as

$$F(s) = \int_{-\infty}^{\infty} f(t)\mathrm{e}^{-\mathrm{j}2\pi st}\mathrm{d}t = \int_{-\infty}^{\infty} f(t)\cos(2\pi st)\mathrm{d}t - \mathrm{j}\int_{-\infty}^{\infty} f(t)\sin(2\pi st)\mathrm{d}t \tag{6.25}$$

Transform $f(t)$ to two parts of the parity's sum as

$$\begin{aligned} F(s) = &\int_{-\infty}^{\infty} f_e(t)\cos(2\pi st)\mathrm{d}t + \int_{-\infty}^{\infty} f_o(t)\cos(2\pi st)\mathrm{d}t \\ &- \mathrm{j}\int_{-\infty}^{\infty} f_e(t)\sin(2\pi st)\mathrm{d}t - \mathrm{j}\int_{-\infty}^{\infty} f_o(t)\sin(2\pi st)\mathrm{d}t \end{aligned} \tag{6.26}$$

The second and third terms are the infinite integrals of the odd and even function products, and the two results are zero, such that the Fourier transform is reduced to.

$$F(s) = \int_{-\infty}^{\infty} f_e(t)\cos(2\pi st)\mathrm{d}t - \mathrm{j}\int_{-\infty}^{\infty} f_o(t)\sin(2\pi st)\mathrm{d}t = F_e(s) + \mathrm{j}F_o(s) \tag{6.27}$$

We obtain the symmetry of the Fourier transform: (1) the even function component is transformed into the even component; (2) the odd function component is transformed into the odd function component; (3) the odd function component introduces the coefficient $-$ j; (4) even function components do not introduce coefficients.

3. Imaginary and Real Components

If a complex function is expressed as the sum of four parts (odd and even real and odd and even imaginary parts), the following four Fourier transform rules can be established: (1) The real even parts produce real even parts (derived by symmetry (1) and (4)); (2) the real odd parts produce imaginary odd parts (derived by symmetries (2) and (3)); (3) the virtual even parts produce imaginary even parts (derived by symmetry (1) and (4)); (4) the imaginary odd parts produce real odd parts (derived by symmetries (2) and (3)).

Since we usually use the real function to represent the input image, it is important to study the case in which the input function is a real function. Note that the transformation result of the real function has an even real part and an odd imaginary part, which is known as the Hermite function. This function exhibits conjugate symmetry:

$$F(s) = F^*(-s) \tag{6.28}$$

where $*$ represents the complex conjugate.

Table 6.2 presents the symmetry of the Fourier transform.

4. Displacement Theorem

The displacement theorem describes the effect of moving the origin of a function on the transformation. For function $f(t)$,

$$\Im[f(t-a)] = \int_{-\infty}^{\infty} f(t-a)\mathrm{e}^{-\mathrm{j}2\pi st}\mathrm{d}t \tag{6.29}$$

where a is the amount of displacement multiplied by the right side of the abovementioned formula,

$$\mathrm{e}^{\mathrm{j}2\pi sat}\mathrm{e}^{-\mathrm{j}2\pi at} = 1 \tag{6.30}$$

Table 6.2 Symmetrical properties of Fourier transform

$f(t)$	$F(s)$
Odd function	Odd function
Even function	Even function
Real odd function	Real even function
Real odd function	Imaginary even function
Imaginary even function	Imaginary even function
Imaginary odd function	Real even function
Complex odd function	Complex even function
Complex even function	Complex even function
Real function	Hermite function
Imaginary function	Inverse Hermite function
Real even, imaginary even function	Real function
Real odd, imaginary even function	Imaginary function

In this case,

$$\Im[f(t-a)] = \int_{-\infty}^{\infty} f(t-a)\mathrm{e}^{-\mathrm{j}2\pi s(t-a)}\mathrm{e}^{-\mathrm{j}2\pi as}\mathrm{d}t \tag{6.31}$$

The variables are replaced as

$$u = t - a, \mathrm{d}u = \mathrm{d}t$$

We move the second index term beyond the points,

$$\Im[f(t-a)] = \mathrm{e}^{-\mathrm{j}2\pi as}\int_{-\infty}^{\infty} f(u)\mathrm{e}^{-\mathrm{j}2\pi su}\mathrm{d}u = \mathrm{e}^{-\mathrm{j}2\pi as}F(s) \tag{6.32}$$

Thus, the displacement of the function introduces a Fourier transform coefficient in its Fourier transform, and if $a = 0$, the coefficient is unity. The complex coefficient is expressed as

$$\mathrm{e}^{-\mathrm{j}2\pi as} = \cos(2\pi as) - \mathrm{j}\,\sin(2\pi as) \tag{6.33}$$

The complex coefficient has a unit amplitude, and the rotation angle of the complex plane changes as s increases, indicating that the displacement of the function does not change the amplitude of its Fourier transform but changes the energy distribution between the real and imaginary parts, which results in a phase shift proportional to s and the amount of displacement.

5. Convolution Theorem

The convolution of two functions can be expressed as

$$\begin{aligned} y(t) = f_1(t) * f_2(t) &= \int_{-\infty}^{\infty} f_1(\tau) f_2(t-\tau)\,\mathrm{d}\tau \\ = f_2(t) * f_1(t) &= \int_{-\infty}^{\infty} f_2(\tau) f_1(t-\tau)\,\mathrm{d}\tau \end{aligned} \tag{6.34}$$

(1) Time domain convolution theorem. If

$$x(t) \leftrightarrow X(\omega), h(t) \leftrightarrow H(\omega) \tag{6.35}$$

then

$$y(t) = x(t) * h(t) \leftrightarrow Y(\omega) = X(\omega)H(\omega) \tag{6.36}$$

This theorem shows that the convolution operation in the time domain is transformed into a product operation in the frequency domain.

(2) Frequency domain convolution theorem. If

$$f(t) \leftrightarrow F(\omega), g(t) \leftrightarrow G(\omega) \tag{6.37}$$

then

$$f(t)g(t) \leftrightarrow \frac{1}{2\pi} F(\omega) * G(\omega) \tag{6.38}$$

This theorem shows that the product operation in the time domain is converted to the convolution operation in the frequency domain.

6. Similarity Theorem

The similarity theorem describes the effect of the scale change of the function argument on the Fourier transform.

Changing the scale of an independent variable is expected to broaden or compress a function. For example, by multiplying an independent variable of Eq. (6.10) by a scale factor to extend or compress the function, the Fourier transform can expressed as

$$\Im[f(at)] = \int_{-\infty}^{\infty} f(at)\mathrm{e}^{-\mathrm{j}2\pi st}\mathrm{d}t \tag{6.39}$$

The integration and indices are multiplied by s/a

$$\Im[f(at)] = \frac{1}{a}\int_{-\infty}^{\infty} f(at)\mathrm{e}^{-\mathrm{j}2\pi at(s/a)}a\mathrm{d}t \tag{6.40}$$

The variables are replaced:

$$u = at,\ \mathrm{d}u = a\mathrm{d}t,$$

In this case,

$$\Im[f(at)] = \frac{1}{|a|}\int_{-\infty}^{\infty} f(u)\mathrm{e}^{-\mathrm{j}2\pi u(s/a)}\mathrm{d}u = \frac{1}{|a|}F(\frac{s}{a}) \tag{6.41}$$

The similarity theorem indicates that a "narrow" function has a "wide" Fourier transform, and vice versa.

7. Rayleigh Theorem

A key function is a nonzero function only in a finite interval. For this type of function, the total energy can be evaluated. The energy of the function is defined as

$$E = \int_{-\infty}^{\infty} |f(t)|^2\mathrm{d}t \tag{6.42}$$

The condition is that the integral exists, Formula (6.43) holds for the instantaneous function, and the energy is a parameter that reflects the total "size" of the function. According to the Rayleigh theorem,

$$\int_{-\infty}^{\infty} |f(t)|^2\mathrm{d}t = \int_{-\infty}^{\infty} |f(s)|^2\mathrm{d}s \tag{6.43}$$

i.e., the transform function has the same energy as that of the original function.

The Rayleigh theorem is consistent with the similarity theorem: If a function narrows, the amplitude does not change, and its energy decreases. According to the similarity theorem, compressing a function is equivalent to widening its transformation while decreasing its amplitude to ensure that the energy in the two domains is equal.

6.2 Special Treatment of Fourier Transform

The Fourier transform described in Sect. 6.1 is a one-dimensional continuous transform; however, the image is two-dimensional discrete. To apply the Fourier transform to image processing, we must solve the above two-dimensional and discrete problems (two-dimensional, discrete). This section describes how to discretize the Fourier transform. First, the discrete Fourier transform is sampled to obtain the mathematical formula of the discrete Fourier transform and its corresponding matrix form. According to the matrix analysis, the computational complexity of the discrete Fourier transform exceeds the expectation, which renders the conversion of the spatial and frequency domains highly challenging. The conversion is expected to be more complex than even the product operation instead of the convolution operation, which limits the idea of converting the space domain convolution into the frequency domain for the product operation. The fast Fourier transform (FFT) [2] is used to solve the problem of computational complexity of the discrete Fourier transform, and the Fourier transform is introduced into the engineering application of image processing. This section describes the principle and process of fast Fourier transform, which can help readers understand the fast Fourier transform.

6.2.1 Discrete Fourier Transform

If both time and frequency are discretized, the Fourier transform of Eq. 6.10 is

$$G_n = G(n\Delta s) = \sum_{k=-N/2}^{N/2} g(k\Delta t)\mathrm{e}^{-\mathrm{j}2\pi(n\Delta s)k\Delta t}\Delta t = \frac{T}{N}\sum_{k=-N/2}^{N/2} g_k\mathrm{e}^{-\mathrm{j}2\pi(n/N)k} \tag{6.44}$$

Because $T = N\Delta t$, the inverse transform can be expressed as

$$g_k = g(k\Delta t) = \sum_{n=-\infty}^{\infty} G(n\Delta s)\mathrm{e}^{-\mathrm{j}2\pi(n\Delta s)k\Delta t}\Delta s = \frac{1}{T}\sum_{n=-\infty}^{\infty} G_n\mathrm{e}^{\mathrm{j}2\pi(\frac{k}{N})n} \tag{6.45}$$

Among the set of coefficients $\{G_n\}$ of $g(k\Delta t)$, only n is small and nonzero.

If $\{f_k\}$ is a sequence of length N, for instance, associated with a continuous function of equal spacing sampling, the discrete Fourier transform (DFT) is sequence $\{F_n\}$:

$$F_n = \frac{1}{\sqrt{N}}\sum_{k=0}^{N-1} f_k\mathrm{e}^{-\mathrm{j}2\pi(\frac{n}{N})k} \tag{6.46}$$

The inverse DFT is:

$$f_k = \frac{1}{\sqrt{N}} \sum_{k=0}^{N-1} F_n e^{j2\pi(\frac{k}{N})n} = \frac{1}{\sqrt{N}} \sum_{k=0}^{N-1} F_n W_N^{nk} \tag{6.47}$$

DFT calculations involve a large number of complex multiplications and additions. According to the definition of the DFT (6.45), it is necessary to sum the data of the input N point and complex coefficient in succession. Each time an F_n value is calculated, N complex multiplication and $N - 1$ complex additions must be performed. A total of N $F_n s$ need to be calculated, and thus, the calculation of an F_n value to calculate the workload must be multiplied by N to calculate the workload of all DFTs. Thus, $N \times N = N^2$ complex multiplications and $N \times (N - 1)$ complex additions must be performed. In actual data processing, N can be quite large. For example, $N = 2^{10} = 1024$. All DFTs require $N^2 = 2^{20} = 1{,}048{,}576$ complex multiplications and $N \times (N - 1) = 1{,}047{,}552$ complex additions. In addition, the input data and intermediate computing data must be stored, and the general required storage capacity and number of operations are proportional to the calculation time. When N is large, the computer directly calculating the DFT is expected to consume a considerable amount of time, and the storage capacity of the entire computer may be occupied. Consequently, real-time data cannot be managed, and this process is highly uneconomical. Consequently, the Fourier transform is challenging to apply to actual text image processing.

6.2.2 Fast Fourier Transform

FFT is a fast algorithm, and its basic principles and calculation formula correspond to the DFT. The FFT increases the computational speed of the DFT by $N/\log_2 N$ times, and thus, the processing of many signals can be coordinated with the overall system speed. The application involves several processed data points and the simulation study of the system in real-time data processing. Before the FFT, in the fields of communication and radar, the speed and cost of digital signal processing are not comparable to those of analog systems. FFT can enhance the competitiveness of digital systems to analyze the spectrum, providing a novel opportunity for the extensive application of digital processing signals.

1. Fast Fourier Transform Method

According to the DFT, we set $W_N^{nk} = e^{-nj2\pi k/N}$. In this case,

$$F_n = \frac{1}{\sqrt{N}} \sum_{k=0}^{N-1} f_k W_N^{nk} \tag{6.48}$$

where W_N^{nk} is the rotation factor. The matrix form of the DFT can be transformed to

$$\begin{bmatrix} F(0) \\ F(1) \\ \vdots \\ F(N-1) \end{bmatrix} = \begin{bmatrix} W^{0\times 0} & W^{1\times 0} & \dots & W^{(N-1)\times 0} \\ W^{0\times 1} & W^{1\times 1} & \dots & W^{(N-1)\times 1} \\ \vdots & \vdots & \ddots & \vdots \\ W^{0\times (N-1)} & W^{1\times (N-1)} & \dots & W^{(N-1)\times (N-1)} \end{bmatrix} \begin{bmatrix} f(0) \\ f(1) \\ \vdots \\ f(N-1) \end{bmatrix} \tag{6.49}$$

The DFT has a fast algorithm, which is based on the rotation factor and has periodic and symmetric characteristics.

(1) Periodicity of W_N^{nk}

$W_N^{nk} = \exp(-\mathrm{j}\frac{2\pi}{N}nk)$ is a complex exponential periodic sequence that has periodic characteristics for ordinal numbers n and k, and its periodicity is N. W_N^{nk} has the following characteristics:

$$W_N^{nk} = W_N^{(n+N)k} = W_N^{n(k+rN)} = W_N^{(nk+rN)} \tag{6.50}$$

$k = 0, 1, 2, \ldots, N-1$; $n = 0, 1, 2, \ldots, N-1$; r represents positive integers.

For example, for $N = 8$,

$$W_8^9 = W_8^{1+8} = W_8^1, \quad W_8^{42} = W_8^{2+5\times 8} = W_8^2 \tag{6.51}$$

Although $nk = 8 \times 8 = 64$ coefficients exist, because of periodicity, only eight coefficients: W_8^0, W_8^1, ···, W_8^7 are independent, and the remaining terms are only the repetitions of the eight coefficients.

(2) Symmetry of W_N^{nk}

Because $W_N^0 = 1$, $W_N^{\frac{N}{2}} = -1$,

$$W_N^{(nk+\frac{N}{2})} = W_N^{nk} + W_N^{\frac{N}{2}} = -W_N^{nk} \tag{6.52}$$

The circle decomposed until the sequence of length N is subdivided into $N/2$ 2-point sequences. This method is recycled to decompose the N-point DFT into $N/2$ 2-point DFT operations, and the computational complexity is considerably decreased.

The proof is as follows (Fig. 6.2):

Let N be a positive integer power of 2, i.e., $N = 2^n$, where n is a positive integer, such that M is a positive integer, and $N = 2M$. Let $N = 2M$ be in the discrete Fourier formula, which can be written in the following form:

$$F_n = \sum_{k=0}^{2M-1} f_k W_N^{nk} = \sum_{k=0}^{M-1} f_{2k} W_{2M}^{2nk} + \sum_{k=0}^{M-1} f_{2k+1} W_{2M}^{n(2k+1)} \tag{6.53}$$

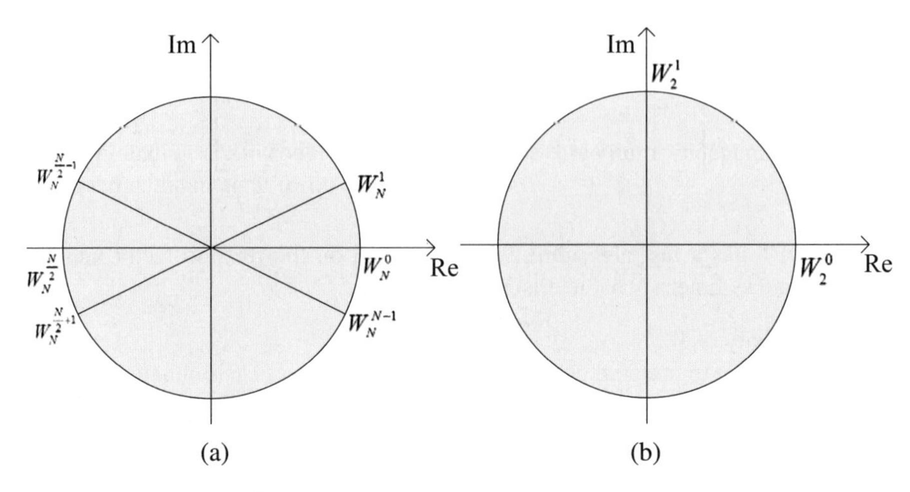

Fig. 6.2 Symmetry of W_N^{nk} (**a**) and final remaining item (**b**)

According to the periodicity of the rotation factor W_N^{nk},

$$W_{2M}^{2nk} = W_M^{nk} \tag{6.54}$$

This formula can be transformed to

$$F_n = \sum_{k=0}^{M-1} f_{2k} W_M^{nk} + \sum_{k=0}^{M-1} f_{2k+1} W_M^{nk} W_{2M}^{n} \tag{6.55}$$

expressed as

$$F_n^e = \sum_{k=0}^{M-1} f_{2k} W_M^{nk}, \quad F_n^o = \sum_{k=0}^{M-1} f_{2k+1} W_M^{nk} \tag{6.56}$$

where u, $x = 0, 1, 2, \ldots, N-1$. In this case,

$$F_n = F_n^e + F_n^o W_{2M}^n \tag{6.57}$$

Considering the symmetry and periodicity of W,

$$W_M^{n+M} = W_M^n, \quad W_{2M}^{n+M} = -W_{2M}^n \tag{6.58}$$

Therefore,

$$F_{n+M} = F_n^e - F_n^o W_{2M}^n \tag{6.59}$$

Thus, a discrete Fourier transform of an N-point can be decomposed into discrete Fourier transforms of two N/2 short sequences, i.e., discrete Fourier transforms F_n^e and F_n^o, which are decomposed into even and odd sequences. Next, through the decomposition of the recursive operation, FFT can be implemented.

Because a large number of similar coefficients exist, many unnecessary duplications exist in the DFT calculation. The FFT solution simplifies these repeated calculations to enable rapid calculation.

3. Comparison of Fast Fourier Transform Advantages

The direct DFT algorithm needs N^2 complex multiplications and $N(N-1)$ complex arithmetic additions.

FFT is performed for the selection of grouping operations, which change the length of the DFT calculation into M classes. Each level has only $N/2$ butterfly operations. Even if the butterfly structure is not simplified, each individual butterfly structure requires only 2 multiplication operations. Therefore, the FFT needs to perform only $M(N/2) \times 2$ complex multiplications, and

$$M = \log_2 N, N = 2^M \tag{6.60}$$

In the M-class butterfly process, the total computing workload for the FFT is.

$$\text{Number of multiplications}: M(N/2) \times 2 = N \log_2 N$$
$$\text{Number of additions}: MN = N \log_2 N$$

Table 6.3 compares the complex multiplier workload for the FFT and direct DFT. When $N = 1024$, the FFT computing efficiency is nearly 100 times higher than that of the DFT.

Table 6.3 Comparison of FFT and DFT workloads

N	M	Direct DFT (N^2)/times	FFT ($N \log_2 N$)/times	Improved ratio $\frac{N}{\log_2 N}$
8	3	64	24	2.7
32	5	1024	160	6.4
64	6	4096	384	10.7
128	7	16,384	896	18.3
256	8	65,536	2048	32
512	9	262,144	4608	56.9
1024	10	1,048,576	10,240	102.4

6.3 Fourier Transform in Pixel Processing

The two sections describe the property of a one-dimensional Fourier transform and its discretization process. However, images are two-dimensional. To ensure application to digital image processing, it is necessary to extend the Fourier transform to two-dimensional discrete space. In this section, from the two-dimensional continuous Fourier transform, the mathematical formula and its properties of the two-dimensional discrete Fourier transform are derived, and the relationship between the two-dimensional discrete Fourier transform and one-dimensional Fourier transform is clarified to ensure that the one-dimensional fast Fourier transform can be easily extended to two-dimensional space. In addition, the main task of applying Fourier transform to image processing is resolved. This section also describes the application of the Fourier transform in image processing.

6.3.1 Two-Dimensional Continuous Fourier Transform

The two-dimensional Fourier transform and its inverse transform are defined as

$$F(u,v)=\int_{-\infty}^{\infty}\int_{-\infty}^{\infty} f(x,y)\mathrm{e}^{-\mathrm{j}2\pi(ux+vy)}\mathrm{d}x\mathrm{d}y \tag{6.61}$$

with

$$f(x,y)=\int_{-\infty}^{\infty}\int_{-\infty}^{\infty} F(u,v)\mathrm{e}^{\mathrm{j}2\pi(ux+vy)}\mathrm{d}u\mathrm{d}v \tag{6.62}$$

where $f(x, y)$ is an image with spectrum $F(u, v)$. In general, $F(u, v)$ represents the complex value function of two real frequency variables u and v. The variable u corresponds to the x-axis, and the frequency domain v corresponds to the y-axis [3, 4].

6.3.2 Two-Dimensional Discrete Fourier Transform

If $g(i, k)$ is an $N \times N$ array (similar to an equidistant rectangular grid for a two-dimensional continuous function of the sample), its two-dimensional discrete Fourier transform is

$$G(m,n) = \frac{1}{N}\sum_{i=0}^{N-1}\sum_{k=0}^{N-1} g(i,k)\mathrm{e}^{-\mathrm{j}2\pi\left(m\frac{i}{N}+n\frac{k}{N}\right)} \tag{6.63}$$

The IDFT is

$$g(i,k) = \frac{1}{N}\sum_{m=0}^{N-1}\sum_{n=0}^{N-1} G(m,n)\mathrm{e}^{\mathrm{j}2\pi\left(i\frac{m}{N}+k\frac{n}{N}\right)} \tag{6.64}$$

As in the one-dimensional case, the discrete Fourier transform is similar to a continuous Fourier transform, a special case of a two-dimensional discrete Fourier transform of a bandwidth-limited function sampled on a rectangular grid.

6.3.3 Matrix Representation

The DFT is represented by a matrix as

$$\boldsymbol{G} = \boldsymbol{F}\,\boldsymbol{g}\,\boldsymbol{F} \tag{6.65}$$

with

$$\boldsymbol{F} = [f_{ik}] = \left[\frac{1}{\sqrt{N}}\mathrm{e}^{-\mathrm{j}2\pi ik/N}\right] \tag{6.66}$$

being an $N \times N$ complex coefficient kernel matrix.

$\boldsymbol{F}$ is a unitary matrix, i.e., the inverse of the matrix is its complex conjugate transpose $[\boldsymbol{F}^{-1} = (\boldsymbol{F}^*)^{\mathrm{T}}]$. To obtain an inverse of the unitary matrix, we can simply exchange the rows and columns of the position and change each element of the imaginary symbol.

Note that when we compute two-dimensional convolution, we usually stack each row into a large column vector and use a large cyclic matrix. However, because the two-dimensional Fourier transform kernel function can be decomposed into row and column operations and $\boldsymbol{F}$ is a unitary matrix, we do not need to perform this operation in calculating a two-dimensional Fourier transform.

6.3.4 Properties of Two-Dimensional Fourier Transform

Table 6.4 summarizes the properties of the two-dimensional Fourier transform. Note that there are several properties of two-dimensional Fourier transforms that do not correspond to a one-dimensional Fourier transform. An example of a two-dimensional image that can be decomposed into the product of the one-dimensional

Table 6.4 Properties of two-dimensional Fourier transform

Property	Spatial domain	Frequency domain
Addition theorem	$f(x, y) + g(x, y)$	$F(u, v) + G(u, v)$
Similarity theorem	$f(ax, by)$	$\frac{1}{\lvert ab\rvert} F(\frac{u}{a}, \frac{v}{b})$
Displacement theorem	$f(x - a, y - b)$	$e^{-j2\pi(au+bv)} F(u, v)$
Convolution theorem	$f(x, y) * g(x, y)$	$F(u, v)G(u, v)$
Separable product	$f(x)g(y)$	$F(u)G(v)$
Differential	$(\frac{\partial}{\partial x})^m (\frac{\partial}{\partial y})^n f(x, y)$	$(j2\pi u)^m (j2\pi v)^n F(u, v)$
Rotation	$f(x\cos\theta, -x\sin\theta + y\cos\theta)$	$F(u\cos\theta + v\sin\theta, -u\sin\theta + v\cos\theta)$
Laplace operator	$\nabla^2 f(x, y) = \left(\frac{\partial^2}{\partial x^2} + \frac{\partial^2}{\partial y^2}\right) f(x, y)$	$-4\pi^2(u^2 + v^2)F(u, v)$
Rayleigh theorem	$\int_{-\infty}^{\infty}\int_{-\infty}^{\infty} \lvert f(x, y)\rvert^2 dxdy$	$\int_{-\infty}^{\infty}\int_{-\infty}^{\infty} \lvert F(u, v)\rvert^2 dudv$

component. Another example is the rotational property, which is useful in computer tomography (CAT) technology, as discussed in Chap. 15.

The Laplace operator is an omnidirectional second-order differential operator that is usually used for edge detection and edge enhancement. When using a Laplace operator for a function, $u^2 + v^2$ is multiplied over the spectrum. According to the convolution theorem, the Laplace operator corresponds to a linear system whose transfer function increases with the square of the frequency.

1. Separation

Suppose

$$f(x, y) = f_1(x) f_2(y) \tag{6.67}$$

In this case,

$$F(u, v) = \int_{-\infty}^{\infty}\int_{-\infty}^{\infty} f_1(x) f_2(y) e^{-j2\pi(ux+vy)} dxdy \tag{6.68}$$

and thus,

$$F(u, v) = \int_{-\infty}^{\infty} f_1(x) e^{-j2\pi ux} dx \int_{-\infty}^{\infty} f_2(y) e^{-j2\pi vy} dy = F_1(u) F_2(v) \tag{6.69}$$

Thus, if a two-dimensional image can be decomposed into two one-dimensional component functions, its spectrum can also be decomposed into two one-dimensional component functions.

For example, for a two-dimensional elliptical Gaussian function,

$$e^{-(x^2/2\sigma_x^2+y^2/2\sigma_y^2)} = e^{-x^2/2\sigma_x^2} e^{-y^2/2\sigma_y^2} \tag{6.70}$$

This term can be decomposed into the product of two one-dimensional Gaussian functions. If the standard variance of the two factors is the same,

$$e^{-(x^2/2\sigma_x^2+y^2/2\sigma_y^2)} = e^{-x^2/2\sigma_x^2} e^{-y^2/2\sigma_y^2} \tag{6.71}$$

This term corresponds to a circular Gaussian function. This function is extremely useful in optical system analysis because it has circular symmetry and can be decomposed into a product of one-dimensional functions.

Using this property, the two-dimensional discrete Fourier transform can be decomposed into a quadratic one-dimensional FFT transform to realize the fast Fourier transform of the image (two-dimensional) to enable the use of the Fourier transform in image processing. To this end, we first obtain $f(x, y)$ from $f(x, y)$ through the Fourier transform. Subsequently, Fourier transform is performed on columns, and we can obtain the Fourier transform result of $f(x, y)$, that is, $F(u, v)$.

2. Similarity

The similarity theorem can be generalized to a two-dimensional case:

$$\Im\{f(a_1x+b_1y, a_2x+b_2y)\} = \int_{-\infty}^{\infty}\int_{-\infty}^{\infty} f(a_1x+b_1y, a_2x+b_2y)e^{-j2\pi(ux+vy)}dxdy \tag{6.72}$$

Variable replacement is performed as

$$w = a_1x+b_1y, z = a_2x+b_2y \tag{6.73}$$

In this case,

$$\begin{cases} x = A_1w + B_1z, dx = A_1dw + B_1dz, \\ y = A_2w + B_2z, dy = A_2dw + B_2dz \end{cases} \tag{6.74}$$

and

$$\begin{aligned} A_1 &= \frac{b_2}{a_1b_2 - a_2b_1}, B_1 = \frac{-b_1}{a_1b_2 - a_2b_1} \\ A_2 &= \frac{-a_2}{a_1b_2 - a_2b_1}, B_2 = \frac{a_1}{a_1b_2 - a_2b_1} \end{aligned} \tag{6.75}$$

The Fourier transform is

$$\begin{aligned}&\Im\{f(a_1x+b_1y,a_2x+b_2y)\}\\&=\int_{-\infty}^{\infty}\int_{-\infty}^{\infty}f(w,z)\mathrm{e}^{-\mathrm{j}2\pi\{(A_1u+A_2v)w+(B_1u+B_2v)z\}}\mathrm{d}z\mathrm{d}w(A_1B_2+A_2B_1)\\&=(A_1B_2+A_2B_1)F(A_1u+A_2v,B_1u+B_2v)\end{aligned}\tag{6.76}$$

3. Rotation

According to the two-dimensional similarity theorem, if $f(x, y)$ rotates by angle θ, the spectrum of $f(x, y)$ also rotates by the same angle. If

$$a_1=\cos\theta,b_1=\sin\theta,a_2=-\sin\theta,b_2=\cos\theta\tag{6.77}$$

then

$$A_1=\cos\theta,A_2=\sin\theta,B_1=-\sin\theta,B_2=\cos\theta\tag{6.78}$$

and

$$\Im\{f(x\cos\theta+y\sin\theta,-x\sin\theta+y\cos\theta)\}=F(u\cos\theta+v\sin\theta,-u\sin\theta+v\cos\theta)\tag{6.79}$$

4. Projection

Suppose that we project a two-dimensional function $f(x, y)$ onto the x-axis to obtain a one-dimensional function:

$$p(x)=\int_{-\infty}^{\infty}f(x,y)\mathrm{d}y\tag{6.80}$$

The (one-dimensional) Fourier transform of $p(x)$ is

$$P(u)=\int_{-\infty}^{\infty}\int_{-\infty}^{\infty}f(x,y)\mathrm{d}y\mathrm{e}^{-\mathrm{j}2\pi ux}\mathrm{d}x\tag{6.81}$$

However, $P(u)$ can be expressed as

$$P(u)=\int_{-\infty}^{\infty}\int_{-\infty}^{\infty}f(x,y)\mathrm{e}^{-\mathrm{j}2\pi(ux+0y)}\mathrm{d}x\mathrm{d}y=F(u,0)\tag{6.82}$$

Therefore, the transformation of $f(x, y)$ on the x-axis is the value of $F(u, v)$ on the u-axis. Combining the rotation, we can determine that the Fourier transform of $F(u, y)$ that is projected on a straight line at an angle θ to the x-axis is equal to the value of $F(u, v)$ along a straight line with an angle θ of the axis. Projection properties are the basis for system identification (Chap. 11) and computed tomography (Chap. 5) using line expansion functions.

6.4 Application of Fourier Transform for Image Pixel Processing

The frequency of an image is an indicator of the severity of the grayscale in the image, which is the gradient of the grayscale in the plane space. The physical meaning of the Fourier transform is to transform the image gray distribution function into the image frequency distribution function. The inverse Fourier transform transforms the frequency distribution function of the image into the grayscale distribution function.

6.4.1 Fourier Transform and Reconstruction

In the actual application of the two-dimensional discrete Fourier transform, the amplitude and phase spectra represent different meanings. We can consider that the phase spectrum contains the texture information of the image, and the amplitude spectrum contains the light contrast information of the image. In many cases, the phase spectrum preserves important characteristics of the signal, while the amplitude spectrum does not [5].

Figure 6.3 shows the input remote sensing image (original), Fig. 6.4a, b show the amplitude and phase spectra, respectively. Next, we reconstruct the image using the phase and amplitude spectra. Figure 6.4c, d show the amplitude and phase spectrum reconstruction, respectively. According to the reconstructed image, the amplitude information map is considerably different from the original image, and the phase information map can preserve most of the texture features of the original image and indicate the outline of the original image. Thus, many important features of the original images are reflected in the phase information map compared to the amplitude information graph.

Fig. 6.3 Original image

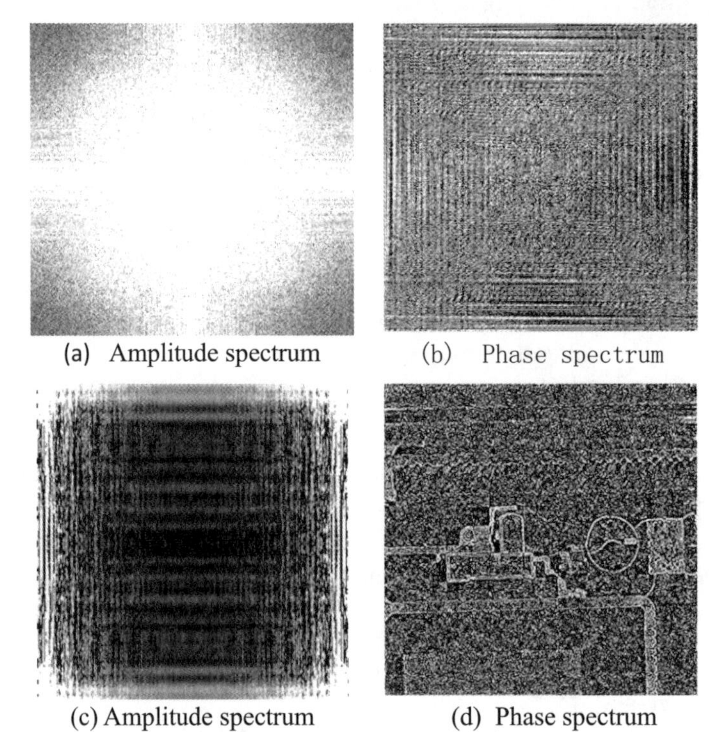

Fig. 6.4 Application of the Fourier transform

6.4.2 Remote Sensing Image Registration

The image registration method based on domain transform is an important remote sensing image processing method. The ratio of the image, rotation and pan variables can be reflected in the Fourier transform domain, and the frequency domain exhibits a high resistance to noise interference.

Assuming that there exists only a displacement relationship between the two images f_1 and f_2, the translations on the x-axis and y-axis are x_0 and y_0, respectively.

$$f_2(x, y) = f_1(x - x_0, y - y_0) \tag{6.83}$$

The relationship between the Fourier transforms F_1 and F_2 of f_1 and f_2

$$F_2(u, v) = \mathrm{e}^{-\mathrm{j}2\pi(ux_0+vy_0)} \times F_1(u, v) \tag{6.84}$$

The cross power spectrum of the two images in the corresponding frequency domain is

$$\frac{F_1(u, v)F_2^*(u, v)}{\left|F_1(u, v)F_2^*(u, v)\right|} = \mathrm{e}^{-\mathrm{j}2\pi(ux_0+vy_0)} \tag{6.85}$$

F_2^* is the complex conjugate of F_2. According to the translation theory, the phase difference between the cross-power spectrum equals the pan amount between the images. After the cross-power spectrum inverse transformation, a pulse function $\delta(x - x_0, y - y_0)$ can be obtained. The subfunction exhibits notable spikes at the offset position, and the value at other locations is close to zero. According to this principle, we can obtain the offset between two images.

Since this algorithm works only for image registration in pan situations, effects of translating, rotating and scaling the triple factor may occur on the two images that need to be registered. Therefore, a complete image registration algorithm must be able to overcome the effects of translation, rotation and scaling. When I_1 and I_2 are two images to be registered,

$$I_2(x, y) = I_1\left((x, y)\begin{bmatrix} \cos\theta_0 & -\sin\theta_0 \\ \sin\theta_0 & \cos\theta_0 \end{bmatrix} s + (\Delta x, \Delta y)\right) \tag{6.86}$$

The Fourier transform is converted to the polar coordinate system:

$$\hat{I}_2 = \mathrm{e}^{\,\mathrm{j}(\omega_x \Delta x+\omega_y \Delta y)} s^{-2} \hat{I}_1(s^{-1}r, \theta + \theta_0) \tag{6.87}$$

where $\hat{I}_1$ and $\hat{I}_2$ are the expressions of I_1 and I_2 in the frequency domain, respectively. Therefore, the relationship between the two images can be established as

$$M_2(\log r, \theta) = s^{-2} M_1(\log r - \log s, \theta + \theta_0) \tag{6.88}$$

In this way, the scaling and rotation factors between the two images are determined (by θ and s, respectively). The coordinates of the polar coordinate system are transformed into the traditional image coordinate system, and image registration is completed.

Figure 6.5 shows the two remote sensing images to be registered. The two images exhibit a certain overlap, and translation, rotation, and zoom relationships exist between the two images. According to the registration algorithm, to achieve image registration, the translation, rotation, and zoom factor can be determined. Figure 6.6 shows the images after registration.

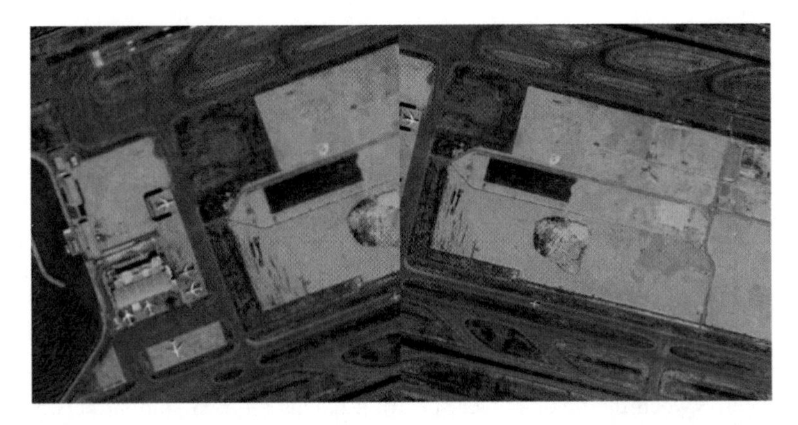

Fig. 6.5 Images to be registered

Fig. 6.6 Images after registration

6.4.3 Remote Sensing Image De-Striping

Significant stripes can be observed in remote sensing images of marine color water temperature scanners (COCTS), such as MODIS and "Ocean One" B (HY-1B) satellites, as shown in Fig. 6.7a. This phenomenon is mainly caused by the difference in the response of different detected elements in the detector. Therefore, the essence of the de-striping operation of the image is to correct the pixel radiance of the image. The traditional method of image stripping is divided into the spatial and transform domains. The spatial domain method is aimed at normalizing the image pixel value. Although this task can be easily accomplished, it changes the true reflectivity of the feature. In the transform domain method, the Fourier transform for the image is performed, its gray value change feature in the frequency domain is analyzed, the image in the frequency domain is filtered through the corresponding template, and the image is converted to the spatial domain through the inverse Fourier transform [6, 7].

As shown in Fig. 6.7a, the image has horizontal stripes. This phenomenon is reflected in Fig. 6.7b, in which the energy is higher along the y-axis direction (i.e., $x = 0$). In this case, $x = 0$ indicates that the lateral gray value in the spatial domain does not exhibit any change, whereas the horizontal stripe indicates that the horizontal gray value is 255 or 0. Therefore, in the template shown in Fig. 6.7c, the mask image is processed at $x = 0$ to process the frequency domain image. In general, the template operation in the frequency domain directly takes 0 value from the mask. However, in this case, to preserve the original high-frequency components, the mean in the mask position on the adjacent left and right elements is set rather than directly taking the 0 value. Figure 6.7d shows the results of the final removal of the band. The filtering effect is high, which demonstrates the significance of the Fourier transform for remote sensing image processing.

6.4.4 Synthetic Aperture Radar (SAR) Imaging Processing

The SAR image generation process is different from that for the general optical remote sensing image. The classic imaging algorithm is the range Doppler (RD) algorithm. As shown in Fig. 6.8, this method mainly includes three steps: distance compression, distance migration correction, and azimuth compression. Among these steps, distance filtering and azimuth compression require matching filtering, and this process is performed in the frequency domain. Therefore, the image needs to be Fourier transformed. However, this Fourier transform is different from the two-dimensional spatial Fourier transform. Readers can refer to the literature and books on SAR imaging algorithms for detailed procedures.

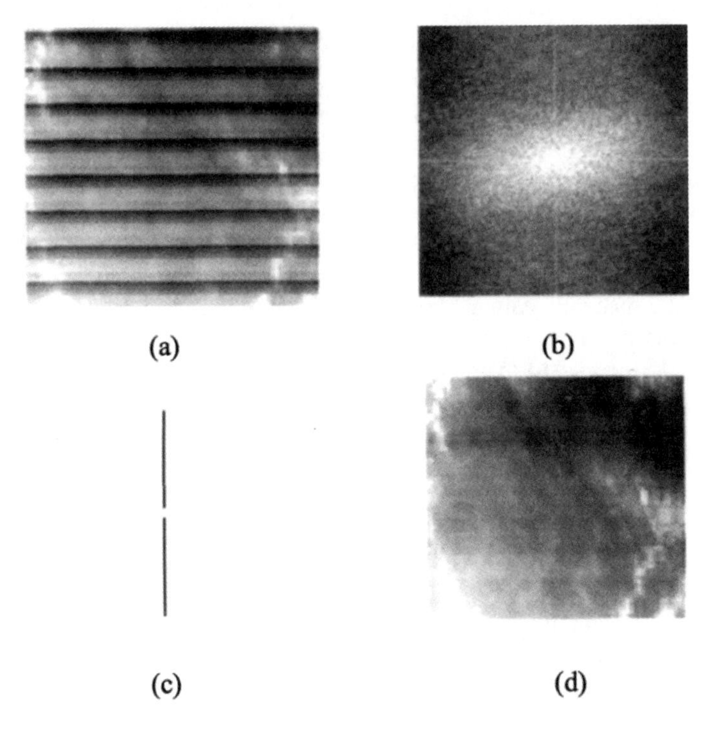

Fig. 6.7 Example of remote sensing image de-striping by Fourier transform

6.5 Conclusion

Fourier transform is a linear integral transformation that establishes a unique correspondence between the complex function of the time (or spatial) domain and frequency domain. The focus of this chapter is to clarify the implementation of the Fourier transform on remote sensing images. The contents of this chapter and mathematical formulas are summarized in Fig. 6.9 and Table 6.5. First, we identify the mathematical basis of Fourier transform and clarify how to transform the Fourier series into the Fourier transform. Next, we describe the discrete Fourier transform and two-dimensional Fourier transform and their application in image processing; Finally, a sample application of the Fourier transform to image processing is described.

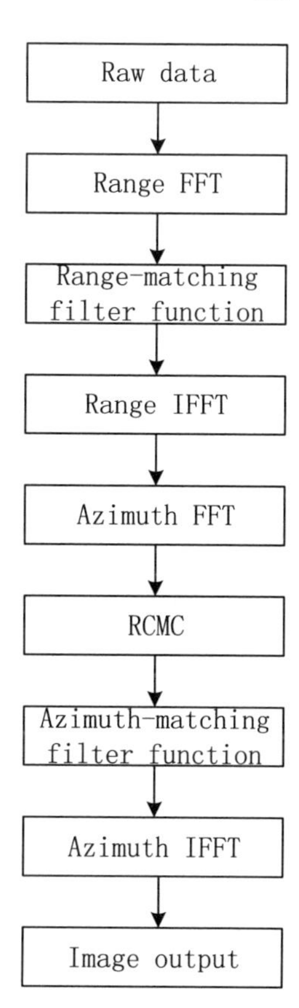

Fig. 6.8 Process flow of the range Doppler imaging algorithm

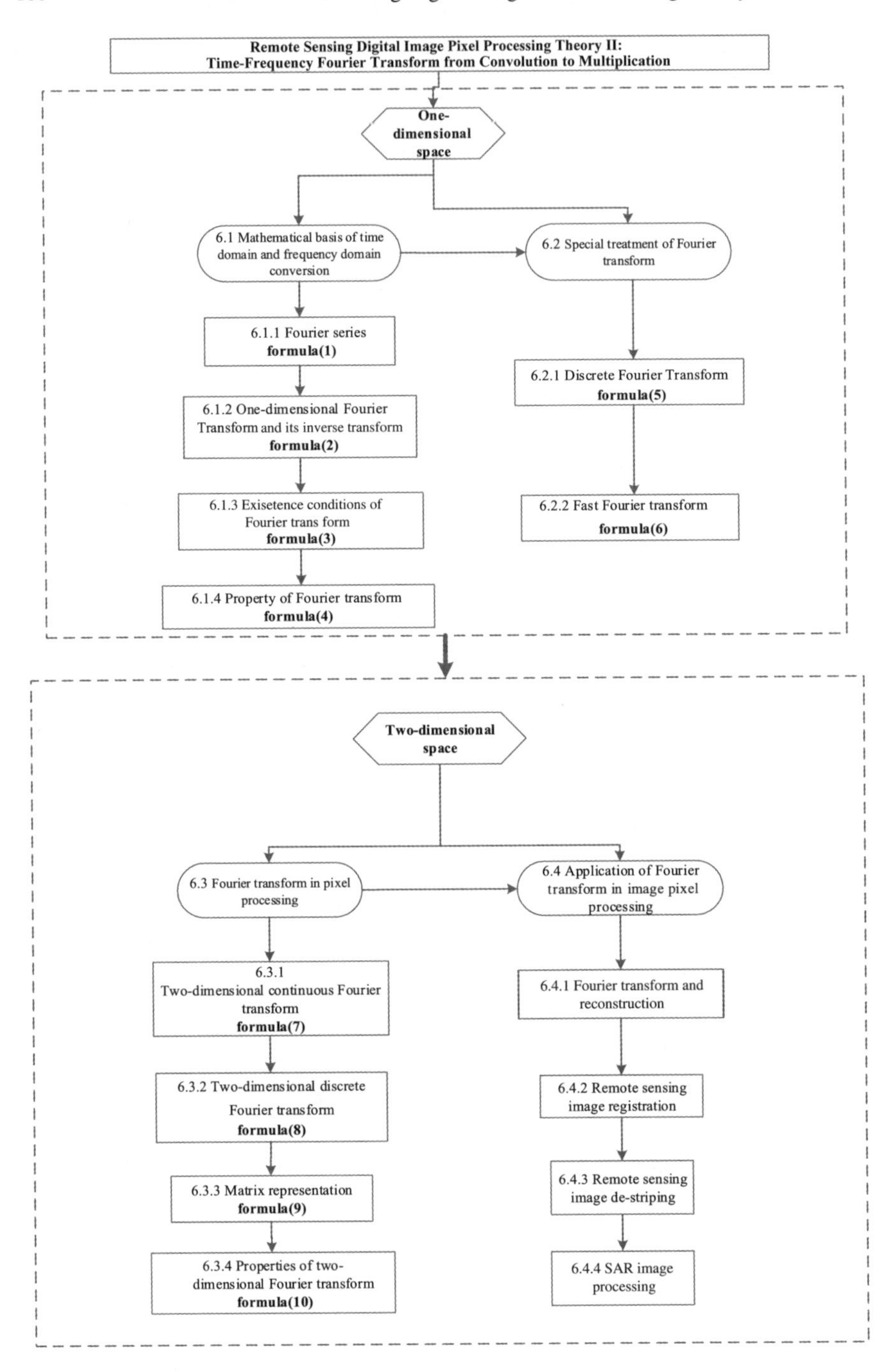

Fig. 6.9 Summary of this chapter's contents

Table 6.5 Summary of mathematical formulas in this chapter

formula (1)	$f(t)=\frac{a_0}{2}+\sum_{n=0}^{\infty}a_n\cos\left(2\pi\frac{n}{T}t\right)+\sum_{n=0}^{\infty}b_n\sin\left(2\pi\frac{n}{T}t\right)$	6.1.1
formula (2)	$F(s)=\Im[f(t)]=\int_{-\infty}^{\infty}f(t)\mathrm{e}^{-\mathrm{j}2\pi st}\mathrm{d}t$	6.1.2
formula (3)	$f(ax,by)$	6.1.3
formula (4)	$\Im[f(t-a)]=\mathrm{e}^{-\mathrm{j}2\pi as}\int_{-\infty}^{\infty}f(u)\mathrm{e}^{-\mathrm{j}2\pi su}\mathrm{d}u=\mathrm{e}^{-\mathrm{j}2\pi as}F(s)$ $y(t)=x(t)*h(t)\leftrightarrow Y(\omega)=X(\omega)H(\omega)$ $\Im[f(at)]=\frac{1}{\lvert a\rvert}\int_{-\infty}^{\infty}f(u)\mathrm{e}^{-\mathrm{j}2\pi u(s/a)}\mathrm{d}u=\frac{1}{\lvert a\rvert}F(\frac{s}{a})$	6.1.4
formula (5)	$F_n=\frac{1}{\sqrt{N}}\sum_{k=0}^{N-1}f_k\mathrm{e}^{-\mathrm{j}2\pi(\frac{n}{N})k}$	6.2.1
formula (6)	$F_n=\frac{1}{\sqrt{N}}\sum_{k=0}^{N-1}f_kW_N^{nk}$	6.2.2
formula (7)	$F(u,v)=\int_{-\infty}^{\infty}\int_{-\infty}^{\infty}f(x,y)\mathrm{e}^{-\mathrm{j}2\pi(ux+vy)}\mathrm{d}x\mathrm{d}y$	6.3.1
formula (8)	$G(m,n)=\frac{1}{N}\sum_{i=0}^{N-1}\sum_{k=0}^{N-1}g(i,k)\mathrm{e}^{-\mathrm{j}2\pi(m\frac{i}{N}+n\frac{k}{N})}$	6.3.2
formula (9)	$\boldsymbol{G}=\boldsymbol{F}\boldsymbol{g}\boldsymbol{F},\ \boldsymbol{F}=[f_{ik}]=\left[\frac{1}{\sqrt{N}}\mathrm{e}^{-\mathrm{j}2\pi ik/N}\right]$	6.3.3
formula (10)	$\Im\{f(a_1x+b_1y,a_2x+b_2y)\}$ $=\int_{-\infty}^{\infty}\int_{-\infty}^{\infty}f(w,z)\mathrm{e}^{-\mathrm{j}2\pi(A_1u+A_2v)w+(B_1u+B_2v)z\}}\mathrm{d}z\mathrm{d}w(A_1B_2+A_2B_1)$ $=(A_1B_2+A_2B_1)F(A_1u+A_2v,B_1u+B_2v)$	6.3.4

References

1. Carleson L. On convergence and growth of partial sums of Fourier series. Acta Mathematica. 1996(116).
2. Bergland GD. A guided tour of the fast Fourier transform. Spectrum, IEEE Trans. 1969(6).
3. Moody A, Johnson DM. Land-surface phenologies from AVHRR using the discrete Fourier transform. Remote Sens Environ. 2001(75).
4. Candan C, Kutay M A, Ozaktas H M. The discrete fractional Fourier transform. Signal Processing, IEEE Transactions, 2000(48).
5. Pan W, Qin K, Chen Y. An adaptable-multilayer fractional Fourier transform approach for image registration. Pattern Anal Mach Intell IEEE Trans Pattern Anal Mach Intell. 2009(31).
6. Xue Y, Ma J, Lai J, et al. Destriping methods for HY-1B satellite images based on Fourier transdrom. Spacecaft Recovery Remote Sens. 2012(33) (In Chinese).
7. Mhabary Z, Levi O, Small E, et al. An exact and efficient 3D reconstruction method from captured light-fields using the fractional Fourier transform. In: SPIE commercial + scientific sensing and imaging. International society for optics and photonics. 2016: 986708-986708-9.

Chapter 7
Remote Sensing Digital Image Pixel Processing Theory III: Frequency Domain Filtering

Chapter Guidance When the information of the space–time domain is converted to the frequency domain, periodic rules of spatial information can be found: the gradually changing edges represent the low-frequency information, and the remote sensing details and mutated edges correspond to the high-frequency information. According to different production requirements, the information of different frequencies can be retained in the frequency domain, and inverse transformation of the real image in the space–time domain can be performed. In this manner, the required remote sensing information can be obtained, and the unnecessary information can be eliminated. This process is known as filtering, and it can be implemented through various types of filters. This method must be selected as the basis for remote sensing image processing based on different types of information.

This chapter characterizes filters as linear and nonlinear filters. In addition to several basic linear filters, such as the basic operator type in the spatial domain and frequency domain, adaptive filters, such as the Wiener filter and matched detector, exist. In terms of nonlinear filters, this chapter discusses the space operator and homomorphic filter.

The structure of this chapter is shown in Fig. 7.1. The specific contents are as follows: Sect. 7.1 Filters and Classification: this part describes the periodic expression of remote sensing time–space domain information and relationship between frequency domain and spatial domain filtering; Sect. 7.2 Linear Filter: this part discusses the relationship between the spatial domain filter operator and transfer function response of the basic frequency-domain filter; Sect. 7.3 Optimal Linear Filter: this part described the optimal criteria to design the Wiener filter and detector design; Sect. 7.4 Nonlinear Filter: we discuss several nonlinear filtering methods such as the correlation operator, median filtering and homomorphic filtering from the spatial domain perspective.

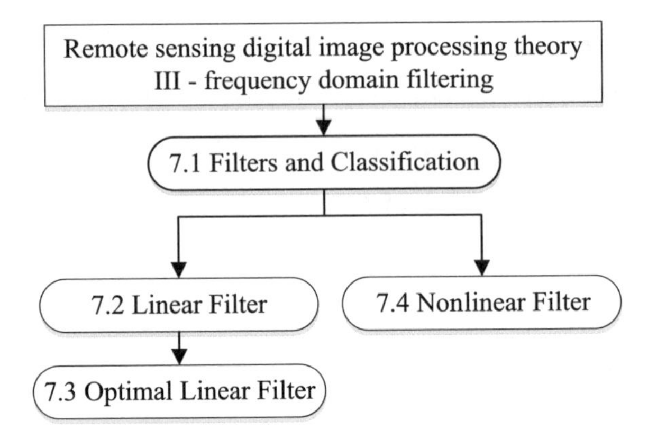

Fig. 7.1 Logical framework of this chapter

7.1 Filters and Classification

This section presents the relationship between the periodic expression of the remote sensing time–space domain information and frequency domain and spatial domain filtering. The filter classification and relationship between the filter and spatial domain provide a framework and theoretical basis for the discussion presented in subsequent sections.

7.1.1 Filters Classification

First, we describe linear and nonlinear systems. For a particular system, input $x_1(t)$ produces output $y_1(t)$:

$$x_1(t) \rightarrow y_1(t) \tag{7.1}$$

Another input $x_2(t)$ produces output $y_2(t)$:

$$x_2(t) \rightarrow y_2(t) \tag{7.2}$$

This system is linear if and only if it satisfies

$$x_1(t) + x_2(t) \rightarrow y_1(t) + y_2(t) \tag{7.3}$$

A linear system exhibits the following characteristics: The sum of the previous two signals is equal to the sum of the previous two outputs. Any system that does not satisfy this constraint is nonlinear.

Filters can be classified from many different perspectives. According to the difference in signal processing, filters can be divided into analog and digital filters.

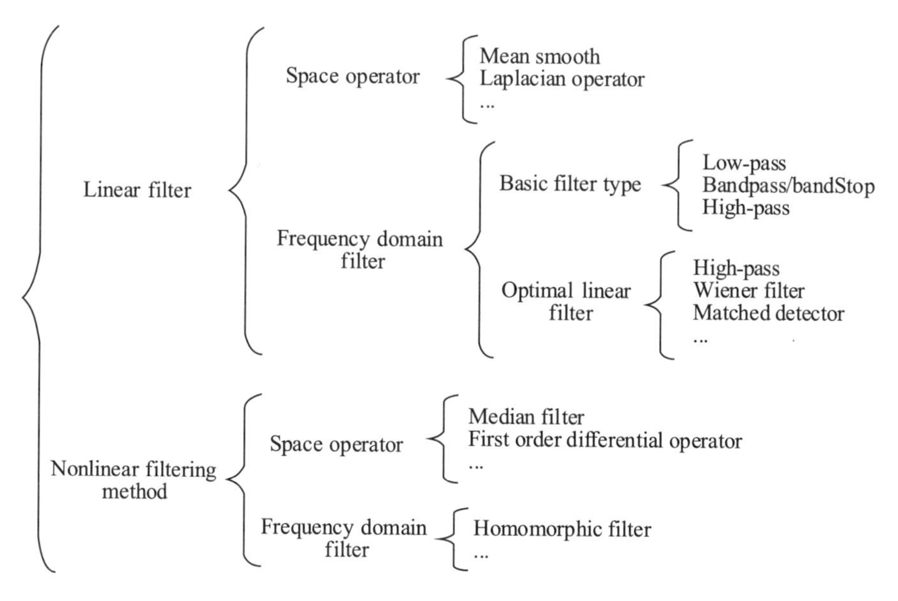

Fig. 7.2 Filter classification

Depending on whether the output of the filter is based on the linear function of the input, filters can be divided into linear and nonlinear filters. According to the filter processing of different frequency bands, filters can be divided into low-pass filters, bandpass/band rejection filters, and high-pass filters [1]. The logic of the filter classification used in this chapter is shown in Fig. 7.2.

7.1.2 Relationship Between Spatial and Frequency Domain Filtering

In the image filtering process, the choice of filtering in the spatial or frequency domain is a fundamental problem. The most basic link between these domains is established considering the convolution theorem. Convolution refers to the process of moving the template of the image pixel by pixel in the image and performing the specified number of calculations for each pixel. In this form, the discrete convolution of two functions $f(x, y)$ and $h(x, y)$ of size $M \times N$ is expressed as $f(x, y) * h(x, y)$, and defined as follows:

$$f(x, y)*h(x, y) = \frac{1}{MN} \sum_{m=0}^{M-1} \sum_{n=0}^{N-1} f(m, n)h(x - m, y - n) \tag{7.4}$$

The negative sign in the formula only indicates that the function h mirrors the symmetry of the origin, which is self-contained in convolution.

$F(u, v)$ and $H(u, v)$ represent the Fourier transform of $f(x, y)$ and $h(x, y)$, respectively, and the convolution definition of $f(x, y) * h(x, y)$ and $F(u, v)H(u, v)$ is composed of Fourier transform pairs expressed as

$$f(x, y) * h(x, y) \Leftrightarrow F(u, v)H(u, v) \tag{7.5}$$

The double arrow indicates that the left expression (spatial domain convolution) can be obtained by performing an inverse Fourier transformation based on the right expression. In contrast, the right expression can be obtained by performing a forward Fourier transformation based on the left expression. A similar result is that the convolution in the frequency domain is reduced to the multiplication of the spatial domain, and vice versa:

$$f(x, y)h(x, y) \Leftrightarrow F(u, v) * H(u, v) \tag{7.6}$$

These two conclusions constitute the convolution theorem.

Before introducing the link between the spatial and frequency domains, we introduce the concept of an impulse function. The impulse function with an intensity of (x_0, y_0) in A is denoted by $A\delta(x - x_0, y - y_0)$ and defined as follows:

$$\sum_{x=0}^{M-1}\sum_{y=0}^{N-1} s(x, y)A\delta(x - x_0, y - y_0) = As(x_0, y_0) \tag{7.7}$$

This equation shows that the sum of the product of function $s(x, y)$ and impulse function $A\delta(x - x_0, y - y_0)$ is equal to the value of $s(x, y)$ at (x_0, y_0), corresponding to impact strength A. The upper limit of the function is defined by assuming $A\delta(x - x_0, y - y_0)$ as an image of size $M \times M$. Image values are available only at (x_0, y_0), and all other values are zero and represent the lower limits of the sum.

By setting f or g in Eq. 7.4 as the impulse function and using the definition in Eq. 7.7 to perform certain processing, we can obtain the convolution of the impulse function "copy" corresponding to the impulse position on the value of this function. This feature is known as the "filter" feature of the convolution function. This aspect is significant for a unit impulse at the origin, expressed as $\delta(x, y)$:

$$\sum_{x=0}^{M-1}\sum_{y=0}^{N-1} s(x, y)\delta(x, y) = s(0, 0) \tag{7.8}$$

Using this simple tool, interesting and useful links between the spatial and frequency domain filtering can be established.

According to the Fourier transform formula, the Fourier transform of a unit impulse at the origin can be determined:

$$F(u,v)=\frac{1}{MN}\sum_{x=0}^{M-1}\sum_{y=0}^{N-1}\delta(x,y)\mathrm{e}^{-\mathrm{j}2\pi(ux/M+vy/N)}=\frac{1}{MN} \tag{7.9}$$

The second step pertains to Formula 7.8. Thus, it can be concluded that the Fourier transform of the impulse function at the origin of the domain is a real constant (which means that the phase angle is 0). If the impulse is placed arbitrarily, the transformation is expected to contain more complex components. The amplitude is the same; however, the nonzero phase angle after the transformation is expected to cause pulse translation.

Assuming that $f(x,y)=\delta(x,y)$ and performing the convolution defined in Formulas 7.4 and 7.7 can be used to obtain

$$f(x,y)*h(x,y)=\frac{1}{MN}\sum_{m=0}^{M-1}\sum_{n=0}^{N-1}\delta(m,n)h(x-m,y-n)=\frac{1}{MN}h(x,y) \tag{7.10}$$

where the attention variables are m and n, and the last step pertains to Eq. 7.8. Formulas 7.5 and 7.10 are combined to obtain

$$\begin{cases} f(x,y)*h(x,y)\Leftrightarrow F(u,v)H(u,v),\\ \delta(x,y)*h(x,y)\Leftrightarrow \Im[\delta(x,y)]H(u,v),\\ h(x,y)\Leftrightarrow H(u,v)\end{cases} \tag{7.11}$$

Using only the properties of the impulse function and convolution theorem, the filters in the spatial and frequency domains can be determined to form Fourier transform pairs. Thus, the filter in the frequency domain can be obtained by performing an inverse Fourier transform based on the filter in the spatial domain, and vice versa [2].

If the two filters are of the same size, filtering is usually more intuitive and effective in the frequency domain, whereas the spatial domain is suitable for smaller filters. Filtering based on Gaussian functions is of particular importance because their shapes are easy to determine, and the Fourier transform and inverse transform of the Gaussian function are real Gaussian functions. These functions are discussed here to limit the scope of variables and simplify the symbol representation.

The frequency domain is denoted by $H(u)$, and the Gaussian filter function is defined as $H(u)=A\mathrm{e}^{-u^2/2\sigma^2}$.

The relevant spatial domain filter is $h(x)=\sqrt{2\pi}\sigma A\mathrm{e}^{-2\pi^2\sigma^2x^2}$.

These two equations represent an important conclusion, expressed in two aspects. First, these expressions form a Fourier transform pair, and the components are real Gaussian functions, which is helpful in the analysis. Second, these functions have the following influence on each other: when $H(u)$ has an extremely wide profile (larger σ), $h(x)$ has an extremely narrow profile, and vice versa. When σ approaches infinity, $H(u)$ and $h(x)$ tend to a constant function and impulse function, respectively [3].

In the frequency domain, the correspondence between the frequency component of the image and visual effect of the image is intuitive. Certain image enhancement tasks are more difficult to represent and analyze in the spatial domains, but they can be expressed and analyzed easily in the frequency domain. Airspace filtering has several advantages in terms of concrete implementation and hardware design.

Certain differences exist between the airspace technology and frequency domain technology. For example, airspace technology, using either point operation or template operation, is based only on the nature of certain pixels. The frequency domain technique uses the data of all the pixels in the image every time and has global properties, which can better reflect the overall characteristics of the image, such as the overall contrast and average gray value [4].

7.2 Linear Filter

This section focuses on the transfer function response of the spatial domain filter operator and fundamental domain of the frequency domain. In the space operator section, the basic theory and properties of the Laplacian operator are introduced. In the frequency domain filtering section, the transfer function and impulse response of various filtering types are discussed.

7.2.1 Space Operator

In space domain template operation, mean smoothing operators and Laplacian operators belong to the linear filtering method. The mean smoothing operator uses the pixel values in the vicinity of the target cell to obtain the gray value of the point. The Laplacian operator is a second-order differential operator, a scalar, and an isotropic operation sensitive to grayscale mutation. Let $\nabla^2 f$ be a Laplace operator:

$$\nabla^2 f = \frac{\partial^2 f}{\partial x^2} + \frac{\partial^2 f}{\partial y^2} \tag{7.12}$$

For discrete digital image $f(x, y)$, the first derivative is

$$\begin{cases} \frac{\partial f(i,j)}{\partial x} = \Delta_x f(i, j) = f(i, j) - f(i-1, j), \\ \frac{\partial f(i,j)}{\partial y} = \Delta_y f(i, j) = f(i, j) - f(i, j-1) \end{cases} \tag{7.13}$$

The second-order partial derivative is

$$\frac{\partial^2 f(i, j)}{\partial x^2} = \Delta_x f(i+1, j) - \Delta_x f(i, j) = f(i+1, j) + f(i-1, j) - 2f(i, j),$$

$$\frac{\partial^2 f(i,j)}{\partial y^2} = \Delta_y f(i,j+1) - \Delta_y f(i,j) = f(i,j+1) + f(i,j-1) - 2f(i,j) \tag{7.14}$$

Therefore, the Laplace operator $\nabla^2 f$ is

$$\nabla^2 f = \frac{\partial^2 f}{\partial x^2} + \frac{\partial^2 f}{\partial y^2} = f(i-1,j) + f(i+1,j) + f(i,j+1) + f(i,j-1) - 4f(i,j) \tag{7.15}$$

To correct the image blur caused by the diffusion phenomenon, the following formula can be used for sharpening:

$$g(i,j) = f(i,j) - k\tau \nabla^2 f(i,j) \tag{7.16}$$

where $k\tau$ is the coefficient associated with the diffusion effect. The value of the coefficient must be reasonable. If the value is extremely large, the edge of the image will exhibit an overshoot. If the value is extremely small, the sharpening effect is not significant.

If $k\tau = 1$, the transformation formula is

$$g(i,j) = 5f(i,j) - f(i-1,j) - f(i+1,j) - f(i,j+1) - f(i,j-1) \tag{7.17}$$

The template is

$$\begin{bmatrix} 0 & -1 & 0 \\ -1 & 5 & -1 \\ 0 & -1 & 0 \end{bmatrix}$$

This Laplace sharpening operation can be converted into a template operation. Commonly used Laplace operator sharpening templates are

$$\begin{bmatrix} 0 & 1 & 0 \\ 1 & -4 & 1 \\ 0 & 1 & 0 \end{bmatrix}, \begin{bmatrix} 1 & 1 & 1 \\ 1 & -8 & 1 \\ 1 & 1 & 1 \end{bmatrix}, \begin{bmatrix} 0 & -1 & 0 \\ -1 & 4 & -1 \\ 0 & -1 & 0 \end{bmatrix}, \begin{bmatrix} -1 & -1 & -1 \\ -1 & 8 & -1 \\ -1 & -1 & -1 \end{bmatrix}$$

The Laplacian operator has two limitations: information loss of the edge occurs, and the Laplace operator corresponds to a second-order difference, which intensifies the image noise by two times. The advantage is that the operator is isotropic, that is, it exhibits rotation invariance [5]. Figure 7.3 compares the filtering effects of three versions of the Laplacian operator.

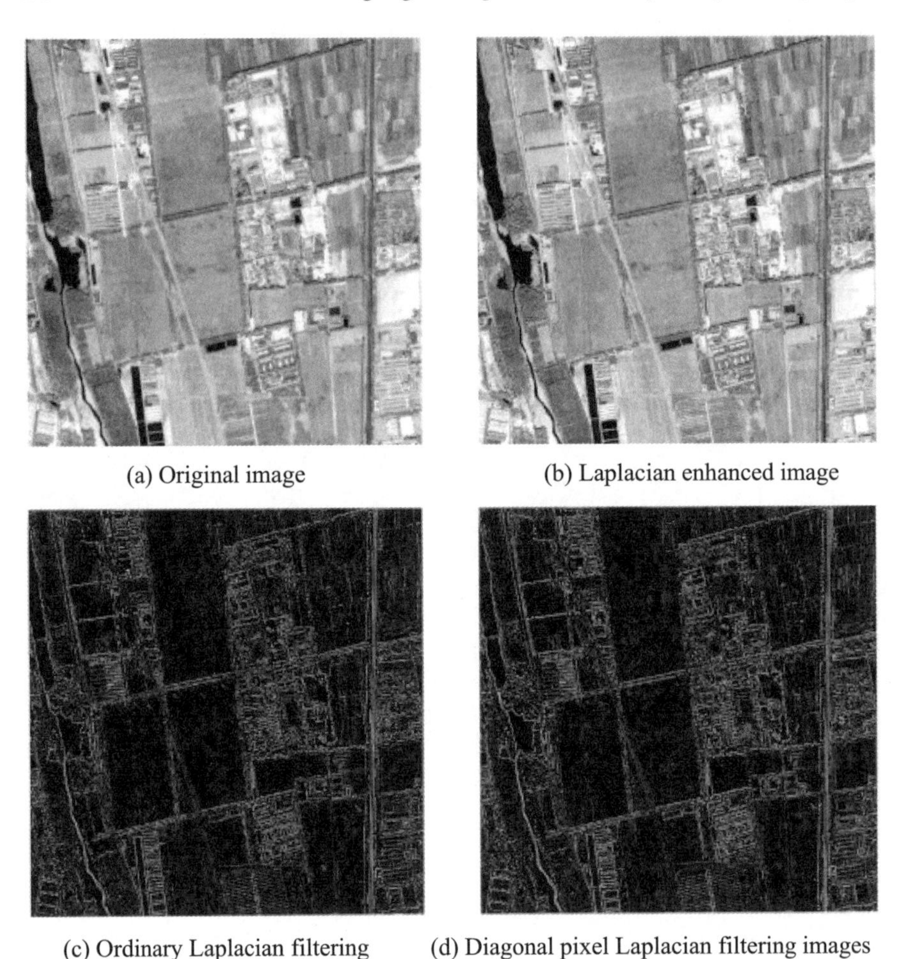

(a) Original image
(b) Laplacian enhanced image
(c) Ordinary Laplacian filtering
(d) Diagonal pixel Laplacian filtering images

Fig. 7.3 Comparison of Laplace operator filtering effects (GF1 full-color band image)

The algebraic coefficient of the Laplace enhancement template operator is 1. The overall energy of the image obtained by filtering the template does not change. For the digital image of any pixel gray value x_i, a 3×3 template is

$$\begin{bmatrix} 0 & a & 0 \\ b & c & d \\ 0 & e & 0 \end{bmatrix}$$

When the template moves on the image, pixel x_i coincides with five coefficients a, b, c, d, and e, and the contribution of the sum of the gray values of the whole image is $x_i(a + b + c + d + e)$. Regardless of the edge, the image is obtained after filtering the overall gray value $\sum x_i(a + b + c + d + e)$, and the sum of the gray value of the original image is $\sum x_i$; therefore, $= a + b + c + d + e = 1$. The sum of the coefficients of the template operator is 1, which is the necessary and sufficient condition to ensure the energy of the filtered digital image.

The Laplacian edge extraction operator, as well as the Roberts, Prewitt, Sobel and other operators, which are discussed later, are based on the mathematical principle that the derivative or partial derivative can be discretely differentiated in digital image processing. Therefore, the sum of the coefficients of the template must be zero and can be extracted when the grayscale changes. Because the energy of the image is mainly concentrated in the background part, the rate of change is small. The edge of the filter operator is filtered out after filtering the energy concentration of the location. Therefore, the edge extraction image is filtered by the edge extraction operator.

7.2.2 Basic Filters in the Frequency Domain

The four basic types of filters in the frequency domain are shown in Fig. 7.4.

1. Ideal Low-Pass Filter

In filter design, the objective of the ideal low-pass filter is to ensure that frequencies less than D_0 are filtered, and frequencies greater of D_0 are completely passed. D_0 is the cutoff frequency, defined as follows:

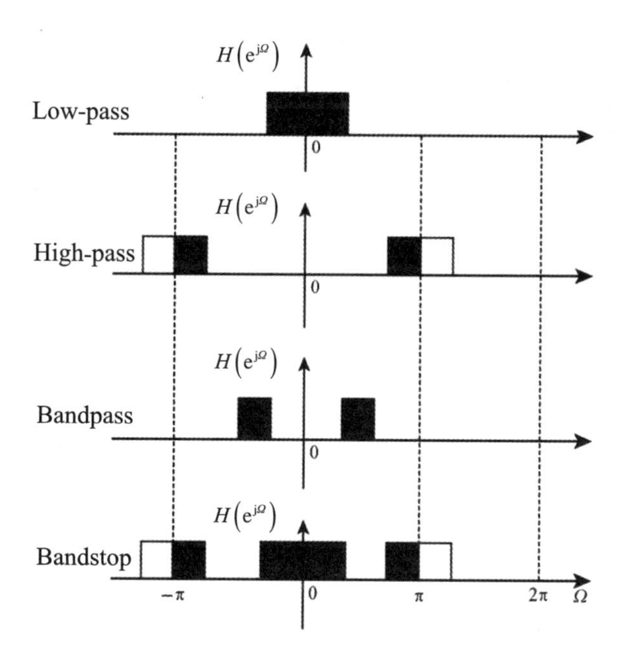

Fig. 7.4 2π distribution of the four filters

$$H(u, v) = \begin{cases} 1, & D(u, v) \le D_0, \\ 0, & D(u, v) > D_0 \end{cases} \tag{7.18}$$

$D(u, v)$ represents the distance from point (u, v) to the origin of the frequency plane:

$$D(u, v) = \left[(u - M/2)^2 + (v - N/2)^2\right]^{\frac{1}{2}} \tag{7.19}$$

M and N denote the frequency range after discretization, and ($M/2$, $N/2$) represents the intersection positions of 1/2 of the discrete frequency ranges u and v.

The energy of the ideal low-pass filter signal is analyzed. The total signal energy P_{T} is

$$P_{\mathrm{T}} = \sum_{u=0}^{M-1} \sum_{v=0}^{N-1} P(u, v) \tag{7.20}$$

If the transformation is centered on a circle, r is the radius of the circle that contains β percent of the energy:

$$\beta = 100\left[\sum_{u} \sum_{v} P(u, v)/P_{\mathrm{T}}\right] \tag{7.21}$$

where $(u^2 + v^2)^{\frac{1}{2}} < r$.

Since the real part $R(u, v)$ and imaginary part $I(u, v)$ of the Fourier transform decrease rapidly with increasing frequencies u and v, the energy decreases rapidly with increasing frequency. Therefore, the energy in the frequency domain plane is concentrated with reference to the frequency of a small circle. When D_0 increases, the energy rapidly decays. The high-frequency part carries less energy but contains rich borders and details. Therefore, the cutoff frequency D_0 decreases. Although the brightness is sufficient (due to the small energy loss), the image is blurred. As the filter radius increases, the amount of high-frequency detail information increases, and the degree of blurring decreases.

2. Ideal High-Pass Filter

A two-dimensional high-pass filter is defined as follows:

$$H(u, v) = \begin{cases} 0 & D(u, v) \le D_0, \\ 1 & D(u, v) > D_0 \end{cases} \tag{7.22}$$

where D_0 is the cutoff frequency length measured from the midpoint of the frequency rectangle. The perspective and profile are shown in Fig. 7.5. In contrast to the ideal low-pass filter, this filter completely removes all the spectral components in the circle with radius D_0. The spectrum outside the circle can pass through without loss.

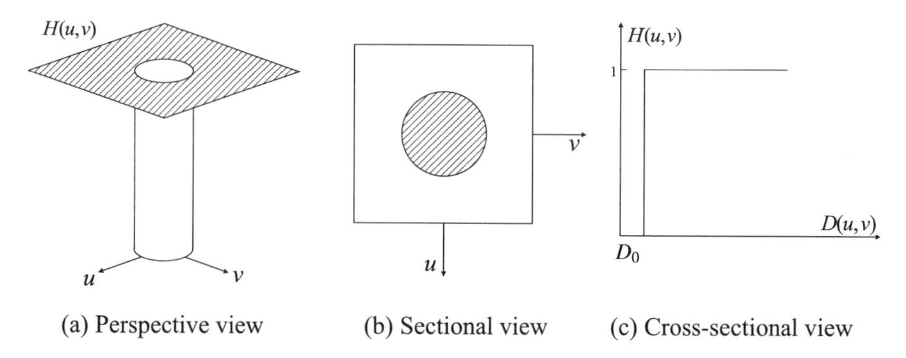

Fig. 7.5 Ideal high-pass filter

Figures 7.5a–c show the perspective, sectional, and cross-sectional view of a typical ideal high-pass filter, respectively.

3. Ideal Bandpass Filter

Assuming that a filter is to be implemented by convolution, it only allows the energy between f_1 and $f_2(f_1 < f_2)$ to pass through, and the required transfer function is

$$G(s) = \begin{cases} 1, & f_1 \leq |s| \leq f_2, \\ 0, & \text{Others} \end{cases} \tag{7.23}$$

$G(s)$ is a pair of rectangular pulses that can be convoluted by rectangular pulses and impulses. Let

$$s_0 = \frac{1}{2}(f_1 + f_2) \text{ and } \Delta s = f_2 - f_1 \tag{7.24}$$

The transfer function of the ideal bandpass filter is

$$G(s) = \Pi\left(\frac{s}{\Delta s}\right) * [\delta(s - s_0) + \delta(s + s_0)] \tag{7.25}$$

Using this transfer function, the impulse response of the ideal bandpass filter can be expressed as follows:

$$g(t) = \Delta s \frac{\sin(\pi \Delta st)}{\pi \Delta st} 2\cos(2\pi s_0 t) = 2\Delta s \frac{\sin(\pi \Delta st)}{\pi \Delta st} \cos(2\pi s_0 t) \tag{7.26}$$

Since $\Delta s < s_0$, Eq. 7.26 describes a cosine wave with a frequency s_0 surrounded by sin $(x)/x$. This impulse response is shown in Fig. 7.6. The number of cosine periods contained between the two zero-crossing points of the envelope depends on the relationship between s_0 and S. If s_0 is fixed and Δs decreases (i.e., narrow passband), a more cosine period will exist between the two zero crossings of the

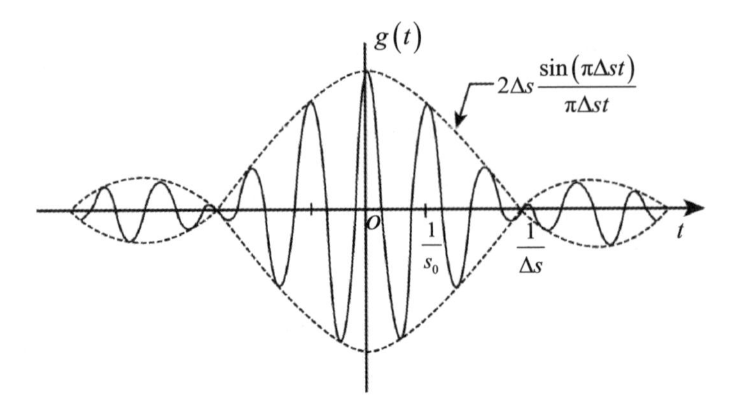

Fig. 7.6 Impulsive response of ideal band-stop filter

envelope. When Δs tends to 0, the impulse response tends to a cosine function. In this extreme case, the convolution operation represents a cross-correlation between the input signal and cosine signal with frequency s_0.

4. Ideal Band-Stop Filter

The ideal band-stop filter blocks the frequency between f_1 and f_2 and allows other frequencies to pass. The transfer function can be expressed as follows:

$$G(s) = \begin{cases} 0, & f_1 \le |s| \le f_2, \\ 1, & \text{Others} \end{cases} \tag{7.27}$$

For convenience, let s_0 be the center frequency, and Δs denote the width of the blocked band. The transfer function can be expressed as 1 (Fig. 7.6).

Subtracting the transfer function of the bandpass filter yields

$$G(s) = 1 - \Pi\left(\frac{s}{\Delta s}\right) * [\delta(s - s_0) + \delta(s + s_0)] \tag{7.28}$$

It can be concluded that the impulse response is

$$g(t) = \delta(t) - 2\Delta s\frac{\sin(\pi\Delta st)}{\pi\Delta st}\cos(2\pi s_0 t) \tag{7.29}$$

The nature of this response varies with the bandwidth and center frequency, similar to bandpass filters. If Δs is extremely small, the band-stop filter is a "notch filter".

In addition to the ideal filter, there exist Butterworth filters, exponential filters, ladder filters, and Gaussian filters. However, these filters are not discussed.

7.3 Optimal Linear Filter

This section focuses on the design of the Wiener filter and matched detector with the relevant optimal criteria. The design principle of the two filters is based on the model, followed by the determination of the optimal criterion, derivation in the frequency domain, and solution of the transfer function.

7.3.1 Wiener Filter

The Wiener filter is a classic linear noise reduction filter, and the mean square error for the function equation must be minimized. The model of the Wiener filter is shown in Fig. 7.7 [6]. Assuming that the observed signal $h(t)$ is composed of $s(t)$ and additive noise signal $n(t)$, the Wiener filter is designed to obtain a model that minimizes the noise signal and restores the useful signal. The impulse response is $h(t)$, and the filter output is $y(t)$. The Wiener filter design is selected such that $y(t)$ ensures that $s(t)$ approaches the impulse response $y(t)$.

1. Optimal Criterion for Wiener Filter

The signal error at the output of the filter is defined as

$$e(t) = s(t) - y(t) \tag{7.30}$$

The difference between the actual output and desired output is a function of time. If the impulse response $h(t)$ is appropriately selected, the error signal is expected to be small. The mean square error is a measure of the mean error, i.e.,

$$\text{MSE} = \varepsilon\left\{e^2(t)\right\} = \int_{-\infty}^{+\infty} e^2(t)\,\mathrm{d}t \tag{7.31}$$

The upper sign is established considering the fact that the error signal is a linear combination of random traversal and is thus also an ergodic random variable. The mean square error defined in 7.31 is expanded:

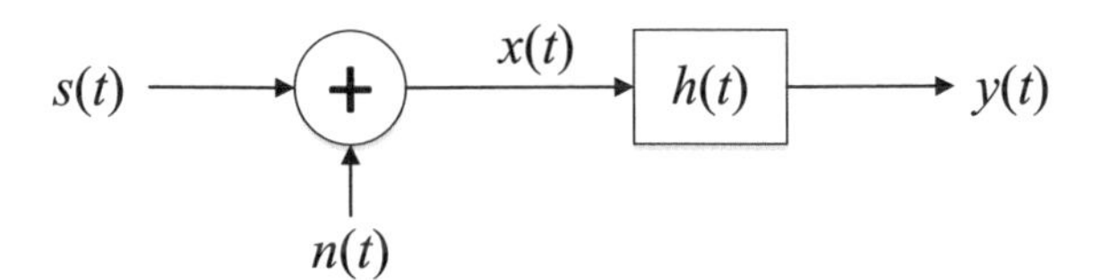

Fig. 7.7 Wiener filter model

$$\begin{aligned}\mathrm{MSE} &= \varepsilon\left\{e^2(t)\right\} = \varepsilon\left\{[s(t) - y(t)]^2\right\} = \varepsilon\left\{s^2(t) - 2s(t)y(t) + y^2(t)\right\} \\ &= \varepsilon\left\{s^2(t)\right\} - 2\varepsilon\{s(t)y(t)\} + \varepsilon\left\{y^2(t)\right\} = T_1 + T_2 + T_3\end{aligned} \tag{7.32}$$

T_1, T_2 and T_3 are introduced to consider these three items independently. T_1 is expressed in the integral form:

$$T_1 = \varepsilon\left\{s^2(t)\right\} = \int_{-\infty}^{\infty} s^2(t)\,\mathrm{d}t = R_s(0) \tag{7.33}$$

This term represents the value of the autocorrelation function of $s(t)$ at $\tau = 0$, and thus, the value is known at the beginning. If $y(t)$ is expressed as a convolution of $x(t)$ and $h(t)$, the second term can be expanded as

$$T_2 = -2\varepsilon\left\{s(t)\int_{-\infty}^{\infty} h(\tau)x(t-\tau)\,\mathrm{d}\tau\right\} \tag{7.34}$$

Since the expected operator is the integral of time, the order of the above formula is modified to obtain

$$T_2 = -2\int_{-\infty}^{\infty} h(\tau)\varepsilon\{s(t)x(t-\tau)\}\,\mathrm{d}\tau \tag{7.35}$$

where the expectation of the integral is the cross-correlation function of $s(t)$ and $x(t)$:

$$T_2 = -2\int_{-\infty}^{\infty} h(\tau)R_{xs}(\tau)\,\mathrm{d}\tau \tag{7.36}$$

T_3 is expanded as the expectation of two convolution products:

$$\begin{aligned}T_3 &= \varepsilon\left\{\int_{-\infty}^{\infty} h(\tau)x(t-\tau)\mathrm{d}\tau \int_{-\infty}^{\infty} h(u)x(t-u)\mathrm{d}u\right\} \\ &= \int_{-\infty}^{\infty}\int_{-\infty}^{\infty} h(\tau)h(u)\varepsilon\{x(t-\tau)x(t-u)\}\,\mathrm{d}\tau\mathrm{d}u\end{aligned} \tag{7.37}$$

If the variable is replaced ($v = t - u$) in the expected operator, the term is

$$\varepsilon\{x(t-\tau)x(t-u)\} = \varepsilon\{x(\ \ + u - \tau)x(v)\} \tag{7.38}$$

This term represents the value of the autocorrelation function of $x(t)$ at $u-t$. The third item is expressed as

$$T_3 = \int_{-\infty}^{\infty}\int_{-\infty}^{\infty} h(\tau)h(u)R_x(u-\tau)\,\mathrm{d}\tau\mathrm{d}u \tag{7.39}$$

The MSE can be defined as

$$\mathrm{MSE} = R_s(0) - 2\int_{-\infty}^{\infty} h(\tau)R_{xs}(\tau)\,\mathrm{d}\tau + \int_{-\infty}^{\infty}\int_{-\infty}^{\infty} h(\tau)h(u)R_x(u-\tau)\,\mathrm{d}\tau\mathrm{d}u \tag{7.40}$$

2. Mean Square Error Analysis

$h_0(t)$ is used to denote a particular function that minimizes the MSE. In general, the difference between arbitrary $h(t)$ and $h_0(t)$ values is different. A function $g(t)$ is defined to represent this difference, i.e.,

$$h(t) = h_0(t) + g(t) \tag{7.41}$$

where $h(t)$ is an arbitrarily selected suboptimal impulse response function, and $g(t)$ is a function chosen to satisfy the equation. The expression of $h(t)$ is substituted into Formula 7.40 to yield

$$\begin{aligned}\mathrm{MSE} = {} & R_s(0) - 2\int_{-\infty}^{\infty} [h_0(\tau) + g(\tau)]R_{xs}(\tau)\,\mathrm{d}\tau \\ & + \int_{-\infty}^{\infty}\int_{-\infty}^{\infty} [h_0(\tau) + g(\tau)][h_0(u) + g(u)]R_x(u-\tau)\,\mathrm{d}\tau\mathrm{d}u\end{aligned} \tag{7.42}$$

This formula is expanded to obtain 7 terms:

$$\begin{aligned}\mathrm{MSE} = {} & R_s(0) - 2\int_{-\infty}^{\infty} h_0(\tau)R_{xs}(\tau)\,\mathrm{d}\tau + \int_{-\infty}^{\infty}\int_{-\infty}^{\infty} h_0(t)h_0(u)R_x(u-\tau)\,\mathrm{d}\tau\mathrm{d}u \\ & + \int_{-\infty}^{\infty}\int_{-\infty}^{\infty} h_0(\tau)g(u)R_x(u-\tau)\,\mathrm{d}\tau\mathrm{d}u + \int_{-\infty}^{\infty}\int_{-\infty}^{\infty} h_0(u)g(\tau)R_x(u-\tau)\,\mathrm{d}\tau\mathrm{d}u \\ & - 2\int_{-\infty}^{\infty} g(\tau)R_{xs}(\tau)\,\mathrm{d}\tau + \int_{-\infty}^{\infty}\int_{-\infty}^{\infty} g(\tau)g(u)R_x(u-\tau)\,\mathrm{d}\tau\mathrm{d}u\end{aligned} \tag{7.43}$$

Comparison of the first three terms of this formula and the first three terms of Eq. 7.40 indicates that the sum of these three terms (denoted by MSE_0) represents the mean square error generated using the optimal impulse response function. Since the autocorrelation function $R(u-\tau)$ is an even function, the fourth and fifth terms of the above equation are equal, and they are combined with the sixth term as

$$\mathrm{MSE}=\mathrm{MSE}_0+2\int_{-\infty}^{\infty} g(u)\left[\int_{-\infty}^{\infty} h_0(\tau)R_x(u-\tau)\mathrm{d}\tau - R_{xs}(u)\right]\mathrm{d}u + \int_{-\infty}^{\infty}\int_{-\infty}^{\infty} g(u)g(\tau)R_x(u-\tau)\mathrm{d}u\mathrm{d}\tau = \mathrm{MSE}_0 + T_4 + T_5 \tag{7.44}$$

T_4 and T_5 are introduced in the interest of conciseness. We prove that T_5 is nonnegative. The autocorrelation function R is written in the integral form:

$$T_5=\int_{-\infty}^{\infty}\int_{-\infty}^{\infty} g(u)g(\tau)\int_{-\infty}^{\infty} x(t-\tau)x(t-u)\mathrm{d}t\mathrm{d}u\mathrm{d}\tau \tag{7.45}$$

The integral of points is changed to obtain

$$T_5=\int_{-\infty}^{\infty}\int_{-\infty}^{\infty} g(u)x(t-u)\mathrm{d}u\int_{-\infty}^{\infty} g(\tau)x(t-\tau)\mathrm{d}\tau\mathrm{d}t \tag{7.46}$$

If the convolution function of $z(t)$ is defined in terms of $g(t)$ and $x(t)$, the equation can be expressed as

$$T_5=\int_{-\infty}^{\infty} z^2(t)\mathrm{d}t \geq 0 \tag{7.47}$$

Next, we consider the MSE:

$$\mathrm{MSE}=\mathrm{MSE}_0+2\int_{-\infty}^{\infty} g(u)\left[\int_{-\infty}^{\infty} h_0(\tau)R_x(u-\tau)\mathrm{d}\tau - R_{xs}(u)\right]\mathrm{d}u + T_5 \tag{7.48}$$

where MSE_0 is the mean square error under the optimal condition, and T_5 is independent of h_0 and nonnegative. The MSE must be minimized to satisfy the conditions. One approach is to ensure that the square brackets have zero values for all u values. This aspect eliminates the T_4 term to ensure that $\mathrm{MSE}_0 \leq \mathrm{MSE}$.

The following discussion proves the necessity. Assume that for certain u values, the square brackets in the value are nonzero. Since $g(u)$ is an arbitrary function, when the item in the parentheses is a positive value, $g(u)$ can be a negative number with a greater absolute value, and several values of T may be large negative numbers and vice versa. In this case, the MSE is smaller than MSE_0, which violates the definition. Therefore, it can be asserted that the bracketed item must be zero. In other words,

$$R_{xs}(\tau) = \int_{-\infty}^{\infty} h_0(u) R_x(u - \tau)\,\mathrm{d}u \tag{7.49}$$

This aspect is a necessary condition to minimize the variance. It is easy to conclude that the above equation is also a sufficient condition for optimizing the filter. Thus, Eq. 7.48 becomes

$$\mathrm{MSE} = \mathrm{MSE}_0 + T_5, \quad T_5 \geq 0 \tag{7.50}$$

For any linear system, the cross-correlation between the input and output can be expressed as

$$R_{xy}(\tau) = h(u) * R_x(u) \tag{7.51}$$

where $R_x(u)$ is the autocorrelation function of the input signal. The right side of Eq. 7.43 is a convolutional integral that can be written as

$$R_{xs}(\tau) = h_0(u) * R_x(u) = R_{xy}(\tau) \tag{7.52}$$

In this scenario, the optimal impulse response is associated with the autocorrelation of the input signal and cross-correlation between the input signal and desired signal. The Fourier transform is implemented on both sides of the above equation

$$P_{xs}(s) = H_0(s) P_x(s) = P_{xy}(s) \tag{7.53}$$

In other words,

$$H_0(s) = \frac{P_{xs}(s)}{P_x(s)} \tag{7.54}$$

3. Wiener Filter Design

Based on the above analysis, the steps to design the Wiener filter can be summarized as follows:

(1) Digitize the sample of the input signal $s(t)$; (2) Find the autocorrelation of the input sample and obtain an estimate of $R_x(\tau)$; (3) Calculate the Fourier transform of $R_x(\tau)$ to obtain $P_x(s)$; (4) Digitize a sample of the input signal in the absence of

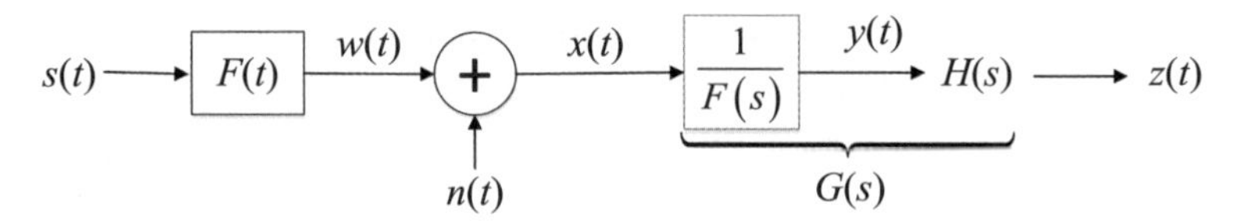

Fig. 7.8 Wiener deconvolution

noise; (5) Determine the signal samples (no noise) and input sample cross-correlation to estimate $R_{xs}(\tau)$; (6) Calculate $P_{xs}(s)$ considering the $R_{xs}(\tau)$ Fourier transform; (7) Calculate the transfer function of the optimal filter as $H_0(s) = \frac{P_{xs}(s)}{P_x(s)}$; (8) If the filter must be subjected to convolution, calculate the Fourier transform of $H_0(s)$ to obtain the impulse response $h_0(t)$ of the optimal linear estimator.

4. Wiener Deconvolution

Due to the inertia of the hardware system, the desired signal $s(t)$ is first degraded by a linear system with an impulse response of $f(t)$, and its output is contaminated by the noise source $n(t)$ to form the observed signal $x(T)$. Deconvolution is performed before Wiener filtering (Fig. 7.8).

Our aim is to design a linear filter $g(t)$ that can simultaneously decelerate the unwanted impulse response $f(t)$ and suppress noise $n(t)$. In the figure, $g(t)$ is composed of a deconvolution filter and a Wiener filter with impulse response $h_0(t)$. Therefore, the spectrum of the observed signal is

$$X(s) = W(s) + N(s) = F(s)S(s) + N(s) \tag{7.55}$$

Furthermore, assuming that $F(s)$ is nonzero, the spectrum sent to the Wiener filter is

$$Y(s) = \frac{X(s)}{F(s)} = S(s) + \frac{N(s)}{F(s)} = S(s) + K(s) \tag{7.56}$$

In the presence of noise and irrelevant signals, the transfer function of the Wiener filter is

$$H_0(s) = \frac{P_s(s)}{P_s(s) + P_k(s)} = \frac{|S(s)|^2}{|S(s)|^2 + \left|\frac{N(s)}{F(s)}\right|^2} \tag{7.57}$$

Thus, the transfer function of the optimal mean square deconvolution filter is

$$G(s) = \frac{H_0(s)}{F(s)} = \frac{1}{F(s)}\left[\frac{P_s(s)}{P_s(s) + P_k(s)}\right] = \frac{F^*(s)P_s(s)}{|F(s)|^2 P_s(s) + P_n(s)} \tag{7.58}$$

7.3.2 Matched Detector

The model of the matched detector is shown in Fig. 7.9. The observed signal $x(t)$ is composed of the contamination of signal $m(t)$ by additive noise $n(t)$. $x(t)$ is input to the impulse response $k(t)$ for the linear filter to obtain output $y(t)$.

The output of the filter can be used to monitor if $m(t)$ has been added. The model can be expressed as

$$y(t) = [m(t) + n(t)] * k(t) = m(t) * k(t) + n(t) * k(t) \tag{7.59}$$

The model is equivalent to the system shown in Fig. 7.10. In other words, the addition of $m(t)$ and $n(t)$ before or after passing through the filter does not influence the whole system [7].

The two component outputs are defined as

$$u(t)=m(t)*k(t)和v(t)=n(t)*k(t) \tag{7.60}$$

$u(t)$ and $v(t)$ denote the signal and noise after filtering, respectively.

1. Optimal Criteria

The ratio of the average signal-to-noise power at the output to the value of zero is set as a measure of the filter performance, i.e.,

$$\rho = \frac{\varepsilon\{u^2(0)\}}{\varepsilon\{v^2(0)\}} \tag{7.61}$$

The prototype signal $m(t)$ is often a narrow function centered on the origin. We want the output power to be maximized at $t = 0$ (i.e., the moment at which the signal appears), and before and after that, when the signal does not exist, the output amplitude is reasonably small. According to the invariant nature, if the signal $m(t - t_1)$

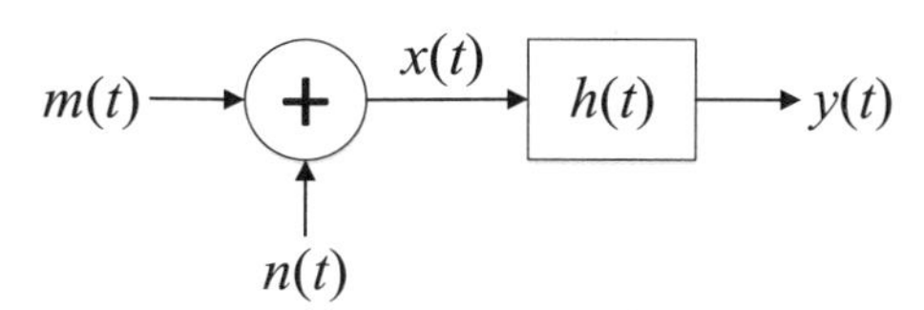

Fig. 7.9 Matched detector model

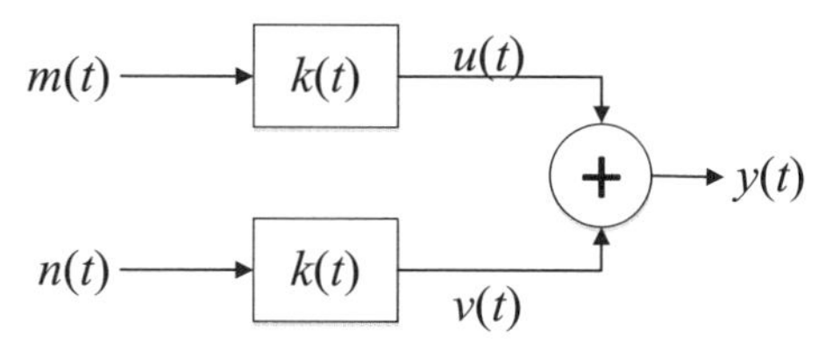

Fig. 7.10 Equivalent filter model for Fig. 7.9

arrives at any other time t_1, the output amplitude of the filter increases at time t_1, which means that the signal appears. If ρ is large, the magnitude of the output $y(t)$ depends to a large extent on whether $m(t)$ is present, and it is insensitive to the fluctuation of noise $n(t)$. Therefore, ρ is maximized to optimization the criteria. Since $u(t)$ is deterministic, it is possible to remove the desired operator in the molecule and express the above formula as

$$\rho = \frac{u^2(0)}{\varepsilon\{v^2(t)\}} = \frac{[m(t) * k(t)]^2}{\varepsilon\{[n(t) * k(t)]^2\}} = \frac{\left[\Im^{-1}\{M(s) * K(s)\}\right]^2}{\varepsilon\{[n(t) * k(t)]^2\}} = \frac{\rho_n}{\rho_d} \tag{7.62}$$

After the introduction of ρ_n and ρ_d, the numerator and denominator can be separately considered. First, the denominator is presented as two volumes:

$$\rho_d = \varepsilon\left\{\int_{-\infty}^{\infty} k(q)n(t-q)\mathrm{d}q \int_{-\infty}^{\infty} k(\tau)n(t-\tau)\mathrm{d}\tau\right\} \tag{7.63}$$

Since the expectation is the integral of time and impulse response $k(t)$ is not a random signal, we can change the order as

$$\rho_d = \int_{-\infty}^{\infty}\int_{-\infty}^{\infty} k(q)k(\tau)\varepsilon\{n(t-q)n(t-\tau)\}\mathrm{d}q\mathrm{d}\tau \tag{7.64}$$

The expectation of the integral pertains to the autocorrelation function $R_n(\tau - q)$ of the noise, which is the inverse Fourier transform of the noise power spectrum $P_n(s)$; therefore,

$$\varepsilon\{n(t-q)n(t-\tau)\} = R_n(\tau - q) = \int_{-\infty}^{\infty} P_n(s)\mathrm{e}^{\mathrm{j}2\pi s(\tau-q)}\mathrm{d}s \tag{7.65}$$

Therefore, the denominator of ρ can be expressed as

$$\rho_d = \int_{-\infty}^{\infty}\int_{-\infty}^{\infty} k(q)k(\tau)\int_{-\infty}^{\infty} P_n(s)\mathrm{e}^{\mathrm{j}2\pi s(\tau-q)}\mathrm{d}s\mathrm{d}q\mathrm{d}\tau \tag{7.66}$$

By decomposing the exponential terms and changing the order of points, we obtain

$$\rho_d = \int_{-\infty}^{\infty} P_n(s)\left[\int_{-\infty}^{\infty} k(q)\mathrm{e}^{-\mathrm{j}2\pi sq}\mathrm{d}q \int_{-\infty}^{\infty} k(\tau)\mathrm{e}^{\mathrm{j}2\pi s\tau}\mathrm{d}\tau\right]\mathrm{d}s \tag{7.67}$$

The terms in square brackets are the product of two inverse Fourier transformations, $K(s)$ and $K(-s)$. Moreover, since the impulse response function is a real function, the transfer function $K(s)$ is of the Hermite type, and $K(-s) = K^*(s)$. Therefore, the items in square brackets can be simplified as

$$K(s)K(-s) = K(s)K^*(s) = |K(s)|^2 \tag{7.68}$$

Substituting this equation into ρ and expanding the Fourier transform term in the molecule yields the following signal-to-noise power ratio:

$$\rho = \frac{\left[\int_{-\infty}^{\infty} K(s)M(s)\mathrm{d}s\right]^2}{\int_{-\infty}^{\infty} |K(s)|^2 P_n(s)\mathrm{d}s} \tag{7.69}$$

This expression must be maximized. Using the Schwarzian inequality principle, it is easy to deduce

$$\rho \leq \int_{-\infty}^{\infty} \frac{|M(s)|^2}{P_n(s)}\mathrm{d}s \tag{7.70}$$

and thus

$$\rho_{\max} = \int_{-\infty}^{\infty} \frac{|M(s)|^2}{P_n(s)}\mathrm{d}s \tag{7.71}$$

which is the necessary condition for maximizing ρ.

2. Transfer Function

The optimal transfer function is obtained by establishing a form and proving that it maximizes. Assume that the optimal transfer function is

$$K_o(s) = C\frac{M^*(s)}{P_n(s)} \tag{7.72}$$

where C is an arbitrary constant. This form is substituted into the general form of ρ:

$$\rho = \frac{\left|\int_{-\infty}^{\infty} C\frac{M^*(s)}{P_n(s)}M(s)\mathrm{d}s\right|^2}{\int_{-\infty}^{\infty} C^2\frac{M(s)^*M(s)}{P_n(s)^*P_n(s)}P_n(s)\mathrm{d}s} \tag{7.73}$$

Because $P_n(s)$ is a real function, $P_n(s)^* = P_n(s)$, and the numerator is the square of the denominator. Therefore, ρ can be simplified:

$$\rho = \int_{-\infty}^{\infty} \frac{|M(s)|^2}{P_n(s)} \mathrm{d}s = \rho_{\max} \tag{7.74}$$

This expression satisfies the necessary conditions for optimization. This aspect indicates that the transfer function set in the formula can maximize the signal-to-noise ratio power ratio at the output of the filter. The magnitude of the transfer function is proportional to the ratio of the amplitude of the signal to the noise power and is a function of the frequency. A constant C is present because our initial effort is to maximize the signal-to-noise ratio at the output.

7.4 Nonlinear Filter

This section introduces the two main categories of nonlinear filtering methods: from the perspective of the spatial domain (Roberts operator, Prewitt operator, Sobel operator, and median filter) and frequency domain (homomorphic filtering).

7.4.1 Space Operator

Commonly used first-order differential operators are the Roberts, Prewitt, and Sobel operators. In image processing, the first-order differential is achieved using the gradient method. For function $f(x, y)$, the gradient on (x, y) is defined by the following two-dimensional column vector:

$$\nabla f(x, y) = \begin{bmatrix} G_x \\ G_y \end{bmatrix} = \begin{bmatrix} \frac{\partial f}{\partial x} \\ \frac{\partial f}{\partial y} \end{bmatrix} \tag{7.75}$$

Formula (7.12) indicates the size and orientation of the gradient vector.

For discrete digital image processing, the size of the gradient is commonly considered, referred to as "gradient" herein. The first-order partial derivative is expressed by the first-order difference approximation:

$$G_x = f(x+1, y) - f(x, y), \quad G_y = f(x, y+1) - f(x, y) \tag{7.76}$$

When this term is evaluated for the whole image, the amount of computation is extremely large. Therefore, in practice, the following expression is used to approximate the modulus of the gradient:

$$|\nabla f| = |G_x| + |G_y| \tag{7.77}$$

This simplified gradient calculation method avoids the complicated operation of squares and prescriptions; however, anisotropy is introduced [8].

1. Roberts Operator

The Roberts operator template is expressed as

$$T_1 = \begin{vmatrix} -1 & 0 \\ 0 & 1 \end{vmatrix}, T_2 = \begin{vmatrix} 0 & 1 \\ -1 & 0 \end{vmatrix}$$

In this approach, the cross-differential is used instead of the derivative, expressed as

$$\begin{aligned} G_x &= f(x+1, y+1) - f(x, y), \\ G_y &= f(x, y+1) - f(x+1, y) \end{aligned} \tag{7.78}$$

The first and second operators have a strong echo on the edge near $45°$ and $-45°$, respectively. G_x and G_y are obtained by filtering the images with these two templates. The final Roberts cross-gradient image is $|G_x| + |G_y|$.

2. Prewitt Operator

The Prewitt operator is an edge template operator, which was proposed by Prewitt in 1970, and is composed of ideal edge subimages extending in horizontal and vertical directions. For the digital image $f(x, y)$, the Prewitt operator is defined as follows:

$$\begin{aligned} G_x &= f(x+1, y-1) + f(x+1, y) + f(x+1, y+1) \\ &\quad + f(x-1, y-1) + f(x-1, y) + f(x-1, y+1), \\ G_y &= f(x-1, y+1) + f(x, y+1) + f(x+1, y+1) \\ &\quad + f(x-1, y-1) + f(x, y-1) + f(x+1, y-1) \end{aligned} \tag{7.79}$$

The operator considers the relationship of the neighborhood points more than the Roberts operator. The template size increases from 2×2 to 3×3, and the template is expressed as

$$T_1 = \begin{vmatrix} -1 & -1 & -1 \\ 0 & 0 & 0 \\ 1 & 1 & 1 \end{vmatrix}, T_2 = \begin{vmatrix} -1 & 0 & 1 \\ -1 & 0 & 1 \\ -1 & 0 & 1 \end{vmatrix} \tag{7.80}$$

3. Sobel Operator

The Sobel operator uses the gradient value of the neighborhood of the pixel to calculate the gradient of the pixel and sets it according to a certain threshold. Equation (7.77) represents the gradient of the image. If d is zero, the image has a longitudinal edge.

The Sobel operator is characterized by a symmetric first-order difference, a 3×3 operator template that is composed of two convolution cores. The template is expressed as

$$T_1 = \begin{vmatrix} -1 & -2 & -1 \\ 0 & 0 & 0 \\ 1 & 2 & 1 \end{vmatrix}, \ T_2 = \begin{vmatrix} -1 & 0 & 1 \\ -2 & 0 & 2 \\ -1 & 0 & 1 \end{vmatrix} \tag{7.81}$$

Figure 7.11 shows the result of the processing of different operators. Usually, one operator checks that the vertical edge has the largest response, while the other operators checks the horizontal edge for the maximum response. The maximum value of the two convolutions is set as the output value of the pixel, and the result of the operation is an edge image.

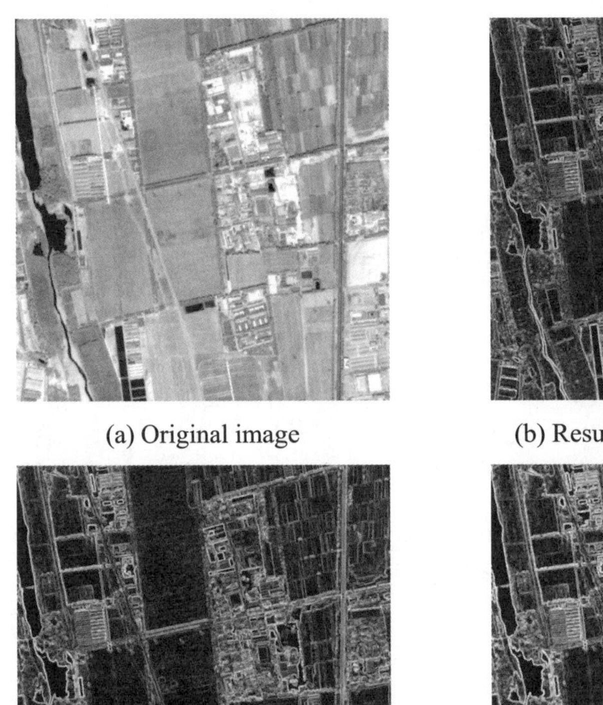

(a) Original image

(b) Results of Roberts operator

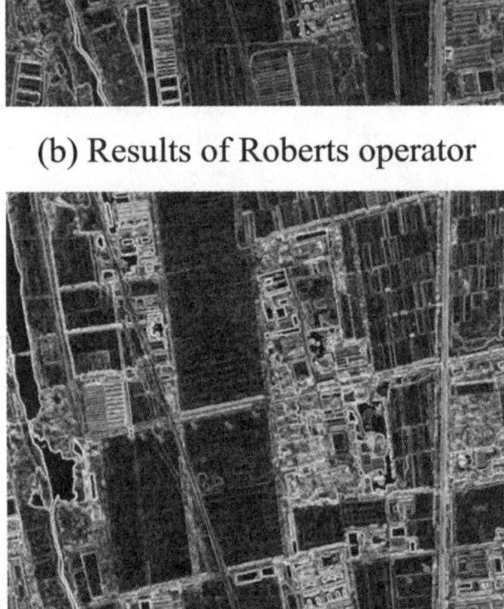

(c) Results of Prewitt operator

(d) Results of Sobel operator

Fig. 7.11 Comparison of multiple operator processing results

4. Median Filter

Median filtering is a nonlinear digital filter technique that is often used to remove noise from images or other signals. The median filter is particularly effective for the removal of salt and pepper noise.

Since the median filter involves the process of convolution of the template window, as the size of the template window increases, the pixel value within the image gradually increases owing the influence of the surrounding pixels, and the whole image becomes increasingly blurred. Therefore, for the median filter, the window size must be appropriately selected [9]. Figure 7.12 compares the processing of the median filter and mean filter for the same image.

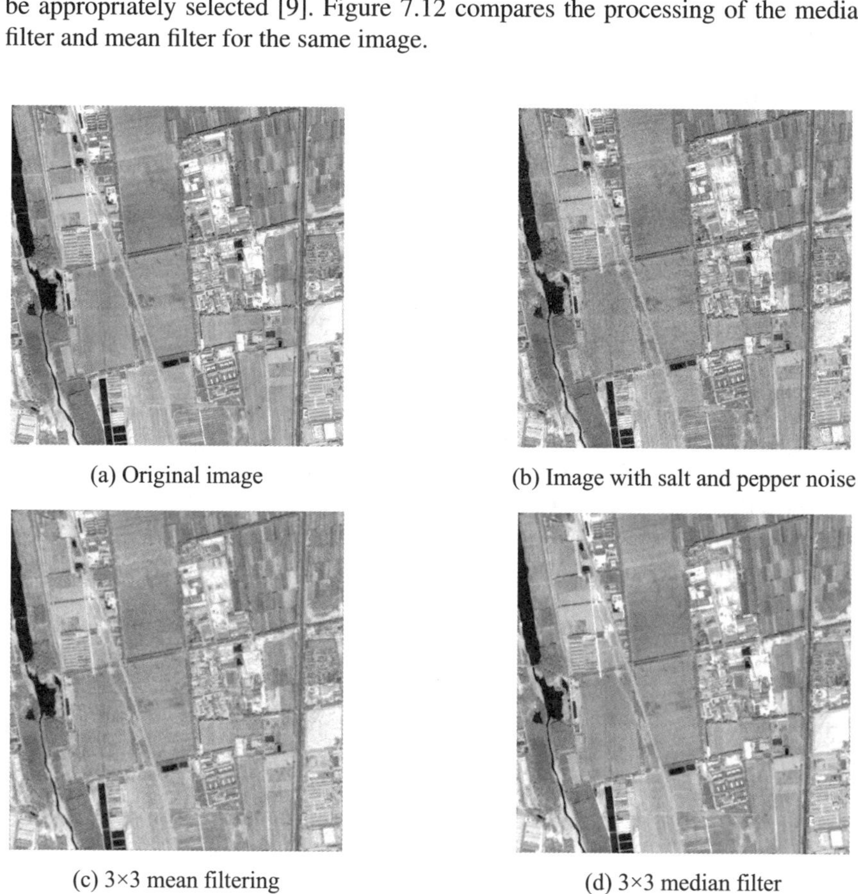

(a) Original image

(b) Image with salt and pepper noise

(c) 3×3 mean filtering

(d) 3×3 median filter

Fig. 7.12 Comparison of median filtering and mean filtering of the same image

7.4.2 Homomorphic Filtering

Homomorphic filtering is based on the image illumination/reflectivity model as the basis for frequency domain processing, and the compression brightness range and enhanced contrast are used to enhance the image quality. Using this method, image processing can mimic the human eye in terms of the brightness response of the nonlinear characteristics to avoid performing the direct Fourier transform of the image distortion.

Homomorphic filters are nonlinear image filtering methods that decrease low frequencies and increase high frequencies, thereby reducing illumination variations and sharpening edges or details. The homomorphic filtering method can be described by the following procedure.

An image $f(x, y)$ can be represented [10] as the product of the irradiation component $i(x, y)$ and reflection component $r(x, y)$

$$f(x, y) = i(x, y) \cdot r(x, y) \tag{7.82}$$

Natural algorithm is applied to both sides of the above formula:

$$\ln f(x, y) = \ln i(x, y) + \ln r(x, y) \tag{7.83}$$

The Fourier transform is applied to this formula.

$$F(u, v) = I(u, v) + R(u, v) \tag{7.84}$$

In this manner, we implement a Fourier transform of the high-frequency and low-frequency components of different filtering functions $H(u, v)$, weaken the low-frequency components and enhance the high-frequency components. After processing, $F(u, v)$ can be obtained:

$$H(u, v) \cdot F(u, v) = H(u, v) \cdot I(u, v) + H(u, v) \cdot R(u, v) \tag{7.85}$$

Reverse transform to the airspace is obtained:

$$h_f(x, y) = h_i(x, y) + h_r(x, y) \tag{7.86}$$

This formula indicates that the enhanced image is composed of the corresponding lighting components and reflection components of the two parts. On both sides of Formula 7.84, we apply the index:

$$g(x, y) = \exp\left|h_f(x, y)\right| = \exp|h_i(x, y)| \cdot \exp|h_r(x, y)| \tag{7.87}$$

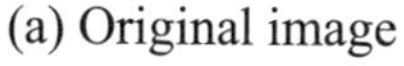

(a) Original image (b) Homomorphic filtered image

Fig. 7.13 Comparison of the results of homomorphic filtering (GF1 full-color band image)

Figure 7.13 shows that remote sensing images with different spatial resolutions are differently processed using the homomorphic filter [11]. If the illumination in the image is uniform, the homomorphic filtering effect is not significant. Instead, homomorphic filtering helps show the details of the darkness in the image [12].

7.5 Summary

Filtering is a key basic step in remote sensing image processing. The filtering performance directly affects the final accuracy of remote sensing image processing. The logic of the most important mathematical basis in this chapter is shown in Fig. 7.14 and Table 7.1.

This chapter introduces filters, which can be classified as linear or nonlinear using Eq. (7.3).

As the image is different from the general signal, after determining the classification (spatial or frequency domain; linear or nonlinear), it is necessary to clarify the intrinsic relationship between the two domains. The relationship between the spatial and frequency domains is established using the Fourier transform, and the two domains form a set of Fourier transform pairs.

This chapter discusses the spatial operators pertaining to the linear filtering method. In image processing, the second-order differential is achieved by defining $\nabla^2 f$. The resulting Laplacian operator is an effective tool for edge detection.

Among linear filtering methods, this chapter discusses the basic filtering method pertaining to the frequency domain. In the frequency domain, the basic filtering model is implemented through the filter transform function to images of low frequencies, high frequencies and other frequency ranges to realize targeted filtering.

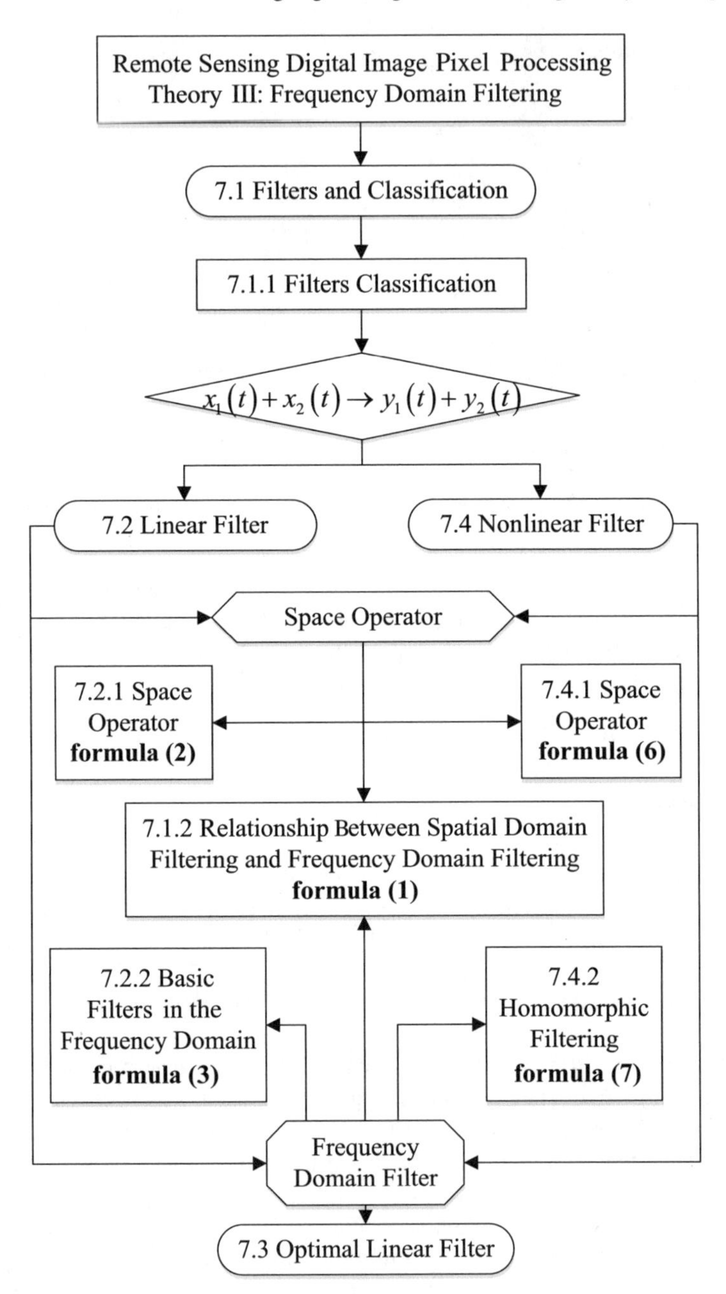

Fig. 7.14 Summary of this chapter

Table 7.1 Summary of the mathematical formulas in this chapter

formula (1)	$f(x,y)*h(x,y) \Leftrightarrow F(u,v)H(u,v)$	7.1.2
formula (2)	$\nabla^2 f = \frac{\partial^2 f}{\partial x^2} + \frac{\partial^2 f}{\partial y^2} = f(i-1,j)+f(i+1,j)+f(i,j+1)+f(i,j-1)-4f(i,j)$	7.2.1
formula (3)	$D(u,v) = \left[(u-M/2)^2+(v-N/2)^2\right]^{\frac{1}{2}}$	7.2.2
formula (4)	$MSE = \varepsilon\{e^2(t)\} = \int_{-\infty}^{+\infty} e^2(t)\mathrm{d}t$	7.2.3
formula (5)	$\rho = \varepsilon\{u^2(0)\}/\varepsilon\{v^2(0)\}$	7.2.3
formula (6)	$\nabla f(x,y) = \begin{bmatrix} G_x \\ G_y \end{bmatrix} = \begin{bmatrix} \frac{\partial f}{\partial x} \\ \frac{\partial f}{\partial y} \end{bmatrix}$	7.3.1
formula (7)	$f(x,y) = i(x,y) \cdot r(x,y)$	7.3.2

The novelty of this chapter is that it introduces two types of optimal linear filters. The design of the Wiener filter is based on the time domain definition and frequency domain design, and the mean square error is considered to determine the optimal filter conditions. In the design of the matched detector, the optimum conditions are determined by discussing the average signal-to-noise power P.

In terms of nonlinear filtering methods, this chapter introduces the traditional spatial operator and homomorphic filtering method. The homomorphic filtering method decreases the low frequency and increases the high frequency, thereby decreasing the illumination change and sharpening the edge or detail of the nonlinear image filtering method, through the time domain definition and frequency domain design.

This chapter highlights the basic principles pertaining to filtering methods and can be valuable for remote sensing digital image processing manuals. Filtering is aimed at enhancing the useful signal and enhancing the image, albeit with strong subjectivity.

References

1. Castleman KR, Selzer RH, Blankenhorn D H. Vessel edge detection in angiogram: an application of the wiener filter. In: Aggarwal JK, Digital signal processing. No. Hollywood, CA: Point lobos Press. 1979.
2. Helbert D, Carre P, Andres E. 3 —D discrete analytical ridgelet transform. IEEE Trans Image Proc. 2006.
3. Goldstein JS, Reed IS, Seharf LL. A multistage representation of the wiener filter based on orthogonal projections. IEEE Trans Inf Theory. 1998.
4. Duncan D, Johnson R, Molnar B, et al. Association between neighborhood safety and overweight status among urban adolescents. BMC Public Health. 2009;9(1).
5. Middleton D. One new classes of matched filters. Inf Theory IEEE Trans. 1960.

6. Jong KD, Albin M, Skärbäck E, et al. Area-aggregated assessments of perceived environmental attributes may overcome single-source bias in studies of green environments and health: results from a cross-sectional survey in southern Sweden. Environ Health. 2011;10(1):4.
7. Hu, J, Sun J, Shao X, et al. Image edge extraction method and prospects. Comput Eng Appl. 2004;40(14):70-73. (In Chinese).
8. Auber TG, Kornprobst TP. Mathematical problems in image processing: partial differential equations and the calculus of variations. New York: Springer. 2002.
9. Sun T, Neuvo Y. Detail-preserving median based filters in image processing. Pattern Recogn Lett. 1994(15).
10. Wen S, You ZS. A Performance optimized algorithm of spatial domain homomorphic-filtering. Appl Res Comput. 2000(17).
11. Sha W, Zhisheng Y. Performance-optimized homomorphic filtering spatial domain algorithm. J Comput Appl. 2000(17). (In Chinese).
12. Cerin E, Barnett A, Sit CH, et al. Measuring walking within and outside the neighborhood in Chinese elders: reliability and validity. BMC Public Health. 2011;11(1):851.

Chapter 8
Remote Sensing Digital Image Pixel Processing Theory IV: Time Domain Sampling

Chapter Guidance The continuous spatial information of the natural world must be transformed into the discrete information needed by the computer, that is, time series information. Therefore, the sampling process is a prerequisite for computer expression analyses. The core of this chapter is the sampling theorem, the analysis of which is based on time-frequency domain conversion, described in Chap. 6, and frequency domain filtering theory, described in Chap. 7.

The logical framework of this chapter is shown in Fig. 8.1. The specific contents are as follows: Sect. 8.1 Sampling and interpolation: This section describes the theoretical basis for the discretization of continuous information. Section 8.2 Propagation of a two-dimensional image of the sampling theorem, which can be directly used for two-dimensional image discretization. Section 8.3 Spectrum analysis and interception to analyze the resulting error; Sect. 8.4 Aliasing error and linear filtering to analyze the root causes of the aliasing and error correction method. Section 8.5 presents the theory and methodological basis of continuous-discrete information conversion.

How can the continuous function be sampled to ensure that the original function is fully restored? To what extent does the sample cause loss of information, and what is the nature of this loss? What is the effect of sampling on the spectrum of the function? To address these questions, the concept of sampling is described in the following sections.

8.1 Sampling and Interpolation

This section focuses on the origin of the Shah sampling theorem. First, we describe the Shannon function, which is an infinite impulse sequence of unit magnitudes that appear along the x-axis. Since sampling and quantization are performed according to the specific time and frequency of the signal, the Shah function is extended to the specific time and frequency units such that there exists an impulse sequence with the time interval τ and frequency interval $1/\tau$.

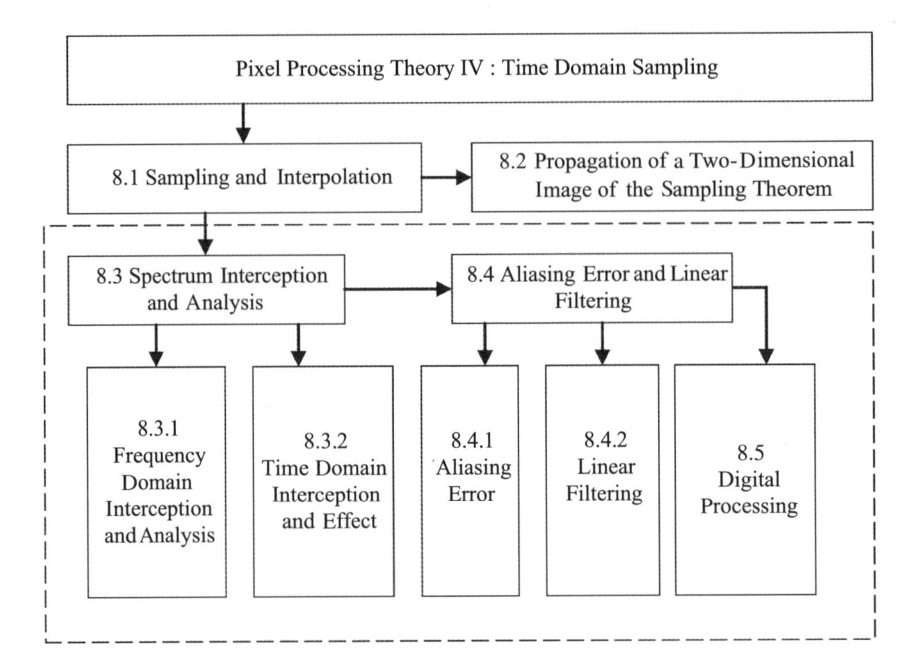

Fig. 8.1 Logical diagram of this chapter

Both the time and frequency domains are sampled based on the Shah function. Time domain processing involves multiplying the original function and impulse sequence function. Frequency domain processing pertains to the convolution of the impulse sequence function and original function that is defined in the frequency domain through the Fourier transform. The sampling theorem can be obtained by analyzing the sampling process [1].

8.1.1 Sampling Interpolation Function (Impulse Sequence): Shah Function

The Shah function models the sampling process:

$$\mathrm{III}(x) = \sum_{n=-\infty}^{\infty} \delta[x - n] \tag{8.1}$$

$\mathrm{III}(x)$ is the unit amplitude impulse sequence that appears along the x-axis interval. The Fourier transform of the Shah function is itself:

$$\Im\{\mathrm{III}(x)\} = \mathrm{III}(s) \tag{8.2}$$

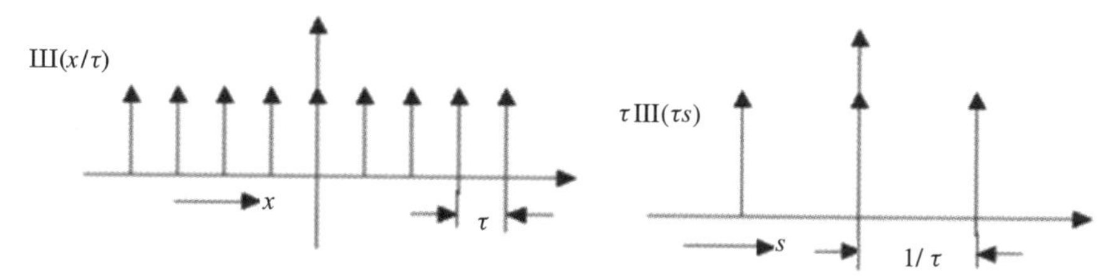

Fig. 8.2 Shah function and its spectrum

Using this function as a model for sampling successive functions, this conclusion is extended to the specific time and frequency, and $\text{III}(x/\tau)$ is the unit amplitude impulse sequence separated by τ along the x-axis. As shown in Fig. 8.2, the similarity theorem is substituted in Eq. (8.2)

$$\Im\{\text{III}(x/\tau)\} = \tau * \text{III}(\tau s) \tag{8.3}$$

According to the similarity theorem, the impulse function exhibits the following property:

$$\delta(ax) = \frac{1}{|a|}\delta(x) \tag{8.4}$$

Since $\text{III}(x)$ is an infinite sequence of impulses that are arranged at equal intervals, this property is also observed in extension or contraction. Specifically,

$$\text{III}(ax) = \sum_{n=-\infty}^{\infty} \delta(ax - n) = \sum_{n=-\infty}^{\infty} \delta\left[a\left(x - \frac{n}{a}\right)\right] \tag{8.5}$$

and thus

$$\text{III}(ax) = \frac{1}{|a|}\sum_{n=-\infty}^{\infty} \delta\left(x - \frac{n}{a}\right) \tag{8.6}$$

If $a = \frac{1}{\tau}$,

$$\text{III}\left(\frac{x}{\tau}\right) = \tau \sum_{n=-\infty}^{\infty} \delta(x - n\tau) \tag{8.7}$$

i.e., the impulse interval is τ. The impulse interval is τ, and not the unit spacing. When the amplitude of the impulse intensity is multiplied by coefficient τ, the Fourier transform is

$$\Im\{\text{III}(x/\tau)\} = \tau\text{III}(\tau s) = \sum_{n=-\infty}^{\infty} \delta(s - n/\tau) \tag{8.8}$$

The last two equations show that in the time domain, an impulse sequence with an intensity τ and interval τ is generated in the frequency domain with an interval $1/\tau$ of the unit impulse sequence. The two sides of Eq. 8.7 can be divided by τ simultaneously to obtain the time-domain impulse sequence of the unit intensity, corresponding to the intensity $1/\tau$ of the impulse sequence in the frequency domain.

8.1.2 Sampling with Shah Function

Assume that the bandwidth of the continuous function $f(x)$ is s_0, i.e.,

$$F(s) = 0, \quad |s| \geq s_0 \tag{8.9}$$

where $F(s)$ is the spectral function, as shown in Fig. 8.3a, b. When sampling at equal intervals τ, the values $f(x)$ are considered only at $x = n\tau$, and no values are considered elsewhere. The sampling process can be modeled simply by multiplying $\text{III}(x/\tau)$ with $f(x)$ to obtain the resulting function $g(x)$. This process sets the function value between the sampling points as zero, and the impulse strength at the sampling point preserves the value of the function. The functions after sampling are shown in Fig. 8.3c, d.

Figure 8.3 shows the model representation of the sampling process in the time and frequency domains [2]. In the time domain, the function of the original continuous function multiplied by the Shah function is sampled. This process is transformed into the frequency domain, which is equivalent to convolving the spectra of the original and impulse functions. The convolution is a function of the sampling function. Thus, it can be concluded that the product in the time domain corresponds to the convolution in the frequency domain.

The time domain shown in Fig. 8.3a is

$$g(x) = f(x)\text{III}\left(\frac{x}{\tau}\right) = f(x)\tau \sum_{n=-\infty}^{\infty} \delta(x - n\tau) = \tau \sum_{n=-\infty}^{\infty} f(nx)\delta(x - n\tau) \tag{8.10}$$

The frequency domain shown in Fig. 8.3b is

$$G(s) = F(s) * \sum_{n=-\infty}^{\infty} \delta(s - n/\tau) = \sum_{n=-\infty}^{\infty} F(s - n/\tau) \tag{8.11}$$

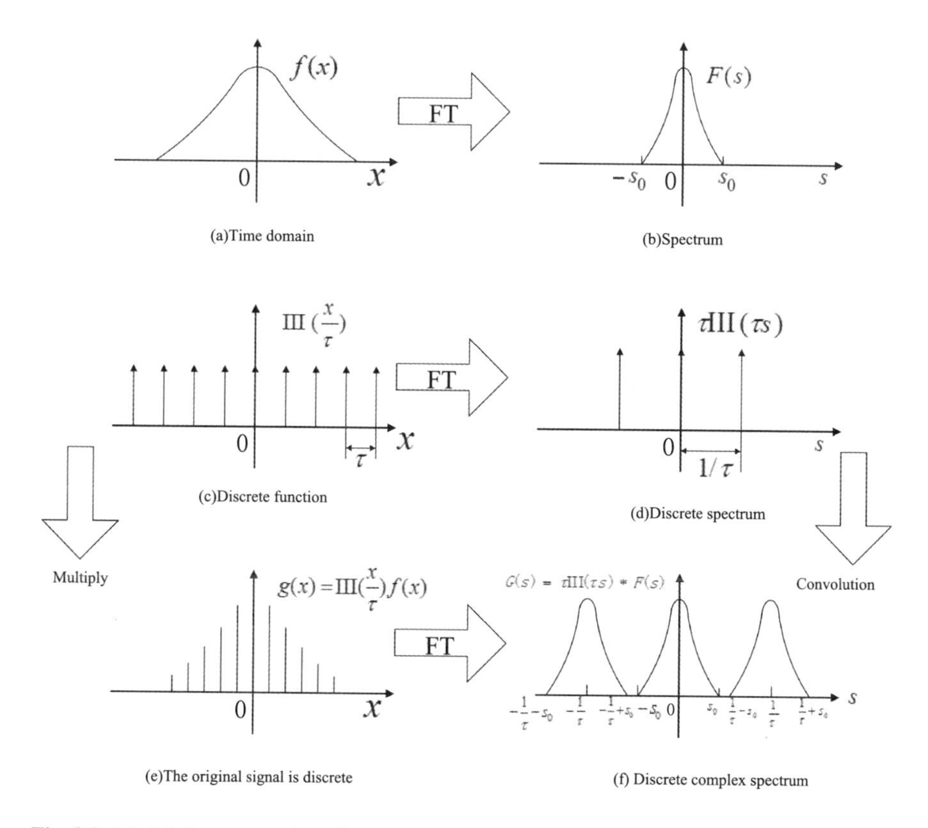

Fig. 8.3 Model characterization of the sampling process in the time and frequency domains

8.1.3 Sampling and Interpolation

The conversion of analog images to digital images inevitably leads to loss of information. How can the restoration of the original image be maximized? We compare the spectral $G(s)$ after sampling and spectrum $F(s)$ before sampling. Values of $G(s)$ are obtained by the convolution of $F(s)$ and the impulse sequence spectrum, resulting in a replica of $F(s)$, a periodic function with a period of $1/\tau$. As shown in Fig. 8.3f, $G(s)$ contains a replica of the infinite spectrum $F(s)$ from negative infinity to positive infinity on the s axis.

1. Sampling

The information $f(x)$ of function $f(x)$ is lost. How can the original function $f(x)$ be restored from the sampling point? If $F(s)$ can be obtained from $G(s)$, $f(x)$ can be derived from $g(x)$. To this end, we maintain the spectrum at the center of the center and eliminate all other copies of $F(s)$ (Fig. 8.4). In particular, we multiply $G(s)$ by the square function $\mathrm{II}\,(s/2s_1)$,

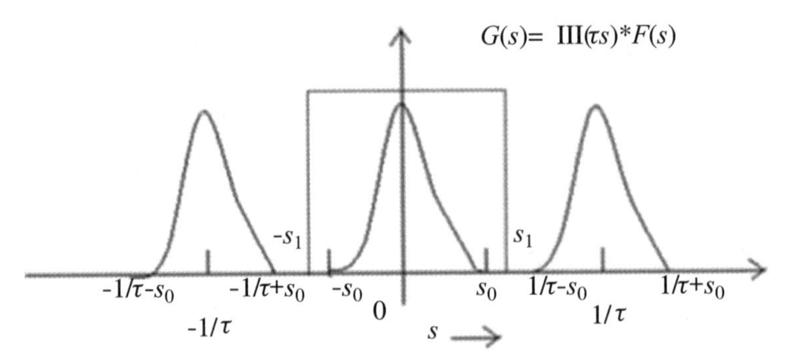

Fig. 8.4 Restoration of the original function $F(s) = G(s)\Pi(s/2s_1)$

where s_0 is the cutoff frequency (bandwidth), s_1 is the intercept window width to restore the original image, and

$$s_0 \le s_1 \le 1/\tau - s_0 \tag{8.12}$$

Thus,

$$G(s)\Pi(s/2s_1) = F(s) \tag{8.13}$$

The spectrum of $f(x)$ is recovered from the spectrum of the sampled signal $g(x)$. Subsequently, the initial function can be expressed as

$$f(x) = \Im^{-1}\{F(s)\} = \Im^{-1}\{G(s)\Pi(s/2s_1)\} \tag{8.14}$$

Due to $\Im\{\Pi(x)\} = \frac{\sin(\pi s)}{\pi s} \Rightarrow \Im^{-1}\{\Pi(s)\} = \frac{\sin(\pi\tau)}{\pi\tau} = \mathrm{sinc}(\pi\tau)$, and $\Im\{f(ax)\} = \frac{1}{|a|}F(s/a) \Rightarrow |a|\Im\{f(ax)\} = F(s/a)$

If $a = 2s_1$,

$$2s_1\Im\{f(2s_1x)\} = F(s/2s_1) \tag{8.15}$$

Consequently,

$$\Im^{-1}\{\Pi(s/2s_1)\} = 2s_1\frac{\sin(\pi\tau)}{\pi\tau}|\tau = 2s_1x = 2s_1\frac{\sin(2\pi s_1x)}{2\pi s_1x} \tag{8.16}$$

The convolution theorem is applied to the right side:

$$f(x) = g(x) * 2s_1\frac{\sin(2\pi\ s_1x)}{2\pi\ s_1x} \tag{8.17}$$

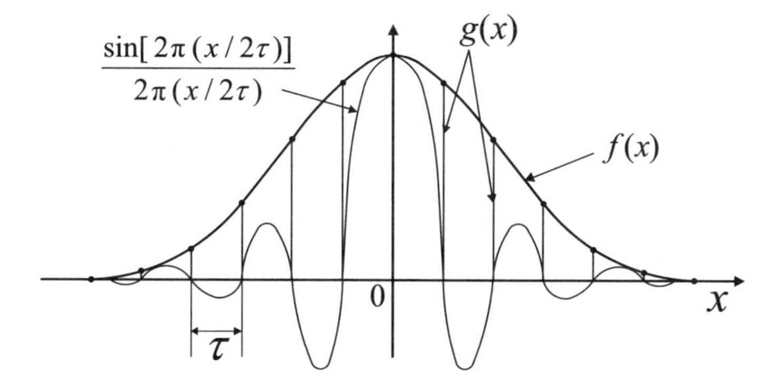

Fig. 8.5 Interpolation

Converting the interpolation function derived by $g(x)$ and Eq. 8.15 is equivalent to copying a narrow function $\sin x/x$ at each sample point, as shown in Fig. 8.5. Equation 8.15 ensures that the sum of the functions $\sin x/x$, which overlap each other, can accurately restore the original function.

Figure 8.5 depicts the situation at time $s_1 = 1/2\tau$. If the reciprocal of the sampling interval is significantly greater than the cutoff frequency s_0, there exists a considerable degree of freedom in the frequency selection of function $\sin x/x$: any value s_1 can be set between s_0 and $1/\tau - s_0$. For convenience, s_1 can be placed at the midpoint [3]

$$s_1 = 1/2\tau \tag{8.18}$$

Therefore, the interpolation function is $\frac{1}{\tau}\frac{\sin(\pi\frac{x}{\tau})}{\pi\frac{x}{\tau}}$, and the spectrum is shown in Fig. 8.6.

In other words, from $g(x)$ to reconstruction $f(x)$, only the function after sampling in the form of an sinc $x = \sin x/x$ interpolation function can be convoluted.

To ensure that the sampled signal can faithfully preserve the information of the original analog signal, the frequency of the sampled signal must be at least twice the

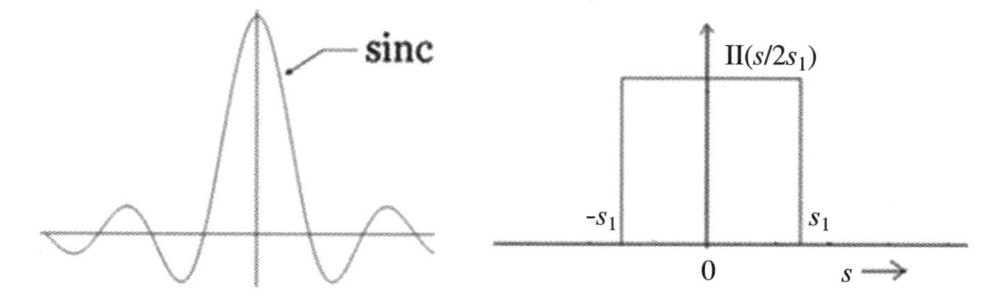

Fig. 8.6 Function sinc and its spectra

highest frequency component of the original signal. This aspect is the basic rule of sampling, known as the sampling theorem.

$$\tau \leq 1/2s_0 \tag{8.19}$$

The sampling theorem, also known as the Shannon sampling law or Nyquist sampling law, is an important basic conclusion in the theory of information, especially in communication and signal processing (E.T. Whittaker' theory of statistics, 1915). The key contributors to this theorem are Claude Shannon, Harry Nyquist and VA Kotelnikov.

From the signal processing viewpoint, the sampling theorem describes two processes: sampling, which transforms a continuous time signal into a discrete time signal; and signal reconstruction, which is the process of transforming discrete signals into a continuous signal [4].

For a continuous signal $x(t)$, the spectrum is $X(f)$, and for the sampling period T_s, the signal $x_s(nT_s)$ is considered. The spectrum $X(f)$ and sampling period T_s meet the following conditions:

(1) Spectrum $X(f)$ is a finite spectrum; when $|f| \geq f_c$ (cutoff frequency), $X(f) = 0$.
(2) When $T_s \leq \frac{1}{2f_s}$ or $2f_c \leq f_s(f_s = \frac{1}{T_s})$, the continuous signal is uniquely determined. The continuous signal is

$$f(x) = \sum_{n=-\infty}^{+\infty} x_s(nT_s) \frac{\sin\left[\frac{\pi}{T_s}(t - nT_s)\right]}{\frac{\pi}{T_s(t-nT_s)}} \tag{8.20}$$

In the formula, $n = 0, \pm 1, \pm 2, \ldots$, f_c is the highest frequency of the signal that can be recognized during the sampling interval, known as the cutoff frequency or Ny quist frequency.

The sampling theorem states that, in general, a continuous signal $x(t)$ with a finite spectrum $X(f)$ is sampled. When the sampling frequency $f_s \geq 2f_c$ is sampled, the sampled signal $x_s(nT_s)$ can be recovered without distortion until the original signal $x(t)$.

2. Sampling Examples

Figure 8.3 shows the relationship between the time domain and frequency domain sampling. The following text describes the different sampling intervals in the frequency domain and their effect on the restoration of the original function [5]. Figure 8.7 shows the spectrum of the sampled Shah function, and Fig. 8.8 shows spectrum $F(s)$ of the original function $f(x)$.

Sample in the frequency domain, i.e.,

$$G(s) = F(s) * \sum_{n=-\infty}^{\infty} f\left(s - \frac{n}{\tau}\right) = \sum_{n=-\infty}^{\infty} F\left(s - \frac{n}{\tau}\right)$$

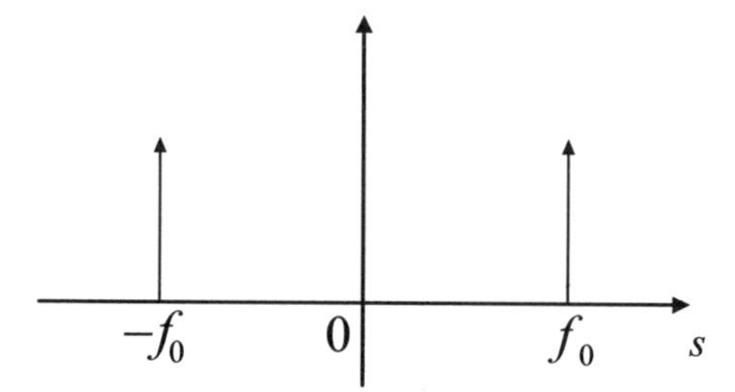

Fig. 8.7 Spectrum of the Shah function

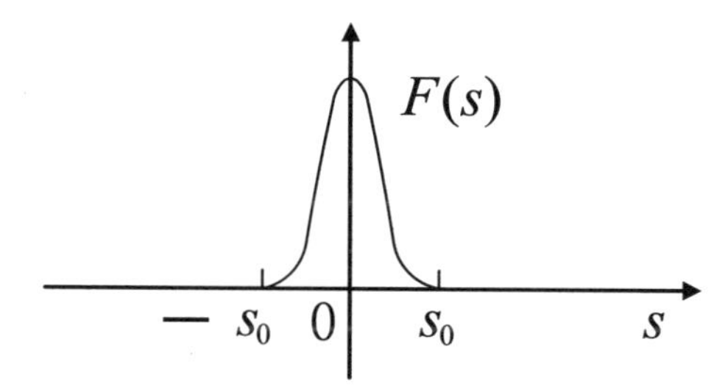

Fig. 8.8 Spectrum of $f(x)$

Case 1: oversampling.

When $1/\tau = 2s_0$, Fig. 8.9 shows the spectrum of oversampling function $g(x)$. There exists a certain distance between the replicas of $F(s)$ in the convolution process (Fig. 8.10). The spectrum is not aliased and can be completely preserved at the origin, where the sampling can completely restore the original function.

Case 2: critical sampling.

When $1/\tau = 2s_0$, Fig. 8.11 shows the spectrum of function $g(x)$ at the time of the critical sampling. The spectra are not aliased in the convolution process (Fig. 8.12). The

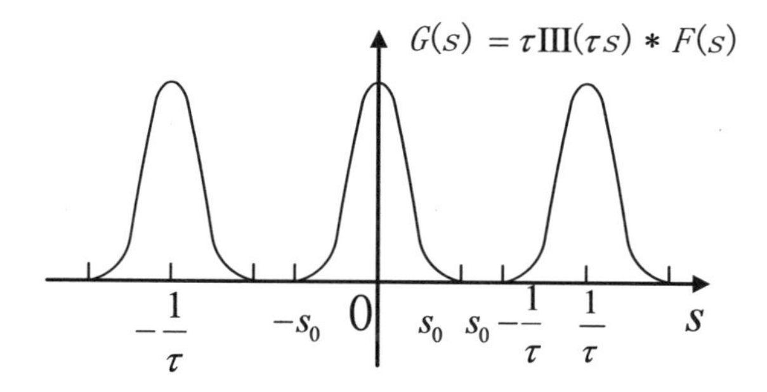

Fig. 8.9 Spectrum of oversampling function $g(x)$

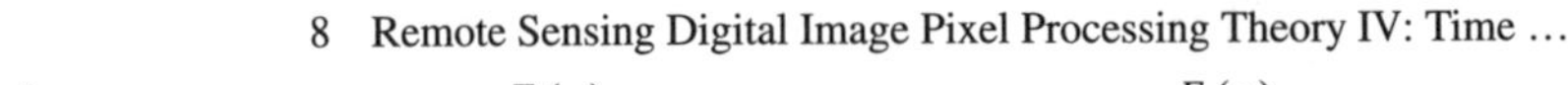

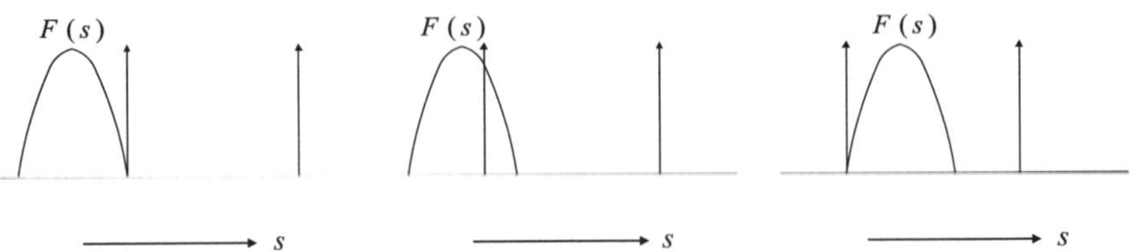

Fig. 8.10 Oversampling process in the frequency domain

spectrum at the origin is completely preserved. This sampling satisfies the sampling theorem and can also restore the original function. However, if the sampling point is not at the amplitude, the magnitude is equal to the sampling amplitude at the discrete sample point, and the maximum amplitude point of the periodic function is the $\pm T/4$ argument point. Thus, the actual amplitude decreases, and the initial phase of the periodic function changes. In the extreme case, when the sampling time is the zero-crossing point of the periodic function, all the sampling points are considered zero-crossing zero signals, corresponding to no signal [6]. Therefore, the Shannon sampling theorem requires that the sampling frequency is greater than and not equal to twice the cutoff frequency.

Case 3: undersampling

When $1/\tau < 2s_0$, Fig. 8.13 shows the addition of the overlays between the replicas of function $g(x)$ after undersampling, and the replicas of $F(s)$ are added (Fig. 8.14).

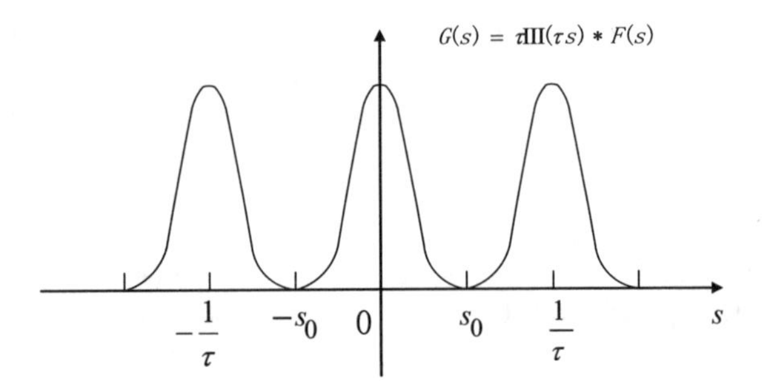

Fig. 8.11 Spectrum of the critical sampling function $g(x)$

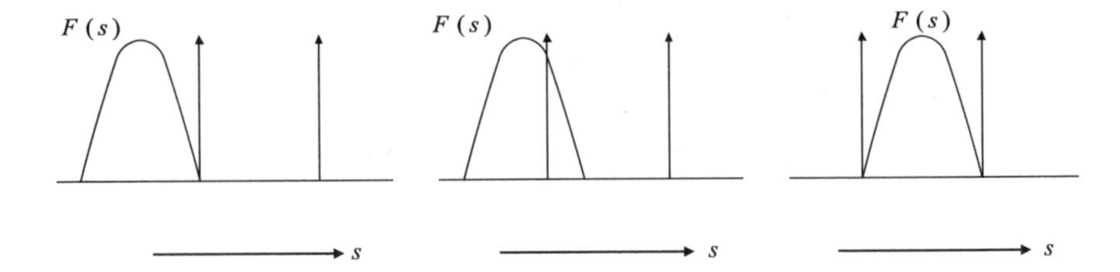

Fig. 8.12 Convolution process of critical sampling in the frequency domain

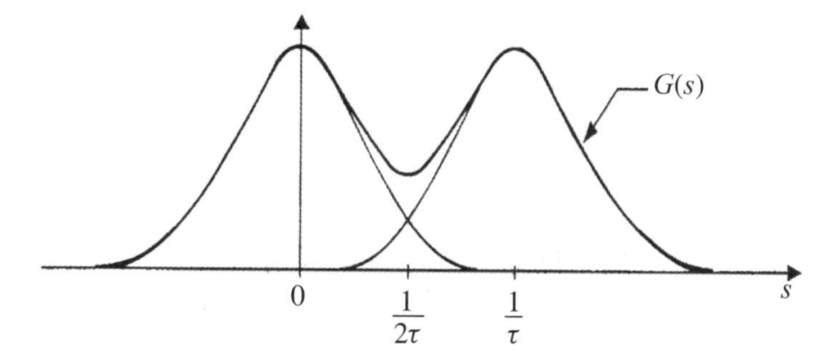

Fig. 8.13 Spectrum of the undersampling function g(x) (aliasing)

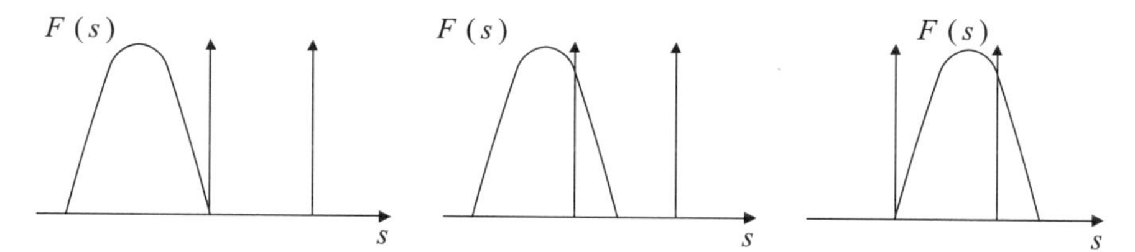

Fig. 8.14 Convolution process in the frequency domain

The figure graphically shows how the high-frequency information is aliased to low-frequency information, and the original function cannot be restored [7].

8.2 Propagation of a Two-Dimensional Image of the Sampling Theorem

With the theoretical support of the sampling theorem of one-dimensional signals, the remote sensing image processing described in this book can extend the sampling theorem to two-dimensional images. The digitization of two-dimensional images includes sampling and quantization. First, the two-dimensional image is sampled according to the sampling theorem of the two-dimensional image. The simulated image is discretized into pixels in time and space after sampling. The next step is to convert the gray value obtained after sampling from analog to discrete, which is known as the quantization of image gray. Sampling affects the degree of reproduction of the details of the image; a greater interval corresponds to a greater loss of detail. If the degree of resolution of the details of the image are quantified, a higher number of quantization bits corresponds to a higher degree of detail resolution. A higher brightness of each pixel corresponds to a more accurate image reproduction. Therefore, different conditions must be set to choose different levels of quantification.

8.2.1 Generalization of the Sampling Theorem

According to the duality of the Fourier transform, the frequency sampling theorem of a time-limited signal can be derived. In addition, according to the basic characteristics of the actual signal and sampling control methods to obtain the sampling theorem of other applications, such as image processing and other technologies widely used in multidimensional signal sampling, communication plays a key role in down sampling [8], that is, bandpass signal sampling. The sampling theorem of bandpass signals indicates that the signal can be reconstructed with a sampling frequency lower than the Nyquist frequency.

1. Frequency Domain Sampling Theorem

Let the signal $f(t)$ represent a finite time signal, that is $|t| \geq t_c$, where $f(t)$ is zero. If the frequency spectrum of the frequency interval pair $f_s \leq 1/2t_c$ is sampled in the frequency domain $f(t)$, it is assumed that the Z transform of an absolutely homogeneous aperiodic sequence $x(n)$ is

$$X(Z) = \sum_{n=-\infty}^{\infty} x(n)Z^{-n} \tag{8.21}$$

Since $x(n)$ is absolutely summable, its Fourier transform exists and is continuous. The Z transform convergence domain includes a unit circle. In this manner, for $X(Z)$ in unit circle N (aliquot sampling), $\widetilde{X}(k)$ can be obtained:

$$\widetilde{X}(k) = X(Z)\Big|_{z=W_N^{-k}} = \sum_{n=-\infty}^{\infty} x(n)W_N^{nk} \tag{8.22}$$

For inverse transformation $\widetilde{X}(k)$ to $\widetilde{X}_N(n)$, the following expression can be obtained:

$$\begin{aligned}
\widetilde{X}_N(n) &= \text{IDFS}\left[\widetilde{X}(k)\right] = \tfrac{1}{N}\sum_{k=0}^{N-1} \widetilde{X}(k)W_N^{-nk} \\
&= \frac{1}{N}\sum_{k=0}^{N-1}\left[\sum_{m=-\infty}^{+\infty} x(m)W_N^{-mk}\right]W_N^{-nk} \\
&= \sum_{m=-\infty}^{+\infty} x(m)\left[\frac{1}{N}\sum_{k=0}^{N-1} W_n^{(m-n)k}\right] \\
&= \sum_{r=-\infty}^{\infty} x(n+rN)
\end{aligned} \tag{8.23}$$

The resulting periodic sequence $\widetilde{X}_N(n)$ is a periodic extension of the aperiodic sequence $x(n)$. The cycle corresponds to the frequency domain sampling points, N.

Therefore, time domain sampling is based on the frequency domain cycle extension, and the same frequency domain sampling is associated with the time domain extension.

Therefore, when $x(n)$ is an infinite sequence, aliasing distortion is performed. When $x(n)$ is a finite length sequence, the length is M. If $N \geq M$, no distortion occurs; if $N < M$, aliasing distortion occurs.

2. Multidimensional Signal Sampling Theorem

Let the signal $f(t_1, t_2, \ldots, t_n)$ be a function of n real variables whose dimension Fourier integral $g(y_1, y_2, \ldots, y_n)$ exists. The corresponding value is zero outside the n dimensional square of the origin symmetry, that is, when $|y_k > w_k|$, $k = 1, 2, \ldots, n$, $g(y_1, y_2, \ldots, y_n) = 0$,

$$f(t_1, t_2, \ldots, t_n) = \sum_{m_1=-\infty}^{+\infty} \cdots \sum_{m_n=-\infty}^{+\infty} f\left(\frac{\pi m_1}{\omega_1}, \frac{\pi m_2}{\omega_2}, \ldots, \frac{\pi m_n}{\omega_n}\right) \frac{\sin(\omega_n t_n - m_n \pi)}{\omega_n t_n - m_n \pi} \tag{8.24}$$

3. Sampling Theorem of Bandpass Signals

If $f(t)$ is the bandpass signal, the pass band is (f_1, f_2), and the sampling frequency f_s must satisfy the following conditions to recover from the sampling signal $f(t)$:

$$\frac{2f_1}{N-1} \geq f_s \geq \frac{2f_2}{N}, N \text{ is any positive integer.}$$

The reconstructed signal $f(t)$ is

$$f(t) = 2WT_s \sum_{n=-\infty}^{+\infty} f(nT_s) \frac{\sin \pi W(t - nT_s)}{\pi W(t - nT_s)} \cos 2\pi f_0(t - nT_s) \tag{8.25}$$

$f_0 = \frac{f_1 + f_2}{2}$ is the center frequency, $W = f_2 - f_1$ is the bandwidth, and $T_s = \frac{1}{f_s}$ is the sampling interval.

The theorem states that for bandpass signals, the bandpass can be used with a sampling frequency lower than the Nyquist frequency. The specific applications can be defined as follows:

(1) Select f_s to satisfy $|f_s - 2W| \leq \varepsilon$, ε is the preselected small positive integer;
(2) Preliminary estimates to satisfy $\frac{2f_1}{N-1} \geq f_s \geq \frac{2f_2}{N}$ with N being the upper limit;
(3) The selected N^* value is the maximum positive integer that does not exceed $\frac{f_1}{f_2 - f_1}$.

8.2.2 Digitization of Two-Dimensional Signals (Images)

The process of one-dimensional signal sampling is described. Subsequently, the sampling in one-dimensional space is extended to that of two-dimensional images, and the sampling theorem is suitable for two-dimensional space.

1. Sampling and Amplitude Quantization

Digital digitization is aimed at digitizing the spatial domain and analog quantity on the Z axis; that is, the digitization of the image refers to the digitization of the signal brightness and image density. The digitization of the spatial domain is known as horizontal digitization, and the digitization of the degree of shading is known as digitization in the vertical or depth direction. Digitization includes two processes: sampling and quantization.

The digital space in the domain, as shown in Fig. 8.15, is a one-dimensional timing signal in the time axis of the simultaneous sampling extended to the x-axis and y-axis. The sampling points are known as pixels. Image digitization refers to the use of discrete pixels at the location or near the location of the gray value to represent the original continuous image.

The simulated image is sampled and discretized into pixels in time and space. The process of converting the gray value obtained from the analog quantity to the discrete amount is known as the quantization of the image grayscale. This process is based on different conditions, and the accuracy of the digital image must be ensured to alleviate the amount of redundancy.

2. Two-Dimensional Sampling Theorem

As shown in Fig. 8.16, the x- and y-axes are equally spaced and sampled by Δx and Δy. The mathematical model can be described by the Shah function of the two-dimensional sampling grid.

To understand the type of frequency components of the sample image and difference between these frequency components and those of the original image, the Fourier

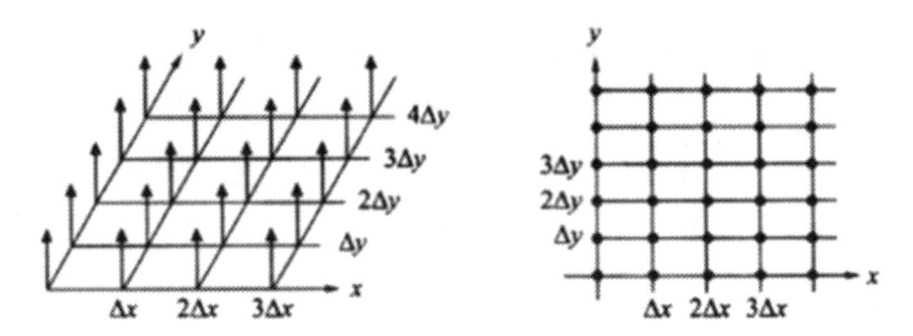

Fig. 8.15 Two-dimensional sampling (rectangular array of pixels)

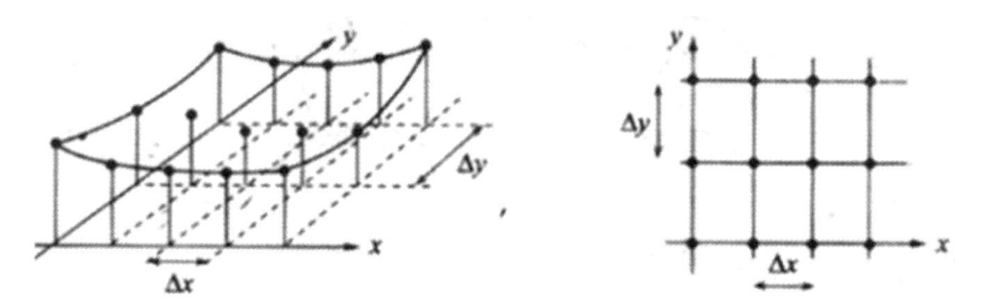

Fig. 8.16 Two-dimensional sampling

transform can be obtained. The result is the same as that of solving a frequency characteristic. As shown in Fig. 8.16, on the spatial frequency axis, the original continuous image spectrum $f(\mu, \upsilon)$ repeats every $1/\Delta x$ and $1/\Delta y$, which is known as the inverse grid.

The mathematical model is used to express the two-dimensional sampling function (Fig. 8.17) and can be expressed as

$$g_s(x, y) = \mathrm{comb}\left(\frac{x}{X}\right)\mathrm{comb}\left(\frac{y}{Y}\right)g(X, Y) \tag{8.26}$$

The comb function can be regarded as the generalization of the unit impinging function in two-dimensional space, which is the set of functions δ. The product is the distribution of any function on the x and y planes. The δ functions of X and Y are multiplied by the function.

The result of multiplying the δ function with the δ function is still a δ function, except that the "size" of the δ function is modulated by the function value at that δ function. In other words, the volume under each δ function is proportional to the value of the point function.

Using the convolution theorem and Fourier transform of the comb function, the spectrum of the sampling function can be derived.

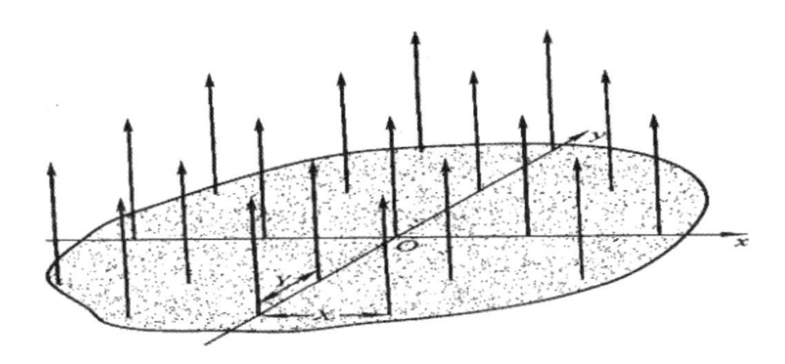

Fig. 8.17 Two-dimensional sampling function

$$
\begin{aligned}
g_s(f_x, f_y) &= F\left\{\mathrm{comb}\left(\frac{x}{X}\right)\mathrm{comb}\left(\frac{y}{Y}\right)\right\} * G(f_x, f_y) \\
&= XY\mathrm{comb}(Xf_x)\mathrm{comb}(Yf_y) * G(f_x, f_y) \\
&= \sum_{n=-\infty}^{\infty}\sum_{m=-\infty}^{\infty} \delta\left(f_x - \frac{n}{X}, f_y - \frac{m}{Y}\right) * G(f_x, f_y) \\
&= \sum_{n=-\infty}^{\infty}\sum_{m=-\infty}^{\infty} G\left(f_x - \frac{n}{X}, f_y - \frac{m}{Y}\right)
\end{aligned} \tag{8.27}
$$

In the case of the Nyquist sampling interval, the spectrum of the original function can be obtained by multiplying the square function of the origin by widths $2B_x$ and $2B_y$. This operation in the frequency domain eliminates several spectral components, often referred to as "filtering"

$$
G_s(f_x, f_y)\,\mathrm{rect}\left(\frac{f_x}{2 * B_x}\right)\mathrm{rect}\left(\frac{f_y}{2 * B_y}\right) = G(f_x, f_y) \tag{8.28}
$$

where the rect function represents a rectangular pulse.

According to the convolution theorem, in the spatial domain,

$$
g_s(x, y) * h(x, y) = g(x, y)
$$

On the left side, the two factors are simplified

$$
\begin{aligned}
g_s(x, y) &= \mathrm{comb}\left(\frac{x}{X}\right)\mathrm{comb}\left(\frac{y}{Y}\right)g(x, y) \\
&= XY\sum_{n=-\infty}^{\infty}\sum_{m=-\infty}^{\infty} \delta(x - nX, y - mY) * g(nX, mY)
\end{aligned} \tag{8.29}
$$

$$
h(x, y) = F\left\{\mathrm{rect}\left(\frac{f_x}{2B_x}\right)\mathrm{rect}\left(\frac{f_y}{2B_y}\right)\right\} = 4B_xB_y\,\mathrm{sinc}\,(2B_xx)\,\mathrm{sinc}\,(2B_yy) \tag{8.30}
$$

The result of the final convolution, the original function, is

$$
g(x, y) = 4B_xB_yXY\sum_{n=-\infty}^{\infty}\sum_{m=-\infty}^{\infty} g(nX, mY)\,\mathrm{sinc}\,[2B_x(x - nX)]\,\mathrm{sinc}\left[2B_y(y - mY)\right] \tag{8.31}
$$

If the maximum allowable sampling interval is considered, i.e., $1/2B_x$, and $1/2B_y$,

$$g(x, y) = \sum_{n=-\infty}^{\infty} \sum_{m=-\infty}^{\infty} g\left(\frac{n}{2B_x}, \frac{m}{2B_y}\right) \operatorname{sinc}\left[2B_x\left(x - \frac{n}{2B_x}\right)\right] \operatorname{sinc}\left[2B_y\left(y - \frac{m}{2B_y}\right)\right] \tag{8.32}$$

For this two-dimensional image signal, the sampling theorem of a one-dimensional signal holds, that is, a continuous image with a small change is obtained, which has a higher spatial frequency component, and the digital image exhibits a wide interval As shown in Fig. 8.18, only the fundamental part near the origin can be restored to the original image, and the overlapping of the spectrum, that is, the location at which aliasing occurs, occurs in other shapes, which is not the same as the original continuous image. At this time, after sampling, the image produces a strip, and the original continuous image and no visible aliasing yield the wrong information.

In the vertical, horizontal or arbitrary directions, when the same changes occur in the law of the image, the spectrum shown in Fig. 8.19 is obtained, with the radius of the circle being R. The image is sampled at equal intervals $\Delta x = \Delta y$ in the x and y directions. The sampling theorem for this square sample image is described in the following section.

When the spatial frequency of up to R continuous images is considered, and the sampling interval is $\Delta x = \Delta y$ with the square spacing,

$$\Delta x = \Delta y \leq \frac{1}{2R} \tag{8.33}$$

If this condition is not established, the wrong sampling would occur.

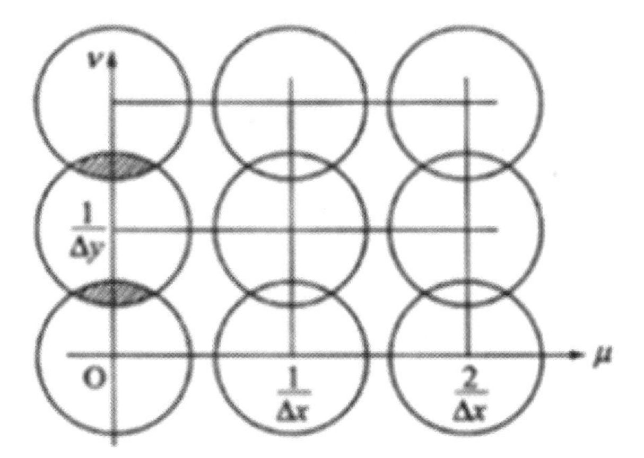

Fig. 8.18 Generation of aliasing

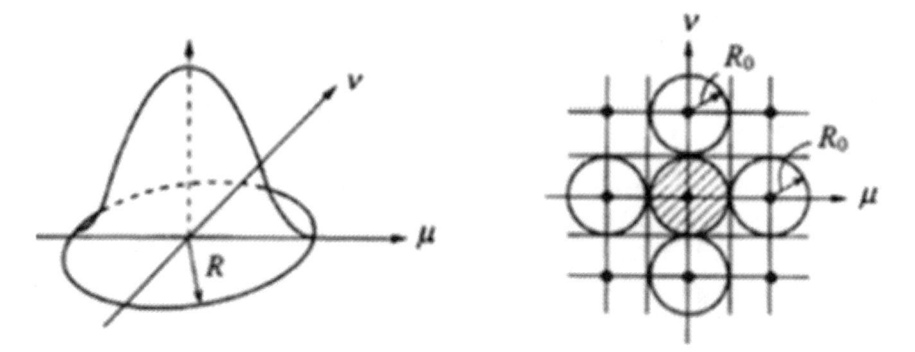

Fig. 8.19 Spectrum and sampling conditions for sampled images

8.3 Spectrum Interception and Analysis

Signal sampling inevitably leads to the interception effect [9], which corresponds to the sampling theorem decision. According to the sampling theorem, $F(s)$ must be a band-limited signal with a frequency component other than the cutoff frequency s_0 of 0, which is the intercept effect.

8.3.1 Frequency Domain Interception and Analysis

Assume that a signal $f(t)$ is represented by N sampling points fixed at a pitch Δt, as shown in Fig. 8.20. The total sampling interval width is

$$T = N * \Delta t \tag{8.34}$$

where T is the width of the intercept window. The signal outside the window to set as 0.

To use the $f(t)$ sampling value to calculate the point on spectrum $F(s)$, the Fourier transform must be obtained. First, we determine the number of points used to calculate the spectrum, spacing of the sampling points, and corresponding spectral range.

Since the sampled signal contains N independent measurements, the calculation of N points on the spectrum is reasonable. The calculation of additional points leads to redundancy, and counting fewer points would render the relevant information unusable. Therefore, in a general procedure for calculating a Fourier transform, it is necessary to calculate N (complex) points on the spectrum from N (complex)

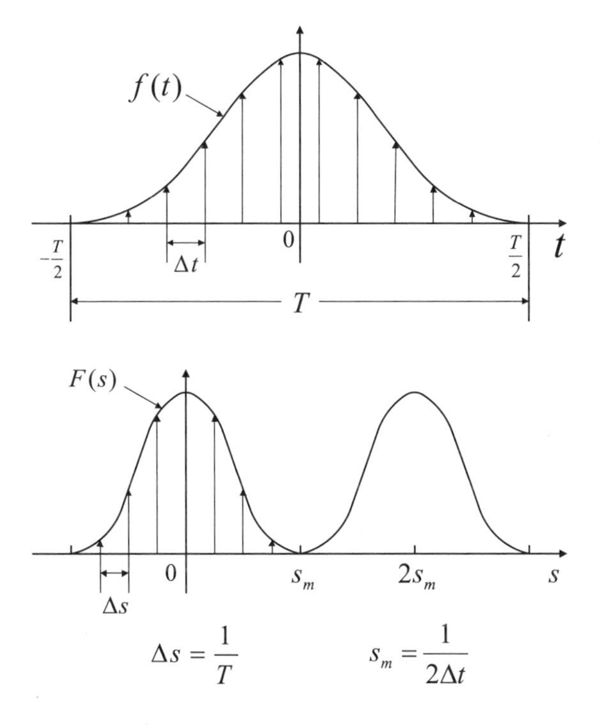

Fig. 8.20 Spectrum calculation

sampling points. For convenience, the calculated points are generally distributed at equal intervals on the s-axis, as shown in Fig. 8.20.

1. Interception in the Frequency Domain

Because $f(t)$ is sampled at sampling interval Δt, the spectral period is $1/\Delta t$. Therefore, only one cycle of coverage $F(s)$ must be calculated. The N sampling points are evenly distributed over a period in which the center is located at the origin $F(s)$. The range of the calculated points is

$$-1/2\Delta t \leq s \leq 1/2\Delta t \tag{8.35}$$

In one cycle in $F(s)$, only N equal spacing sampling points are set. Therefore,

$$N\Delta s = 1/\Delta t \tag{8.36}$$

where

$$\Delta s = 1/N\Delta t = 1/T \tag{8.37}$$

where Δs is the sampling interval in the frequency domain. The optimal approach to calculate the $f(t)$ spectrum is to calculate the equidistant point defined by Eq. 8.43, where the frequency ranges from $-s_m$ to s_m. In this case,

Table 8.1 Summary of sampling and interception parameters

Parameter	Area	Relationship
Number of sampling points	Time domain and frequency domain	$N = \frac{T}{\Delta t} = \frac{2s_m}{\Delta s}$
Sampling interval	Time domain	$\Delta t = \frac{T}{N} = \frac{1}{2s_m}$
Sampling interval	Frequency domain	$\Delta s = \frac{2s_m}{N} = \frac{1}{T}$
Truncated window width	Time domain	$T = N\Delta t$
Maximum computable frequency	Frequency domain	$s_m = \frac{1}{2\Delta t} = \frac{1}{2}N\Delta s$

$$s_m = 1/2\Delta t \tag{8.38}$$

The highest frequency s_m that can be calculated is inversely proportional to the sampling interval Δt in the time domain. The degree of the calculated spectrum is the sampling interval Δs in the frequency domain, which is inversely proportional to the width T of the intercept window in the time domain [10].

2. Spectrum Calculation

The sampling interval in one domain determines the width of the interception window in another domain. To calculate the high-frequency components in the spectrum, the interval must be carefully sampled in the time domain. In addition, to ensure a high resolution (small) in the spectrum, it is necessary to use a large intercept window in the time domain, even if the time domain function is extremely narrow. The relationship between the samples and intercepted parameters in the time and frequency domains is described in Table 8.1.

To calculate $f(t)$ as a complex function, N real values and N imaginary values are transformed to generate N real values and N virtual values in the frequency spectrum. If $f(t)$ is a real function, N real values and N zeros (imaginary parts) produce $N/2$ real numbers and $N/2$ imaginary numbers in the right half of the spectrum. Since $F(s)$ is of the Hermite type, the left half of the spectrum is a mirror image of the right half. Thus, from the information content perspective, the $N/2$ real values and $N/2$ dummy values of the left half of the spectrum are redundant. In both cases, the number of unconstrained sample points is the same in both domains.

8.3.2 Time Domain Interception and Effect

The finite length of the time domain inevitably leads to the infinite length of the frequency domain, resulting in the truncated spectrum not existing. Thus, the dense

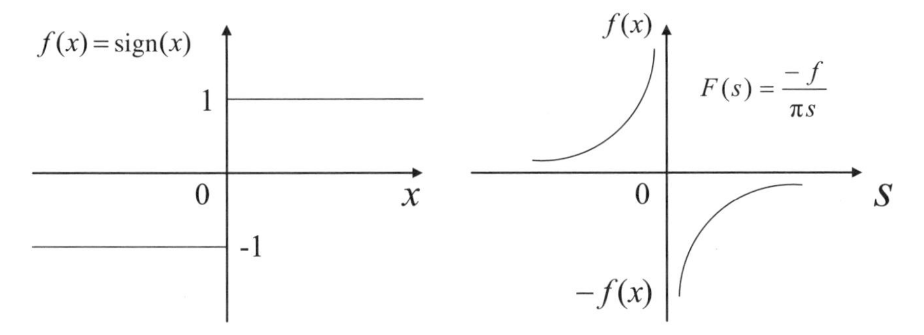

Fig. 8.21 Step function and its frequency spectrum

sampling of the original signal cannot be completely reverted, and the infinite domain of the time domain exists. Consequently, the time domain sampling cannot be cut off. In the limited amount of computer calculations, interception and interception effects are unavoidable [11]. Owing to interception, the calculated spectrum is different from the actual spectrum. Similar to the sampling pitch, it is preferable to choose to intercept the window width to produce the exact result. The following example illustrates the effect of interception.

Consider the function $\text{sign}(x)$ shown in Fig. 8.21. To calculate the spectrum, the function must be intercepted within a finite interval T. Since the function extends to infinity with a constant magnitude, it appears that this example is sensitive to interception.

The interception window can be expressed as $\text{II}(x/T)$, and the center of the window is zero. If we use the intercept window with a width T to intercept the function, as shown in Fig. 8.22, the resulting function is

$$\begin{aligned} g(x) = f(x)\,\text{II}(x/T) &= \text{II}\left(\frac{x - T/4}{2 \times T/4}\right) - \text{II}\left(\frac{x + T/4}{2 \times T/4}\right) \\ &= \text{II}\left(\frac{x}{T/2} - \frac{1}{2}\right) - \text{II}\left(\frac{x}{T/2} + \frac{1}{2}\right) \end{aligned} \tag{8.39}$$

The positive and negative $T/4$ values on the cell indicate the movement of the center of the window, as shown in Fig. 8.23. Since the intercepted function is an odd rectangular pulse pair, it can be expressed as

$$g(x) = \text{II}\left(\frac{x}{T/2}\right) * \left[\delta\left(x - \frac{T}{4}\right) - \delta\left(x + \frac{T}{4}\right)\right] \tag{8.40}$$

In other words, window $\text{II}\left(\frac{x}{T/2}\right)$ is convolved with two pulses to obtain $g(x)$. Equation 8.40 is transformed to obtain the spectrum of the clipped edge function:

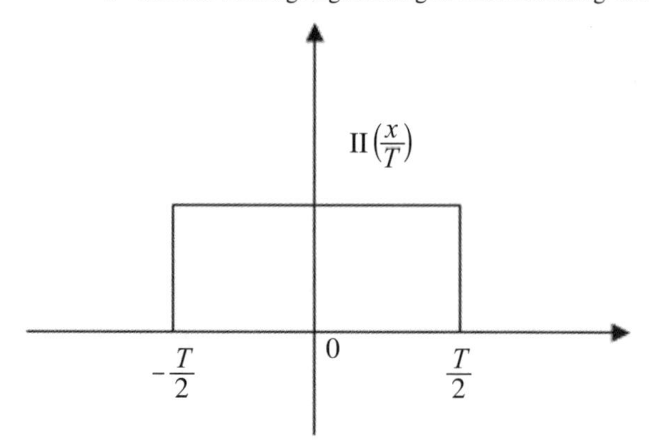

Fig. 8.22 Interception window in the time domain

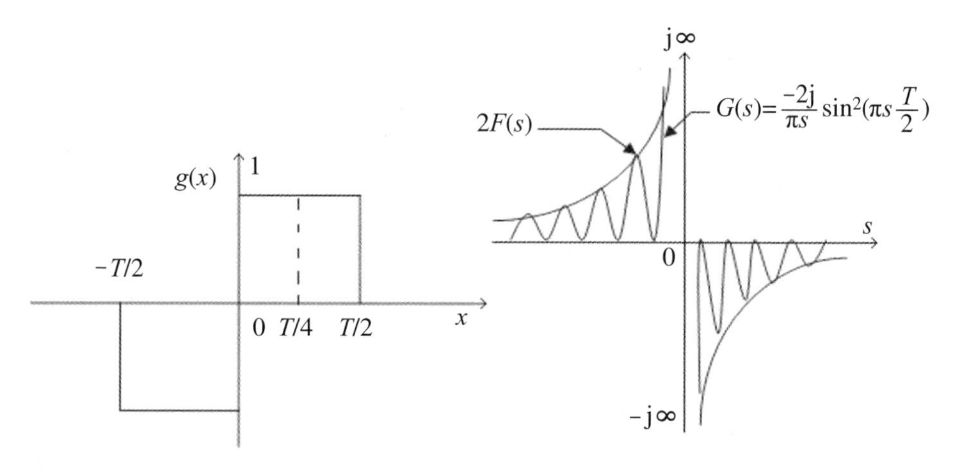

Fig. 8.23 Intercepted step function and its frequency spectrum

$$G(s)\Im\{g(x)\} = \Im\left\{\mathrm{II}\left(\frac{x}{T/2}\right)\right\}\Im\left\{\delta\left(x - \frac{T}{4}\right) - \delta\left(x + \frac{T}{4}\right)\right\} \tag{8.41}$$

where

$$\begin{aligned}\Im\left\{\mathrm{II}\left(\frac{x}{T/2}\right)\right\} &= \int_{-\infty}^{\infty} \mathrm{II}(2x/T)\mathrm{e}^{-\mathrm{j}2\pi sx}\mathrm{d}x \\ &= \int_{-\frac{\pi}{4}}^{\frac{\pi}{4}} \mathrm{e}^{-\mathrm{j}2\pi sx}\frac{\mathrm{d}(-\mathrm{j}2\pi sx)}{-\mathrm{j}2\pi s} = \frac{\mathrm{e}^{-\mathrm{j}2\pi s}}{\mathrm{j}2\pi s}\Big|_{-\frac{T}{4}}^{\frac{T}{4}} = \frac{\sin(\pi s\frac{T}{2})}{\pi s}\end{aligned} \tag{8.42}$$

$$\Im\left\{\delta\left(x-\frac{T}{4}\right)-\delta\left(x+\frac{T}{4}\right)\right\}=\int_{-\infty}^{\infty}\left[\delta\left(x-\frac{T}{4}\right)-\delta\left(x+\frac{T}{4}\right)\right]\mathrm{e}^{-\mathrm{j}2\pi sx}\mathrm{d}x$$

$$=\frac{\mathrm{e}^{-\mathrm{j}2\pi s\frac{T}{4}}-\mathrm{e}^{\mathrm{j}2\pi s\frac{T}{4}}}{\mathrm{j}2}(\mathrm{j}2)=-\mathrm{j}\sin\left(\pi s\frac{T}{2}\right) \quad (8.43)$$

Therefore,

$$G(s)=-\mathrm{j}\sin(\pi s\frac{T}{2})\frac{\sin\left(\pi s\frac{T}{2}\right)}{\pi s} \quad (8.44)$$

Consequently,

$$G(s)=\frac{-2\mathrm{j}}{\pi s}\sin^2\left(\frac{\pi sT}{2}\right)=2F(s)\left[\frac{1}{2}-\frac{1}{2}\cos(\pi sT)\right]=\frac{-\mathrm{j}}{\pi s}[1-\cos(\pi sT)] \quad (8.45)$$

The graph is shown in Fig. 8.24. The frequency of the intercepted signal is twice the sinusoid enveloped by the required spectrum *F(s)*. The large change in the spectrum pertains to interception [12]. This case involves a large modification of the original function.

Since our objective is to calculate the points on *G*(*s*), it is necessary to identify where these points will fall on the *G*(*s*) sine curve. The sampling points on *G*(*s*) are identified at the following discrete frequencies:

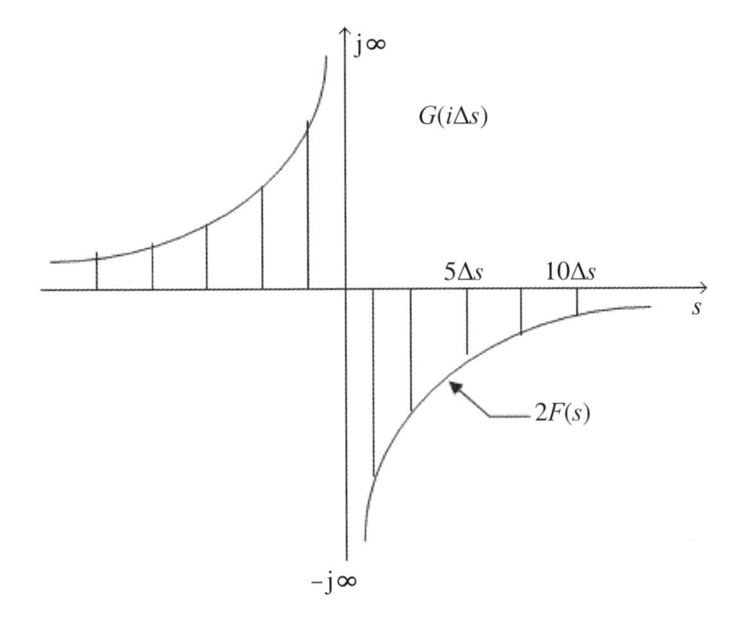

Fig. 8.24 Computation spectrum of the step function

$$s_i = i\Delta s = \frac{i}{T}, i = 0, 1, 2, ..., \frac{N}{2} \tag{8.46}$$

The calculated value is

$$G(s_i) = 2F(s_i)\left[\frac{1}{2} - \frac{1}{2}\cos(i\pi)\right] \tag{8.47}$$

The cosine term takes a value of 1 and -1 when i is even and odd, respectively. Therefore,

$$G(s_i) = \begin{cases} 2F(s_i), & i = 2n + 1, n = 0, 1, \ldots \\ 0, & i = 2n, n = 0, 1, \ldots \end{cases} \tag{8.48}$$

as shown in Fig. 8.24. Note the result of the interception in the above example: The odd points are correct (although the size is twice the normal size), and the even points are zero. This finding shows that interception redistributes energy between odd and even points.

The target result can be obtained by convolving $G(i\Delta s)$ with a narrow triangular local averaging filter such as [1/4, 1/2, 1/4]. This value is equivalent to the result of multiplying the clipped edge with a window function of the form $2\text{sinc}\,x$. Thus, discontinuities occurring at $\pm$ $T/2$ can be prevented to avoid interception.

8.4 Aliasing Error and Linear Filtering

This section is aimed at clarifying the inevitability of aliasing errors. According to the sampling theorem, aliasing can be completely avoided by selecting a suitable sampling interval for a function with a limited bandwidth [13]. Interception destroys the limited bandwidth, and aliasing occurs in digital processing. A wider time domain corresponds to a narrower frequency domain. Therefore, the limited frequency domain required by the sampling theory corresponds to an infinitely large time domain. The aliasing error is naturally unavoidable in this case. Both oversampling and frequency domain filtering can mitigate the effects of aliasing errors to a certain extent, and double oversampling can satisfy these requirements.

8.4.1 *Aliasing Error*

Assume that $\tau > 1/2s_0$. When the $F(s)$ replicas are repeated to form $G(s)$, the replicas are superimposed. If the interpolation function is used, it is impossible to accurately recover $f(x)$, as shown in Fig. 8.25, because

$$G(s)\,\Pi\left(\frac{s}{2s_1}\right) \neq F(s) \tag{8.49}$$

1. Effect of Overlapping Spectral Replicas

The energy above spectrum s_1 is folded below and added to the spectrum. The difference in this energy is known as aliasing, and the difference between the interpolated functions and $f(x)$ is known as the aliasing error.

In general, the frequency above has more energy, and more energy is folded back to the spectrum, corresponding to a higher mixed aliasing error [14]. Note: When $f(x)$ is an even function, $F(s)$ is also an even function, and the aliasing effect enhances the energy in the spectrum. If the function is an odd function, the opposite effect is attained, and the energy in the spectrum decreases. If $f(x)$ is nonodd or noneven, the aliasing increases the even function part and decreases the odd function part, rendering the function practically more likely to be even than its spectrum.

Unfortunately, this phenomenon does not occur owing to the interception process. Assuming that a function with a limited bandwidth is intercepted for a finite length T, the process can be modeled by multiplying the function by a rectangular pulse of width T, equivalent to the sin x/x function, whose spectrum corresponds to infinitely persistent convolution in the frequency domain. Since the convolution of the two functions cannot be narrower than any of the functions, it can be concluded that the spectrum of the intercepted function is infinite in the frequency domain. A wider time domain corresponds to a narrower frequency domain, and the sampling domain of the required time domain function for the frequency domain is limited, which means that the requirements of the time domain function are limited. When the time domain is infinite, the truncation error occurs, and the time domain is truncated to convolution of the infinite sinc function, resulting in the bandwidth of the frequency domain being no longer limited [15]. Although aliasing cannot be completely avoided, the resulting error can be limited and decreased to an acceptable level of practical application.

The following example shows how to set an upper limit for the aliasing error and choose the digitization parameters to achieve the required accuracy if aliasing cannot be prevented.

The system can be identified by calculating the response spectrum of a rectangular wave, as shown in Fig. 8.26. If $f(t)$ is the input pulse and $g(t)$ is the system output, the transfer function is

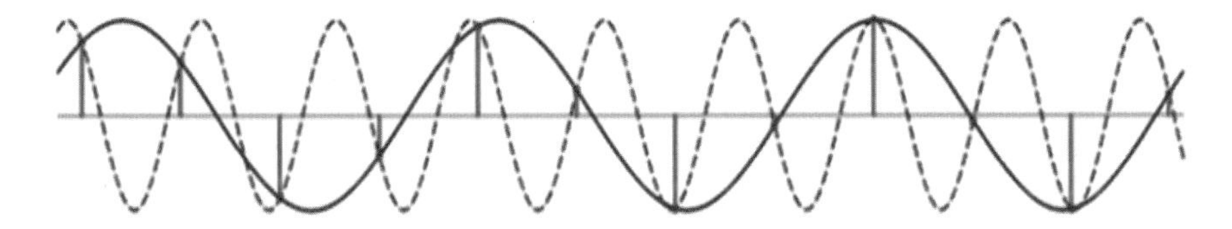

Fig. 8.25 Loss of function due to undersampling

$$H(s) = \frac{G(s)}{F(s)} \tag{8.50}$$

Assume that in this example, the system is a low-pass filter whose output is a rectangular wave with slightly rounded corners.

To estimate Eq. 8.50 through digital calculations, $f(t)$ and $g(t)$ must be digitized, and their spectrum must be determined. The sampling interval Δt and sampling time range T must be chosen to obtain a high spectral resolution to ensure that a sufficiently small aliasing error is obtained. To this end, a measure of the spectral resolution and aliasing error must be defined, and these two quantities must be linked to the sampling parameters. In this manner, N, T and Δt can be appropriately selected.

Figure 8.27 shows the input signal and spectrum. Since $F(s)$ extends from negative infinity to positive infinity, no values of Δt can completely avoid aliasing. $F(s)$ in the envelope of the form $1/s$ ensures that the peak value of the function tends to zero as the frequency increases. If the sinusoidal fluctuation of the function is neglected, only its envelope is considered, and the maximum spectrum that may be aliased is noted. The intensity appears at frequency s_m. This frequency corresponds to the worst case of aliasing, and the measure of aliasing error is defined as the ratio of $F(s_m)$ to $F(0)$. Since the value of $F(0)$ is equal to 1 and the envelope of $F(s)$ is $1/2\pi as$, the upper bound of aliasing can be considered equal to

$$A \leq \frac{1}{2\pi a s_0} = \frac{2\Delta t}{2\pi a} = \frac{\Delta t}{\pi a} \tag{8.51}$$

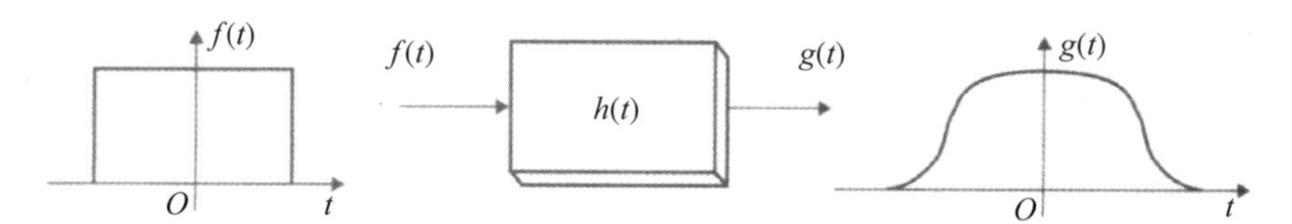

Fig. 8.26 Linear system identification

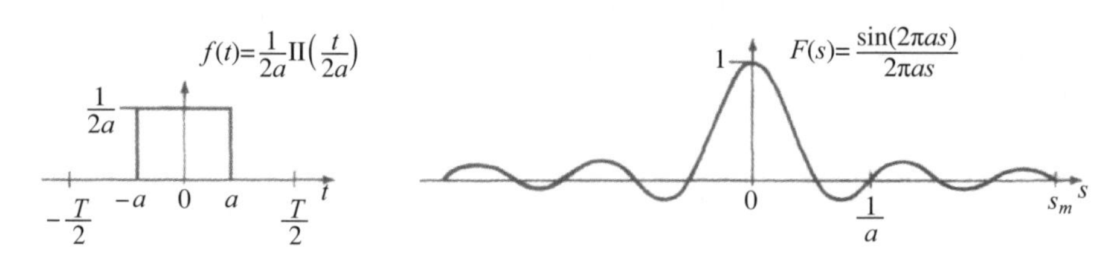

Fig. 8.27 Input signal and its spectrum

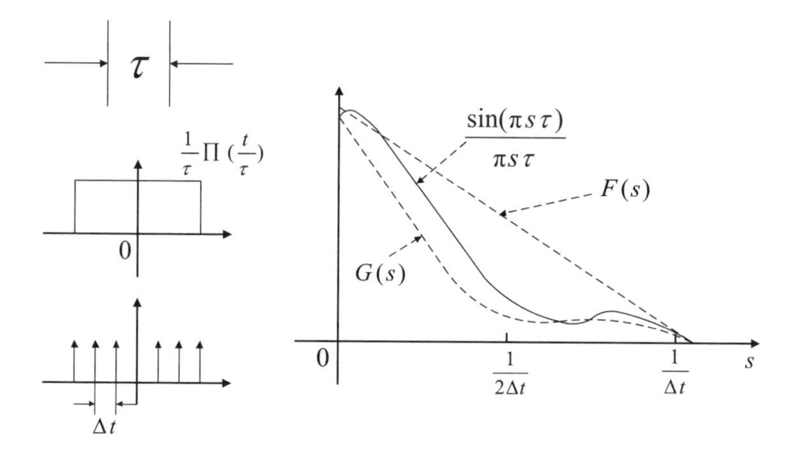

Fig. 8.28 Use of a rectangular aperture to decrease aliasing

The upper bound of the aliasing error defined by this equation is proportional to Δt but not to T. In this way, as long as Δt is sufficiently smaller than pulse width $2a$, the aliasing error can be made as small as necessary.

2. Alias Control

Two parameters can be used to control the aliasing that can cause loss of interesting information in the graph: sampling aperture and sampling interval. This section describes how to decrease the aliasing error through the anti-aliasing filter.

Figure 8.28 shows the method of using a rectangular sampling aperture to alleviate aliasing. The aperture width is twice the sampling pitch [16]. Therefore, the first zero crossing of the transfer function occurs at $f_N = \frac{1}{2\Delta t}$. In this case, the energy above f_N (the part that causes aliasing) can be considerably decreased.

The triangular sampling aperture shown in Fig. 8.29 has a width of four sample points, and the first zero crossing occurs at f_N. Since the frequency spectrum decreases more rapidly as the spectrum increases, compared to the rectangular pulses, aliasing can be alleviated more effectively. However, similar to the rectangular pulse, the energy in $F(s)$ below the f_N part is decreased.

Aliasing can be attributed to the large gap between pixels in the sensor chip of the CCD camera. The sampling aperture is extremely narrow to realize anti-aliasing, and high-frequency information cannot be removed before sampling. Moreover, the camera is slightly defocused, and thus, the lens acts as an anti-aliasing filter.

In addition, continuous functions cannot be severely distorted after digital processing. Oversampling is a viable solution. Decreasing the sampling interval can ensure that f_N lies beyond the frequency of interest in the spectrum. In this manner, even if aliasing pollutes the high-frequency part of the spectrum, the data of interest are not or only slightly affected. According to experience, two oversampling operations are sufficient for most applications.

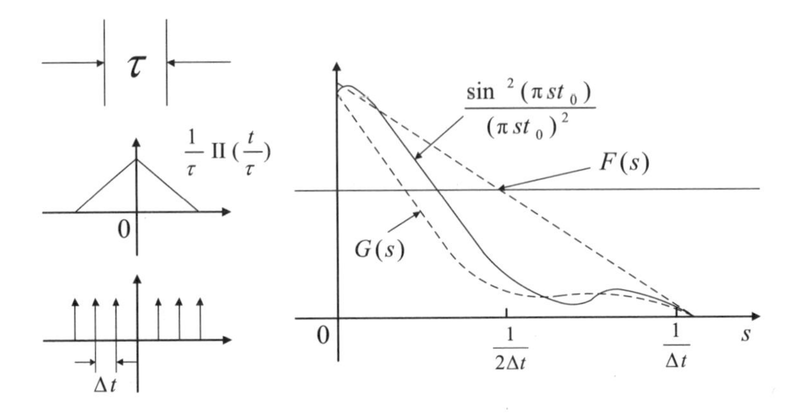

Fig. 8.29 Use of a triangle aperture to decrease aliasing

8.4.2 Linear Filtering

Linear filtering can be achieved in two ways. First, the filtering operation shown in Fig. 8.30 can be implemented by digitally convoluting the sampled functions $f(t)$ and $h(t)$ to generate $g(t)$. Second, through the numerical integration of the Fourier transform algorithm, $f(t)$ and $h(t)$ can be converted to the frequency domain, and the multiplication of the output spectrum $G(s)$ can be performed. The inverse transformation is the output signal.

If one or both convolutional input signals have a small duration, the digitized convolution algorithm is computationally simple. Otherwise, the efficient Fourier transform algorithm renders the second method more practical. This section compares the two methods with aliasing and interception errors.

1. Convolution Filtering

As mentioned, the sampling of $f(t)$ and $h(t)$ renders the spectrum cyclical. If both signals are sampled at the same pitch Δt, the spectra will have the same period $1/\Delta t$. The convolution of the signal after sampling is aimed at multiplying the two spectra in the frequency domain to obtain $G(s)$, which is the periodic function of frequency Δt. In interpolating $g(t)$, as discussed, the spectrum is reduced to a replica at the origin.

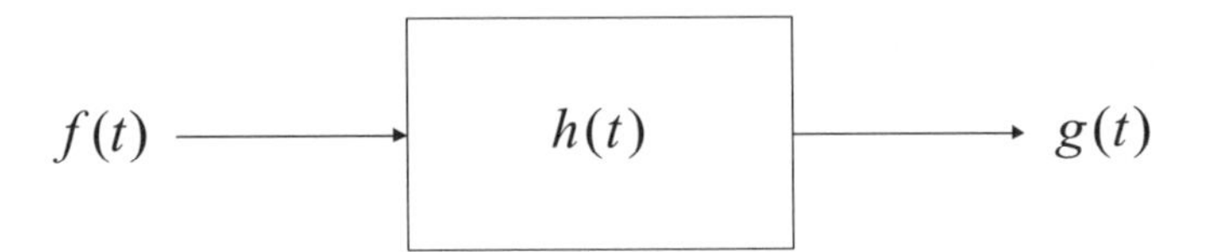

Fig. 8.30 Linear system

If $f(t)$ and $h(t)$ are functions of a limited spectral bandwidth and cutoff below $s = 1/2\Delta t$, $g(t)$ will be spectral bandwidth-limited, and thus, it can be recovered through interpolation. However, interception destroys the finiteness of the spectrum, and thus, a certain degree of aliasing is inevitable. This aliasing can be directly expressed in terms of $g(t)$. In summary, digital convolution does not produce effects other than those of sampling, interception, and interpolation [17].

Common convolution filters include smoothing filters, sharpening filters, first-order differential filters, and Roberts, Prewitt, and Sobel filters.

2. Frequency Domain Filtering

Figure 8.31 illustrates the calculation of a Fourier transform. The input signal $f(t)$ is sampled to obtain $x(t)$, which has a continuous periodic spectrum. In calculating the Fourier transform of $x(t)$, the equally spaced points on the main period of the periodic spectrum are calculated, as shown in the figure.

We calculate the N points with the same pitch between $-1/2\ \Delta t$ and $1/2\ \Delta t$. $Y(s)$ represents the calculated spectrum as it is spectrum $X(s)$ of $x(t)$.

Since $Y(s)$ is sampled, its inverse transform $Y(t)$ is a continuous (not sampled) and infinite persistence period function. The calculated spectrum $Y(s)$ is not $x(t)$ or the spectrum of unrecorded $f(t)$. Instead, $Y(s)$ is the spectrum of a continuous periodic function with period T. All the sampling points of $x(t)$ lie exactly in the main period of $y(t)$, which is the original function $f(t)$ of the sampling function $x(t)$.

If the inverse transformation is performed digitally, it is possible to recover $x(t)$ from $Y(s)$ and interpolate $x(t)$ to recover $f(t)$. In this case, the fact that $Y(s)$ corresponds to a periodic function does not lead to an adverse effect. Nevertheless, it is challenging to modify the spectrum to achieve digital filtering.

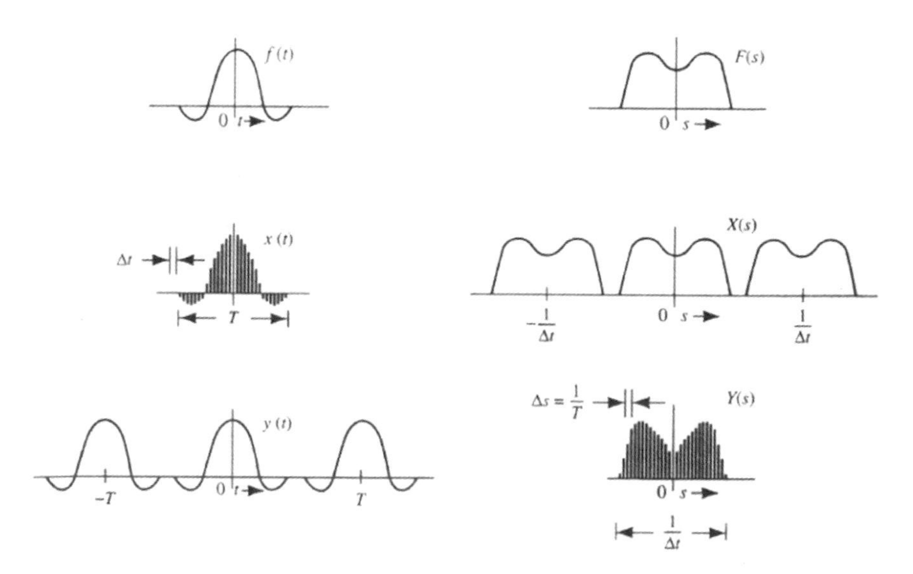

Fig. 8.31 Frequency domain filtering

The replication spectrum overlap is discussed in the following text. It is assumed that frequency domain filtering is performed by multiplying $Y(s)$ and transfer function $H(s)$. This process convolutes $y(t)$ with the impulse response $h(t)$. Since $y(t)$ is periodic, the convolution will move the adjacent period to the main period in the vicinity.

If $h(t)$ is narrow and $y(t)$ is approximately constant near $t = T/2$, the overlap of this adjacent period will have only a small effect. However, if $x(t)$ is not equal at both ends of the intercept window, $y(t)$ is discontinuous. In this case, $h(t)$ for convolution pertains to the discontinuous fuzzy pasting in the cutting window at both ends of the "artificial" traces.

Although the blurring effect at both ends of the intercept window cannot be completely avoided, the effect can be reduced to a tolerable level by (a) making the intercept window wider than the important part of the signal to retain the information of interest (b) adjusting $x(t)$ t to ensure that both ends of the intercept window have the same amplitude; in this case, when a periodic function is obtained, there will be no or only minor discontinuities. This aspect can be realized by multiplying the intercepted function by a specific window function. This function is the unit amplitude in most of the window and is reduced to zero at both ends.

The fuzzy effect at both ends of the intercept window encountered in the frequency domain filtering method corresponds to the aliasing caused by the sampling in the time domain. When the calculated spectrum is used to achieve linear filtering, the effect of interception must be quantitatively analyzed.

Common filters in the frequency domain are ideal low-pass, bandpass, band-stop, and n-order Butterworth filters.

8.5 Digital Processing

This section describes the image digitization process. The signal is intercepted in the time domain and sampled in the frequency domain. It is necessary to consider the sampling interval according to the sampling theorem and use the filter function to weaken the interception effect and aliasing error. Finally, the discrete image is converted to a digital image by interpolation, which is the digital processing flow of the image.

In this section, we only digitize a function and refactor it without processing it. The discussion begins with the continuous function f shown in Fig. 8.32. This function has a triangular amplitude spectrum and a random phase.

When this signal is digitized, it must be intercepted within a finite interval T. The intercept window $\text{II}(t/T)$ and its spectrum are shown in Fig. 8.33. The intercepted function and its frequency spectrum are shown. Intercepting $f(t)$ convolves the spectrum with a narrow sin x/x function.

The digitizer is equivalent to a framework with finite width of the sampling aperture at each sampling point. The signal is averaged over the aperture, and the local

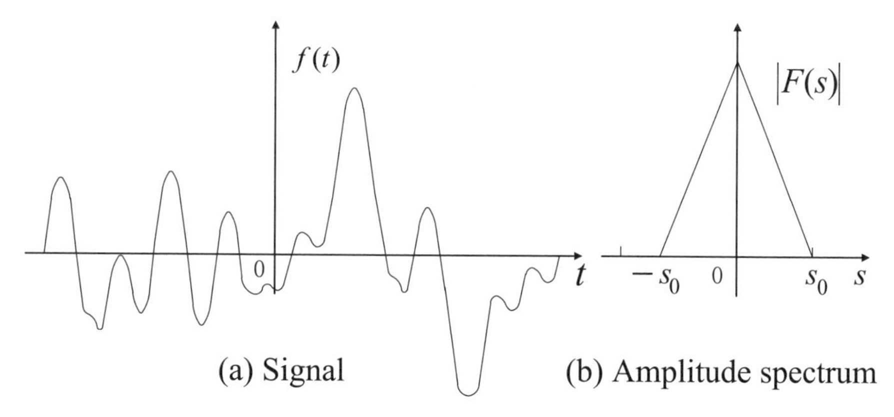

Fig. 8.32 Signals and their spectra

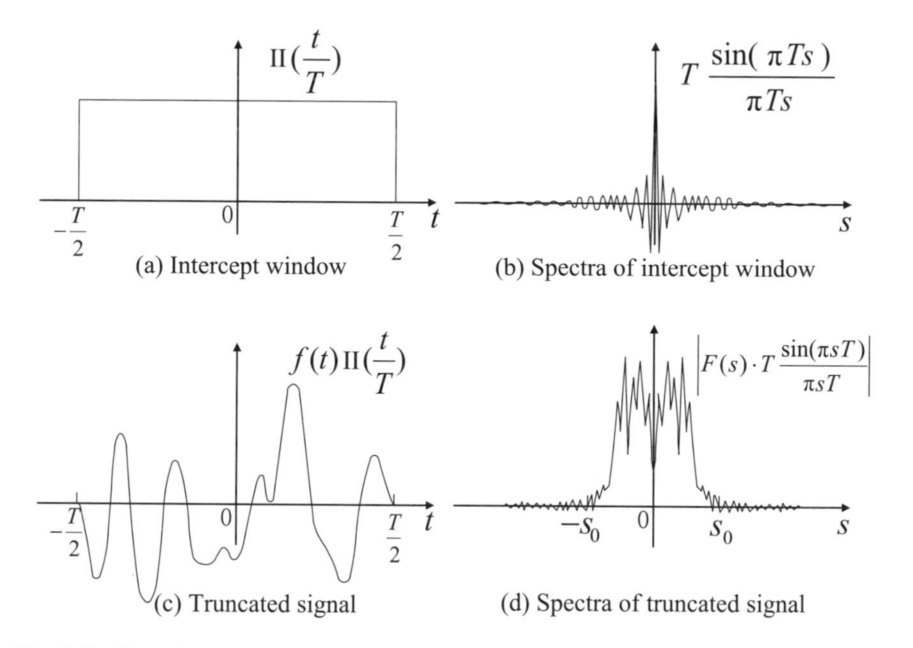

Fig. 8.33 Signal interception

average can be modeled as convolution with the appropriate sampling aperture function. For the image digitizer, the aperture function is a model of the spatial sensitivity of the scanning point. The electronic signal is commonly used with a circuit with a definite integration time, as shown in Fig. 8.34. A narrow rectangular pulse of width t is used as a model for the function of the sampling aperture. As shown in Fig. 8.34a, the convolution of the clipped signal with a sampled aperture function is equivalent to multiplying its spectrum by a fairly wide sin x/x function. If the sampling window

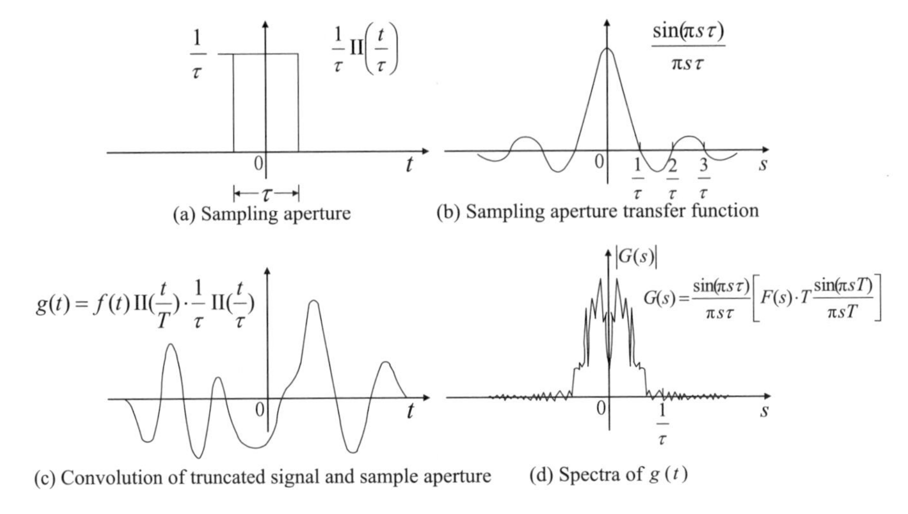

Fig. 8.34 Convolution with sampling aperture function

is a Gaussian function, the spectrum of the intercepted signal is multiplied by a wide Gaussian function. In both cases, the sampling aperture decreases the high-frequency energy in the signal. In Fig. 8.34, when the frequency exceeds $s = 1/\tau$, the polarity of the energy is reversed.

The sampling process is shown in Fig. 8.35. After the sampled signal is smoothed, the intercepted signal is multiplied by $\mathrm{III}(t/\Delta t)$ to obtain the sampled result. As shown in Fig. 8.35d, sampling causes the spectrum of the signal to be periodic because it duplicates the original spectrum at interval $1/\Delta t$.

The intent is to restore $f(t)$ by interpolation of the sampled signal. Figure 8.36 shows the interpolation performed by convolving a triangular pulse with the sampled function. In this figure, the width of the triangular pulse is $2t_0$. Convolving the sampled function with the interpolation function is equivalent to multiplying its spectrum by a function of the form $\sin^2 x/x^2$. As the function decreases with increasing frequency, all replicas tend to be zero except for the master replica at $s=0$.

The ideal interpolation function is sin x/x, which multiplies the function spectrum by the center pulse at $s=0$.

If $h(t)$ denotes the function obtained by interpolating the sampled function,

$$h(t) = \left(\left\{\left[f(t)\ \mathrm{II}\ \frac{t}{T}\right] * \frac{1}{\tau}\mathrm{II}\left(\frac{t}{\tau}\right)\right\} \mathrm{III}\left(\frac{t}{\Delta t}\right)\right) * \frac{1}{t_0}\Lambda\left(\frac{t}{t_0}\right) \tag{8.52}$$

The spectrum is

$$H(s) = \left(\left\{\left[F(s) * T\frac{\sin(\pi s\tau)}{\pi s\tau}\right]\frac{\sin(\pi s\tau)}{\pi s\tau}\right\} * \Delta t\,\mathrm{III}(s\Delta t)\right)\left[\frac{\sin(\pi s t_0)}{\pi s t_0}\right]^2 \tag{8.53}$$

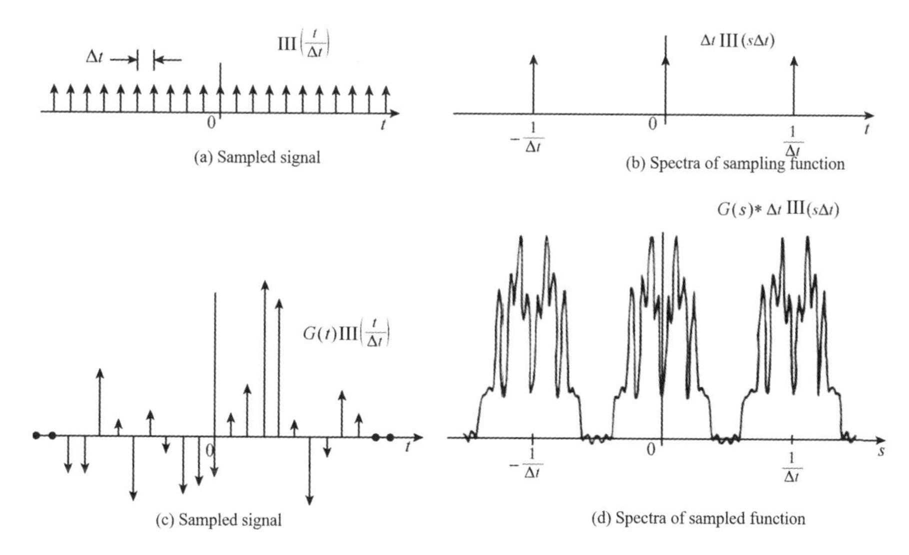

Fig. 8.35 Sampled signal

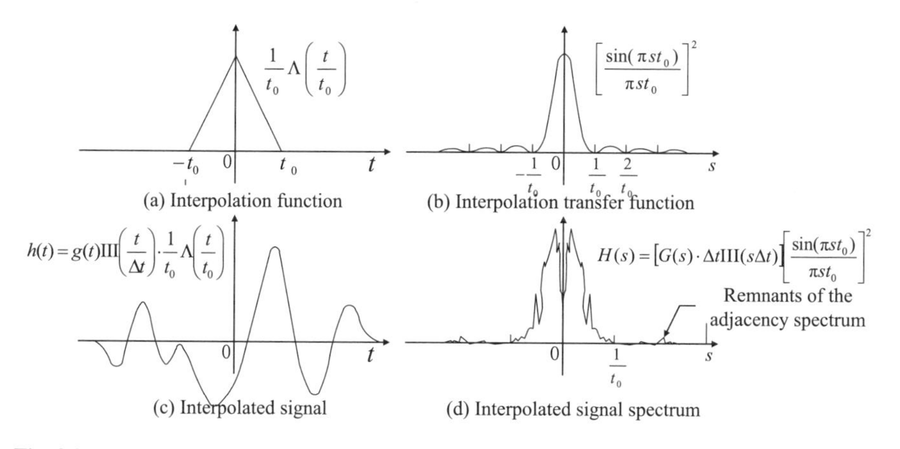

Fig. 8.36 Interpolating the sampled function

The problem is to identify the degree of influence on the digital processing. Although the sampling pore and interpolation functions can be arbitrarily selected, an appropriate relationship must be maintained between them. For example, the width of the sampling aperture τ must be approximately equal to the sampling pitch Δt. In addition, for linear interpolation, $t_0 = \Delta t$.

As shown in Fig. 8.36, the sampling aperture tends to decrease the high-frequency energy in the spectrum, which alleviates aliasing. If the sampling aperture has a negative transfer function, the polarity of the high-frequency energy is reversed.

Sampling causes the spectrum to become periodic, which can lead to aliasing of the energy above the folding frequency.

8.6 Summary

The core of this chapter is the Shannon sampling theorem and its extension. The sampling theorem provides an important theoretical basis in the framework of digital image processing. The general process of sampling is to input a continuous signal, select the sampling interval, discretize the signal, and obtain the discrete signal spectrum through the window function truncation signal and inverse Fourier transform to obtain the original signal. The discrete process can be abstracted as the Shah function multiplied by the original signal. The key to the introduction of the sampling theorem is to control the spectral aliasing.

In this chapter, the sampling theorem is extended to the frequency sampling theorem and two-dimensional sampling theorem. The theoretical model of the two-dimensional sampling theorem is consistent with the digital image. It is convenient to use the sampling theorem to sample and obtain the signal of the two-dimensional image. This aspect is of significance in remote sensing, medicine, earth science and other applications. An important mathematical basis of this chapter is the Fourier transform, which can be used to derive the original signal spectrum, or the discrete signal inverse transform, which can be used to obtain the original signal.

The contents of this chapterand mathematical formulas aresummarized in Fig. 8.37 and Table 8.2, respectively.

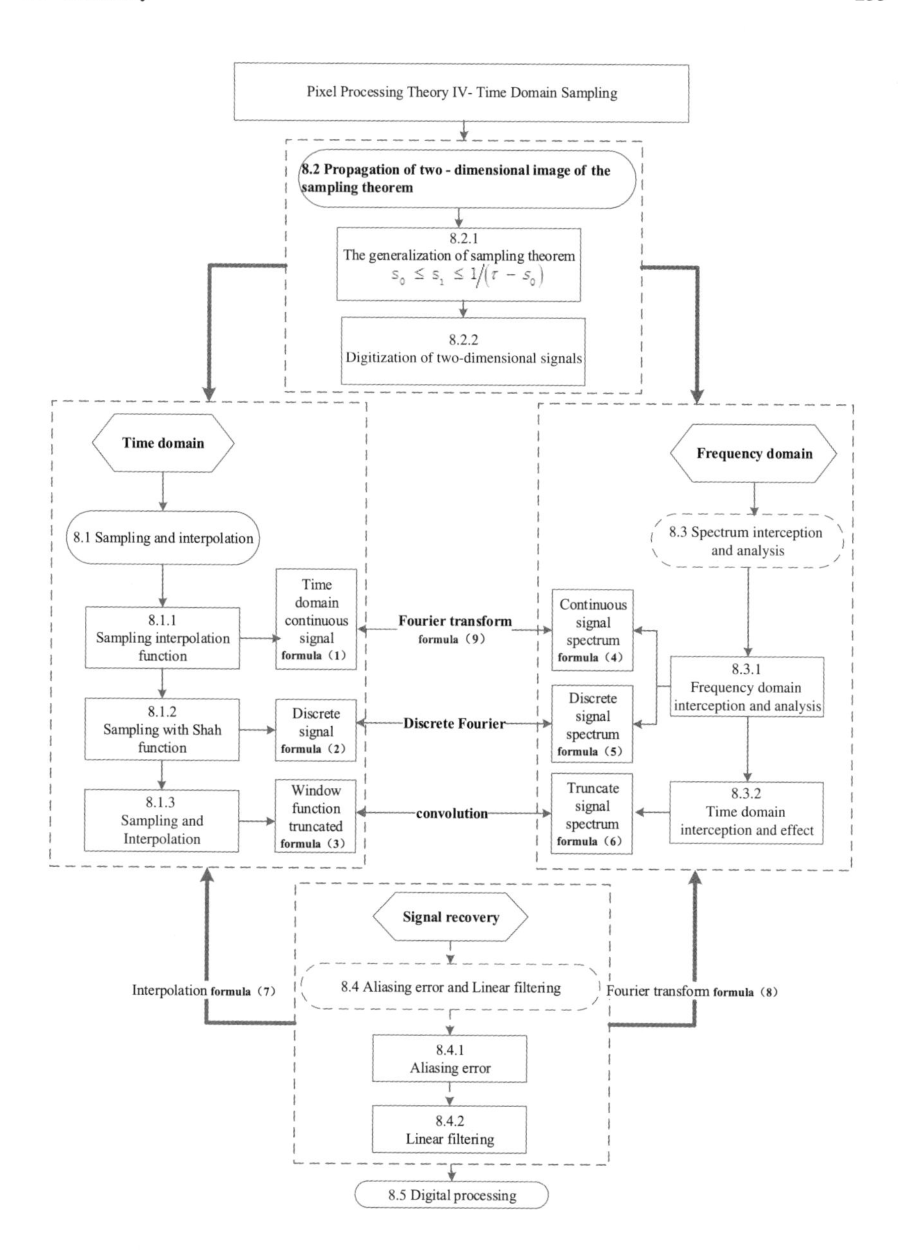

Fig. 8.37 Summary of the contents of this chapter

Table 8.2 Summary of the formulas in this chapter

formula (1)	$y = g(t)$	8.1.3
formula (2)	$\text{III}(\frac{x}{\tau}) = \tau \sum_{n=-\infty}^{\infty} \delta(x - n\tau)$	8.1.1
formula (3)	$F(s) = G(s)$	8.1.3
formula (4)	$\hat{X}(j\Omega) = \int_{-\infty}^{\infty} \hat{x}(t) e^{-j\Omega t} dt = \int_{-\infty}^{\infty} x(t) p_\delta(t) e^{-j\Omega t} dt$	8.2.1
formula (5)	$G(s)\Im\{g(x)\} = \Im\{\text{II}(\frac{x}{T/2})\}\Im\{\delta(x - \frac{T}{4}) - \delta(x + \frac{T}{4})\}$	8.3.2
formula (6)	$f(x) = g(x) * 2s_1 \frac{\sin(2\pi s_1 x)}{2\pi s_1 x}$	8.1.3

References

1. Xia WJ, Wang F. Understanding frequency domain sampling theorem in digital signal processing. Sci Technol Innovation Herald. 2012(18). (In Chinese).
2. Chen AL. Theoretical proof and validation of frequency domain sampling theorem. China New Telecommun. 2012(14). (In Chinese).
3. Qin T, Li YF, Yang SZ. The development of multiwavelet sampling theorem. Acta Mathematica Sinica. 2012(55). (In Chinese).
4. Yan BB, Shi MM, Lv JF. Spectral analysis reconstruction of original continuous signal. China E-commerce. 2011(8). (In Chinese).
5. Zhang CW, Xiang Q, Qin KY. Research on band-limited signal sampling theory of linear regular transform domain. Acta Electronica Sinica. 2010(1). (In Chinese).
6. Pan JX, Mo HY. Insufficient pixel-based artifact processing method. Chin J Stereology Image Anal. 2008(13). (In Chinese).
7. Liu Y, Li XH, Wang YL. Undersampling-based frequency measurement method. J Electron Measur Instrum. 2010(1). (In Chinese).
8. Tang YY, Yang SZ, Cheng ZX. Approximate sampling theorem of two-dimensional continuous signal. Appl Math Mech. 2003(24). (In Chinese).
9. Wang S, Chen MX, Wang AM. Anti-aliasing non-uniform periodic sampling and spectrum analysis method. Signal Process. 2005(21). (In Chinese).
10. Ye ZF, Ding ZZ. Spectral alias-free sampling and signal fully reconfigurable sampling. J Data Acquisition Process. 2005(20). (In Chinese).
11. Rong YT, Chen ZL. An interpolation algorithm based on Shannon sampling theorem. J Northwest Text Inst. 2000(14). (In Chinese).
12. Marks IRJ. Introduction to Shannon sampling and interpolation theory. Berlin: Springer. 2012.
13. Marks IRJ. Advanced topics in Shannon sampling and interpolation theory. New York: Springer. 1993.
14. Olevskii A, Ulanovskii A, On multi-dimensional sampling and interpolation. Anal Math Phys. 2012.2(2).
15. Hu Y. Truncation error analysis of sampling theorem. J Graduate Sch Chin Acad Sci. 2008(25). (In Chinese).
16. Osgood B, Siripuram A, Wu W. Discrete sampling and interpolation: universal sampling sets for discrete bandlimited spaces. IEEE Trans Inf Theory. 2012;58(7).
17. Wang C Y, Hou Z X, He K, et al. JPEG-based image coding algorithm at low bit rates with down-sampling and interpolation. In: The international conference on wireless communications, networking and mobile computing. 2018.

Chapter 9
Basics of Remote Sensing Digital Image Pixel Transformation I: Space–Time Equivalent Orthogonal Basis

Chapter Guidance According to the Fourier reciprocal frequency domain transformation described in Chap. 6, remote sensing digital image transformation represents a two-dimensional matrix transformation. Any transformation is an orthogonal unitary transformation [1], that is, the transformation matrix is full rank. In other words, a cell matrix unit transformation corresponds to that before and after, and transformation and inverse transformation are performed. Through transformation, any image can be transformed into a linear combination of a set of uncorrelated orthogonal basis images whose coefficients are not zero. To determine the merits of the transformation, the number of nonzero coefficients can be compared: a smaller number of coefficients corresponds to more efficient operation and simpler storage. This aspect is the mathematical essence of image processing, transmission, storage and compression and summarizes the theory of linear representation of the basis function–basis vector–basis image. This chapter is the focus of the book, specifically, the basis of orthogonal transforms of different forms of remote sensing digital images.

The contents of this chapter are as follows: Sect. 9.1 Linear transformation: This section describes the basic properties of linear transformation and theoretical basis of orthogonal transform; Sect. 9.2 Basis function, basis vector and basis image: This section describes the process of transforming the basis function to the basis vector and basis vector, thereby obtaining the basis image. Sect. 9.3 Instance transformation. The derivation of the Haar and Hadamard transformations strengthens the theory of the basis function, basis vector and basis image. Section 9.4 Based on the derivation of the Hadamard transformation, we present a sample application to help the reader understand the process of transformation. Figure 9.1 illustrates the logical framework of this chapter.

The discrete images in linear space are divided into two steps: vector algebraic transformation and matrix expression.

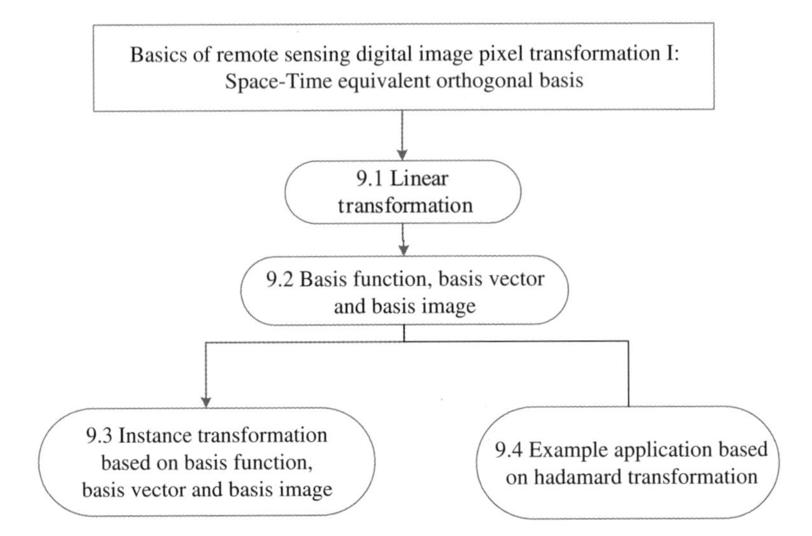

Fig. 9.1 Logical diagram of this chapter

9.1 Linear Transformation

This section presents the definition of the linear space and basic properties of linear transformation. Linear transformation is a mapping from one vector space V to another vector space W while maintaining the addition and multiplication operations. Linear transformation is performed in linear space. Subsequently, from the concept of one-dimensional discrete linear transformation and unitary transformation, the two-dimensional discrete linear transformation process is described, and the transformation of arbitrary discrete images is performed for the linear combination of orthogonal basis images.

9.1.1 Linear Space

Definition 1 To establish an operation between the elements in V, i.e., $\forall\alpha, \beta \in V, \exists! \gamma \in V$ corresponding to V, with $\gamma = \alpha + \beta$ referred to as an adder. Symbols "∀", "∃", and "!" represent "for any", "exist", and "the only one," respectively.

Definition 2 To perform a computation between the elements in sets V and P, i.e., c is a value of complex field C. $\forall k \in \Pi$, $\alpha \in c$, $\exists! \eta \in c$ corresponding to this value, defined as [alpha] k. The product (multiplication) is defined as $\eta = k * \alpha$. Multiplication symbol * may be omitted if the above two operations satisfy the following eight axioms:

(1) $\forall\alpha, \beta \in V, \alpha + \beta = \beta + \alpha.$

(2) $\forall \alpha, \beta, \gamma \in V, \alpha + (\beta + \gamma) = (\alpha + \beta) + \gamma$.
(3) $\forall \alpha \in c, \exists 0 \in V$, and thus, $\alpha + 0 = \alpha$. (existence of the additive identity; 0 known as the additive identity or zero element)
(4) $\forall \alpha \in c, \exists \beta \in V$, and thus, $\alpha + \beta = 0$ (presence of additive inverse; for α and β, the additive inverse is $-\alpha$)
(5) $\forall \alpha \in c, 1 * \alpha = \alpha$.
(6) $\forall k, l \in c, \exists \alpha \in V$, and thus, $(kl)\,\alpha = k(l\alpha)$.
(7) $\forall k, l \in c, \exists \alpha \in V$, and thus, $(k + l)\,\alpha = k\alpha + l\alpha$.
(8) $\forall k \in c, \exists \alpha, \beta \in V$, and thus, $k\,(\alpha + \beta) = k\alpha + k\beta$.

V corresponds to the linear space P, referred to as $V(P)$, and defines the linear elements in a certain amount of space. The operation that satisfies the abovementioned eight axioms is known as a linear operation. If P is a real number field R, V is a real linear space; if P is a complex field C, V is a complex linear space.

9.1.2 Linear Transformation

1. Definition of Linear Transformation

Definition 3 The linear transformation T in space V refers to a linear transformation. If any $\alpha, \beta, k \in P$, $T(\alpha + \beta) = T(\alpha) + T(\beta)$; $T(k\alpha) = kT(\alpha)$. Thus, $T: V \to V$, and T is the linear transformation. If and only if for any $\alpha \in V$, $T(\alpha) = S(\alpha)$, the two linear transformations S and T are equal.

2. Basic Properties of the Linear Transformation

Assume that T is a linear transformation, α is any vector in a plane k, and a is a real number. The basic properties of the linear transformation are as follows:

Property 1 *Linear transformation refers to the process of changing the zero vector in linear space to a zero vector of the image, i.e., if* α *is a linear transformation of* V, $\sigma(0) = 0$. *Note: In* $\sigma(k\alpha) = k\sigma(\alpha)$, *set* $k = 0$.

Property 2 *Linear transformation refers to the process of changing the negative vector in linear space to a negative vector of images, i.e., if* α *is a linear transformation of* V, $\sigma(-\alpha) = -\sigma(\alpha)$. *Note: In* $\sigma(k\alpha) = k\sigma(\alpha)$, *set* $k = -1$.

Property 3 α *is a linear transformation of* V, *and the necessary and sufficient conditions of* σ *are* $\forall \alpha, \beta \in V, k_1, k_2 \in F$, so $\sigma(k_1\alpha + k_2\beta) = k_1\sigma(\alpha) + k_2\sigma(\beta)$.

Proof If $\sigma(k_1\alpha + k_2\beta) = k_1\sigma(\alpha) + k_2\sigma(\beta)$, and $k_1 = k_2 = 1$, $\sigma(\alpha + \beta) = \sigma(\alpha) + \sigma(\beta)$. If $k_2 = 0$, $(k_1\alpha) = k_1\sigma(\alpha)$. Therefore, α is a linear transform of V, and the adequacy is proven

Property 4 *Linear transformation ensures that the linear combination of linear relationships is unchanged, i.e., if* α *is a linear transformation of* V, *and* $\beta = k_1\alpha_1 + k_2\alpha_2 + \cdots + k_r\alpha_r$, $\sigma(\beta) = k_1\sigma(\alpha_1) + k_2\sigma(\alpha_2) + \cdots + k_r\sigma(\alpha_r)$.

Proof If $\alpha_1, \alpha_2, \ldots, \alpha_r$ represent a linear correlation, $\sigma(\alpha_1)$, $\sigma(\alpha_2)$. $\sigma(\alpha_r)$ are also linearly related.

3. Basic Operations of Linear Transformation

Definition 4 T_1, T_2, T_3 are linear transformations on *V*, and the following three basic operations can be defined:

(1) The sum of linear transformations $T = T_1 + T_2$ is defined as $T(\alpha) = (T_1 + T_2)(\alpha) = T_1(\alpha) + T_2(\alpha), \forall \alpha \in V$
(2) The product of linear transformation $T = T_1 T_2$ is defined as $T(\alpha) = T_1(T_2(\alpha)), \forall \alpha \in A$
(3) The quantitative multiplication of linear transformation $T = kT_1$ is defined as $T(\alpha) = k(T_1(\alpha)), \forall \alpha \in V, k \in P$
(4) Inverse transformation of linear transformation

Definition 5 If *I* is the unit linear transformation of linear space *V*, *T* is the linear transformation. If there is one linear transformation that makes $TS = ST = I$, we define linear transformation *T* as reversible. *S* is the inverse transformation of *T*, defined as T^{-1}.

9.1.3 One-Dimensional Linear Transformation

Definition If $\boldsymbol{x}$ is a vector of *N*, and $\boldsymbol{T}$ is a matrix of *N*,

$$y_i = \sum_{j=0}^{N-1} t_{i,j} x_j \text{ or } \boldsymbol{y} = \boldsymbol{T}\boldsymbol{x} \tag{9.1}$$

This expression represents a linear transformation of vector $\boldsymbol{x}$ (where $i = 0, 1, 2, \ldots, N-1$). Matrix $\boldsymbol{T}$ is also known as the nuclear matrix of this transformation. Notably, in terms nuclear and convolution of nuclear, nuclear has different meanings.

The conversion result is another *N*-dimensional vector $\boldsymbol{y}$. This transformation, referred to as a linear transformation $\boldsymbol{y}$, consists of the input element and first order. Each element of y_i is the inner product of the input vector $\boldsymbol{x}$ and $\boldsymbol{T}$.

A simple example of a linear transformation is a rotation vector of the two-dimensional coordinate system, where vector $\boldsymbol{x}$ represents the rotation angle θ.

$$\begin{bmatrix} y_1 \\ y_2 \end{bmatrix} = \begin{bmatrix} \cos\theta & -\sin\theta \\ \sin\theta & \cos\theta \end{bmatrix} \begin{bmatrix} x_1 \\ x_2 \end{bmatrix} \tag{9.2}$$

If $\boldsymbol{T}$ is nonsingular, the original vector can be recovered through the inverse transform:

$$\boldsymbol{x} = \boldsymbol{T}^{-1}\boldsymbol{y} \tag{9.3}$$

Each element of $\boldsymbol{x}$ is the inner product of $\boldsymbol{y}$ and a certain row of $\boldsymbol{T}^{-1}$. For the previous example, this aspect is equivalent to rotating the same angle in the opposite direction.

9.1.4 Unitary Transformation

The basic linear expression is strictly reversible and satisfies certain orthogonal conditions, i.e., it corresponds to a unitary transform.

If $\boldsymbol{T}$ is a unitary matrix,

$$\boldsymbol{T}^{-1} = \boldsymbol{T}^{*\mathrm{T}} \text{and } \boldsymbol{T}\boldsymbol{T}^{*\mathrm{T}} = \boldsymbol{T}^{*\mathrm{T}}\boldsymbol{T} = \boldsymbol{I} \tag{9.4}$$

where each element $\boldsymbol{T}^*$ represents the complex conjugate, and T represents transposition. If $\boldsymbol{T}$ is a unitary matrix and all elements are real numbers, it is an orthogonal matrix and satisfies

$$\boldsymbol{T}^{-1} = \boldsymbol{T}^{\mathrm{T}} \text{and } \boldsymbol{T}\boldsymbol{T}^{\mathrm{T}} = \boldsymbol{T}^{\mathrm{T}}\boldsymbol{T} = \boldsymbol{I} \tag{9.5}$$

Element (i, j) is the inner product of Rows i and j of $\boldsymbol{T}$. As indicated in Formula 9.5, when $i = j$, the inner product is 1; otherwise, the product is 0. Therefore, the rows of $\boldsymbol{T}$ represent a set of orthogonal vectors.

In general, the transformation matrix $\boldsymbol{T}$ is nonsingular (i.e., the rank of $\boldsymbol{T}$ is N), which makes the transformation reversible. Thus, all rows of $\boldsymbol{T}$ form an orthogonal basis (a set of orthogonal basis vectors or unit vectors) of an N-dimensional vector space.

Thus, a linear unitary transformation produces a vector $\boldsymbol{y}$ with N transform coefficients, each of which is the inner product of the input vector $\boldsymbol{x}$ and a row of the transformation matrix $\boldsymbol{T}$. The inverse transformation is similar to the computation, and a set of inner products is generated by the transform coefficient vector and row of the inverse transformation matrix.

Positive transformations are often considered a decomposition process: the signal vector is decomposed into its elementary components, which are naturally expressed in the form of a basis vector. The transformation factor specifies the amount of each component in the original signal. The inverse transform is treated as a synthetic process, that is, components are added to synthesize the original vector. The transform coefficients specify the number of individual components added to accurately and completely reconstruct the input vector.

The transformed vector is a representation of the original vector. Since this vector has the same number of elements as the original vector (i.e., the same degree of freedom), and the original vector can be recovered without error, it can be treated as another representation of the original vector.

9.1.5 Two-Dimensional Discrete Linear Transformation

For a two-dimensional case, an N-dimensional matrix $\boldsymbol{F}$ is transformed into another N-dimensional matrix $\boldsymbol{G}$, and the general form of the linear transformation is

$$\boldsymbol{G}_{m,n} = \sum_{i=0}^{N-1}\sum_{k=0}^{N-1} \boldsymbol{F}_{i,k}(i,k,m,n) f(i,k) \tag{9.6}$$

where i, k, m, and n are discrete variables in the range 0 to N-1, and $\boldsymbol{F}(i, k, m, n)$ is the transformed kernel function. As shown in Fig. 9.2, the $\boldsymbol{F}(i, k, m, n)$ matrix can be considered a block of $N^2 \times M^2$ elements, and each row has N blocks. For N rows, each block is an $N \times M$ matrix. The block is indexed by m and n, and the elements of each block (submatrix) are indexed by i and k.

If $\boldsymbol{F}(i, k, m, n)$ can be decomposed, the component functions in the row direction and column direction are the component functions of the product, i.e., if

$$\boldsymbol{F}(i,k,m,n) = \boldsymbol{T}_r(i,m)\boldsymbol{T}_c(k,n) \tag{9.7}$$

the transformation is separable. In other words, the transformation can be performed in two steps: a row operation, followed by a column (or vice versa) operation, that is,

$$\boldsymbol{G}_{m,n} = \sum_{i=0}^{N-1}\left[\sum_{k=0}^{N-1} \boldsymbol{F}_{i,k}\boldsymbol{T}_c(k,n)\right]\boldsymbol{T}_r(i,m) f(i,k) \tag{9.8}$$

Furthermore, if the two component functions are the same, this transformation can be termed symmetric (not to be confused with the symmetric matrix). In this case,

$$\begin{array}{c c} & \begin{array}{ccc} n=1 & n=2 & n=N \end{array} \\ \begin{array}{c} m=1 \\ m=2 \\ \\ m=M \end{array} & \begin{pmatrix} [\ ] & [\ ] & \cdots & [\ ] \\ [\ ] & [\ ] & \cdots & [\ ] \\ \vdots & \vdots & \ddots & \vdots \\ [\ ] & [\ ] & \cdots & [\ ] \end{pmatrix} \end{array}$$

Fig. 9.2 Nuclear matrix

$$F(i,k,m,n)=T(i,m)T_r(i,m) \tag{9.9}$$

Formula 9.8 can be expressed as

$$G_{m,n}=\sum_{i=0}^{N-1}T(i,m)\left[\sum_{k=0}^{N-1}F_{i,k}T(k,n)f(i,k)\right]\text{or}G=TFT \tag{9.10}$$

where T is a unitary matrix, known as the transformed kernel matrix. In this chapter, we use this representation to indicate a common, separable, symmetric unitary transformation.

Inverse transform

$$F=T^{-1}GT^{-1}=T^{*t}GT^{*t} \tag{9.11}$$

can accurately restore F.

9.2 Basis Function, Basis Vector and Basis Image

In this section, we describe the process of transforming the basis function to the basis vector to obtain the basis image, and mathematically express the transformation relationship between the basis function, basis vector and basis image.

The link between discrete transformations (discrete Fourier transform, discrete cosine transform, discrete Walsh transform, and discrete Hadamard transform) can be determined by the concepts of the basis function, basis vector and basis image. The difference between the various linear transformations is only the basis function [2]. Different functions correspond to different transformations.

9.2.1 Basis Function

The basis function is a specific set of rules to generate basis vectors. The basis function is assumed to be f, and the basis vectors can be generated by the basis functions f_i, i.e.,

$$f_i=\begin{bmatrix}a_{i1}\\ \dots\\ a_{iN}\end{bmatrix} \tag{9.12}$$

f_i satisfies 9.1 when

$$f_i^{\mathrm{T}} f_j = \begin{cases} 1, & i = j \\ 0, & i \neq j \end{cases} \tag{9.13}$$

The Fourier transform uses a complex exponential as the basis of its prototype function, and the basis of the function is different frequencies. Similarly, the cosine, Haar and Hadamard transformations satisfy the definition of the basis function.

9.2.2 Basis Vector

A unit vector is a basis vector, also known as a unit vector. Using the basis vector, a simple one-dimensional discrete transformation can be performed. Suppose there exist n orthogonal vectors

$$\boldsymbol{a}_1 = \begin{bmatrix} a_{11} \\ a_{21} \\ \vdots \\ a_{n1} \end{bmatrix}, \quad \boldsymbol{a}_2 = \begin{bmatrix} a_{12} \\ a_{22} \\ \vdots \\ a_{n2} \end{bmatrix}, \quad \cdots, \quad \boldsymbol{a}_n = \begin{bmatrix} a_{1n} \\ a_{2n} \\ \vdots \\ a_{nn} \end{bmatrix} \tag{9.14}$$

$$\sum_{k=1}^{n} a_{ki} a_{kj} = \begin{cases} C, & i = j \\ 0, & i \neq j \end{cases} \tag{9.15}$$

When $C = 1$, the matrix is normalized orthogonal. That is, each vector is a unit vector, and the matrix of the basis vector satisfying this equation is

$$\boldsymbol{A} = \begin{bmatrix} a_{11} & a_{12} & \cdots & a_{1n} \\ a_{21} & a_{22} & \cdots & a_{2n} \\ \vdots & \vdots & & \vdots \\ a_{n1} & a_{n2} & \cdots & a_{nn} \end{bmatrix} \tag{9.16}$$

satisfying

$$\boldsymbol{A}^{\mathrm{T}} \boldsymbol{A} = \boldsymbol{A}\boldsymbol{A}^{\mathrm{T}} = \boldsymbol{I} \tag{9.17}$$

For $N \times N$ matrix $\boldsymbol{A}$, there exist N scalars λ_k $(K = 0,\ldots, N-1)$, and thus, $|\boldsymbol{A} - \lambda_k \boldsymbol{I}| = 0$, where λ_k is the eigenvalue of the matrix. In addition, N satisfies $\boldsymbol{A}\boldsymbol{v}_k = \lambda_k \boldsymbol{v}_k$. Vector $\boldsymbol{v}_k$ is known as the characteristic vector of $\boldsymbol{A}$. The eigenvectors are $N \times 1$-dimensional, and each $\boldsymbol{v}_k$ corresponds to an eigenvalue λ_k. These feature vectors form an orthogonal basis set. The orthogonal basis set is the basis vector, and its number is equal to rank r of the eigenvector matrix. When the matrix is full rank, that is, $r = N$, the number of basis vectors is N.

Any vector $\boldsymbol{F}_i$ can be weighted by k_i to reconstruct the value, as shown in Formula 9.18

$$\boldsymbol{F}_i = \begin{bmatrix} x_1 \\ x_2 \\ \cdots \\ x_n \end{bmatrix} = \sum_{i=1}^{n} k_i \boldsymbol{a}_i \tag{9.18}$$

In image processing, the operation is performed using a two-dimensional matrix ($m \times n$). The computation of linear algebra is a one-dimensional vector when $n = 1$ or $m = 1$, and the mathematics is directed to a single number (dimension number 1×1, generally not marked) of the process. These entities have a close relationship, as shown in Eq. 9.19. If the variable is a function of respective units and satisfies this relationship when the dimension is 1×1, it can be expressed as a basis function, also known as a common function in higher mathematics.

$$\underset{\text{Image}}{\begin{bmatrix} a_{11} & \cdots & a_{1n} \\ \vdots & \ddots & \vdots \\ a_{m1} & \cdots & a_{mn} \end{bmatrix}} \overset{\text{if } n=1}{\Rightarrow} \underset{\text{Vector}}{\begin{bmatrix} a_{11} \\ \vdots \\ a_{m1} \end{bmatrix}} \overset{m=1}{\Rightarrow} \underset{\text{Number}}{\begin{bmatrix} a_{11} \end{bmatrix}_{1*1}} = a_{11} \tag{9.19}$$

9.2.3 Basis Image

Basis image $\boldsymbol{F}_{i,j}$ can be generated by the outer product of any two sets of basis vectors $\boldsymbol{f}_i$, i.e.,

$$\boldsymbol{F}_{i,j} = \boldsymbol{f}_i \boldsymbol{f}_j^{\mathrm{T}} = \begin{bmatrix} a_{i1} \\ \cdots \\ a_{iN} \end{bmatrix} \begin{bmatrix} a_{j1} & \dots & a_{jN} \end{bmatrix} = \begin{bmatrix} a_{i1}a_{j1} & \dots & a_{i1}a_{jN} \\ \dots & \dots & \dots \\ a_{iN}a_{j1} & \dots & a_{iN}a_{jN} \end{bmatrix} \tag{9.20}$$

When the basis vector $\boldsymbol{f}_i$ is $N \times 1$, the basis image is N^2.

Any image $X(i, j)$ can be described through the basis image $\boldsymbol{F}_{i,j}$ and weighted to achieve reconfiguration, i.e.,

$$X(i, j) = \begin{bmatrix} x_{11} & \dots & x_{1N} \\ \dots & \dots & \dots \\ x_{N1} & \dots & x_{NN} \end{bmatrix} = \sum_{i,j=1,\dots,N} k_{ij} \boldsymbol{F}_{ij} \tag{9.21}$$

where k_{ij} is not equal to 0, and a smaller number corresponds to more effective results of image conversion. This value is an objective evaluation of the image conversion efficiency. In digital image processing, the computer only stores functions, k values, and indices i and j. When the image has N^2 pixels and N is sufficiently large, if the nonzero number is m, the computer storage unit consists of 3 m weighting units and a basis function f unit. When m is considerably smaller than N, the size of the image transformation decreases by an order of magnitude.

9.3 Instance Transformation Based on the Basis Function, Basis Vector and Basis Image

In this section, based on the theory of linear transformation and the understanding of the basis function, basis vector and basis image, we consider the Haar and Hadamard transformations to illustrate the process.

9.3.1 Haar Transformation

The Haar transformation is a symmetric and separable unitary transformation, which uses the Haar function as a basis function. In this case, $N = 2^n$, where n is an integer.

The Fourier transform basis function includes different frequencies, and the Haar functions in the scale (width) and position are different. Therefore, the dimensions and position of the Haar transformation have dual properties, evident in their basis functions [3]. This attribute establishes the basis for the wavelet transform discussed in the next chapter.

1. Basis Function Definition

Since the Haar function changes in both scale and position, a double indexing mechanism must exist. The Haar function defined in the [0, 1] interval is as follows. For an integer $0 \leq k \leq N - 1, p$ and q are uniquely determined:

$$k = 2^p + q - 1 \tag{9.22}$$

In this configuration, k is not only a function of p and q, but p and q are functions of k. For any $k > 0$, p is the maximum power of 2 in $2^p \leq k$, and $q - 1$ is the remainder.

The Haar function is defined as

$$h_0(x) = \frac{1}{\sqrt{N}} \tag{9.23}$$

and

$$h_k(x) = \frac{1}{\sqrt{N}} \begin{cases} 2^{p/2}, & \frac{q-1}{2^p} \le x < \frac{q-\frac{1}{2}}{2^p} \\ -2^{p/2}, & \frac{q-\frac{1}{2}}{2^p} \le x < \frac{q}{2^p} \\ 0, & \text{others} \end{cases} \tag{9.24}$$

For $i = 0, 1, \cdots, N-1$, if $x = i/N$, one can generate a set of basis functions, except that $k = 0$ is a constant, and each basis function has a separate rectangular pulse pair. The changes in scale (width) and position are shown in Fig. 9.3. The index p specifies the scale, and q determines the amount of translation (position).

2. Basis Vector Matrix

According to Formula 9.24, when N = 8 and $k = 0$, $\sqrt{8}h_0(x) = 1$.

When $k = 1$, $2^p \le 1$, and thus, $p = 0$, $q = k - 2^p + 1 = 1$, and

$$h_1(x) = \frac{1}{\sqrt{8}} \begin{cases} 1, & 0 \le x < \frac{4}{8} \\ -1, & \frac{4}{8} \le x < 1 \end{cases} \tag{9.25}$$

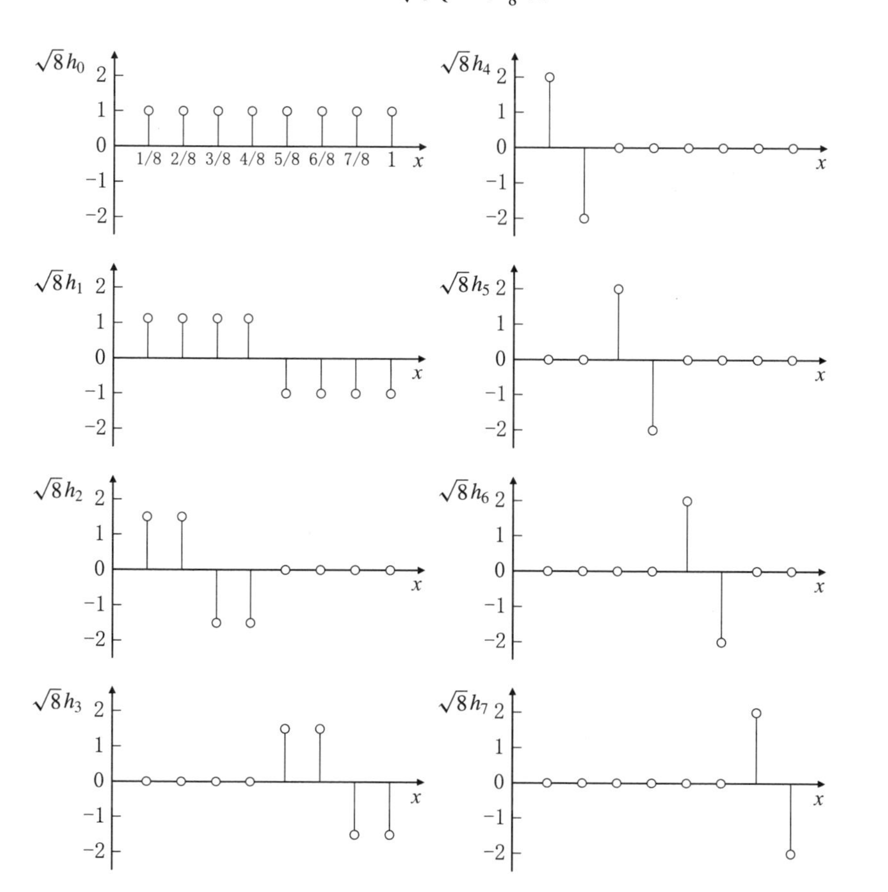

Fig. 9.3 Haar transformation basis function

When $k = 2$, $2^p \le 2$, and thus, $p = 1, q = k - 2^p + 1 = 1$, and

$$h_2(x) = \frac{1}{\sqrt{8}} \begin{cases} \sqrt{2}, & 0 \le x < \frac{2}{8} \\ -\sqrt{2}, & \frac{2}{8} \le x < \frac{4}{8} \\ 0, & \frac{4}{8} \le x < 1 \end{cases} \tag{9.26}$$

When $k = 3$, $2^p \le 3$, and thus, $p = 1, q = k - 2^p + 1 = 2$, and

$$h_3(x) = \frac{1}{\sqrt{8}} \begin{cases} \sqrt{2}, & \frac{4}{8} \le x < \frac{6}{8} \\ -\sqrt{2}, & \frac{6}{8} \le x < 1 \\ 0, & 0 \le x < \frac{4}{8} \end{cases} \tag{9.27}$$

When $k = 4$, $2^p \le 4$, and thus, $p = 2, q = k - 2^p + 1 = 1$, and

$$\sqrt{8}h_4(x) = \begin{cases} 2, & 0 \le x < \frac{1}{8} \\ -2, & \frac{1}{8} \le x < \frac{2}{8} \\ 0, & \frac{2}{8} \le x < 1 \end{cases} \tag{9.28}$$

When $k = 5$, $2^p \le 5$, and thus, $p = 2, q = k - 2^p + 1 = 2$, and

$$\sqrt{8}h_5(x) = \begin{cases} 2, & \frac{2}{8} \le x < \frac{3}{8} \\ -2, & \frac{3}{8} \le x < \frac{4}{8} \\ 0, & 0 \le x < \frac{2}{8}, \frac{4}{8} \le x < 1 \end{cases} \tag{9.29}$$

When $k = 6$, $2^p \le 6$, and thus, $p = 2, q = k - 2^p + 1 = 3$, and

$$\sqrt{8}h_6(x) = \begin{cases} 2, & \frac{4}{8} \le x < \frac{5}{8} \\ -2, & \frac{5}{8} \le x < \frac{6}{8} \\ 0, & 0 \le x < \frac{4}{8}, \frac{6}{8} \le x < 1 \end{cases} \tag{9.30}$$

When $k = 7$, $2^p \le 7$, and thus, $p = 2, q = k - 2^p + 1 = 4$, and

$$\sqrt{8}h_7(x) = \begin{cases} 2, & \frac{6}{8} \le x < \frac{7}{8} \\ -2, & \frac{7}{8} \le x < 1 \\ 0, & 0 \le x < \frac{6}{8} \end{cases} \tag{9.31}$$

The result is derived from the above formula to obtain the results shown in Fig. 9.3. The 8×8 unitary core matrix of the Haar transformation is

$$
\boldsymbol{H}_r = \frac{1}{\sqrt{8}} \begin{bmatrix} 1 & 1 & 1 & 1 & 1 & 1 & 1 & 1 \\ 1 & 1 & 1 & 1 & -1 & -1 & -1 & -1 \\ \sqrt{2} & \sqrt{2} & -\sqrt{2} & -\sqrt{2} & 0 & 0 & 0 & 0 \\ 0 & 0 & 0 & 0 & \sqrt{2} & \sqrt{2} & -\sqrt{2} & -\sqrt{2} \\ 2 & -2 & 0 & 0 & 0 & 0 & 0 & 0 \\ 0 & 0 & 2 & -2 & 0 & 0 & 0 & 0 \\ 0 & 0 & 0 & 0 & 2 & -2 & 0 & 0 \\ 0 & 0 & 0 & 0 & 0 & 0 & 2 & -2 \end{bmatrix} \tag{9.32}
$$

3. Basis Image

Each row in formula (9.32) is a basis vector, and any two rows of the core matrix can generate corresponding basis images. Figure 9.4a shows all base images when $N = 8$. Because the number of basis vectors is $N = 8$, the number of group images is $N^2 = 64$. The lower right quadrant can be used to search for small features of different positions in the image. All basis images when $N = 2$ and $N = 4$ are shown in Figs. 9.4b, c, respectively.

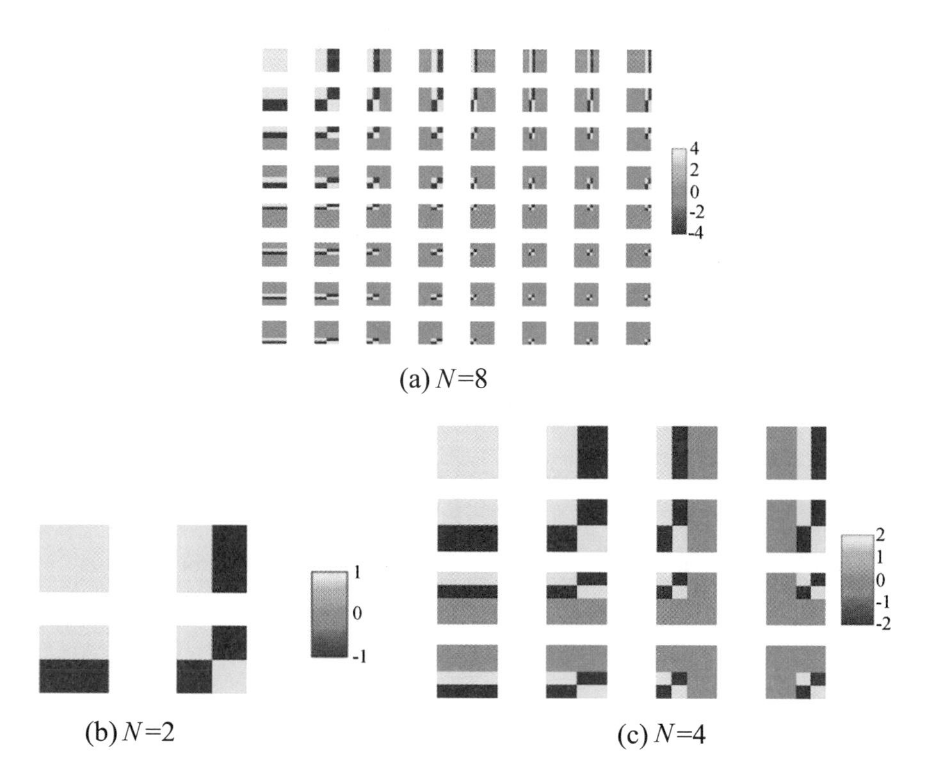

Fig. 9.4 Haar transformation of the basis image

9.3.2 Hadamard Transformation

1. Basis Function Definition

Hadamard transformation is a generalized Fourier transform; specifically, it is a symmetrical, separable unitary transformation, and its core matrix contains only two elements: + 1 and −1. In this case, $N = 2^n$, where n is an integer [4].

When $N = 2$, the kernel matrix is

$$\frac{1}{\sqrt{2}}\boldsymbol{H}_2 = \frac{1}{\sqrt{2}}\begin{bmatrix} 1 & 1 \\ 1 & -1 \end{bmatrix}$$

When $N > 2$, the kernel matrix can be generated as a block matrix:

$$\frac{1}{\sqrt{N}}\boldsymbol{H}_N = \frac{1}{\sqrt{N}}\begin{bmatrix} \boldsymbol{H}_{N/2} & \boldsymbol{H}_{N/2} \\ \boldsymbol{H}_{N/2} & -\boldsymbol{H}_{N/2} \end{bmatrix} \tag{9.33}$$

When $N = 2^n$, if the $N^{-1/2}$ factor appears on the front matrix, the matrix contains only one element, and thus, the conversion calculation amount is small.

2. Basis Vector Matrix

The Hadamard transformation matrix is mainly in the form of a point 2^k transition matrix $N = 2^k$, in which the smallest unit matrix is a 2×2 Hadamard transformation matrix. The following equations present the generation of two points, four points and eight Hadamard transformation matrix steps.

When $N = 1$,

$$W_1 = 1$$

When $N = 2$, the two-point Hadamard transformation basis vector matrix is

$$\boldsymbol{W}_2 = \frac{1}{\sqrt{2}}\begin{bmatrix} 1 & 1 \\ 1 & -1 \end{bmatrix} \tag{9.34}$$

When $N = 4$, the four-point Hadamard transformation basis vector matrix is

$$\boldsymbol{W}_4 = \frac{1}{2}\begin{bmatrix} 1 & 1 & 1 & 1 \\ 1 & -1 & 1 & -1 \\ 1 & 1 & -1 & -1 \\ 1 & -1 & -1 & 1 \end{bmatrix} \tag{9.35}$$

When $N = 8$, the Hadamard transformation basis vector matrix is

$$W_8 = \frac{1}{2\sqrt{2}}\begin{bmatrix} W_4 & W_4 \\ W_4 & -W_4 \end{bmatrix} = \frac{1}{2\sqrt{2}}\begin{bmatrix} 1 & 1 & 1 & 1 & 1 & 1 & 1 & 1 \\ 1 & -1 & 1 & -1 & 1 & -1 & 1 & -1 \\ 1 & 1 & -1 & -1 & 1 & 1 & -1 & -1 \\ 1 & -1 & -1 & 1 & 1 & -1 & -1 & 1 \\ 1 & 1 & 1 & 1 & -1 & -1 & -1 & -1 \\ 1 & -1 & 1 & -1 & -1 & 1 & -1 & 1 \\ 1 & 1 & -1 & -1 & -1 & -1 & 1 & 1 \\ 1 & -1 & -1 & 1 & -1 & 1 & 1 & -1 \end{bmatrix} \tag{9.36}$$

3. Basis Image

According to the relationship between the basis functions, basis vectors and basis images described in Sect. 9.2, when $N = 2$, 4, and 8, according to formulas (9.34), (9.35), and (9.36), any of the outer products of two row vectors can form $N^2 = 4$, $N^2 = 16$, and $N^2 = 64$ basis images, respectively.

9.4 Sample Application Based on Hadamard Transformation

This section is based on the principle of Hadamard transformation derivation presented in the last chapter. Two-dimensional and four-dimensional transformations are considered as examples to illustrate the application of the Hadamard transformation [5].

Example 1 A two-dimensional image is defined as

$$X = \begin{bmatrix} 1 & 2 \\ 3 & 4 \end{bmatrix} \tag{9.37}$$

The objective is to describe this image with basis images.

Solution: For a two-dimensional image, the kernel matrix of Hadamard transformation is

$$H_2 = \frac{1}{\sqrt{2}}\begin{bmatrix} H_1 & H_1 \\ H_1 & -H_1 \end{bmatrix} = \frac{1}{\sqrt{2}}\begin{bmatrix} 1 & 1 \\ 1 & -1 \end{bmatrix}, H_4 = \frac{1}{2}\begin{bmatrix} H_2 & H_2 \\ H_2 & -H_2 \end{bmatrix} \tag{9.38}$$

The basis image is formed as shown in Fig. 9.5.

The linear transformation is

$$X = AY \tag{9.39}$$

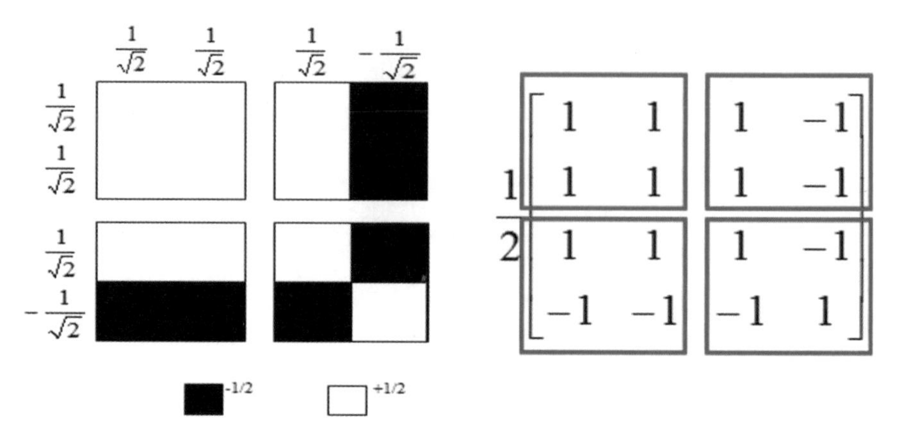

Fig. 9.5 Generation of basis images

The original image is introduced in the transformation formula

$$\begin{bmatrix} 1 \\ 2 \\ 3 \\ 4 \end{bmatrix} = \frac{1}{2}\begin{bmatrix} 1 & 1 & 1 & 1 \\ 1 & -1 & 1 & -1 \\ 1 & 1 & -1 & -1 \\ 1 & -1 & -1 & 1 \end{bmatrix} \boldsymbol{Y} \tag{9.40}$$

In this case,

$$\boldsymbol{Y} = 2\begin{bmatrix} 1 \\ 2 \\ 3 \\ 4 \end{bmatrix} \times \begin{bmatrix} 1 & 1 & 1 & 1 \\ 1 & -1 & 1 & -1 \\ 1 & 1 & -1 & -1 \\ 1 & -1 & -1 & 1 \end{bmatrix}^{-1} \tag{9.41}$$

According to the calculation,

$$\boldsymbol{Y} = \begin{bmatrix} 5 & -1 & -2 & 0 \end{bmatrix}^{\mathrm{T}} \tag{9.42}$$

The original image can be represented by the basis image as follows:

$$\boldsymbol{X} = \begin{bmatrix} 1 & 2 \\ 3 & 4 \end{bmatrix} = \frac{1}{2} \times \left(5\begin{bmatrix} 1 & 1 \\ 1 & 1 \end{bmatrix} - \begin{bmatrix} 1 & -1 \\ 1 & -1 \end{bmatrix} - 2\begin{bmatrix} 1 & 1 \\ -1 & -1 \end{bmatrix}\right) \tag{9.43}$$

as shown in Fig. 9.6.

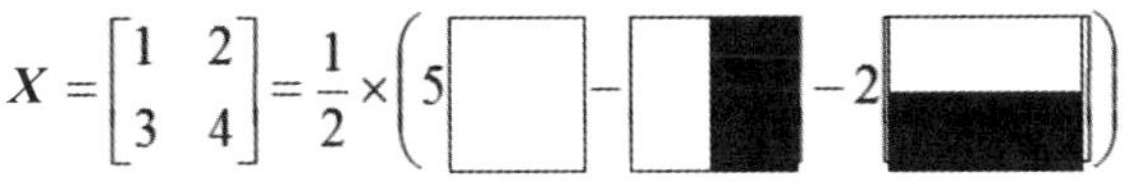

$$X = \begin{bmatrix} 1 & 2 \\ 3 & 4 \end{bmatrix} = \frac{1}{2} \times \left(5 \square - \square - 2 \square \right)$$

Fig. 9.6 Basis image representation of two-dimensional images

Example 2 Describe a four-dimensional image $F = \begin{bmatrix} 1\,2\,3\,4 \\ 1\,2\,3\,4 \\ 1\,2\,3\,4 \\ 1\,2\,3\,4 \end{bmatrix}$ with basis images.

Solution: First, we express the image in the form of an array

$$f = [1\,2\,3\,4\,1\,2\,3\,4\,1\,2\,3\,4\,1\,2\,3\,4]^{\mathrm{T}} \tag{9.44}$$

Next, through the calculation of the two-dimensional image,

$$f = \begin{bmatrix} 1\,2\,3\,4 \\ 1\,2\,3\,4 \\ 1\,2\,3\,4 \\ 1\,2\,3\,4 \end{bmatrix} = \frac{1}{4}\left\{ 10 \begin{bmatrix} 1\,1\,1\,1 \\ 1\,1\,1\,1 \\ 1\,1\,1\,1 \\ 1\,1\,1\,1 \end{bmatrix} - 2 \begin{bmatrix} 1 & -1 & 1 & -1 \\ 1 & -1 & 1 & -1 \\ 1 & -1 & 1 & -1 \\ 1 & -1 & 1 & -1 \end{bmatrix} - 4 \begin{bmatrix} 1 & 1 & -1 & -1 \\ 1 & 1 & -1 & -1 \\ 1 & 1 & -1 & -1 \\ 1 & 1 & -1 & -1 \end{bmatrix} \right\} \tag{9.45}$$

the original image can be represented by the basis image, as shown in Fig. 9.7.

Example 3

Suppose there exist two images: one image involve a large change in the original data, as shown in the left expression of Formula 9.46, and the other image is a raw data distribution that is relatively uniform, as shown in the right expression of Formula 9.46.

$$f_1 = \begin{bmatrix} 1\,3\,3\,1 \\ 1\,3\,3\,1 \\ 1\,3\,3\,1 \\ 1\,3\,3\,1 \end{bmatrix}, f_2 = \begin{bmatrix} 1\,1\,1\,1 \\ 1\,1\,1\,1 \\ 1\,1\,1\,1 \\ 1\,1\,1\,1 \end{bmatrix} \tag{9.46}$$

The objective is to obtain the Hadamard transformation.

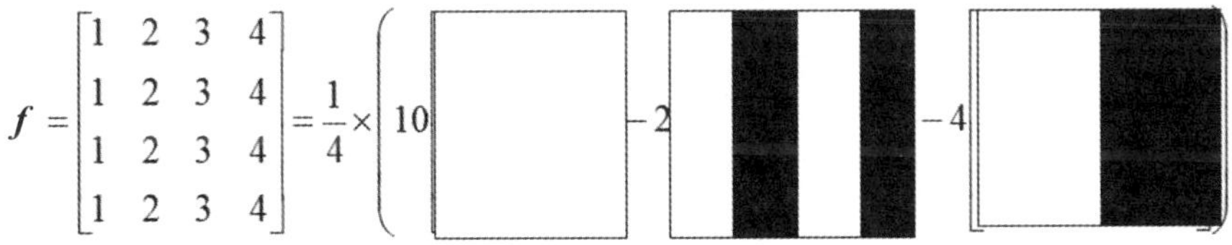

$$f = \begin{bmatrix} 1 & 2 & 3 & 4 \\ 1 & 2 & 3 & 4 \\ 1 & 2 & 3 & 4 \\ 1 & 2 & 3 & 4 \end{bmatrix} = \frac{1}{4} \times \left(10 \square - 2 \square - 4 \square \right)$$

Fig. 9.7 Basis image representation of four-dimensional images

Solution: The two images are obtained using the Hadamard transformation. According to the Hadamard transformation matrix, the basis vector matrix is

$$\boldsymbol{H}_4=\frac{1}{2}\begin{bmatrix} 1 & 1 & 1 & 1 \\ 1 & -1 & 1 & -1 \\ 1 & 1 & -1 & -1 \\ 1 & -1 & -1 & 1 \end{bmatrix} \tag{9.47}$$

The transformation can be expressed as

$$\boldsymbol{W}_1=\frac{1}{4}\begin{bmatrix} 1 & 1 & 1 & 1 \\ 1 & -1 & 1 & -1 \\ 1 & 1 & -1 & -1 \\ 1 & -1 & -1 & 1 \end{bmatrix}\begin{bmatrix} 1 & 3 & 3 & 1 \\ 1 & 3 & 3 & 1 \\ 1 & 3 & 3 & 1 \\ 1 & 3 & 3 & 1 \end{bmatrix}\begin{bmatrix} 1 & 1 & 1 & 1 \\ 1 & -1 & 1 & -1 \\ 1 & 1 & -1 & -1 \\ 1 & -1 & -1 & 1 \end{bmatrix}=\begin{bmatrix} 8 & 0 & 0 & -4 \\ 0 & 0 & 0 & 0 \\ 0 & 0 & 0 & 0 \\ 0 & 0 & 0 & 0 \end{bmatrix} \tag{9.48}$$

$$\boldsymbol{W}_2=\frac{1}{4}\begin{bmatrix} 1 & 1 & 1 & 1 \\ 1 & -1 & 1 & -1 \\ 1 & 1 & -1 & -1 \\ 1 & -1 & -1 & 1 \end{bmatrix}\begin{bmatrix} 1 & 1 & 1 & 1 \\ 1 & 1 & 1 & 1 \\ 1 & 1 & 1 & 1 \\ 1 & 1 & 1 & 1 \end{bmatrix}\begin{bmatrix} 1 & 1 & 1 & 1 \\ 1 & -1 & 1 & -1 \\ 1 & 1 & -1 & -1 \\ 1 & -1 & -1 & 1 \end{bmatrix}=\begin{bmatrix} 4 & 0 & 0 & 0 \\ 0 & 0 & 0 & 0 \\ 0 & 0 & 0 & 0 \\ 0 & 0 & 0 & 0 \end{bmatrix} \tag{9.49}$$

This value shows that for the original images, $N^2 = 16$. After the Hadamard transformation, only two nonzero values remain, and the data are considerably compressed.

Example 4 Example of Hadamard transformation in a remote sensing image; see Fig. 9.8

As shown in Examples 3 and 4, the Hadamard transformation is characterized by energy concentration, the distribution of the original data is more uniform, and the transformed data are concentrated on the corners of the matrix. Thus, the Hadamard transformation can be used to compress the image information. At present, the Hadamard transformation is being widely used in spectral data acquisition, target recognition and classification, weak signal detection and other fields.

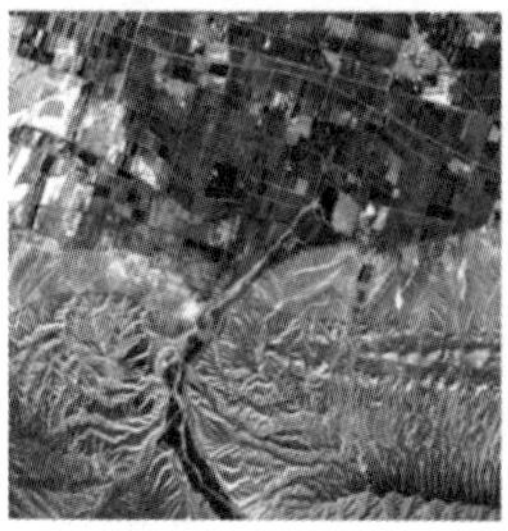

(a) Original image

(b) Hadamard transform

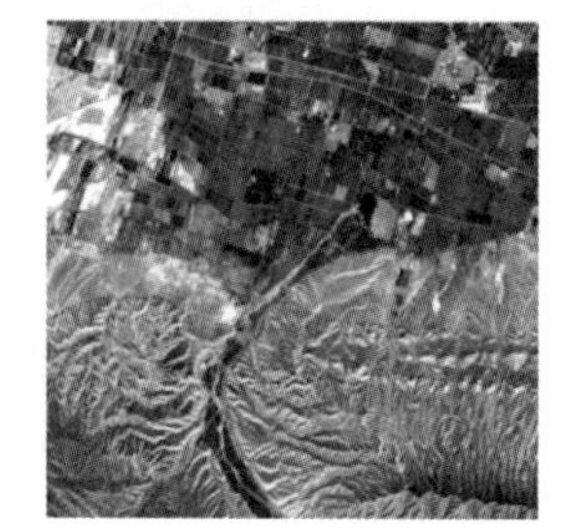

(c) Hadamard inverse transform

Fig. 9.8 Hadamard transformation results

Table 9.1 Advantages and disadvantages of Hadamard transformation

Advantages	Disadvantages
① Only real number operations ② Only addition and subtraction operations and no multiplication ③ Several properties similar to the discrete Fourier transform ④ Forward and reverse conversion forms are similar	① Convergence rate lower than that of the discrete cosine transformation, and thus, the spectrum analysis is inferior ② The number of additions and subtractions is smaller than those of the discrete Fourier transformation and discrete cosine transformation

A Hadamard transformation imaging spectrometer is a type of spectral imager based on the Hadamard transformation and is applied in multichannel detection in optics. The device can encode the spectral information of the image considering the two-dimensional information of the image. The inverse transformation method restores the spectral information. The advantages and disadvantages of the Hadamard transformation are summarized in Table 9.1.

9.5 Summary

The contents of this chapter and mathematical formula are summarized in Fig. 9.9 and Table 9.2, respectively.

The orthogonal transformation in digital image processing is performed in linear space $V(P)$. This chapter first introduces the theoretical basis of linear transformation. Next, we describe the concept and relationship of the basis function, basis vector and basis image. The basis vector can be used to generate the basis vector. The basis images can be generated by the outer product of any two sets of basis vectors, and any image can be weighted by the basis image to achieve refactoring. The relationship among the basis vector ($\boldsymbol{A}$), matrix theory ($\boldsymbol{A}\boldsymbol{A}^{\mathrm{T}} = \boldsymbol{I}$), basis function ($\boldsymbol{T}\boldsymbol{T}^{*\mathrm{T}} = \boldsymbol{I}$) and basis image ($\boldsymbol{F}_{m,n} = \boldsymbol{T}(p, m)\boldsymbol{T}(q, n)$) is the focus of this chapter and the book.

Considering this aspect, the Haar transformation and Hadamard transformation are described. For the Haar transformation, the emphasis is on understanding the basis function $h_0(x) = \frac{1}{\sqrt{N}}$. For the Hadamard transformation, the focus is the recursive matrix $\frac{1}{\sqrt{N}}\boldsymbol{H}_N = \frac{1}{\sqrt{N}}\begin{bmatrix} \boldsymbol{H}_{N/2} & \boldsymbol{H}_{N/2} \\ \boldsymbol{H}_{N/2} & -\boldsymbol{H}_{N/2} \end{bmatrix}$ to determine any $2^k(k \in N)$ Hadamard transformations, which is the basis for its application to digital image processing.

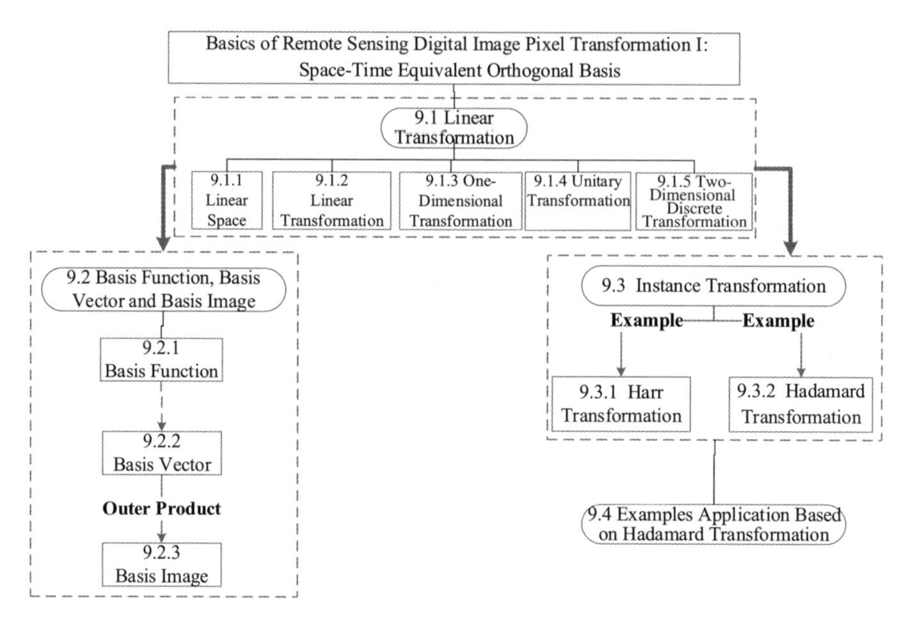

Fig. 9.9 Summary of the contents of this chapter

Table 9.2 Summary of the mathematical formulas in this chapter

formula (1)	$y_i = \sum_{j=0}^{N-1} t_{i,j} x_j$ or $\boldsymbol{y} = \boldsymbol{T x}$	9.1.3
formula (2)	$\boldsymbol{G}_{m,n} = \sum_{i=0}^{N-1} \sum_{k=0}^{N-1} \boldsymbol{F}_{i,k}(i,k,m,n) f(i,k)$	9.1.5
formula (3)	$\boldsymbol{f}_i = \begin{bmatrix} a_{i1} \\ \ldots \\ a_{iN} \end{bmatrix}$	9.2.1
formula (4)	$\boldsymbol{f}_i^{\mathrm{T}} \boldsymbol{f}_j = \begin{cases} 1, & i = j \\ 0, & i \neq j \end{cases}$	9.2.1
formula (5)	$\boldsymbol{F}_{\mathrm{i}} = \begin{bmatrix} x_1 \\ x_2 \\ \cdots \\ x_n \end{bmatrix} = \sum_{i=1}^{n} k_i \boldsymbol{a}_i$	9.2.2
formula (6)	$\boldsymbol{F}_{i,j} = \boldsymbol{f}_i \boldsymbol{f}_j^{\mathrm{T}} = \begin{bmatrix} a_{i1} \\ \ldots \\ a_{iN} \end{bmatrix} \begin{bmatrix} a_{j1} & \ldots & a_{jN} \end{bmatrix} = \begin{bmatrix} a_{i1}a_{j1} & \ldots & a_{i1}a_{jN} \\ \ldots & \ldots & \ldots \\ a_{iN}a_{j1} & \ldots & a_{iN}a_{jN} \end{bmatrix}$	9.2.3

(continued)

Table 9.2 (continued)

formula (7)	$X(i,j)=\begin{bmatrix} x_{11} & \dots & x_{1N} \\ \dots & \dots & \dots \\ x_{N1} & \dots & x_{NN} \end{bmatrix}=\sum\limits_{i,j=1,\dots,N} k_{ij}F_{ij}$	9.2.3
formula (8)	$h_0(x)=\frac{1}{\sqrt{N}}$	9.3.1
formula (9)	$\frac{1}{\sqrt{N}}H_N=\frac{1}{\sqrt{N}}\begin{bmatrix} H_{N/2} & H_{N/2} \\ H_{N/2} & -H_{N/2} \end{bmatrix}$	9.3.2

References

1. Yang Z. Inner product of the orthogonal transformation relationship, symmetric transformation, symmetric transformation. Math Sci. 1993(1). (In Chinese).
2. Hou Z-X, Gao Z-Y, Yang A-P. A JPEG algorithm based on full phase cosine biorthogonal transformation. J Image Graph. 2007(12). (In Chinese).
3. Cheng J, Chen X-X. The mapping transform between Hadamard transform and Haar transform spectral coefficients. J Circuits Syst 2001(6).
4. Zhao F, Li J, Li SH. Semi-fragile watermark algorithm based on Walsh Hadamard transform and convolution coding. J Commun. 2009(30).
5. Qi D. Matrix transformation and its application in image information hiding. J North China Univ Technol. 1999(11). (In Chinese).

Chapter 10
Basics of Remote Sensing Digital Image Pixel Transformation II: Time–Frequency Orthogonal Basis

Chapter Guidance The Fourier transform described in Chap. 6 is based on the global reciprocal domain transform, which can only compress or expand the image globally and is not compatible with compression and expansion. However, real remote sensing images generally contain low-frequency parts (such as mountain ranges) and high-frequency parts (such as steep peaks). How can the low-frequency part of the image be compressed and the high-frequency part be expanded to achieve induction or analysis? There are challenges to each of the reciprocal and high-frequency parts (such as steep peaks), and the time–frequency combined orthogonal basis (subject to the orthogonal basis law described in Chap. 9) theory provides a way to solve such problems. This kind of window operation, which can be performed according to the actual scale and start and end times, is often known as the wavelet transform, and its essence is the coexistence of the scale factor and delay factor under amplitude energy conservation.

A wavelet is a mathematical function defined at a finite interval with an average of zero. Wavelet-based transformations are known as wavelet transforms. The wavelet transform is different from the commonly used Fourier transform, which is usually used in the frequency domain, and it is a local transformation of space (time) and frequency. The signal is subjected to multiscale refinement analysis by stretching and translating operations to achieve time subdivision in the high-frequency part and frequency subdivision in the low-frequency part. This framework is known as the "mathematical microscope". The structure of this chapter is shown in Fig. 10.1. The main contents of this chapter are as follows: Sect. 10.1 Time–frequency analysis and wavelet transform: This section introduces the limitations of the Fourier time–frequency reciprocal domain transformation and time–frequency combination wavelet transform; Sect. 10.2 Continuous wavelet transform and series expansion: This section describes the mathematical connotation of the wavelet transform; Sect. 10.3 Discrete wavelet transform: This section describes the basic method of wavelet transform for discrete signals or images; Sect. 10.4 Wavelet transform application. This section describes the basic idea and example of applying wavelet transform to remote sensing image processing.

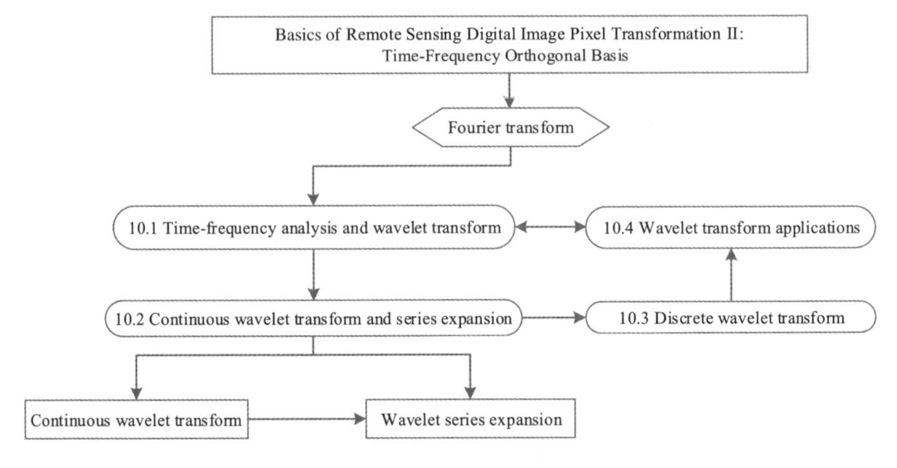

Fig. 10.1 Logical diagram of this chapter

Therefore, this chapter provides an important time–frequency analysis method for remote sensing image processing.

10.1 Time–frequency Analysis and Wavelet Transform

In this section, we introduce the limitations of Fourier transformation in the time–frequency reciprocal domain and significance of the time–frequency combination wavelet transform. The time–frequency combined wavelet transform is a valuable solution to the problem of low-frequency compression and high-frequency expansion of remote sensing images by translation and scale zooming, which can compensate for the deficiency of the global or reciprocal domain transform represented by the Fourier transform in addressing stationary signals.

10.1.1 Time–frequency Analysis

An image can be regarded as a random field, and its gray value forms the space (time) domain. The Fourier transform associates the space (time) domain with the frequency domain to achieve the effect that a dense signal in the spatial domain is spread and analyzed in the frequency domain. The high-frequency component interprets the abrupt part of the signal, such as the edge of the image, while the low-frequency component determines the overall image of the signal, such as the contour of the image. The Fourier transform is a global transformation. For functions that are mostly composed of intervals with zero values, such as the nonstationary signal shown in Fig. 10.2a, the Fourier transform will render the spectrum highly

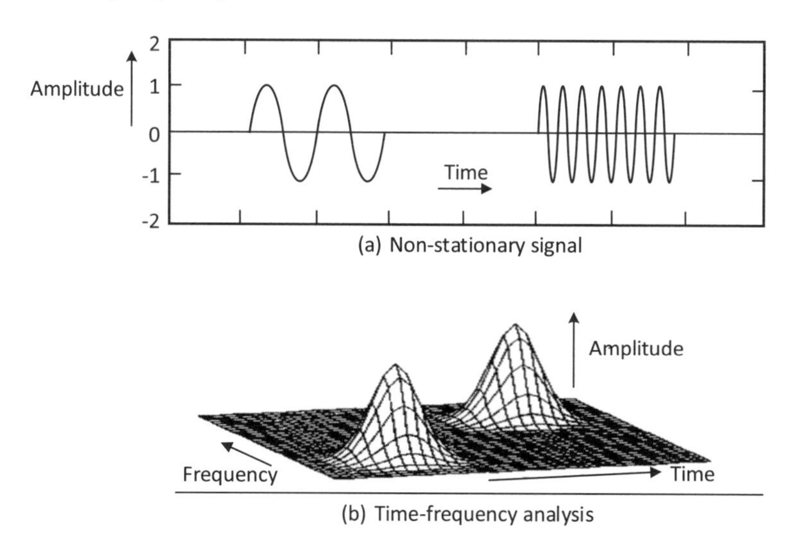

Fig. 10.2 Time–frequency space

confusing. Moreover, the time of the frequency component and its changes with time cannot be obtained. Many natural or artificial signals, such as seismic waves [1], voices, and images, are nonstationary; that is, the statistical nature of the signal is a function of time. Furthermore, the location of the signal mutation is often of interest. Different time processes correspond to the same spectrum. For these cases, the Fourier transform-based analysis method is not effective. Therefore, time–frequency analysis is gaining increasing attention, as shown in Fig. 10.2b.

10.1.2 Wavelet Transform

To solve the problem of the Fourier transform being unable to perform time–frequency analysis and analyze the local range characteristics, in 1946, Gabor (D. Gabor) proposed windowed Fourier transform, that is, the Gabor transformation. The method uses a window function with a certain width in time to realize shifting on the time axis, multiplies the signal to be analyzed and performs the Fourier transform to realize time–frequency analysis. However, the method cannot satisfy the high-frequency narrow window and low-frequency wide window requirements and cannot constitute an orthogonal basis, t. Moreover, the computation involves a large amount of redundancy.

In 1982, J. Morlet abandoned the orthogonal basis function of the Fourier transform, constructed the Morlet wavelet and proposed the concept of wavelet analysis. Wavelet analysis is a linear time–frequency analysis method that can obtain the time information by the translation of the signal and the frequency information by scaling the signal with a variable time–frequency window. This theory promoted related research, and in 1986, Meyer (Y. Meyer) proved the existence of a wavelet orthogonal system. In 1988, I. Aubechies constructed a series of practical compact regular wavelets. In 1989, S. Mallat unified subband coding and pyramid algorithms based on the multiresolution analysis theory and proposed pyramidal decomposition and reconstruction, which is the basis of wavelet analysis in the engineering field. In 1992, R. Coifman et al. proposed the wavelet packet decomposition theory, which can obtain sophisticated high-frequency information.

At present, wavelet transform is being widely used in image processing, edge detection, speech synthesis, seismic exploration, atmospheric turbulence detection, astronomical recognition, machine vision and other fields.

10.2 Continuous Wavelet Transform and Series Expansion

10.2.1 Continuous Wavelet Transform

1. Wavelet Function

If $\Psi(x)$ is a real-valued function, and the spectrum $\Psi(s)$ satisfies the following formula:

$$C_{\Psi} = \int_{-\infty}^{\infty} \frac{|\Psi(s)|^2}{|s|} \mathrm{d}s < \infty \tag{10.1}$$

$\Psi(x)$ is known as the basic wavelet or mother wavelet, and the origin is set as the center. Because s is in the denominator,

$$\Psi(0) = 0 \Rightarrow \int_{-\infty}^{\infty} \Psi(x)\mathrm{d}x = 0 \tag{10.2}$$

A set of wavelet basis functions $\{\Psi_{a,b}(x)\}$ can be generated by translating and stretching the parent wavelet $\Psi(x)$:

$$\Psi_{a,b}(x) = \frac{1}{\sqrt{a}}\Psi\left(\frac{x-b}{a}\right) \tag{10.3}$$

where a is the scaling factor and can change the scale of the wavelet, as shown in Fig. 10.3. b is the translation factor, which can change the position of the wavelet, as shown in Fig. 10.4. a and b are real, and $a > 0$. The front coefficients of Formula 10.4 ensure that the norm of the wavelet basis function is equal and that energy conservation is satisfied.

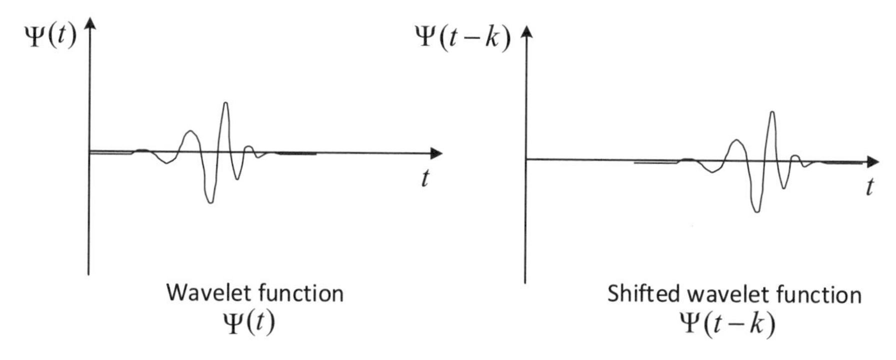

Fig. 10.3 Effect of scaling factors

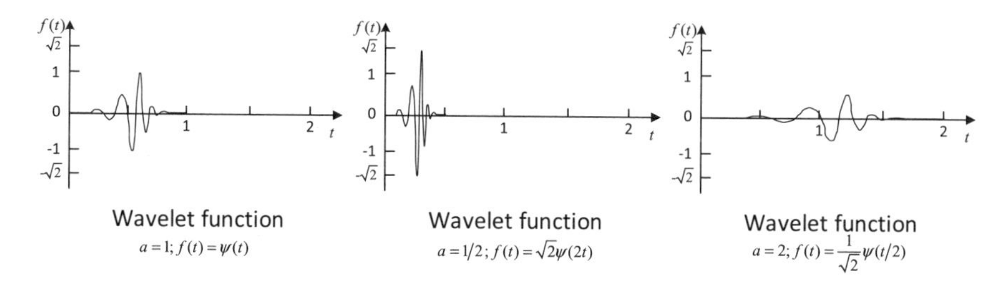

Fig. 10.4 Effect of translational factors

2. Definition of Continuous Wavelet Transform

In 1980, Grossman and Morlet proposed the continuous wavelet transform, which is the product of the signal and wavelet function summed over the entire interval. For example, for the binary function $f(x, y) \in L^2(R^2)$, $L^2(R)$ denotes the square product of the function on the real axis, and its two-dimensional continuous wavelet transform and inverse transformation are

$$W_f(a, b_x, b_y) = \int_{-\infty}^{\infty} \int_{-\infty}^{\infty} f(x, y) \Psi_{a,b_x,b_y}(x, y) \mathrm{d}x \mathrm{d}y \tag{10.4}$$

$$f(x, y) = \frac{1}{C_\Psi} \int_{-\infty}^{\infty} \int_{-\infty}^{\infty} \int_{-\infty}^{\infty} W_f(a, b_x, b_y) \Psi_{a,b_x,b_y}(x, y) \mathrm{d}b_x \mathrm{d}b_y \frac{\mathrm{d}a}{a^3} \tag{10.5}$$

where b_x and b_y represent translations in two dimensions, and the two-dimensional elementary wavelet $\Psi(x, y)$ is expressed as

$$\Psi_{a,b_x,b_y}(x, y) = \frac{1}{|a|} \Psi\left(\frac{x - b_x}{a}, \frac{y - b_y}{a}\right) \tag{10.6}$$

3. Properties of Continuous Wavelet Transform

Property 1 (Superposition)

If f(x), g(x)∈L^2(R), k_1 *and* k_2 *are arbitrary constants*:

$$W_{k_1 f+k_2 g}(a,b)=k_1 W_f(a,b)+k_2 W_g(a,b) \tag{10.7}$$

Property 2 (Translational)

If f(x)∈L^2(R),

$$W_{f(x-x_0)}(a,b)=W_{f(x)}(a,b-x_0) \tag{10.8}$$

Property 3 (Scalability)

If f(x)∈L^2(R),

$$W_{f(\lambda x)}(a,b)=\frac{1}{\sqrt{\lambda}}W_{f(x)}(\lambda a,\lambda b),\ \lambda>0 \tag{10.9}$$

Property 4 (Multiplication principle)

If f(x) *and* g(x)∈L^2(R),

$$\int_0^{+\infty}\int_{-\infty}^{+\infty}\frac{1}{a^2}W_f(a,b)W_g(a,b)\mathrm{d}b\mathrm{d}a=C_\Psi\int_R f(x)g(x)\mathrm{d}x \tag{10.10}$$

and $C_\Psi=\int_0^{+\infty}\frac{|\Psi(\omega)|^2}{\omega}\mathrm{d}\omega$

Property 5 (Inverse formula)

If f(x)∈L^2(R),

$$f(x)=\frac{1}{C_\Psi}\int_0^{+\infty}\int_{-\infty}^{+\infty}\frac{1}{a^2}W_f(a,b)\Psi_{a,b}(x)\mathrm{d}b\mathrm{d}a \tag{10.11}$$

Property 6 (Redundancy)

Because of the correlation of a *and* b, *there is redundancy in the information expression in the continuous wavelet transform. The reconstruction formula is not unique, and the kernel function* $\Psi_{a,b}(x)$ *also has many choices. Although the redundancy can enhance the stability of the signal reconstruction, it increases the difficulty of analyzing and interpreting the results of the wavelet transform.*

10.2.2 Wavelet Series Expansion

Unlike the continuous wavelet transform, the wavelet series expansion specifies the scaling and translation coefficients as integers rather than real numbers.

1. Dyadic Wavelet

The basis function of the dyadic wavelet is obtained by binary scaling (each scaling is a multiple of 2) and binary translation (each translation is a shift of $k/2^j$) of the fundamental wavelet $\Psi(x)$

$$\Psi_{j,k}(x) = 2^{j/2}\Psi(2^j x - k),\ j, k \in Z \tag{10.12}$$

2. Wavelet Series Expansion

If $\{\Psi_{j,k}(x)\}$ is the normative orthogonal basis of $L^2(R)$, i.e.,

$$\langle\Psi_{j,k}, \Psi_{l,m}\rangle = \int_{-\infty}^{\infty} \Psi_{j,k}(x)\Psi_{l,m}(x)dx = \delta_{j,l}\delta_{k,m}, \tag{10.13}$$

$\Psi(x) \in L^2(R)$ is the orthogonal wavelet. Here, j, k, l, m are integers, $\delta_{j,k}$ is the Kronecker δ function, and "$\langle\bullet\rangle$" refers to the inner product.

The following process proves that the dyadic wavelet belongs to the orthogonal wavelet:

$$\begin{aligned}\langle\Psi_{j,k}, \Psi_{l,m}\rangle &= \int_{-\infty}^{\infty} 2^{j/2}\Psi(2^j x-k)2^{l/2}\Psi(2^l x-m)\mathrm{d}x \\ &\overset{令 t=2^j x-k}{=} 2^{-j/2}\cdot 2^{l/2}\int_{-\infty}^{\infty} \Psi(t)\Psi(2^{l-j}(t+k)-m)\mathrm{d}t \overset{j=l}{=} \\ &\int_{-\infty}^{\infty} \Psi(t)\Psi(t+k-m)\mathrm{d}t = \delta_{0,k-m} = \delta_{k,m}\end{aligned} \tag{10.14}$$

When $0 \le t < 1, \Psi(t) \ne 0$, and when $m-k \le t < 1+m-k$, $\Psi(t-(m-k)) \ne 0$. When $k \ne m$, $\Psi(t)\Psi(t-(m-k)) \equiv 0$.

When $j \ne l$ and $l > j$, we set $r = l - j > 0$, i.e.,

$$\langle\Psi_{j,k}, \Psi_{l,m}\rangle = 2^{r/2}\int_{-\infty}^{\infty} \Psi(x)\Psi(2^r x + s))\mathrm{d}x \tag{10.15}$$

According to the definition of $\Psi(x)$, only the case of $0 \le x < 1$ needs to be discussed. We integrate the above formula, i.e.,

$$\left\langle \Psi_{j,k},\ \Psi_{l,m} \right\rangle = 2^{r/2}\int_{-\infty}^{\infty} \Psi(x)\Psi(2^r x+s))\mathrm{d}x = \int_{0}^{1/2} \Psi(2^r x+s)\mathrm{d}x - \int_{1/2}^{1} \Psi(2^r x+s)\mathrm{d}x$$
$$\overset{令 t=2^r x+s}{=} 2^{-r}\int_{s}^{a} \Psi(t)\mathrm{d}t - 2^{-r}\int_{a}^{b} \Psi(t)\mathrm{d}t = 0 \tag{10.16}$$

where $a = s + 2^{r-1}$ and $b = s + 2^r$. Since the intervals $[s, a]$ and $[a, b]$ do not contain the support interval of $\Psi(t)$, the two integrals are equal to zero. The proof is completed.

Any $f(x) \in L^2(R)$ can be expressed as

$$f(x) = \sum_{\mathrm{j}=-\infty}^{\infty} \sum_{\mathrm{k}=-\infty}^{\infty} \mathrm{c}_{\mathrm{j,k}} \overline{\Psi_{j,k}(x)} \tag{10.17}$$

which represents the complex conjugate of the function, and

$$c_{j,k} = \langle f(x), \Psi_{j,k}(x) \rangle = 2^{j/2} \int_{-\infty}^{\infty} f(x)\Psi(2^j x - k)\mathrm{d}x \tag{10.18}$$

Formulas 10.18 and 10.19 determine the wavelet series expansion of $f(x)$ for wavelet $\Psi(x)$

3. Compact Dyadic Wavelet

If $f(x)$ and $\Psi(x)$ are limited to zero outside the [0, 1] interval, the classes of orthogonal wavelet functions mentioned above become compact and can be determined by a single index n:

$$\Psi_n(x) = 2^{j/2}\Psi(2^j x - k) \tag{10.19}$$

where j and k are functions of n, i.e.,

$$n = 2^j + k;\ j = 0, 1, \ldots, k = 0, 1, \ldots, 2^j - 1 \tag{10.20}$$

The corresponding inverse transformation is

$$f(x) = \sum_{n=0}^{\infty} c_n \Psi_n(x) \tag{10.21}$$

The conversion factor is

$$c_n = \langle f(x), \Psi_n(x) \rangle = 2^{\frac{j}{2}} \int_{-\infty}^{\infty} f(x)\Psi(2^j x - k)\mathrm{d}x \tag{10.22}$$

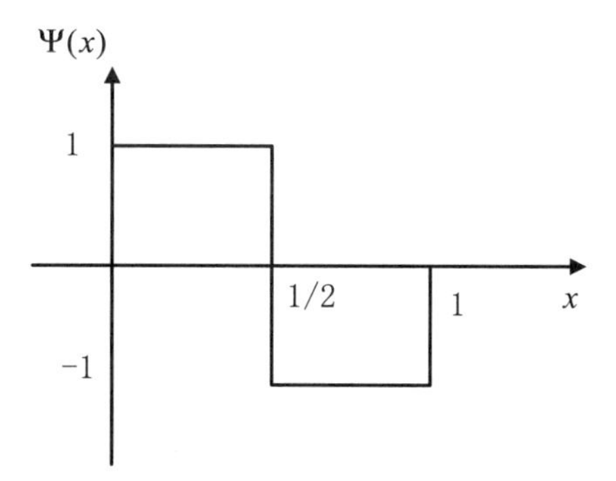

Fig. 10.5 Haar basic wavelet

4. Example of Haar Wavelet Transform

In 1909, A. Haar proposed the Haar wavelet. The basic function of the Haar wavelet is defined in the interval [0, 1), as shown in Fig. 10.5.

The formula is

$$\Psi(x) = \begin{cases} 1, x \in [0, 1/2) \\ -1, x \in [1/2, 1) \end{cases} \tag{10.23}$$

The Haar transform is a compact, dyadic, orthogonal normalized wavelet transform. In $\Psi_n(x) = 2^{\frac{j}{2}}\Psi(2^j x - k) = 2^{j/2}\Psi\left(x - \frac{k}{2^j}/\frac{1}{2^j}\right)$, the width $w = 1/2^j$, amplitude $h = 2^{\frac{j}{2}}$, position $p = k/2^j$, and $n = 2^j + k$. If $N = 8$, $2^j \leq n$, $k = n - 2^j$, and $n = 0,1\ldots, 7$. For $n = 0$–7, the process of the Haar transform is shown in Table 10.1. The corresponding Haar transform basis function is shown in Fig. 10.6.

Table 10.1 Haar transform basis function ($N = 8$)

n	j	k	w	h	p	$\Psi_n(x)$
	$2^j \leq n$	$k = n - 2^j$	$w = \frac{1}{2^j}$	$h = 2^{\frac{j}{2}}$	$p = \frac{k}{2^j}$	
0	–	0	–	–	–	$\Psi_0(x) = 1$
1	0	0	1	1	0	$\Psi_1(x) = \Psi(x)$
2	1	0	$1/2$	$\sqrt{2}$	0	$\Psi_2(x) = \sqrt{2}\Psi(2x)$
3	1	1	$1/2$	$\sqrt{2}$	$1/2$	$\Psi_3(x) = \sqrt{2}\Psi(2x - 1)$
4	2	0	$1/4$	2	0	$\Psi_4(x) = 2\Psi(4x)$
5	2	1	$1/4$	2	$1/4$	$\Psi_5(x) = 2\Psi(4x - 1)$
6	2	2	$1/4$	2	$2/4$	$\Psi_6(x) = 2\Psi(4x - 2)$
7	2	3	$1/4$	2	$3/4$	$\Psi_7(x) = 2\Psi(4x - 3)$

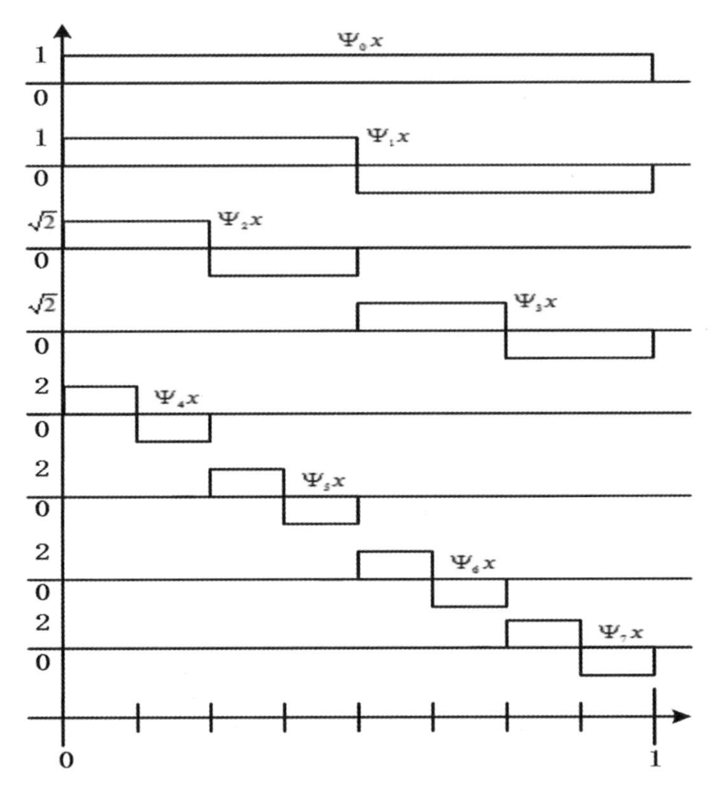

Fig. 10.6 Haar transform basis function ($N = 8$)

10.3 Discrete Wavelet Transform

This section focuses on the basic method for the wavelet transform of discrete signals or images based on the multiresolution technology, such as image pyramid, subband coding, discrete wavelet and Mallat algorithms. The Mallat algorithm is of significance because it makes the wavelet transform truly applicable.

10.3.1 Multiresolution Analysis

Let $\{V_j\}$, $j \in Z$ be a series of subspaces of function space $L^2(R)$ that satisfy

(1) Uniform monotonicity: $\Lambda \subset \Lambda_2 \subset V_1 \subset V_0 \subset \Lambda$
(2) Asymptotic completeness: $I_{j\in Z}\, V_j = \{0\}$; $\Upsilon_{j\in Z}\, V_j = L^2(R)$
(3) Scaling regularity: $f(t) \in V_j \Leftrightarrow f(2^j t) \in V_0,\ j \in Z$
(4) Translation invariance: $f(t) \in V_0 \Leftrightarrow f(t-n) \in V_0, \forall n \in Z$

(5) Existence of orthogonal basis: For $\Phi \in V_0$ such that $\{\Phi(t-n)\}$ is an orthogonal base of V_0, i.e.,

$$V_0 = \text{span}\{\Phi(t-n)\}, n \in Z, \int_R \Phi(t-n)\Phi(t-m)\mathrm{d}t = \delta(n-m) \quad (10.24)$$

$\{V_j\}$, $j \in Z$ is the multiresolution analysis of $L^2(R)$, where $\{V_j\}$, $j \in Z$ and W_j are known as the scale space and wavelet space, respectively. W_j is the orthogonal complement of V_j in V_{j-1}, and $W_i \perp W_j (i \neq j)$, and thus, W_j is termed V_j in the detail space of V_{j-1}, as shown in Fig. 10.7.

Similarly, there exists function $\Phi(t)$ to generate a closed subspace W_0, and thus, the biscale equation of the wavelet function is

$$\Phi(t) = \sqrt{2} \sum_{k=-\infty}^{+\infty} h_0(k)\Phi(2t-k) \quad (10.25)$$

If $\Phi(t)$ is the orthogonal basis of V_0, $\Phi_{j,n}(t) = 2^{-j/2}\Phi(2^{-j}t-n)$ is the orthogonal basis of subspace V_j, and $\Phi(t)$ is known as the scaling function of V_0.

Multiresolution analysis is performed using a scale function, and thus, multiresolution analysis is equivalent to finding the properties of the scale function. The scale function can be constructed using the low-pass filter, and the impulse response of the discrete low-pass filter is known as the scale vector. The scaling functions can replicate by themselves with a half scale of the weight and structure, with a weight of $h_0(k)$, or repeatedly with dimensions of the rectangular pulse function convolution $h_0(k)$, through the numerical calculation shown in Fig. 10.8. The wavelet function based on the high-pass filter implementation and scale of the low-pass filter function can be used as the wavelet function and scale function for generating the function of the next level, and both functions operate to realize signal decomposition, while the filter banks form a framework for decomposition. The figure shows the framework to solve the scale function. The specific solution can be determined using Formula (10.26), and the filter coefficient can be referenced in the work of Daubechies [2].

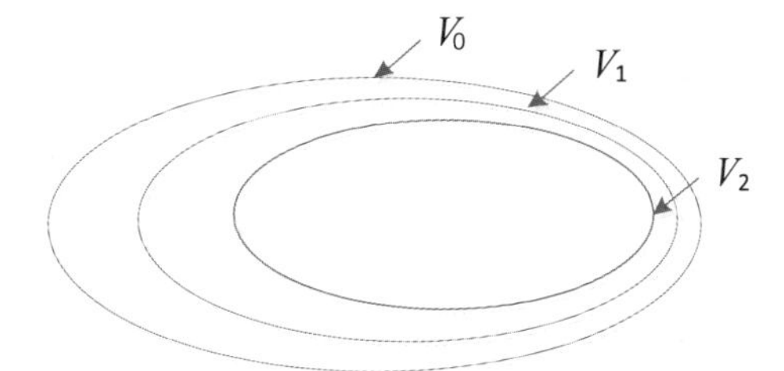

Fig. 10.7 Multiresolution scale space

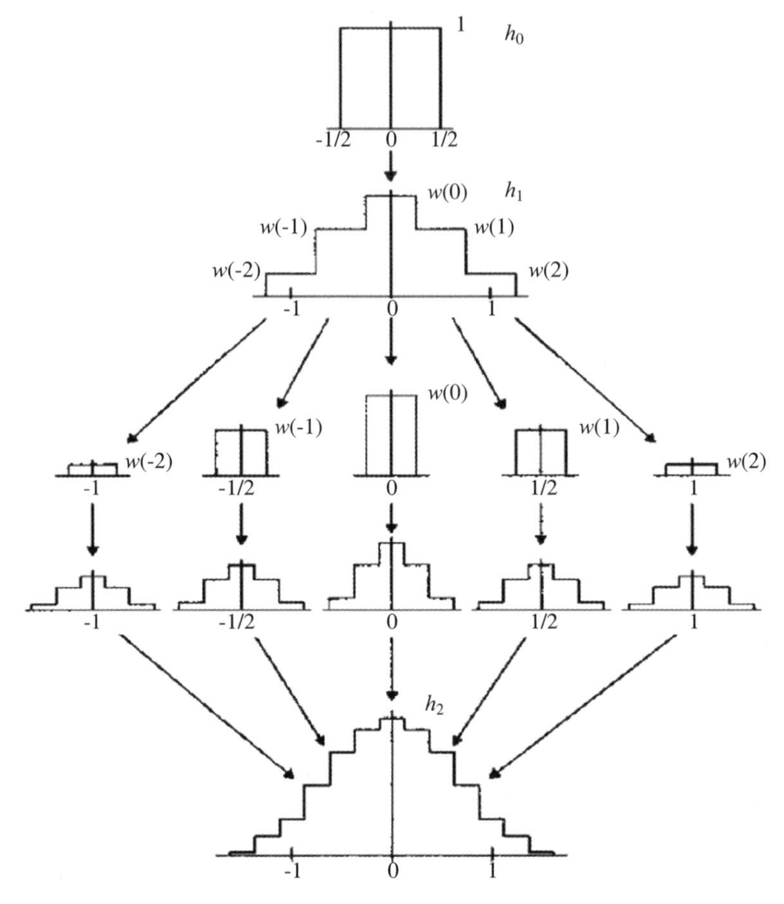

Fig. 10.8 Convolution construction of scale functions [2]

10.3.2 Subband Coding and Mallat Algorithm

The Mallat algorithm [3] is essentially a discrete wavelet transform algorithm that uses double-band coding to build the wavelet transform from the bottom up in an iterative way. Subband coding is intended to decompose the signal into narrowband components and represent the signal in a redundant way with nonreconstructed error [4], as shown in Fig. 10.9.

If the given limit-band signal is set as $f(t)$ and Fourier transform is performed, $f(i\Delta t), i = 0, 1, \cdots, N-1$ can be obtained according to the Nyquist sampling law. First, the discrete signal $f(i\Delta t)$ is passed through a half-band low-pass filter $h_0(i\Delta t)$ to obtain the low-resolution representation $g_0(i\Delta t)$. Subsequently, $f(i\Delta t)$ is passed through a half-band high-pass filter to obtain the high-frequency information $g_1(i\Delta t)$. Next, the lower half-band signal $g_0(i\Delta t)$ is again implemented with a half band subband. Thus, a $N/2$ point high half-band signal is obtained with two

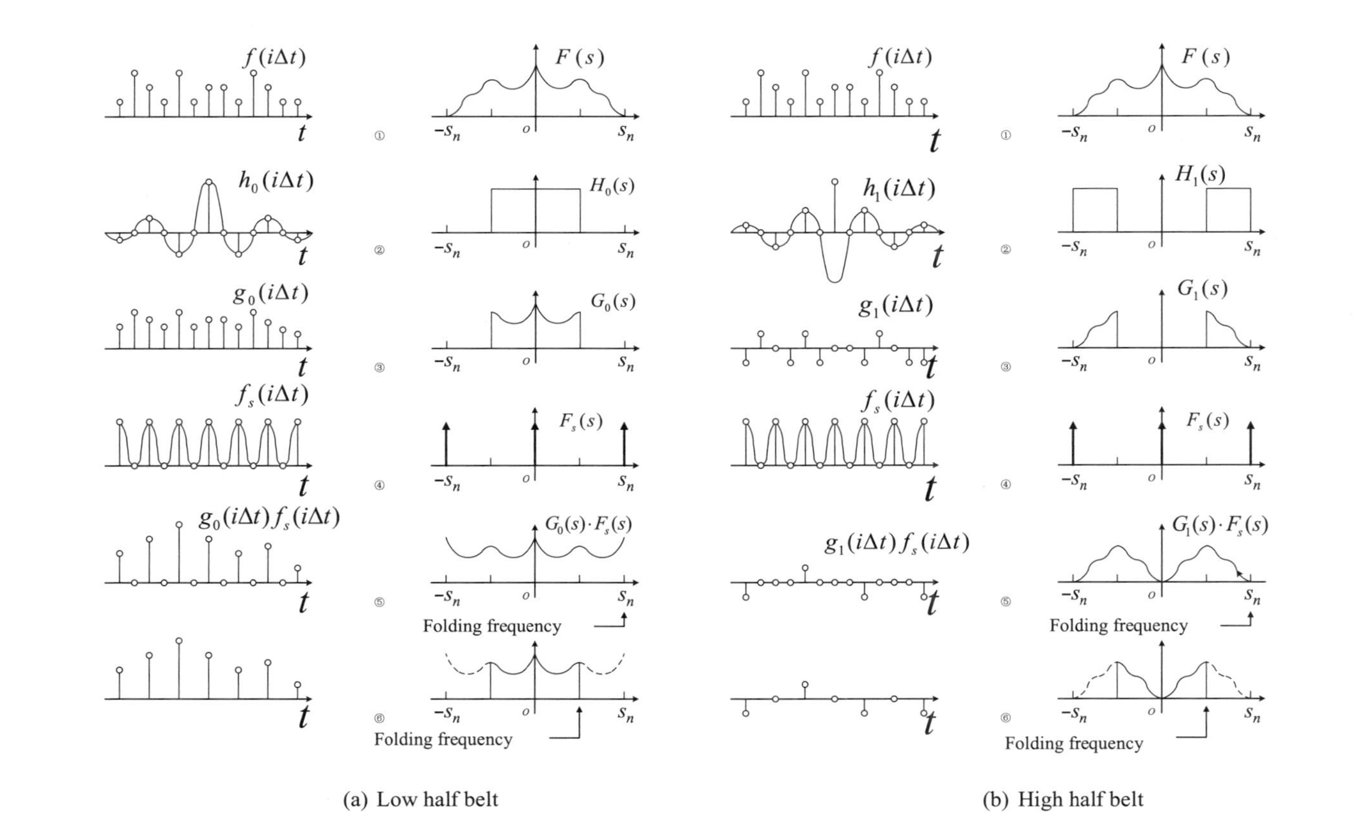

Fig. 10.9 Subband coding, where ① denotes a sampled signal and its band-limited spectrum, ② shows the ideal half-band high/low-pass filter, ③ indicates the high/low-pass filtered signal, ④ shows the interval sampling function, ⑤ denotes odd sampling points to zero, and ⑥ shows discarded odd sampling points

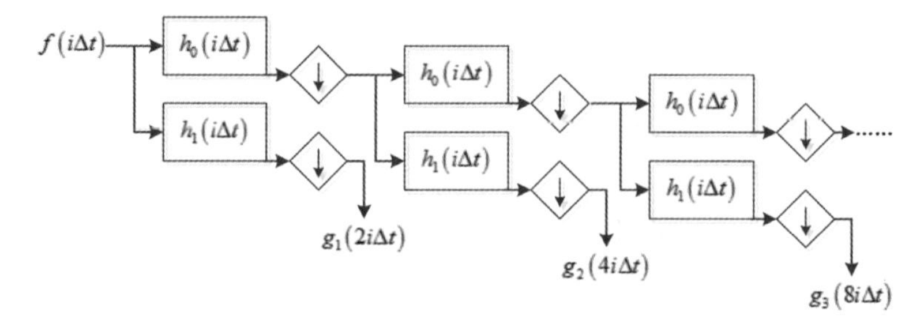

Fig. 10.10 Signal band decomposition process

$N/4$ point subband signals, corresponding to the first and second $\frac{1}{4}$ regions of the interval $[0, s_n]$, respectively, and s_n is the Nyquist sampling frequency. This process is iteratively performed. Each step retains the high half-band signal and encodes the lower half-band signal until a low half-band signal with only one point is obtained. The Mallat algorithm can be convoluted by the signal and filter coefficients and implemented by two decimations. The two decimation processes can be performed by preserving even data and discarding odd data, as shown in Fig. 10.10.

Specifically, for $f \in L^2(R)$, if $P_j f \in V_j$, $Q_j f \in W_j$, then P_j is the projection space of the scale space, Q_j is the projection operator of the wavelet space, and

$$P_{j-1} f = P_j f + Q_j f \tag{10.26}$$

$$f = \lim_{j \to -\infty} P_j f = P_j f + Q_j f + Q_{j-1} f + \cdots = \sum_{j=-\infty}^{+\infty} Q_j f \tag{10.27}$$

We set the scale function for the multiresolution analysis as $\phi(t)$ and discrete sequence as $c_n^0 \in l^2(Z)$. The constructor function $f(t)$ is

$$f(t) = \Sigma_n c_n^0 \phi(t-n) \Rightarrow P_1 f = \Sigma_k c_k^1 \phi_{1k},\ Q_1 f = \Sigma_k d_k^1 \phi_{1k} \tag{10.28}$$

The process of signal decomposition is the process of searching for the relationship of coefficients c_k^j, d_k^j and c_n^{j-1}, i.e.,

$$c_k^j = \Sigma_n h(n-2k) c_n^{j-1},\ d_k^j = \Sigma_n g(n-2k) c_n^{j-1} \tag{10.29}$$

And

$$h(n) = 2^{-1/2} \int \phi(1/2t)\phi(t-n)\mathrm{d}t,\ g(n) = 2^{-1/2} \int \Psi(1/2t)\phi(t-n)\mathrm{d}t \tag{10.30}$$

10.3.3 Image Pyramid

An image pyramid that interprets the image at multiple resolutions is a series of images that are gradually reduced in resolution by the pyramid shape, as shown in Fig. 10.11. This framework is used for machine vision and image compression.

The bottom and top of the pyramid are the high-resolution representation and low-resolution approximation of the image to be processed, respectively. If the image size is $N \times N$, and the pyramid has J levels, the j_{th}-level size is $2^j \times 2^j$, where $0 \leq j \leq J$. The decomposition of a two-dimensional image can be considered a product of two one-dimensional tensors. In the decomposition process, the resolution is usually set as $2^{-j} (j \geq 0)$, the original image is set as the 0th layer, the resolution is 1, and the approximate image under the resolution 2^{-j} is extracted from the underlying image in the formation. $f_0(x, y)$ is the original image, and $f_1(x, y)$ is the approximation image of $f_0(x, y)$, which passes through the low-pass filter and the second selected A_j operator:

$$f_1(x, y) = A_1[f_0(x, y)] \tag{10.31}$$

and

$$f_0^e(x, y) = E_1[f_1(x, y)] \tag{10.32}$$

Therefore, the detail image $d_0(x, y)$ is

$$d_0(x, y) = f_0(x, y) - f_0^e(x, y) \tag{10.33}$$

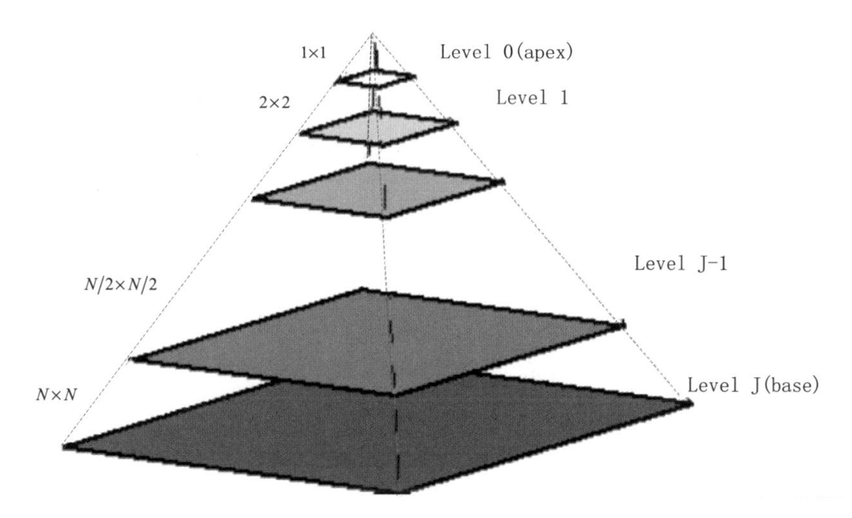

Fig. 10.11 Image pyramid

Therefore, the original image $f_0(x, y)$ is broken down into two parts: the detail image for which the scale is 1 and resolution is 1/2, and the approximation image for which the scale is 1/2 and resolution is 1/2. The original image passes through the selection operator A_j, and the interpolation operator E_j can generate a series of approximation images $f_j(x, y)$ and detail images $d_j(x, y)$.

The distribution of $f_j(x, y)$ is similar to the Gaussian function, in a framework known as the Gaussian pyramid, and the distribution of $d_j(x, y)$ is similar to the Laplacian function, in a framework known as the Laplacian pyramid.

10.3.4 The Principle of Discrete Wavelet Transform

The principle of the one-dimensional wavelet transform is shown in Fig. 10.12, where $x[n]$ is the discrete input signal with length N. $g[n]$ is the low-pass filter. The high-frequency part of the input signal is filtered out, and the low-frequency part is output. $h[n]$ is the high-pass filter. In contrast to the low-pass filter, the low-frequency part is filtered out and the high-frequency part is output. $\downarrow$ is the downsampling filter.

In each run, the low-frequency part of the one-dimensional discrete wavelet transformation is subdivided into high-frequency and low-frequency parts, and the resolution considerably decreases. The α-layer coefficients of the one-dimensional discrete wavelet transform can be expressed as

$$\begin{cases} x_{\alpha,L}[n] = \sum_{k=0}^{K-1} x_{\alpha-1,L}[2n-k]g[k], \\ x_{\alpha,H}[n] = \sum_{k=0}^{K-1} x_{\alpha-1,L}[2n-k]h[k]. \end{cases} \tag{10.34}$$

The mathematical principle of the two-dimensional wavelet transform is derived from the one-dimensional discrete wavelet transform. The situation of separating the

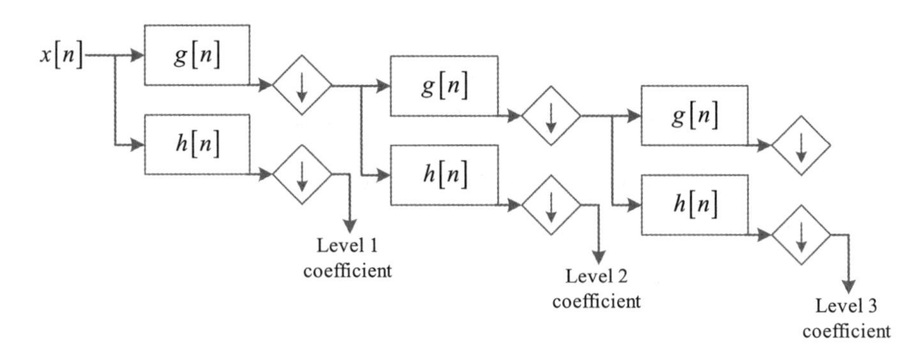

Fig. 10.12 Multidimensional diagram of the one-dimensional discrete wavelet transform

two-dimensional scale function is considered:

$$\phi(x, y) = \phi(x)\phi(y) \tag{10.35}$$

where $\phi(x)$ is a one-dimensional scale function. If $\psi(x)$ is the corresponding wavelet, the following three two-dimensional fundamental wavelet is the basis of transformation, i.e.,

$$\psi^1(x, y) = \phi(x)\psi(y) \;\; \psi^2(x, y) = \psi(x)\phi(y) \;\; \psi^3(x, y) = \psi(x)\psi(y) \tag{10.36}$$

The following set of functions represent the orthogonality of $L^2(R^2)$, where j, l, m, n are integers:

$$\left\{\psi_{j,m,n}^l(x, y)\right\} = \left\{2^j \psi^l(x - 2^j m, y - 2^j n)\right\} \;\; j \geq 0; \; l = 1, 2, 3 \tag{10.37}$$

The processing of the α-layer of the two-dimensional input signal $x[m, n]$ is performed in two steps. First, high-pass, low-pass and downfrequency operations are performed in the n direction:

$$\begin{cases} v_{\alpha,\mathrm{L}}[m, n] = \sum_{k=0}^{K-1} x_{\alpha-1,\mathrm{L}}[m, 2n - k]g[k], \\ v_{\alpha,\mathrm{H}}[m, n] = \sum_{k=0}^{K-1} x_{\alpha-1,\mathrm{L}}[m, 2n - k]h[k]; \end{cases} \tag{10.38}$$

Next, along the m direction, $v_{\alpha,\mathrm{L}}[m, n]$ and $v_{\alpha,\mathrm{H}}[m, n]$ are subjected to high-pass, low-pass and down frequency operations:

$$\begin{cases} x_{\alpha,\mathrm{LL}}[m, n] = \sum_{k=0}^{K-1} v_{\alpha,\mathrm{L}}[2m - k, n]g[k], \\ x_{\alpha,\mathrm{LH}}[m, n] = \sum_{k=0}^{K-1} v_{\alpha,\mathrm{L}}[2m - k, n]h[k], \\ x_{\alpha,\mathrm{HL}}[m, n] = \sum_{k=0}^{K-1} v_{\alpha,\mathrm{H}}[2m - k, n]g[k], \\ x_{\alpha,\mathrm{HH}}[m, n] = \sum_{k=0}^{K-1} v_{\alpha,\mathrm{H}}[2m - k, n]h[k]. \end{cases} \tag{10.39}$$

Finally, four groups of decomposition signals are obtained, and their size is half of the original. The subscript LL represents the low frequency part, LH represents

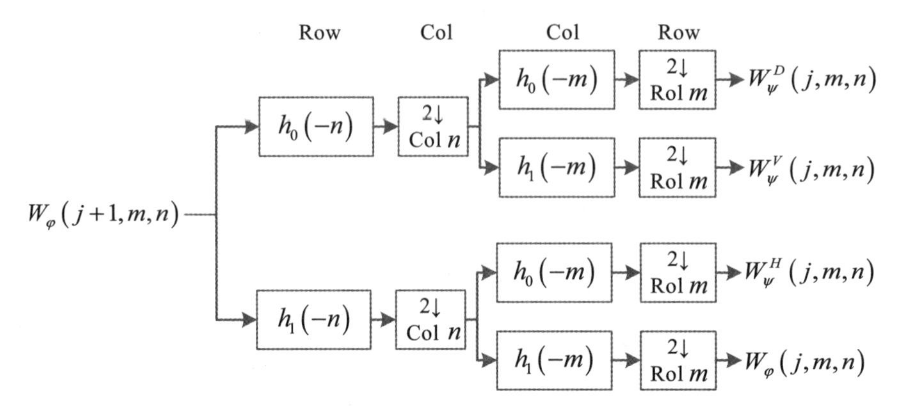

Fig. 10.13 Two-dimensional positive wavelet transform

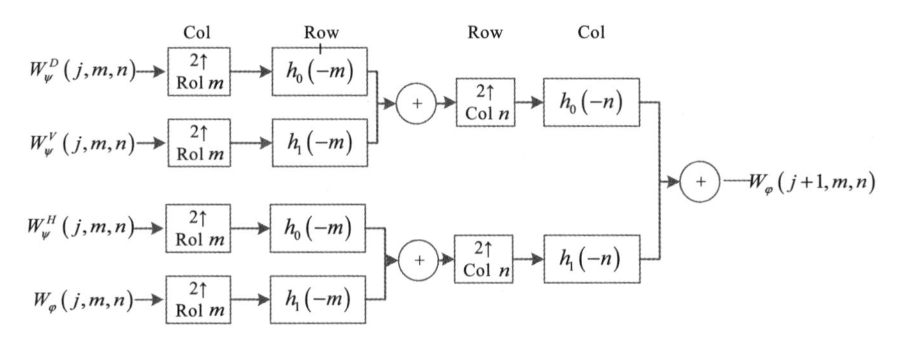

Fig. 10.14 Two-dimensional inverse wavelet transform

the longitudinal high frequency, H L represents the transverse high frequency, and H H represents the diagonal high frequency.

The two-dimensional wavelet transform is shown in Fig. 10.13, and the inverse transformation is shown in Fig. 10.14. In the two figures, j is an integer, indicating hierarchy; m and n are integers, representing rows and columns, respectively; $h_0(x)$ denotes a low-pass filter, and $h_0(x)$ denotes a high-pass filter.

10.4 Wavelet Transform Applications

This section describes the basic idea and examples of applying the wavelet transform to remote sensing image processing. The core of using wavelet transform to denoise remote sensing images is to adjust the high frequency coefficient of the wavelet transform and reconstruct the signal to obtain the image after denoising. The core of remote sensing image fusion is to combine the features of different scales and different frequencies after multiple wavelet transforms and reconstruct the signal to obtain the image after fusion.

10.4.1 Remote Sensing Image Denoising

Assume that the noisy signal $x(n)$ is expressed as

$$x(n) = f(n) + \sigma e(n) \tag{10.40}$$

where $f(n)$ is the original signal, $e(n)$ is Gaussian white noise $N(0, 1)$, and σ is the noise intensity. The wavelet transform denoising process is as follows [5]:

(1) First, the noise signal is decomposed by the wavelet. An orthogonal wavelet basis and number of multiscale decomposition layers A are selected, and orthogonal wavelet transform is performed on $x(n)$ to obtain the high frequency coefficient W_k^j $(1 \leq j \leq J, 1 \leq k \leq N)$ of the j_{th} layer wavelet transform, where N is the number of wavelet coefficients.

(2) The high-frequency wavelet coefficients are adjusted using the soft threshold method.

① Estimate the noise intensity σ empirically:

$$\sigma = \frac{1}{0.6745} \cdot \frac{1}{N} \cdot \sum_{1}^{N} \left| W_k^j \right|, 1 \leq j \leq J \tag{10.41}$$

② Calculate the general threshold T_1.

For a standard Gaussian variable with independent and identically distributed distributions, the probability that the maximum value is less than T_1 approaches 1 as N increases. Since Gaussian white noise $e(n)$ has an independent and identical distribution, M wavelet coefficients are obtained by the orthogonal wavelet basis decomposition of the J layer, which have an independent distribution. The probability that the maximum amplitude in these noise wavelet coefficients is lower than T_1 is close to 1. The general threshold T_1 can be calculated using the following formula:

$$T_1 = \sigma \sqrt{2 \ln(M)} \tag{10.42}$$

③ Calculate Stein's unbiased risk threshold T_2.

To ensure that the approximation error of the signal is minimal, the threshold must be adapted to the variation in the wavelet coefficients. The risk value corresponding to each threshold can be obtained, and the threshold with the smallest risk value is set as the selection threshold. Therefore, the square of the high-frequency coefficients of a certain wavelet transform is arranged in ascending order: $P = [P_1, P_2 \cdots P_N]$,

$P_1 \leq P_2 \cdots \leq P_N$, where N is the number of wavelet coefficients of the layer. The corresponding risk vector $[r_1, r_2 \cdots r_N]$ is calculated, i.e.,

$$r_i = \frac{(N-2i) + (N-i)P_i + \sum_{k=1}^{N} p_k}{N}, i = 1, 2, \ldots, N$$

Therefore, the minimum value of R is selected as the minimum risk value. The minimum value is r_k. In accordance with the subscript variable k, the corresponding wavelet coefficient square P_k is determined, and the threshold Stein unbiased risk threshold T_2 is derived:

$$T_2 = \sigma \cdot \sqrt{P_k} \tag{10.43}$$

④ Calculate the Stein unbiased deduction threshold T_3 for the test method.

This method pertains to the selection of the optimal predictor threshold. S is the sum of the squares of the N wavelet coefficients, and h is the minimum energy level:

$$S = \sum_{i=1}^{N} W_i^2, g = \frac{S-N}{N}, h = (\log_2 N)^{\frac{3}{2}} \sqrt{N}$$

The method to determine the Stein unbiased risk threshold T_3 is

$$T_3 = \begin{cases} T_1, & g < h \\ \min(T_1, T_2), & g > h \end{cases} \tag{10.44}$$

The soft-threshold method is used to solve the high-frequency coefficients of the wavelet transform under different scale decompositions:

$$\hat{W}_k^j = \begin{cases} sign\left(W_k^j\right)\left(\left|W_k^j\right| - T_3\right), & \left|W_k^j\right| \geq T_3 \\ 0, & \left|W_k^j\right| < T_3 \end{cases} \quad k \in N \tag{10.45}$$

where $\hat{W}_k^j$ is the j_{th} layer estimation wavelet coefficient adjusted by the wavelet soft threshold method, N is the number of wavelet coefficients of the layer, and T_3 is the threshold. The algorithm model is shown in Fig. 10.15

(3) Signal reconstruction. The estimated wavelet transform high-frequency coefficient $\hat{W}_k^j$ and low-frequency wavelet coefficients are decomposed at each scale to reconstruct the estimated value $\hat{x}(n)$ of the original signal.
(4) Evaluation of the denoising effect. The denoising effect is evaluated by SNR and the original signal root mean square error RMSE:

$$\text{SNR} = 10 \log_2 \frac{\sum_n x^2(n)}{\sum_n [x(n) - \hat{x}(n)]^2} \tag{10.46}$$

Fig. 10.15 Soft threshold method to shrink the wavelet coefficient

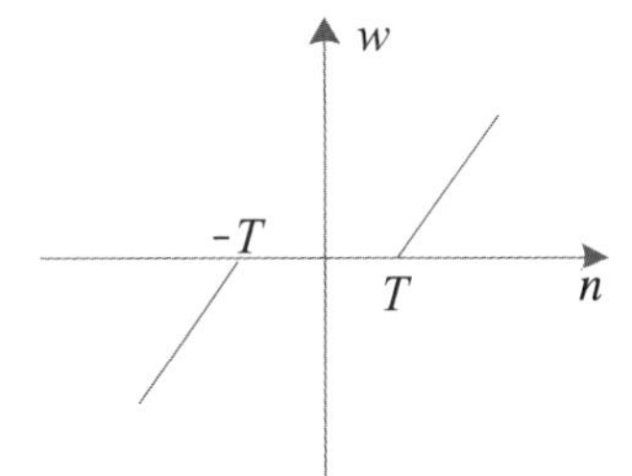

$$\mathrm{RMSE} = \sqrt{\frac{1}{N}\sum_{n}\left[x(n) - \hat{x}(n)\right]^2} \tag{10.47}$$

where $x(n)$ is the original signal, and $\hat{x}(n)$ is the estimated signal after wavelet denoising. If SNR is higher and RMSE is smaller, the estimated signal is more similar to the original signal, and the denoising effect is superior.

10.4.2 Remote Sensing Image Fusion

Wavelet transform refers to a multiscale, multidirectional, and multiresolution decomposition, and the image can be decomposed into a lower resolution of the approximate image and high frequency detail images. Therefore, different scales of spatial features can be used for the pixel-level fusion of multiresolution images obtained by different sensors. The transformation is nonredundant and lossless, the total amount of data after wavelet decomposition remains unchanged, and the correlation of the image differences between adjacent scales is removed, which can fully reflect the local variation characteristics of the original image. Furthermore, through the directivity, the fusion result with a superior visual effect can be obtained.

The wavelet transform is an orthogonal transform; however, the orthogonal filter does not have the linear phase characteristic, and the resulting phase distortion may lead to the distortion of the image edge. Therefore, image fusion is generally performed in biorthogonal wavelet transforms. The steps are shown in Fig. 10.16. The first step is preprocessing, including noise reduction and registration. Next, a biorthogonal wavelet transform is implemented to decompose multiple images. Subsequently, the subgraphs of different levels corresponding to multiple images are fused according to the fusion algorithm, and the wavelet decomposition map after fusion is obtained. Finally, the multiscale image is transformed into a wavelet transform to obtain the final fusion result.

The wavelet fusion rule is divided into two types: single pixel and region-based features. The former type includes the direct substitution or addition of the wavelet coefficients, maximum value selection, and weighted average. The latter type includes the gradient-based method, local variance-based method, and local

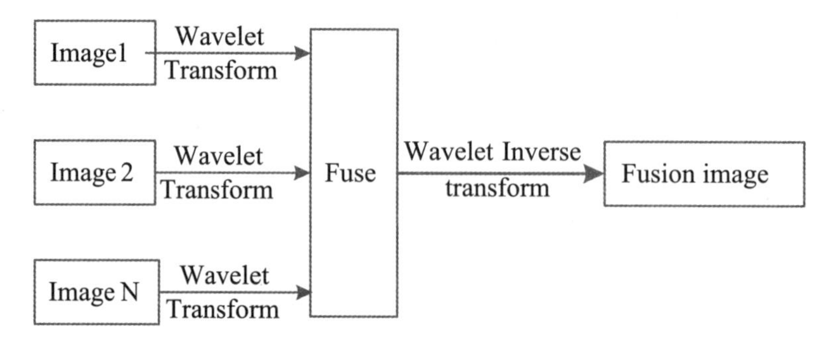

Fig. 10.16 Image fusion algorithm based on wavelet transform

energy-based method. Pixel-based fusion rules exhibit a high degree of sensitivity to the edges, requiring strict image registration. Based on the regional fusion rules, considering the correlation of adjacent pixels and decreasing the sensitivity to the edge, the application is more extensive. The data fusion rules based on the wavelet transform are shown in Table 10.2.

Table 10.2 Comparison of common methods

Image fusion rules based on wavelet transform	Applicable image	Advantage	Disadvantage
Larger absolute value of the coefficient method	The original image is rich in high frequency components, and the brightness contrast is higher	Retains the original image features and contrast	Other features of the image are easily covered
Weighted average method	Wide range of applications	The weight coefficient is adjustable and can eliminate part of the noise, and the loss of the original image information is smaller	Image contrast reduction, and need for grayscale enhancement
Elimination of high frequency noise method	Wide range of applications	High frequency noise is eliminated, the fusion results have a higher contrast, and the original image features can be better retained	Loss of part of the high-frequency information
Double threshold method	The original image is a gray image with a distribution of balanced and high frequency components	Dual threshold is optional, and the algorithm is practical	Regardless of the original image grayscale distribution characteristics, edge jumps may occur

In this context, considerable research has been performed. For example, Qiang et al. used the sym4 wavelet for the wavelet decomposition of high spatial resolution aerial images and low spatial resolution satellite image. By calculating the information entropy and standard deviation, the authors found that the results based on the local variance fusion principle are superior than those based on the weighted average method [6]. Ranchin used the D4 wavelet through the SW, MW, and RWM transformation methods for the image fusion of SPOT images and KVR-1000 images, and the results were superior than those of the BROBEY, HIS and PCA standard fusion methods, with the optimal effect corresponding to the RWM [7]. Li used D8 and biorthogonal B-spline wavelet through SW and SWF transformation methods for Landsat TM images and SPOT panchromatic images. The fusion processes are shown in Figs. 10.17 and 10.18 [8]. Figure 10.18 (a) shows the original SPOT image, (b) shows the original Landsat TM image, (c) shows the HIS fusion result, (d) shows the PCA fusion result, (e) shows the discrete wavelet transform fusion result, and (f) shows the discrete wavelet framework transform fusion result. Compared with the spatial correlation, the fusion result of the wavelet transform is superior to that of standard fusion, and the fusion method of the discrete wavelet frame is superior when the registration precision is low.

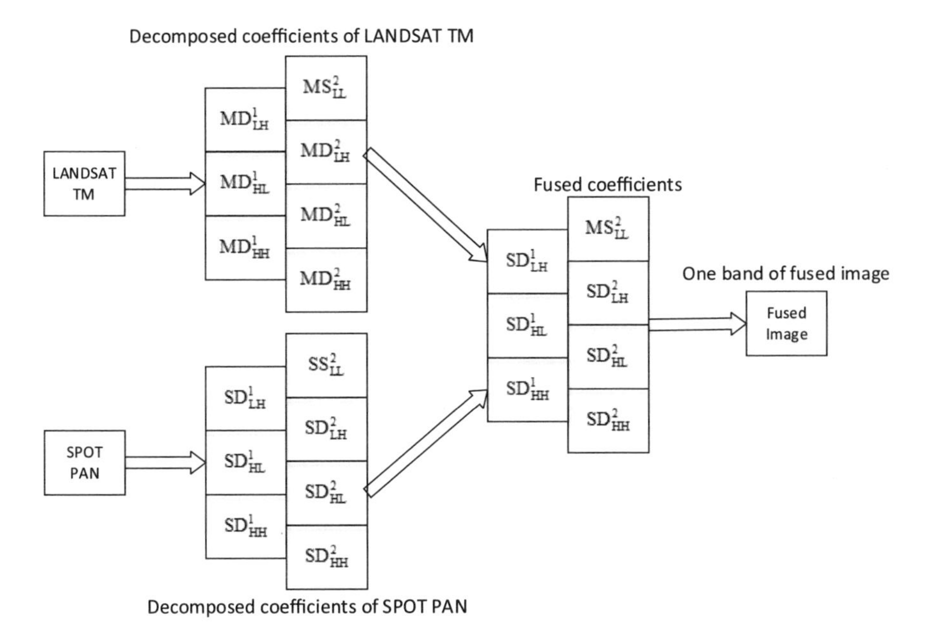

Fig. 10.17 Discrete wavelet frame transform based image fusion

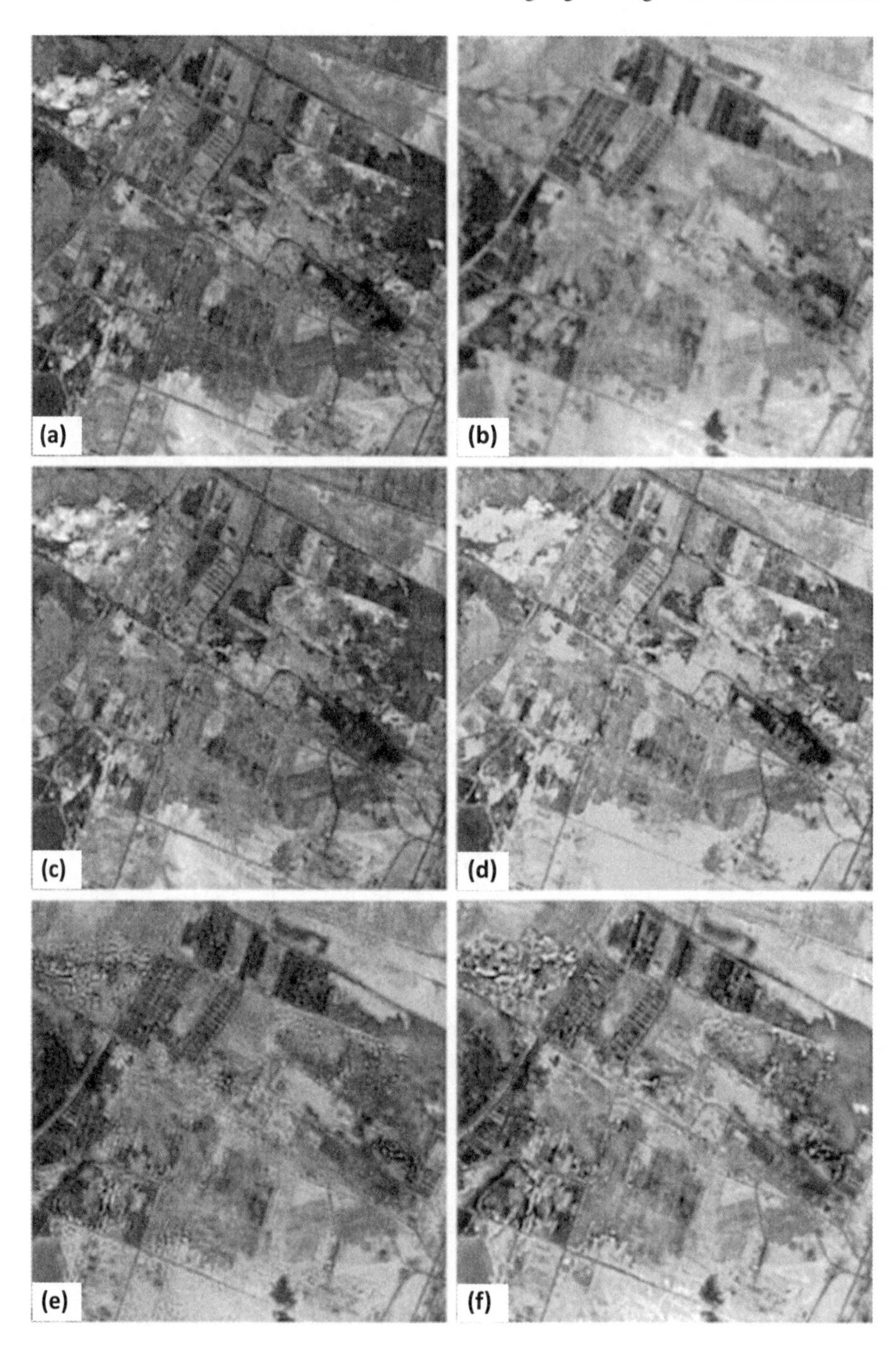

Fig. 10.18 Original image and fusion results

10.5 Summary

The wavelet transform is localized in the time and frequency domains and is more suitable for analyzing frequency-varying signals such as those associated with remote sensing images than Fourier transforms. This chapter describes the time–frequency analysis method from the perspective of continuous wavelet transform to understand its essence. The problem of redundancy of information is solved by implementing the continuous wavelet with series expansion, dyadic wavelet, and tightly packed binary wavelet. The wavelet transform and amount of shift are discrete. Under the guidance of subband coding and multiresolution analysis, the Mallat algorithm is used to rapidly construct a wavelet orthogonal basis. Multiresolution analysis is performed using a scale function, and thus, the multiresolution analysis is equivalent to finding the properties of the scale function. The wavelet transform can be regarded as a set of bandpass filters to filter the signal. Compared with the traditional method, the wavelet transform exhibits notable advantages in remote sensing image denoising and fusion, and the principle relation is shown in Fig. 10.19. The corresponding mathematical formulas are shown in Table 10.3.

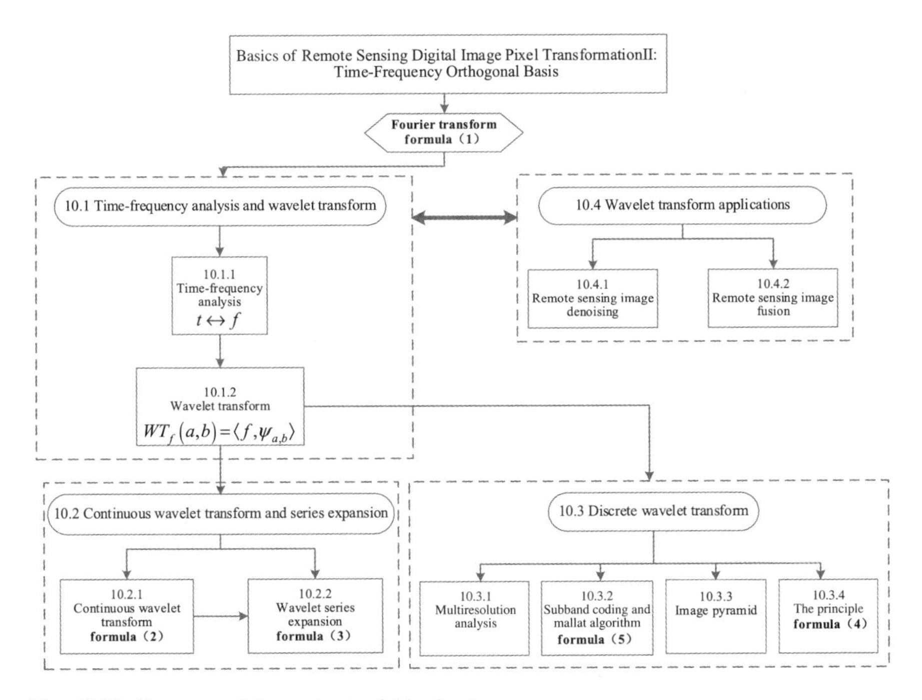

Fig. 10.19 Summary of the contents of this chapter

Table. 10.3 Summary of the formulas in this chapter

formula (1)	$\begin{cases} f(t) = \dfrac{1}{2\pi}\displaystyle\int_{-\infty}^{+\infty} F(\omega)\mathrm{e}^{\mathrm{j}\omega t}\mathrm{d}\omega \\ F(\omega) = \displaystyle\int_{-\infty}^{+\infty} f(t)\mathrm{e}^{-\mathrm{j}\omega t}\mathrm{d}t \end{cases}$	
formula (2)	$W_f(a, b_x, b_y) = \int_{-\infty}^{\infty}\int_{-\infty}^{\infty} f(x, y)\Psi_{a,b_x,b_y}(x, y)\mathrm{d}x\mathrm{d}y$ $f(x, y) = \frac{1}{C_\Psi}\int_0^{\infty}\int_{-\infty}^{\infty}\int_{-\infty}^{\infty} W_f(a, b_x, b_y)\Psi_{a,b_x,b_y}(x, y)\mathrm{d}b_x\mathrm{d}b_y\frac{\mathrm{d}a}{a^3}$	10.2.1
formula (3)	$\Psi_{j,k}(x) = 2^{j/2}\Psi(2^j x - k),\ j, k \in Z$ $\Psi_n(x) = 2^{j/2}\Psi(2^j x - k)$ $n = 2^j + k,\ j = 0, 1, \cdots;\ k = 0, 1, \cdots, 2^j - 1$	10.2.2
formula (4)	$\begin{cases} x_{\alpha,\mathrm{L}}[n] = \displaystyle\sum_{k=0}^{K-1} x_{\alpha-1,\mathrm{L}}[2n-k]g[k] \\ x_{\alpha,\mathrm{H}}[n] = \displaystyle\sum_{k=0}^{K-1} x_{\alpha-1,\mathrm{L}}[2n-k]h[k] \end{cases}$	10.3.4
formula (5)	$\begin{cases} c_k^j = \Sigma_n h(n-2k)c_n^{j-1} \\ d_k^j = \Sigma_n g(n-2k)c_n^{j-1} \end{cases}$	10.3.2

References

1. Meng HY, Liu GZ. Multiscale seismic waveform inversion by wavelet transform. Chin J Geophys. 1999. (In Chinese).
2. Daubechies I. Orthonormal bases of compactly supported wavelets. Commun Pure Appl Math. 1988(41).
3. Mallat S. A theory for multiresolution signal decomposition: the wavelet representation. Geosci Remote Sens IEEE Trans Geosci Remote Sens. 1989.
4. Cheng Y, Zhang X, Wen J. Coding of mosaic image based on wavelet sub-band substitute. PCSPA. 2010.
5. María GA, José LS, Raquel GC, et al. Fusion of multispectral and panchromatic images using improved IHS and PCA mergers based on wavelet decomposition. Geosci Remote Sens IEEE Trans Geosci Remote Sens 2004(42).
6. Qiang ZX, Peng JX, Wang HQ. Remote sensing image fusion based on local deviation of wavelet transform. J Huazhong Univ Sci Tech (Nature Science Edition). 2003(31). (In Chinese)
7. Ranchin T, Wald L. Fusion of high spatial and spectral resolution images: the ARSIS concept and its implementation. Photogram Eng Remote Sens. 2000(66).
8. Li S, James K T, Yaonan W. Using the discrete wavelet frame transform to merge Landsat TM and SPOT panchromatic images. Inf Fusion. 2002.

Part III
Technology and Application

Chapter 11
Technology I of Remote Sensing Digital Image Processing: Noise Reduction and Image Reconstruction

Chapter Guidance The originally recorded remote sensing digital images involve noise and system error after imaging, transmission and processing. Therefore, the first task of image processing is to restore the original image. Only after identifying the original real images can useful feature information in the image be extracted according to the demands. The essence is to remove the digital image processing or input convolution effect and equivalent noise.

The aim of image restoration is to eliminate or alleviate the image quality degradation caused by the image acquisition and transmission process (or degradation) and restore the original image (or real scene) as much as possible [1]. Figure 11.1 shows the framework of the image degradation. The main contents of this chapter are as follows: Sect. 11.1 Error image degradation and two principle elements of degeneration: This section describes the identification method; Sect. 11.2 Overview of mathematical deconvolution operation and nature of the classic restoration filter. This section describes the corresponding processing method; Sect. 11.3 Linear algebra restoration and less restrictive restoration. This section describes the corresponding art method; Sect. 11.4 Multilook and filter restoration. This section describes the common recovery methods for SAR images; Sect. 11.5 Superresolution restoration in remote sensing. This section describes the special frequency methods for variable restoration; Sect. 11.6 Modeling image restoration software implementation. This section describes the software and hardware for software implementation.

Image $f(x, y)$ is blurred by the linear system $h(x, y)$ and superimposed on noise $n(x, y)$, leading to the degraded image $g(x, y)$. The degraded image convolutes with restoration filter $w(x, y)$, resulting in restoration image $\hat{f}(x, y)$. In this chapter, the first part is aimed at analyzing the two elements of the basic model and degraded image restoration model. Next, we introduce the classic methods of image restoration, restoration linear algebra techniques, filtering multilook image restoration and superresolution techniques beyond the cutoff frequency associated with restoration technology. Subsequently, we analyze the software for image model restoration, which is the basis of image restoration.

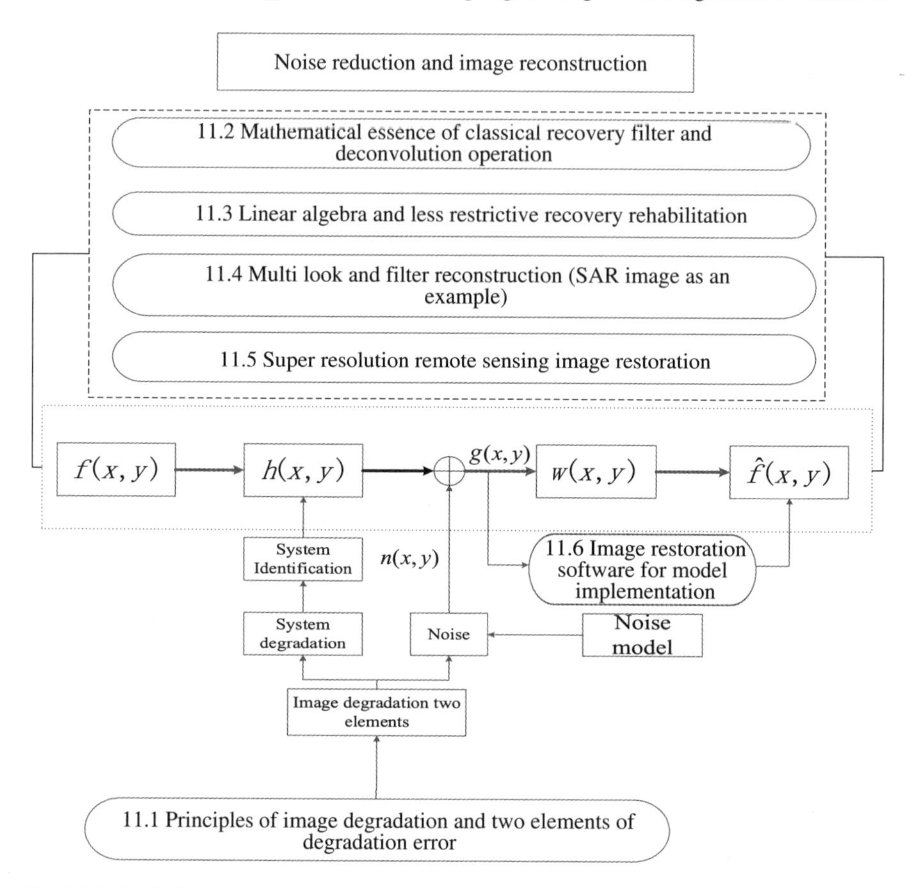

Fig. 11.1 Logical framework

11.1 Principles of Image Degradation and Two Elements of Degradation Error

This section first clarifies the image degradation and image restoration process, analyzes the image degradation mechanism, and clarifies the objectives of image restoration. Subsequently, we analyze the two elements of image degradation. To clarify the identification method, we explain the restoration method for images.

11.1.1 Image Degradation Mechanism

The key objective of image restoration is to establish the degradation model. Image degradation occurs in the modeling process due to the presence of a degenerate system. The original input image transforms into the output degradation image owing to the effect of the degradation system and noise. Assuming that there exists a known black-and-white static degenerate plane image $g(x, y)$, the degradation system is $h(x, y)$, and the noise is $n(x, y)$, the degradation process is as shown in Fig. 11.2.

The relationship between the input and output can be expressed as

$$g(x, y) = h(x, y) * f(x, y) + n(x, y) \tag{11.1}$$

For ease of discussion, we set $n(x, y) = 0$ temporarily, and simplify (11.1) as

$$g(x, y) = h(x, y) * f(x, y) \tag{11.2}$$

After determining function $g(x, y)$ and degradation system h using formula (11.2), $f(x, y)$ can be obtained. Generally, $g(x, y)$ is known, and thus, the core of the restoration processing system is to identify $h(x, y)$. The following assumptions are set for the system degradation:

(1) H is a linear system. According to the characteristics of the linear system, the following relationship exists in the time domain:

$$\begin{aligned} h(x, y) * [k_1 f_1(x, y) + k_2 f_2(x, y)] &= k_1 h(x, y) * f_1(x, y) \\ &\quad + k_2 h(x, y) * f_2(x, y) \\ &= k_1 g_1(x, y) + k_2 g_2(x, y) \end{aligned} \tag{11.3}$$

where $f_1(x, y)$ and $f_2(x, y)$ are the two input image systems; $g_1(x, y)$ and $g_2(x, y)$ are the corresponding output images; and k_1 and k_2 are constants.

(2) System $h(x, y)$ is a spatial location invariant system, that is, for each arbitrary input image $f(x, y)$,

$$h(x, y) * f(x - \alpha, y - \beta) = g(x - \alpha, y - \beta) \tag{11.4}$$

where α and β are the displacements of the spatial position. The response to any point depends only on the value of that point in the image, regardless of the amount of displacement.

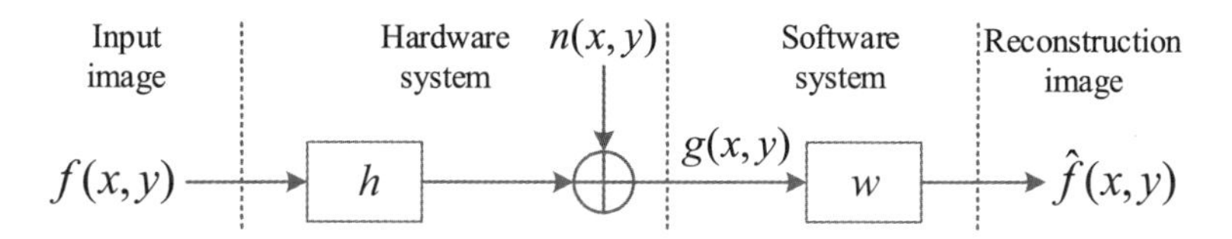

Fig. 11.2 Degraded image restoration process

11.1.2 Two Elements of Image Degradation

1. Image Degradation and System Model Identification Caused by System Convolution

(1) Remote sensing system with image degradation

For remote sensing systems, degradation of the image sensing system can be expressed as

$$H = f_1 \cdot f_2 \cdot A \cdot G \tag{11.5}$$

where f_1 indicates the errors caused by remote sensing platforms, f_2 refers to the errors caused by the imaging system, A refers to atmospheric effects, and G refers to the error caused by the inversion feature.

① Image degradation caused by remote sensing platforms refers to image deformation caused by changes in attitude of aircraft or satellites in flight
② Image degradation based on the imaging system is caused by the strain sensor with different projection imaging modalities, such as panoramic changes or pitch changes caused by the projection.
③ Image degradation caused by atmospheric effects pertains to the influence of the atmosphere on the target wave absorption, scattering, and refraction, which decreases the image quality.
④ Image degradation caused by inversion features pertains to image retrieval processing errors, and the former model error may include error correction radiation and image processing, such as geometric correction. The latter model refers to the complexity of the target feature, and inversion makes it difficult to accurately model the inversion-related errors.

(2) Methods of remote sensing image system identification

① Using the test target for system identification

In many cases, the system transfer function can be directly determined. In the case shown in Fig. 11.2, the impulse response function $h(x, y)$ is unknown and needs to be determined; if there exist an appropriate test signal $f(x, y)$ as shown in Fig. 11.3, Formula (11.2) can be transformed to the frequency domain convolution multiplication relationship. The system transfer function is

$$H(u, v) = \frac{G(u, v)}{F(u, v)} \tag{11.6}$$

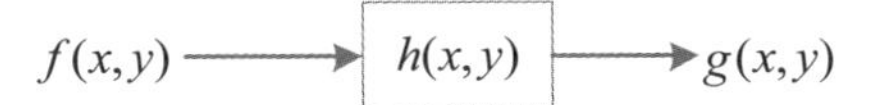

Fig. 11.3 Linear system

In the optimal solution, there are no zero points in $F(u,\ v)$. If zero points exists but $H(u, v)$ is relatively smooth, the equations can still be solved with numerical methods.

② Using the cross-correlation system for identification

As shown in Fig. 11.4, the output of a linear system is correlated with the input. The output spectrum function of the cross-correlator is

$$Z(s) = G(s) * F(s) = H(s)F(s) * F(s) = H(s)P_f(s) \tag{11.7}$$

$P_f(s)$ is the power spectrum of the input signal. If $f(x)$ refers to noncorrelation white noise, the value may be constant; thus, the output of the cross-correlator is the impulse response of the system. In this case, the random noise image can be used as an image input system, which calculates the cross-correlation with the system output, to obtain a point spread function of the system. Furthermore, the spectral output of the cross-correlation function is the system transfer function.

③ Determining the optical transfer function from the spectrum of the degraded image

An image of a complex scene has a relatively smooth amplitude spectrum. If the transfer function that causes the degradation has zero points in the spectral degradation, these zero points will force the image frequency to become zero at a certain frequency. If the position of an appropriate model of the fuzzy function is given, the position of these zero (or near-zero) points in the spatial frequency plane may be used to determine the positional parameters of the blurred optical transfer function.

By taking the logarithm of the power spectrum of the degraded image, the concave amplitude caused by the zero point in the degenerate transfer function can be enhanced. If the zero points are equally spaced, there will be a series of periodic spikes on the logarithm of the power spectrum. The power spectrum of the logarithmic power spectrum, often known as the cepstrum, can be used to determine the exact distance between these spikes, and the zero points of the degradation transfer function can be obtained.

2. Noise-Inducing Image Degradation and Model to Describe Noise

Noise is one of two factors leading to image degradation. Digital image noise pertains to the image acquisition and transfer process, and effectively removing image noise

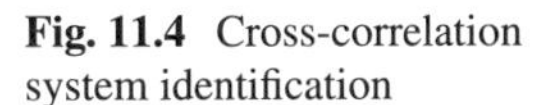
Fig. 11.4 Cross-correlation system identification

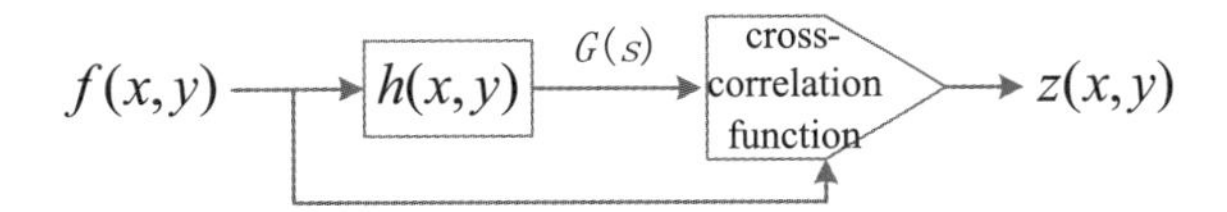

is a key step. Noise modeling is the first step in removing noise, and the mean and variance are used in the statistical sense to describe the noise.

When the digital image signal is a two-dimensional signal, the two-dimensional intensity distribution is $f(x, y)$. The mean of the noise indicates the overall strength of the noise in the image:

$$\overline{n} = E[n(x, y)] = \frac{1}{MN}\sum_{x=1}^{M}\sum_{y=1}^{N} n(x, y) \tag{11.8}$$

The variance of the noise indicates the difference in the intensity distribution of the noise in the image, defined as

$$\sigma_n^2 = E\{[n(x, y) - \overline{n}]^2\} = \frac{1}{MN}\sum_{x=1}^{M}\sum_{y=1}^{N} [n(x, y) - \overline{n}]^2 \tag{11.9}$$

According to the probability distribution, noise can be divided into Gaussian noise, uniform noise, Rayleigh noise, salt and pepper noise, gamma noise and index noise.

(1) Gaussian noise

Gaussian noise is referred to as normal noise (Fig. 11.5), the position of which is constant, i.e., every point has noise, but the amplitude of the noise is random. The random variable z of Gaussian probability density is

$$p(z) = \frac{1}{\sqrt{2\pi}\sigma} e^{\frac{-(z-\mu)^2}{2\sigma^2}} \tag{11.10}$$

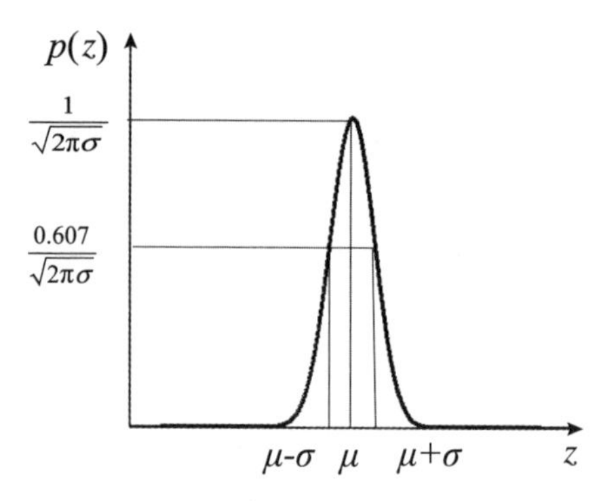

Fig. 11.5 Gaussian noise density function

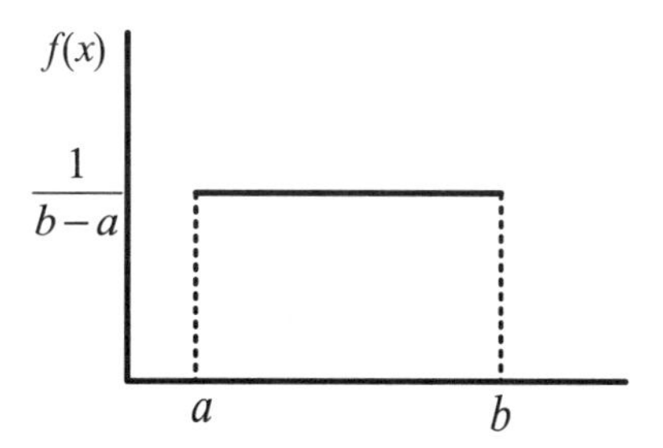

Fig. 11.6 PDF of uniform noise

where z represents a gray value, μ is the average value or a desired value of z, and σ represents the standard deviation. The square of the standard deviation σ^2 is known as the variance of z. When z obeys the Gaussian distribution, its value falls in the range $[(\mu - \sigma), (\mu + \sigma)]$ or $[(\mu - 2\sigma), (\mu + 2\sigma)]$ with a probability of approximately 70% and 95%, respectively.

(2) Uniform noise

The probability density of the uniform noise distribution (Fig. 11.6) is

$$p(z) = \begin{cases} \frac{1}{b-a}, & a < z < b \\ 0, & \text{others} \end{cases} \tag{11.11}$$

The mean and variance of the probability density are defined as

$$\mu = \frac{a+b}{2}, \sigma^2 = \frac{(b-a)^2}{12} \tag{11.12}$$

(3) Salt and pepper noise

The probability density of pepper noise (also known as impulse noise; Fig. 11.7) is

$$p(z) = \begin{cases} p_a, & z = a \\ p_b, & z = b \\ 0, & \text{others} \end{cases} \tag{11.13}$$

If $b > a$, b will appear as a bright spot, and a will appear as a dark spot. If p_a or p_b is zero, the impulse noise is known as a single-stage pulse. If p_a and p_b are not zero or the two values are approximately equal, the value of the impulse noise is randomly distributed, similar to salt and pepper. Impulse noise can be positive or negative. Typically, negative and positive pulses occur as pulsed black dots (pepper point) and white dots (salt point), respectively.

(4) Rayleigh noise

The probability density function of Rayleigh noise (Fig. 11.8) is

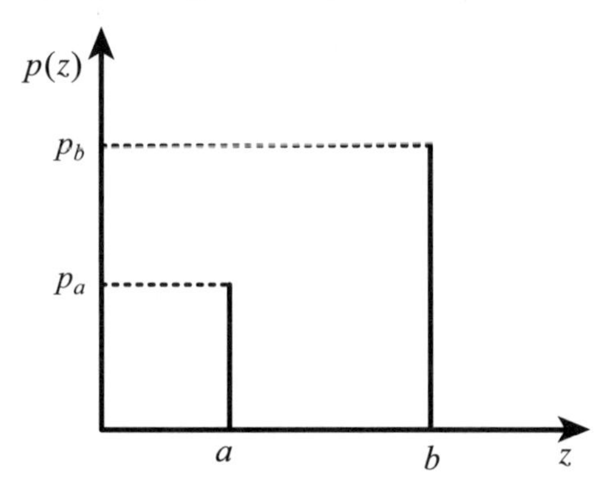

Fig. 11.7 PDF of salt and pepper noise

$$p(z) = \begin{cases} \frac{2}{b}(z-a)e^{-\frac{(z-a)^2}{b}}, & z \geq a \\ 0, & z < a \end{cases} \tag{11.14}$$

The mean and variance of probability density are $\mu = a + \sqrt{\frac{\pi b}{4}}$ and $\sigma^2 = \frac{b(4-\pi)}{4}$, respectively.

(5) Gamma noise

Gamma noise (Fig. 11.9), is also known as Irish noise, with the following probability density function:

$$p(z) = \begin{cases} \frac{a^b z^{b-1}}{(b-1)!}e^{-az}, & z \geq 0 \\ 0, & z < 0 \end{cases} \tag{11.15}$$

Here, $a > 0$; b is a positive integer, and "!" represents a factorial. The density mean and variance are

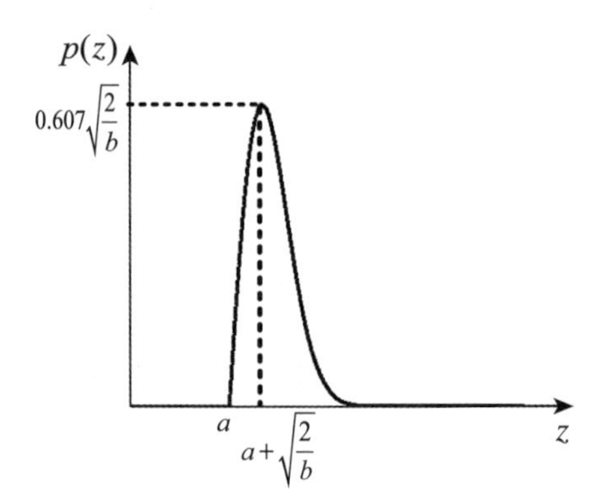

Fig. 11.8 PDF of Rayleigh noise

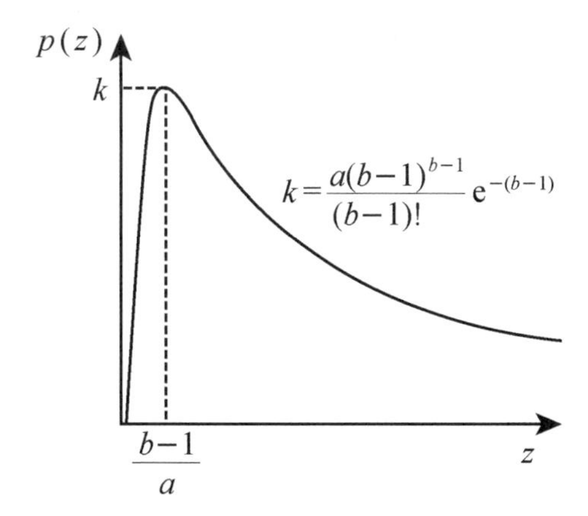

Fig. 11.9 PDF of gamma noise

$$\mu = \frac{b}{a}, \sigma^2 = \frac{b}{a^2}$$

(6) Index noise

The probability density function of index noise (Fig. 11.10) is

$$p(z) = \begin{cases} a\mathrm{e}^{-az}, z \geq 0 \\ 0, \ z < 0 \end{cases} \tag{11.16}$$

where $a > 0$. The expectation and variance of the probability density function are $\mu = \frac{1}{a}$ and $\sigma^2 = \frac{1}{a^2}$, respectively. The probability density function of the exponential distribution is a special case of the Irish probability distribution when $b = 1$.

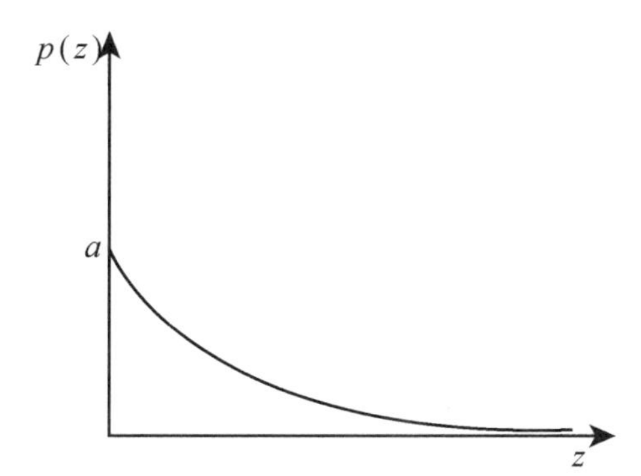

Fig. 11.10 PDF of index noise

11.2 Mathematical Essence of Classic Recovery Filter and Deconvolution Operation

In the previous section, the image degradation mechanism is described, and two elements of image degradation are analyzed. This section explores the nature of the mathematical deconvolution operation and restoration of classic filters and reviews the effective tools in practical image restoration.

11.2.1 Introduction of Classic Recovery Filter

Image deblurring and denoising are the two major problems in image restoration. According to the mathematical tools and focused recovery problems, conventional image restoration methods can be categorized into divided inverse filtering method and algebraically spatial filtering. Image restoration classification can be represented by the model shown in Fig. 11.11.

Inverse filtering methods include the classic inverse filtering method, Wiener filter method and Kalman filter method. In the frequency domain conversion function, the inverse filtering classical transformation function of the reciprocal of image distortion for high SNR images can effectively suppress noise. However, the method cannot be applied at low SNRs.

Algebraic methods can be divided into the pseudoinverse method, singular value decomposition pseudoinverse method, Wiener estimation method and constrained image restoration method.

Spatial filtering methods include the mean filtering, median filtering, nonlinear filtering, laminated filtering, and morphological filtering methods. The nonlinear filter corresponding to these methods can effectively remove the impulse noise

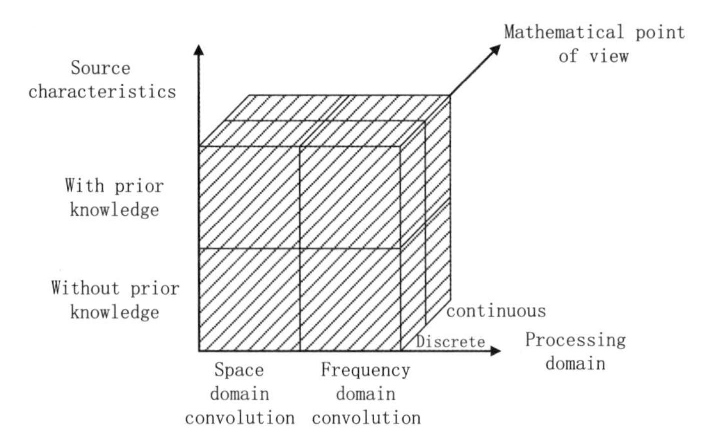

Fig. 11.11 Classification of image restoration methods

while maintaining the details of the image and obtain a satisfactory restoration effect under certain conditions. Therefore, the spatial filtering method is of significance. However, nonlinear filters do not have a complete set of theories and lack a unified design method. Moreover, such methods are generally based on noise and prior knowledge of linear or nonlinear filtering methods. The following text presents the basic principles of several classic filters.

11.2.2 Inverse Filtering

For the image degradation model, the basic principle of inverse filtering (Fig. 11.2), that is, the convolution theorem of the Fourier transform holds:

$$G(u, v) = H(u, v)F(u, v) + N(u, v) \tag{11.17}$$

where $G(u, v)$, $H(u, v)$, $N(u, v)$ and $F(u, v)$ are the Fourier transformations of deteriorated image $g(x, y)$, point spread function $h(x, y)$, noise $n(x, y)$ and original image $f(x, y)$, respectively. According to formula (11.17),

$$F(u, v) = \frac{G(u, v)}{H(u, v)} - \frac{N(u, v)}{H(u, v)} \tag{11.18}$$

If the noise is negligible, the above equation may be approximated by

$$\hat{F}(u, v) = \frac{G(u, v)}{H(u, v)} \tag{11.19}$$

By obtaining the inverse Fourier transform, we can obtain a restored image, i.e.,

$$f(x, y) = \Im^{-1}[F(u, v)] = \Im^{-1}\left[\frac{G(u, v)}{H(u, v)}\right] \tag{11.20}$$

In other words, if the Fourier transform and “filtering” transfer function of the degraded image are known, the original image can be obtained by Fourier transform, and the inverse Fourier transform of the original image can be obtained. Here, $G(u, v)$ is divided by $H(u, v)$, which pertains to inverse filtering, which is the basic principle of inverse filter restoration.

In general, an inverse filtering method cannot accurately estimate the zero points of $H(u, v)$. Instead of performing direct inverse filtering with $1/H(u, v)$, an additional function $M(u, v)$ related to u, v is used. In the time domain, the processing framework is as shown in Fig. 11.2. In the frequency domain, in the absence of zero points and noise, the following formula can be used:

$$M(u, v) = \frac{1}{H(u, v)} \tag{11.21}$$

The model shown in Fig. 11.2 includes the degradation and recovery operations. Degradation and recovery of the overall transfer function can be expressed with *H(u, v)* and *M(u, v)*, respectively:

$$\hat{F}(u, v) = [H(u, v)M(u, v)]F(u, v) \tag{11.22}$$

H(u, v) is known as the input transfer function, and *M(u, v)* corresponds to transfer function processing. *H (u, v)*, and *M(u, v)* are known as output transfer functions. In general, the amplitude of *H(u, v)* decreases rapidly with increasing distance *u*, *v* from the plane origin, while the amplitude variation of the noise term *N*(*u*, *v*) is relatively flat. Far from the origin, the *u*, *v* plane *N(u, v)/H(u, v)* values increase. However, in most cases, image *F*(*u*, *v*) is extremely small, and thus, the lower case is applicable. In this case, noise dominates, and the original image cannot be satisfactorily recovered. This rule described using 1/*H*(*u*, *v*) is effective only when inverse filtering is applied in the origin neighborhood. In other words, *M*(*u*, *v*) must satisfy

$$M(u, v) = \begin{cases} \frac{1}{H(u, v)}, & u^2 + v^2 \le w_0^2 \\ 1, & u^2 + v^2 > w_0^2 \end{cases} \tag{11.23}$$

The selection of w_0^2 must exclude the zero points of *H (u, v)* from this neighborhood.

11.2.3 Wiener Filter

Helstrom used the least mean square error estimation method to propose a Wiener deconvolution filter with the following two-dimensional transfer function. According to the explanation of the Wiener filter in Chap. 7, Formula 11.24 can be obtained:

$$G(u, v) = \frac{H^*(u, v)}{|H(u, v)|^2 + P_n(u, v)/P_f(u, v)} \tag{11.24}$$

where P_f and P_n are the power spectra of the signal and noise, respectively. The following text describes the mathematical nature of Wiener's deconvolution.

The Wiener filter is a least squares filter. The filter is a restoration method that minimizes the mean square error between the original image *f*(*x*, *y*) and its restored image, and it pertains to constraint recovery. In addition to the requirement of understanding the transfer function of the degrading model, it is also necessary to know (at least theoretically) the statistical properties of the noise and correlation between the noise and image. The problem is to choose a suitable transformation matrix $\boldsymbol{Q}$, as choosing different types of $\boldsymbol{Q}$ can yield different types of Wiener filter recovery methods. If the correlation matrices $\boldsymbol{R}_{\mathrm{f}}$ and $\boldsymbol{R}_{\mathrm{n}}$ corresponding to image $f(x, y)$ and noise $n(x, y)$, respectively, are used to represent $\boldsymbol{Q}$, the Wiener filter restoration method can be established.

Suppose $\boldsymbol{R}_{\mathrm{f}}$ and $\boldsymbol{R}_{\mathrm{n}}$ are the correlation matrices of the original image $f(x, y)$ and noise $n(x, y)$, respectively, defined as

$$\left.\begin{aligned} \boldsymbol{R}_{\mathrm{f}} &= E\{\boldsymbol{f}\boldsymbol{f}^{\mathrm{T}}\} \\ \boldsymbol{R}_{\mathrm{n}} &= E\{\boldsymbol{n}\boldsymbol{n}^{\mathrm{T}}\} \end{aligned}\right\} \tag{11.25}$$

where $E\{\cdot\}$ denotes the theoretical expectation operation, and the element ij of $\boldsymbol{R}_{\mathrm{f}}$ is represented by $E\{f_i f_j\}$, which is the correlation between the i-th and j-th elements of f. Similarly, $\boldsymbol{R}_{\mathrm{n}}$ of the ij-th elements is expressed by $E\{n_i\, n_j\}$, which is the correlation between the n i-th and j-th elements. Because elements f and n are real numbers,

$$\left.\begin{aligned} E\{f_i f_j\} &= E\{f_j f_i\} \\ E\{n_i\, n_j\} &= E\{n_j\, n_i\} \end{aligned}\right\}$$

Thus, $\boldsymbol{R}_{\mathrm{f}}$ and $\boldsymbol{R}_{\mathrm{n}}$ are real symmetric matrices. For most functions of the image, the correlation between the pixels does not extend beyond the image distance of 20 to 30 points. Thus, typically, near the main diagonal of the correlation matrix, we will have nonzero elements, which will be zero in the upper right corner and lower left corner of the region.

Assuming that the correlation of any two pixels is related only to the distance between the pixels, regardless of where they are located, $\boldsymbol{R}_{\mathrm{f}}$ and $\boldsymbol{R}_{\mathrm{n}}$ can be approximated as block cycle matrices. Thus, these matrices can be diagonalized using a $\boldsymbol{W}$ matrix of $\boldsymbol{H}$ eigenvectors. Let $\boldsymbol{A}$ and $\boldsymbol{B}$ be the diagonal matrices corresponding to $\boldsymbol{R}_{\mathrm{f}}$ and $\boldsymbol{R}_{\mathrm{n}}$; in this case,

$$\boldsymbol{R}_{\mathrm{f}} = \boldsymbol{W}\boldsymbol{A}\,\boldsymbol{W}^{-1},\ \boldsymbol{R}_{\mathrm{n}} = \boldsymbol{W}\boldsymbol{B}\,\boldsymbol{W}^{-1} \tag{11.26}$$

As mentioned, the elements in matrices $\boldsymbol{A}$ and $\boldsymbol{B}$ are the Fourier transforms of the elements in the correlation matrices $\boldsymbol{R}_{\mathrm{f}}$ and $\boldsymbol{R}_{\mathrm{n}}$, respectively, denoted by $S_{\mathrm{f}}(u, v)$ and $S_{\mathrm{n}}(u, v)$, respectively. According to the information theory, the Fourier transform of the autocorrelation function of the random vector is the power spectral density of the random vector. Thus, $S_{\mathrm{f}}(u, v)$ and $S_{\mathrm{n}}(u, v)$ are the spectral densities of $f(x, y)$ and $n(x, y)$, respectively.

If the selected linear operator $\boldsymbol{Q}$ satisfies the following relationship,

$$\boldsymbol{Q}^{\mathrm{T}}\boldsymbol{Q} = \boldsymbol{R}_{\mathrm{f}}^{-1}\boldsymbol{R}_{\mathrm{n}} \tag{11.27}$$

this formula is substituted into the general formula of the constrained least squares algebraic restoration solution:

$$\hat{f} = \left(\boldsymbol{H}^{\mathrm{T}}\boldsymbol{H} + \boldsymbol{R}_{\mathrm{f}}^{-1}\boldsymbol{R}_{\mathrm{n}}\right)^{-1}\boldsymbol{H}^{\mathrm{T}}\boldsymbol{g} \tag{11.28}$$

The cycle matrix diagonalization and (11.25) lead to

$$\hat{f} = \left(\boldsymbol{W}\boldsymbol{D}^{*}\boldsymbol{D}\boldsymbol{W}^{-1} + \boldsymbol{W}\boldsymbol{A}^{-1}\boldsymbol{B}\boldsymbol{W}^{-1}\right)^{-1}\boldsymbol{W}\boldsymbol{D}^{*}\boldsymbol{W}^{-1}\boldsymbol{g} \tag{11.29}$$

where the diagonal elements of $\boldsymbol{D}$ correspond to the Fourier transform of the block elements in $\boldsymbol{H}$and "*" denotes a conjugate. By multiplying the left side of the equation with $\boldsymbol{W}^{-1}$ and several transformation matrices, the equation can be written as

$$\boldsymbol{W}^{-1}\hat{f} = \left(\boldsymbol{D}^{*}\boldsymbol{D} + \boldsymbol{A}^{-1}\boldsymbol{B}\right)^{-1}\boldsymbol{D}^{*}\boldsymbol{W}^{-1}\boldsymbol{g} \tag{11.30}$$

The elements of this formula can be expressed as

$$\begin{aligned}\hat{F}(u,v) &= \left\{\frac{H^{*}(u,v)}{|H(u,v)|^{2} + \gamma[S_{\mathrm{n}}(u,v)/S_{\mathrm{f}}(u,v)]}\right\}G(u,v) \\ &= \left\{\frac{1}{H(u,v)}\cdot\frac{|H(u,v)|^{2}}{|H(u,v)|^{2} + \gamma[S_{\mathrm{n}}(u,v)/S_{\mathrm{f}}(u,v)]}]\right\}G(u,v)\end{aligned} \tag{11.31}$$

When $\gamma = 1$, the item in braces represents a Wiener filter. When $\gamma = 0$, the items in the braces correspond to the inverse filter. Wiener deconvolution is the best method for deconvolution for the noise transfer function; however, three problems limit the effectiveness of this approach: First, Wiener deconvolution corresponds to the minimum MSE criterion, which not necessarily subjective. Second, the variable space cannot effectively manage the point spread function (PSF), such as a shift-variant nonlinear system. Third, nonstationary processes (image and noise) are not suitable, and the image is often nonstationary.

11.3 Linear Algebra and Less Restrictive Recovery Rehabilitation

This section describes the method of image restoration linear algebra, mainly from two perspectives: with and without constraints. Next, the constraints of constant fuzzy displacement, stationarity of signal and noise and image restoration are defined. These methods are commonly used in practical applications.

11.3.1 Linear Algebra Recovery

The algebra image restoration algorithm, proposed by Andrews and Hunt et al., is based on a discrete degenerate system model:

$$\boldsymbol{g}=\boldsymbol{H}\boldsymbol{f}+\boldsymbol{n} \tag{11.32}$$

where $\boldsymbol{g}$, $\boldsymbol{f}$, and $\boldsymbol{n}$ are the N-dimensional column vectors, and $\boldsymbol{H}$ is a matrix of dimensions $N\times N$.

1. Unconstrained Algebraic Restoration

Rewriting Formula (11.1), the image noise is defined as

$$\boldsymbol{n}=\boldsymbol{g}-\boldsymbol{H}\boldsymbol{f} \tag{11.33}$$

If $N=0$ or the noise is not considered, the following method pertains to the solution of the least squares problem. If $\boldsymbol{e}(\hat{\boldsymbol{f}})$ be the difference between the estimated value $\hat{\boldsymbol{f}}$ and fuzzy vector $\boldsymbol{g}$, Formula (11.32) can be rewritten as

$$\boldsymbol{g}=\boldsymbol{H}\boldsymbol{f}=\boldsymbol{H}\hat{\boldsymbol{f}}+\boldsymbol{e}(\hat{\boldsymbol{f}}) \tag{11.34}$$

The square of the norm of $\boldsymbol{e}(\hat{\boldsymbol{f}})$ is

$$\left\|\boldsymbol{e}(\hat{\boldsymbol{f}})\right\|^2=\boldsymbol{e}(\hat{\boldsymbol{f}})^{\mathrm{T}}\boldsymbol{e}(\hat{\boldsymbol{f}})=\left(\boldsymbol{g}-\boldsymbol{H}\hat{\boldsymbol{f}}\right)^{\mathrm{T}}\left(\boldsymbol{g}-\boldsymbol{H}\hat{\boldsymbol{f}}\right) \tag{11.35}$$

$\left\|\boldsymbol{e}\left(\hat{\boldsymbol{f}}\right)\right\|^2$ can be regarded as a measure of the error term $\boldsymbol{e}\left(\hat{\boldsymbol{f}}\right)$. $\hat{\boldsymbol{f}}$ may be selected such that it minimizes the difference between the results obtained after blur $\boldsymbol{H}$(degradation) and the observed image $\boldsymbol{g}$ in the mean squares sense. If $\boldsymbol{f}$ and the $\boldsymbol{H}$ results are similar, the optimal estimate of $\boldsymbol{f}$ can be obtained.

The unconstrained restoration is the least-squares solution presented in Formula (11.35), i.e., we find a squared error vector corresponding to the minimum value. Moreover, the argument is not bound by any unconstrained restoration.

If the objective function is

$$W(\hat{\boldsymbol{f}})=\left\|\boldsymbol{e}(\hat{\boldsymbol{f}})\right\|^2=\left(\boldsymbol{g}-\boldsymbol{H}\hat{\boldsymbol{f}}\right)^{\mathrm{T}}\left(\boldsymbol{g}-\boldsymbol{H}\hat{\boldsymbol{f}}\right) \tag{11.36}$$

the derivative is set as zero, and

$$\frac{\partial W(\hat{\boldsymbol{f}})}{\partial \boldsymbol{f}}=-2\boldsymbol{H}^{\mathrm{T}}\left(\boldsymbol{g}-\boldsymbol{H}\hat{\boldsymbol{f}}\right)=0 \tag{11.37}$$

We determine

$$\hat{f} = \left(\boldsymbol{H}^{\mathrm{T}}\boldsymbol{H}\right)^{-1}\boldsymbol{H}^{\mathrm{T}}\boldsymbol{g} \tag{11.38}$$

where $(\boldsymbol{H}^{\mathrm{T}}\boldsymbol{H})^{-1}\boldsymbol{H}^{\mathrm{T}}$ is known as the generalized inverse matrix of $\boldsymbol{H}$, since $\boldsymbol{H}$ is an $N \times N$ square matrix. In this case,

$$\hat{f} = \boldsymbol{H}^{-1}\boldsymbol{g} \tag{11.39}$$

Formula (11.39) defines the inverse filter, i.e., the algebraic solution under unconstrained conditions. The fuzzy displacement constant $\boldsymbol{H}$ is a circulant-block matrix. The convolution in the frequency domain can be obtained as

$$\hat{F}(u, v) = \frac{G(u, v)}{H(u, v)} \tag{11.40}$$

If $H(u, v)$ is zero, $\boldsymbol{H}$ is singular, and $\boldsymbol{H}^{-1}$ and $(\boldsymbol{H}^{\mathrm{T}}\boldsymbol{H})^{-1}$ do not exist.

The unconstrained linear algebraic recovery and Wiener filter are different; the latter form corresponds to the smallest mean square error of the reconstructed signal and original signal, and the difference between the squared sum of the difference between the descending image g and estimated image after degradation is minimized. Therefore, the results of the two methods are not necessarily the same.

2. Constrained Algebraic Restoration

Factors that affect image restoration include the noise interference $\boldsymbol{n}$ and transfer function of the imaging system $\boldsymbol{H}$. The latter term includes the effects of optics and electronics in image sensors. If the noise is neglected, in accordance with Formula (11.32), the original image $\boldsymbol{f}$ can be restored using the inverse matrix $\boldsymbol{H}$, i.e.,

$$\boldsymbol{f} = \boldsymbol{H}^{-1}\boldsymbol{g} \tag{11.41}$$

Mathematically, this inverse matrix must exist and be unique. Even if slight disturbance is present in the image restoration of the blurred image, a strong disturbance will be obtained, which cannot be ignored:

$$\boldsymbol{H}^{-1}[\boldsymbol{g} + \varepsilon] = \boldsymbol{f} + \delta \tag{11.42}$$

where ε is an arbitrarily small perturbation, $\delta \gg \varepsilon$. Assuming that the inverse matrix is present and unique, for an imaging system, a digitizer, or a truncation error, the collected digitized image is disturbed. Due to the randomness of noise, the blurred image $\boldsymbol{g}$ may have infinite possible cases, and the recovery becomes challenging. Furthermore, there is a possibility that the inverse matrix $\boldsymbol{H}^{-1}$ is not known but still exists, and $\boldsymbol{f}$ is the approximate solution, which is known as singularity recovery problems. To overcome the pathological nature of the restoration problem, it is often necessary to apply a certain kind of constraint on the operation in the process of recovery to choose one of the many possible outcomes. This restraining method is known as the constrained algebra restoration method.

Under normal circumstances, considering the minimization process in the presence of noise, the equality of both ends of Eq. (11.42) can be expressed as

$$\|\boldsymbol{n}\|^2 = \left\|\boldsymbol{g} - \boldsymbol{H}\hat{\boldsymbol{f}}\right\|^2 \tag{11.43}$$

The restoration with constraints can be realized as follows. Let $\boldsymbol{Q}$ be a linear operator of $\boldsymbol{f}$. The recovery problem corresponds to constraints to minimize $\left\|\boldsymbol{Q}\hat{\boldsymbol{f}}\right\|^2$, as indicated in Formula (11.43). Using the correction function (objective function) of Lagrange multipliers

$$\begin{aligned} W(\hat{\boldsymbol{f}}, \lambda) &= \left\|\boldsymbol{Q}\hat{\boldsymbol{f}}\right\|^2 + \lambda\left\|\boldsymbol{g} - \boldsymbol{H}\hat{\boldsymbol{f}}\right\|^2 - \|\boldsymbol{n}\|^2 \\ &= (\boldsymbol{Q}\hat{\boldsymbol{f}})^{\mathrm{T}}(\boldsymbol{Q}\hat{\boldsymbol{f}}) + \lambda(\boldsymbol{g} - \boldsymbol{H}\hat{\boldsymbol{f}})^{\mathrm{T}}(\boldsymbol{g} - \boldsymbol{H}\hat{\boldsymbol{f}}) - \boldsymbol{n}^{\mathrm{T}}\boldsymbol{n} \end{aligned} \tag{11.44}$$

the partial derivative of $\mathrm{W}\left(\hat{\boldsymbol{f}}, \lambda\right)^2$ on $\hat{\boldsymbol{f}}$ is set as zero:

$$\frac{\partial W(\hat{\boldsymbol{f}}, \lambda)}{\partial \hat{\boldsymbol{f}}} = 2[(\boldsymbol{Q}^{\mathrm{T}}\boldsymbol{Q} + \lambda \boldsymbol{H}^{\mathrm{T}}\boldsymbol{H})\hat{\boldsymbol{f}} - \lambda \boldsymbol{H}^{\mathrm{T}}\boldsymbol{g}] = 0 \tag{11.45}$$

The following expression is obtained:

$$\hat{\boldsymbol{f}} = (\boldsymbol{H}^{\mathrm{T}}\boldsymbol{H} + \gamma \boldsymbol{Q}^{\mathrm{T}}\boldsymbol{Q})^{-1}\boldsymbol{H}^{\mathrm{T}}\boldsymbol{g} \tag{11.46}$$

This expression is the general formula for constraining the least squares algebraic restoration solution, where $\gamma = 1/\lambda$. For example, if the image constraint indicates that after recovery, the energy of $\hat{\boldsymbol{f}}$ and $\boldsymbol{g}$ blurred image remains unchanged, i.e.,

$$\hat{\boldsymbol{f}}^{\mathrm{T}}\hat{\boldsymbol{f}} = \boldsymbol{g}^{\mathrm{T}}\boldsymbol{g} = C \tag{11.47}$$

the method evaluates the minimum value of the objective function $W\left(\hat{\boldsymbol{f}}\right)^2$ and computes the minimum value of $W\left(\hat{\boldsymbol{f}}, \lambda\right)^2$ under the conditions of Formula (11.47), i.e., the auxiliary function

$$W(\hat{\boldsymbol{f}}, \lambda) = W(\hat{\boldsymbol{f}}) + \lambda(\hat{\boldsymbol{f}}^{\mathrm{T}}\hat{\boldsymbol{f}} - C) = (\boldsymbol{g} - \boldsymbol{H}\hat{\boldsymbol{f}})^{\mathrm{T}}(\boldsymbol{g} - \boldsymbol{H}\hat{\boldsymbol{f}}) + \lambda(\hat{\boldsymbol{f}}^{\mathrm{T}}\hat{\boldsymbol{f}} - C) \tag{11.48}$$

Take the derivative of $\hat{\boldsymbol{f}}$ and set it as zero:

$$\frac{\partial W(\hat{\boldsymbol{f}}, \lambda)}{\partial \hat{\boldsymbol{f}}} = -2\boldsymbol{H}^{\mathrm{T}}\boldsymbol{g} + 2\boldsymbol{H}^{\mathrm{T}}\boldsymbol{H}\hat{\boldsymbol{f}} + 2\lambda\hat{\boldsymbol{f}} = 0$$

The following expression is obtained:

$$\hat{f} = (\boldsymbol{H}^{\mathrm{T}}\boldsymbol{H} + \lambda \boldsymbol{I})^{-1}\boldsymbol{H}^{\mathrm{T}}\boldsymbol{g} \tag{11.49}$$

where $\boldsymbol{I}$ is a unit matrix. According to Eqs. (11.46) and (11.49), when $\boldsymbol{Q}=\boldsymbol{I}$ or $\boldsymbol{Q}$ is an orthogonal matrix, the two equations are equal. In operation, turn constant λ until satisfied.

If $\gamma = 0$, and $\boldsymbol{Q} = \boldsymbol{I}$, (11.49) transforms into (11.46); this situation corresponds to a pseudoinverse filter. Furthermore, $\boldsymbol{f}$ and $\boldsymbol{n}$ may be considered random variables, and $\boldsymbol{Q}$ is a signal-to-noise ratio:

$$\boldsymbol{Q} = \boldsymbol{R}_{\mathrm{f}}^{1/2}\boldsymbol{R}_{\mathrm{n}}^{1/2} \tag{11.50}$$

where $\boldsymbol{R}_{\mathrm{f}} = \varepsilon\{\boldsymbol{f}\boldsymbol{f}^{\mathrm{T}}\}$ and $\boldsymbol{R}_{\mathrm{n}} = \varepsilon\{\boldsymbol{n}\boldsymbol{n}^{\mathrm{T}}\}$ are covariance matrices of the signal and noise, respectively. Solution (11.46) is

$$\hat{f} = (\boldsymbol{H}^{\mathrm{T}}\boldsymbol{H} + \gamma \boldsymbol{R}_{\mathrm{f}}^{\mathrm{T}}\boldsymbol{R}_{\mathrm{n}})^{-1}\boldsymbol{H}^{\mathrm{T}}\boldsymbol{g} \tag{11.51}$$

γ is the adjustable parameter. Note that $\gamma = 1$. The classic Wiener filter is obtained, whose mean square error in the original image and restoration images is minimized.

The frequency domain of the Wiener filter is described by minimizing Eq. (11.48) with the criterion of Formula (11.51) using the linear algebraic method. Notably, in the abovementioned method, the filter is the restoration filter that is closest to the original image (in the mean square sense). Although this method can yield the same result faster, the rigor of the optimal filter is not guaranteed.

11.3.2 Less Restrictive Recovery

This section is not limited to the conditions in which "fuzzy displacement is the same, and the signal and noise are stable" and considers a situation mimicking actual situations.

1. Blurred in Space

For astigmatism, coma aberration, and blurring restoration with field curvature, a space coordinate is changed, and the image is degraded through geometrical transformation. The resulting function exhibits spatial invariance. Ordinary space invariant restoration methods are used, and a geometric transformation is performed opposite to the previous inverse transform image and restored to the original format.

2. Blurred with Time

To address phase distortions caused by atmospheric conditions such as unevenness caused by stars, the solution is as follows: first, obtain the objects to be observed and a reference time point for the average power spectrum. Next, the power spectrum is divided by the celestial star point to realize power spectrum deconvolution. The obtained result is the diffraction limit of the power spectrum to estimate the unknown celestial bodies. The inverse transformation of the objects is used to determine the autocorrelation function. To this end, binary and other celestial bodies of interest must be identified.

3. Nonstationary Signal and Noise

The filters discussed earlier in this section assume that the signal and noise are smooth. A stable image means that the local power spectrum is the same or nearly the same across the entire image; however, this is usually not the case. In fact, most images are highly uneven. Many images can be thought of as being composed of relatively flat areas that are locked by a relatively high gradient edge lock, for example, aerial photographs of farmland.

11.4 Multilook and Filter Reconstruction (Considering the Case of SAR)

Synthetic aperture radar (SAR) technology is an important part of remote sensing, and SAR images exhibit several similarities and differences with optical images in terms of the degradation mechanism and restoration principle [8]. In this section, based on the restoration principles and methods of optical imagery, similarity and commonality restoration methods are established based on the characteristics of SAR images. Due to the system imaging principles, SAR images include speckle noise. Speckle noise in SAR images cannot accurately reflect the scattering characteristics of objects, which critically affects the image quality, reduces the ability to extract the target information in the image, and complicates the interpretation of SAR images. This speckle noise requires unique noise suppression, and elimination of speckle noise is an key issue in SAR image processing and main objective of SAR image restoration. The following analysis focuses on SAR speckle noise generation principles and multilook processing and filter processing methods for removing noise in these two categories.

11.4.1 Probability Distribution Function of Speckle Noise in SAR Images

Ideal point target scattered electromagnetic wave, a spherical wave which is an echo, which is equal everywhere on a spherical surface. The actual target feature is often a plurality of points over the target composition, and the signal received by SAR is the vector sum of these ideal point targets. Thus, when the actual target coherent electromagnetic wave radiation that scatters back an echo is not fully determined by the total scattering coefficient of the target feature, but around the scattering coefficient, values exhibit large random fluctuations in the image. The reflection is the speckle noise. Owing to these fluctuations, the SAR image of a target having a uniform scattering coefficient does not have a uniform gradation, and granular entities such as white textured black and white dots are observed.

1. Single-Look SAR Image

The probability distribution of monoscopic image speckle negative exponential distribution for a uniform target scene and image pixel intensity is

$$p(I) = \frac{\exp(-I/\langle I \rangle)}{\langle I \rangle} \tag{11.52}$$

where I is the pixel intensity, and $< I >$ is the variance in the intensity. In terms of the amplitude value A or D expressed in decibels, the relationship of the intensity I is $I = A^2$ and $D = 10\lg I = 10\ln I/\ln 10$. Thus, the intensity probability distribution can be converted directly as

$$p(A) = \frac{2A}{\langle I \rangle} \exp(-A^2/\langle I \rangle) \tag{11.53}$$

$$p(D) = \frac{\exp(\frac{D}{K} - \frac{\exp(D/K)}{\langle I \rangle})}{K \langle I \rangle} \tag{11.54}$$

where $K = 10/\ln 10$. The values are all Rayleigh distributed.

2. Multilook SAR Image

To improve the SNR of the multiview image to be processed, multilook processing is performed for the average of n discrete subimages of the same scene. For n independent incoherent superpositions of subimages to change the speckle noise distribution, the probability distribution of the intensity I is a Gamma distribution, i.e.,

$$p(I) = \frac{n^n I^{n-1}}{(n-1)! \langle I \rangle^n} \exp(-nI/\langle I \rangle) \tag{11.55}$$

$$p(A) = \frac{2n^n A^{2n-1}}{(n-1)! \langle I \rangle^n} \exp(-nA^2/\langle I \rangle) \tag{11.56}$$

$$p(D) = \frac{n^n}{K(n-1)! \langle I \rangle^n} \exp\left(\frac{nD}{K} - \frac{n\exp(D/K)}{\langle I \rangle}\right) \tag{11.57}$$

11.4.2 SAR Image Restoration Methods

SAR image restoration can be understood in terms of the noise of the SAR image. The reason for the restoration is that, under a random field model, it is necessary to restore the real random field X from the observed random field Y.

If the speckle noise in the SAR image cannot be well suppressed, the noise will critically affect the radiation resolution of the image, rendering it difficult to perform postprocessing, such as interpretation and classification of the image. The existing SAR image denoising methods can be divided into two categories: based on the spatial resolution loss of multiview processing, as a pretreatment; based on a filtering algorithm involving statistical characteristics of the noise streaks of SAR images, as a posttreatment.

1. Multiview Processing

A single-vision SAR image refers to an SAR image generated using the entire length of a synthetic aperture. To reduce speckle noise, multiview processing is often performed. Multiview processing refers to reducing the distance or azimuth processor bandwidth, thereby dividing the azimuth spectrum or range direction into individual (or partially overlapping) N segments. Each segment generates a lower resolution image of a single view. Each monoscopic image pixel corresponds to incoherent superposition of the multiview images obtained and averaged to suppress speckle noise and increase the SNR. Pieces of the SAR image are overlaid.

After multiview processing, the radiometric resolution is improved, but multiview processing reduces the signal bandwidth utilization, thereby reducing the spatial resolution. Therefore, the selection process involves a balance of number of SAR images according to the application.

In practice, in multiview processing, there may be certain overlap between adjacent subbands. When the resolution is known, the subband width is given. Set the entire Doppler bandwidth as B_D, the subband width as B_S, and overlapping width of adjacent subbands as B_O, L is defined as

$$L = \frac{B_D - B_S}{B_S - B_O} + 1 \tag{11.58}$$

Under this condition, a greater overlapping width of adjacent subbands corresponds to a larger number, which decreases the speckle noise. A larger overlap width of adjacent subband corresponds to a higher correlation between subimages, which increases the speckle noise. Thus, in conventional multiview processing, the overlapping width of adjacent subband selection is empirical. Figure 11.12 shows the result of SAR image restoration after different views.

2. Filtering

With the development of digital image processing technologies, several airspace filtering algorithms have emerged internationally and have been applied for speckle

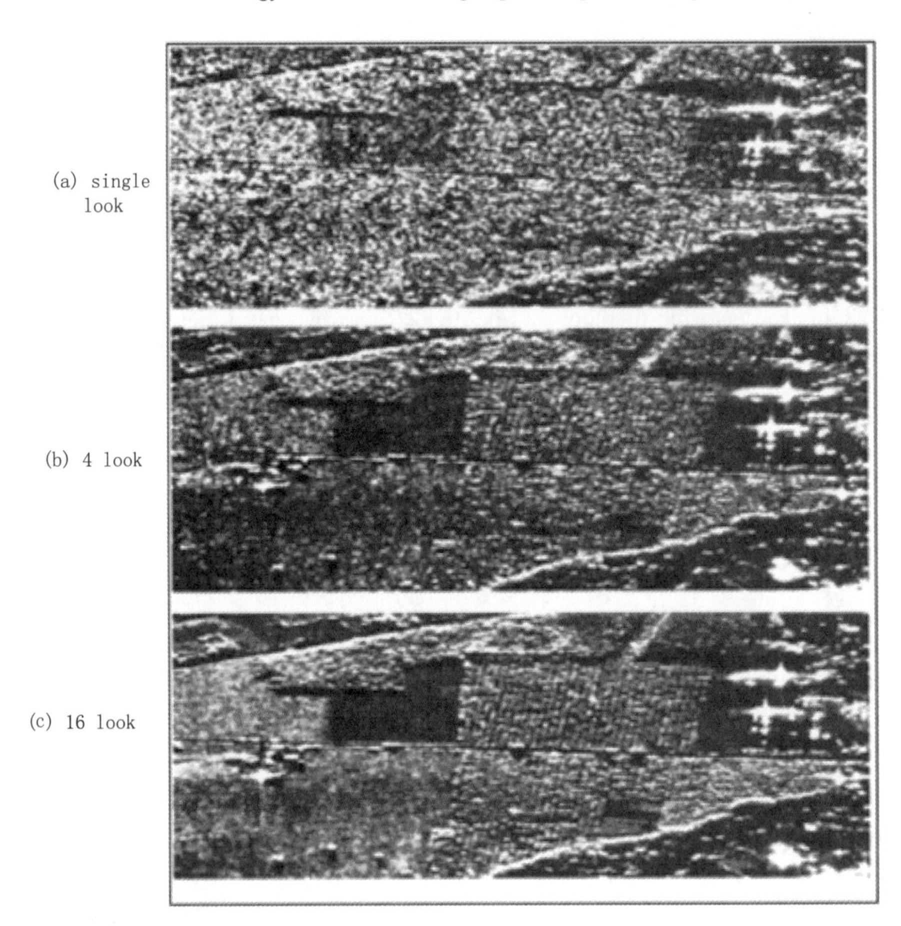

Fig. 11.12 SAR image after multiview processing

noise suppression in SAR images. Among these algorithms, local adaptive filtering is an effective method, which takes a sliding window on the image and filters the pixels in the window to obtain the filter value of the center pixel of the window. In the following text, we describe several typical algorithms.

(1) Classic spatial filter

Classic spatial filters are mean and median filters, which exhibit advantages of simple operation and effective suppression of speckle noise. However, it is difficult to retain the edge information of the image, and the image is blurred, which decreases the resolution.

① Mean filter.

The basic principle of the mean filter is that the filter window is selected such that the number of filter points in the window is an odd number, and the average value of all pixels in the selected slide window is considered the value of the center pixel point, defined as

$$u_i = \text{average}\,(f_j | j \in \Omega_i) \tag{11.59}$$

where u_i is the gray value of the center pixel of the filtered window Ω_i, f_j is the gray value of each pixel of window Ω_i. Although the filter can suppress the speckle noise to a certain extent, it blurs the edge texture information.
② Median filter.

The basic principle of the median filter is that the filter window is usually selected such that the number of filter points in the window is an odd number. Subsequently, all observations in the sliding window are sorted according to their numerical values, and the observation value in the middle position serves as the output of the median filter, defined as

$$u_i = \text{median}\,(f_j | j \in \Omega_i) \tag{11.60}$$

where u_i is the gray value of the center pixel of the filtered window Ω_i, and f_j is the gray value of each pixel of window Ω_i. For SAR images, the median filter can effectively filter out the isolated point noise, which blurs the edges. Information Information such as line and object edges is lost, and the spatial resolution of the image decreases.

(2) Classic adaptive filter

These filters consider adaptive local statistical characteristics of an image, i.e., the filters are a function of the local statistical parameters. Compared with conventional methods, the edge information retention effect is enhanced, smoothing can be adjusted and the edge can be retained through parameter control.

① Lee filter.

In the Lee filter, the multiplicative noise model is

$$A\mathrm{e}^{\mathrm{j}\varphi} = \sum_{k=1}^{N} A_k \mathrm{e}^{\mathrm{j}\varphi_k} \tag{11.61}$$

Linear Taylor expansion is performed using a first-order model, and the least mean square error is used to estimate this linear model to obtain the filtering formula:

$$\hat{R}(t) = I(t)W(t) + I(t)(I - W(t)) \tag{11.62}$$

where $\hat{R}(t)$ is the speckle image value. In Formula 11.58, *R(t)* is the estimated value, *I(t)* is the mean value of the speckle window, and *W(t)* is a weighting function:

$$W(t) = 1 - \frac{C^2}{C_I^2(t)} \tag{11.63}$$

$$W(t) = 1 - \frac{C_u^2}{C_I^2(t)}$$

where C_u and $C_I(t)$ are the standard deviation coefficients of plaque $u(t)$ and image $I(t)$, respectively, and C_u represents the coefficients of variation of the entire normal noise image:

$$C_u = \frac{\sigma_u}{\overline{u}}, C_I(t) = \frac{\sigma_I(t)}{\overline{I}(t)}$$

where σ_u and $\overline{u}$ are the mean value and standard deviation of plaque $u(t)$, and $\sigma_I(t)$ and $\overline{I}(t)$ are the standard deviation and mean value of $I(t)$, respectively.

② Kuan filter.

The Kuan multiplicative noise filter model is associated with the signal plus noise model. The minimum mean square error (MMSE) is used to estimate the model parameters. The Lee filter and filtering formula have the same form but with different weights, i.e.,

$$W(t) = \frac{1 - C_u^2 / C_I^2(t)}{1 + C_u^2} \tag{11.64}$$

③ Frost filter.

The Frost filter is different from the Lee filter and Kuan filter. The Frost filter estimates the expected reflection intensity by convolving the observed image with the impulse response of the SAR system. The SAR response of the system is obtained by the MMSE formula:

$$\varepsilon^2 = E\left|(R(t) - I(t)m(t))^2\right| \tag{11.65}$$

In the formula, $m(t)$ is the impulse response of the SAR system, and $R(t)$ is modeled as an autoregressive process. The autocorrelation function is

$$R_R = \sigma_R^2 \exp(-a|r|) + \overline{R}^2 \tag{11.66}$$

where $\overline{R}$ is the local mean value of the signal, σ_R^2 is the local variance, and a is the autocorrelation coefficient. Corresponding to different ground conditions, the three parameters have different values, and we can obtain an estimate of $m(t)$:

$$m(t) = K_2 a \exp(-a|t|)^2 \tag{11.67}$$

where K_2 is a normalization constant. Through simplification, a simple expression can be obtained for $a = K\left(\sigma_1/\overline{I}\right)^2 = KC_I^2$. The form of the Frost filter is

$$m(t) = K_1 a \exp(-KC_I^2(t_0)|t|) \tag{11.68}$$

where K_1 is a normalization constant. According to this equation, the Frost filter is a weighted mean filter. A pixel farther from the center of the smaller weight percentage point value in the current processing window has less impact on the central pixel, and pixels close to the center point correspond to a larger share of the weight, thus affecting the updated center pixel.

④ Sigma filter.

The sigma filter is based on the following assumptions: the gray value of the pixel in the window is relatively close to that of the center pixel. For the one-dimensional Gaussian distribution, the probability interval of the sample points, 2σ, decreases to 93.5%. Assuming noise obeys a Gaussian distribution, in the window filtering process, we select only the points for which the pixel value in the window falls within the 2σ range. Use the average gray value as an estimate of the center pixel gray value. Other significant pixels are treated as edges without filtering, and σ is the standard deviation of the pixel values of the scan window.

(3) Geometry filter

Considering the plane coordinates plus the gray value of the image as a three-dimensional model, the noise is removed by morphological methods. The edge retention of this filter is due to the local grayscale statistical filter, such as the typical maximum a posteriori (MAP) gamma filter.

Speckle noise filtering is based on the known observation intensity I. The MAP estimation of the desired reflection intensity R under the Bayes criterion is

$$p(R(t)|I(t)) = \frac{p_{\text{speckle}}(I(t)|R(t))p_R(R(t))}{p_I(I(t))} \tag{11.69}$$

$$R_{\text{MAP}} = \max_R(p(R(t)|I(t))) \tag{11.70}$$

In the formula, $p_{\text{speckle}}(R(t)|I(t))$ is the probability density function of the intensity, $p_R(R(t))$ is the prior probability density function of R. Moreover,

$$p(R(t)|I(t)) \propto p_{\text{speckle}}(I(t)|R(t))p_R(R(t))$$
$$= \left(\frac{L}{R(t)}\right)^L \frac{I(t)^{L-1}}{\Gamma(L)} \exp\left[-\frac{LI(t)}{R(t)}\right]\left(\frac{\upsilon(t)}{\mu(t)}\right)^{\upsilon(t)} \frac{R(t)^{\upsilon(t)-1}}{\Gamma(R(t))} \exp\left[-\frac{\upsilon(t)R(t)}{\mu(t)}\right] \tag{11.71}$$

We set $\frac{\mathrm{d}P(R(t)|I(t))}{\mathrm{d}R(t)} = 0$ and obtain the MAP resolution of *R(t)* as the root of the following equation:

$$\frac{R(t)_{\text{MAP}}^2}{\mu(t)} + \left[\frac{L+1}{\nu(t)}\right] R(t)_{\text{MAP}} - \frac{LI(t)}{\nu(t)} = 0 \tag{11.72}$$

where $\mu(t)$ is *R*(*t*) of the local mean value, and $\nu(t)$ is a partial order, which can be estimated from the field of pixels. The abovementioned methods to process an image by fixing a window do not consider metasurrounding structural information. Therefore, in practical applications, these algorithms do not make the boundary or detail features of the image fuzzy, and it is difficult to ensure that the interior of the area is sufficiently smooth.

(4) Hierarchical filtering

Wavelet theory, which is based on the transformation of an image into a series of image information points of different scales, performs the filtering of a low- and high-resolution image, representing low-frequency and high-frequency component noise, respectively, to preserve major thresholding edge information, and then rebuilds the image. This method is a highly complex algorithm but maintains noise suppression and edge effects.

In addition to the conventional filter, morphological filtering and random field model-based filtering method exist.

11.5 Superresolution Remote Sensing Image Restoration

In the process of image restoration, the increase in image resolution is limited by the limit resolution of the optical system [6], and the various image restoration techniques cannot exceed this limit. This section introduces superresolution remote sensing image restoration techniques to define the frequency restoration process parameters.

K. R. Castleman, to overcome the limitations of diffraction limit recovery information technologies, developed superresolution (super resolution) technology, generally used with limited function extrapolation. The key to superresolution technology is that the resolution of the image increases. Therefore, superresolution technology is an image reconstruction technique based on the physical characteristics of the features and high resolution is obtained from the low resolution. There exist two ways to obtain high-resolution remote sensing images: by enhancing the imaging system performance imaging, and through superresolution processing to indirectly obtain high-resolution remote sensing images.

If the imaging system is linear space-invariant, the imaging system and imaging process pertain to low-pass filters [7]. The solution of the original image is restricted (Fig. 11.13), i.e., outside the cutoff frequency, $H(u) = 0$. It is more difficult to find information beyond the cutoff frequency $F(u)$ at a distant location.

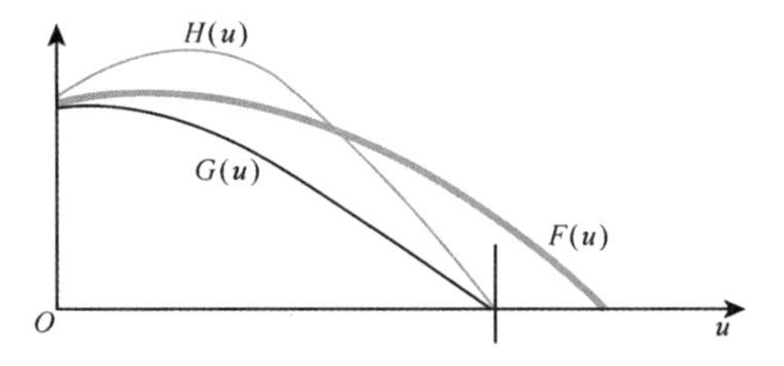

Fig. 11.13 Superresolution principle

$F(u)$ can be estimated to restore information beyond the cutoff frequency.

11.5.1 Single-Image Superresolution Restoration

This technique uses a single image to restore information beyond the diffraction limit of the image, thereby increasing the spatial resolution of the image. Representative methods include analytical continuation, the Harris method, continuous energy degradation, the prolate spheroidal wave function method, linear extrapolation, and the superimposed sinusoidal template method. The Harris method uses the previously learned sampling theorem for reconstruction, that is,

$$F(s) = \left[\mathrm{III}\left(2s^{\mathrm{T}}\right)F(s)\right] * \frac{2\pi\,\sin\left(2\pi\,s^{\mathrm{T}}\right)}{2\pi\,s^{\mathrm{T}}} \tag{11.73}$$

There are two obstacles in this kind of method in actual operation: First, the image pertaining to the point spread function cannot be precisely known, and second, the process is highly sensitive to noise. The desired frequency spectrum band cannot be obtained, and the original image cannot be correctly reconstructed. The disadvantage of this type of method is that more information from multiple images cannot be obtained, and practical application is less likely. Therefore, Andrews and Hunt named this phenomenon the "superresolution myth".

11.5.2 Steps of Superresolution Image Reconstruction

This technique refers to the application of multiple low-resolution remote sensing images with certain mathematical methods (including the spatial and frequency domains) for image reconstruction, resulting in high spatial resolution remote sensing image. Foreign research can be divided into three categories: algebraic interpolation, spatial iteration and frequency domain solution.

1. Algebraic Interpolation

The algebraic interpolation method estimates the desired pixel value based on neighboring known pixels. The problem of interpolating an image from multiple images is a simple sample reconstruction problem. The oversampling above the Nyquist frequency can be completely reconstructed by convolution with the *Sinc* function, whereas undersampling below the Nyquist frequency cannot be reconstructed.

2. Spatial Iteration Method

This method is based on the spatial domain of digital image interpolation through analog sampling and gradual correction of the iterative implementation [2]. Peleg et al. used the analog sampling method and simulated annealing algorithm iteration to enhance the spatial resolution, that is, the initial estimate for the first cell values fluctuates, and the influence of this floating error is identified to determine the optimal fluctuation of the analog sample [3]. Later, Peleg with Irani considered a similar tomography to develop the method of back projection to reconstruct images with high spatial resolution, considering image registration and image degradation [4]. Patti et al. conducted a POCS study of several low-resolution image reconstruction methods for high-resolution images [5]. The algorithm considers the noise and uses blur sensor hardware to avoid relative movement between the sensor and imaging target. All these methods do not consider more general problems, such as the nonuniform sampling of samples and differences in spatial resolution and radiation brightness between image samples.

3. Frequency Domain Deblending

If a single digital image is undersampled, it cannot be reconstructed by algebraic interpolation because the undersampling of the signal causes the signal to alias in the corresponding frequency domain. However, if aliasing can be mitigated by the spectrum of multiple images, the spatial resolution of the corresponding spatial domain image will be improved.

Although superresolution recovery has been established in theory, it is challenging to implement for the following reasons: influence of noise; model accuracy; and difficulty in matching the analytic function in the given section (a small matching error may lead to a large SR error). The latest developments include time domain POCS superresolution reconstruction, time domain MAP superresolution reconstruction, machine learning based on superresolution reconstruction, dictionary-based superresolution reconstruction and video field superresolution reconstruction.

11.6 Image Restoration Software for Model Implementation

The previous section describes the principles and common methods for image restoration. How can computing software and hardware features be used on the basis of the existing principles of the method promptly and efficiently to achieve image restoration? This section describes image restoration software and hardware, in terms of the transform domain and spatial domain. Moreover, the section describes the software and hardware implementation of the restoration methods.

11.6.1 Transform Domain Filtering

If the recovery operation is linear with constant displacement, it can be implemented in the frequency domain by multiplying the Fourier transform with the following steps: first, two-dimensional DFT is performed. The spectrum and transfer function are multiplied point by point, and the inverse of the two-dimensional DF is performed. For an image sized $N \times N$, the Fourier transform requires $(N\log_2 N)^2$ multiplication and addition operations. This method is effective if a reconstructed transfer function in the frequency domain is available.

11.6.2 Large Kernel Convolution

If recovery is a linear and unchanging operation, it can be realized through space convolution. As long as the point spread function and the image are sized $N \times N$, the point spread function recovery and discrete convolution above DFT are numerically equivalent. The convolution of two $N \times N$ arrays requires N^4 operations, considerably exceeding that for the DFT method, which renders the method unsuitable for larger images.

11.6.3 Small Kernel Convolution

Through rational design, a small nucleus $M \times M$ (where $M < N$) sized 9×9 or even 3×3, yields sufficient results close to the complete $N \times N$ convolution kernel. The core with the size of $M \times M$ convolutes the image with the size $N \times N$, with M^2N^2 multiplications. A small kernel can be obtained as follows:

1. Undersampling of the Transfer Function

Using the inverse transformation of the two-dimensional DFT, we can obtain a point spread function with $M \times M$ from a transfer function of size $M \times M$. By undersampling an $N \times N$ transfer function recovery, a size of $M \times M$ can be obtained. The value can be passed through a suitably modulated dimensional compression transfer function from 0 to cover the entire Nyquist zone frequency, while allowing the size of the convolution kernel to be decreased.

To avoid the undersampling process producing aliasing, we may need to smoothly restore the modulation transfer function. In this case, a small core is achieved through the smooth modulation transfer function, which is not consistent with the original foot size modulation transfer function. Smoothing of the modulation transfer function narrows the point spread function, rendering the space more limited, which is equivalent to the field reversed low-pass filter.

2. Kernel Cutoff

A more efficient method of convolution with a small core is to directly truncate the array of point spread functions to an acceptable small size. The multiplication of the dot diffusion function with the square pulse is equivalent to convolving the modulation transfer function with the $\sin x/x$ function. This operation considerably changes the transfer function unless the point spread function is limited in space (i.e., substantially zero in the truncated window). It is possible to judge whether the effect of the truncation is acceptable by calculating the truncated transfer function and comparing it with the original reconstructed modulation transfer function.

The truncation of the kernel degrades the "detail" in the transfer function, as removing the high-frequency component from the spectral function reduces the details in the corresponding image. However, the transfer function of the kernel truncation can generally reflect the shape of the ideal modulation transfer function but cannot reflect the local change.

Similar to the DFT positive transform, the DFT inverse transform is an orthogonal transform. Therefore, each basis function is orthogonal to other basis functions. Regardless of how the remaining nuclear elements are modified, the nuclear function does not lose any details due to cutoff. In other words, in terms of the mean square, the truncated kernel modulation transfer function is the best approximation of the large kernel modulation transfer function.

11.6.4 Kernel Decomposition

Modern image processing systems include special hardware that can perform high-speed convolution on a small nucleus. If the $M \times M$ core is decomposed into a series of smaller (3×3) cores and convolved in sequence, the hardware is involved. Although this method cannot accurately replace any $M \times M$ core, it often has a high approximation effect. There are two methods: singular value decomposition, and SGK decomposition. The former method exhibits several advantages in describing the distribution of matrix data, and the singular value of the image represents the algebraic feature of the image. For any real matrix $\boldsymbol{A}$, there is a unique singular value diagonal matrix $\boldsymbol{r}$. Therefore, the singular value vector corresponding to the real matrix A is also unique, and the singular value can uniquely describe the image. The latter method is an image decomposition method that combines SVD with small-scale nuclear technology. SVD decomposition.

1. SVD Decomposition

Singular value decomposition (SVD), as shown in Fig. 11.14, represents an $M \times M$ matrix with rank R as the sum of R rank-1 $M \times M$ matrices, and each $M \times M$ matrix with rank 1 is the outer product of two M-dimensional eigenvectors and the weighting coefficients when they are added is a singular value of the matrix. Since convolution and integration are interchangeable, this process can be expressed as a sum of R images, where each image is convoluted by one of the $M \times M$ matrices above rank 1.

As each convolution of each line by the first image corresponds to $M \times 1$ convolutions for each column and $1 \times M$ convolutions, the summation of each matrix requires only $2MN^2$ multiplication and addition operations. Thus, convolution can pass all $2RMN^2$ completion operations (if $R < M/2$, it is required than the original $M^2 N^2$ fewer times). If the kernel exhibits circular symmetry, the row below the center row is the same as the row above it, and the rank is not greater than $(M+1)/2$. In addition, in many image systems, these one-dimensional row convolutions and column convolutions can be achieved by high-speed hardware.

As a numerical example, consider a 3×3 convolution kernel:

$$\boldsymbol{F}=\begin{bmatrix} 1 & 2 & 1 \\ 2 & 3 & 2 \\ 1 & 2 & 1 \end{bmatrix} \tag{11.74}$$

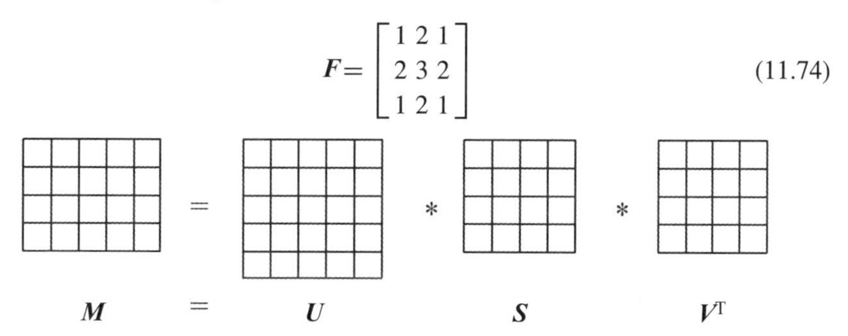

Fig. 11.14 SVD decomposition diagram (*U* and *V* in the figure are orthogonal matrices, and *S* is a diagonal matrix containing singular values)

Since this matrix is square and has symmetry, its singular value decomposition can be simplified, and its unitary matrix is equal, i.e.,

$$U = FF^{\mathrm{T}} = V = F^{\mathrm{T}}F = \begin{bmatrix} 6 & 10 & 6 \\ 10 & 17 & 10 \\ 6 & 10 & 6 \end{bmatrix} \tag{11.75}$$

The eigenvector is

$$u_1 = v_1 = \begin{bmatrix} 0.454 \\ 0.766 \\ 0.454 \end{bmatrix}, u_2 = v_2 = \begin{bmatrix} 0.542 \\ -0.643 \\ 0.542 \end{bmatrix}, u_3 = v_3 = \begin{bmatrix} -0.707 \\ 0 \\ 0.707 \end{bmatrix} \tag{11.76}$$

For a feature value of zero, U and V are matrices with rank 2. The singular value is located

$$\Lambda = U^{\mathrm{T}}FV = \begin{bmatrix} 5.37 & 0 & 0 \\ 0 & -0.372 & 0 \\ 0 & 0 & 0 \end{bmatrix} \tag{11.77}$$

on the diagonal, and the SVD sum can be expressed as

$$F = \sum\nolimits_{j=1}^{2} \Lambda_{j,j u_j v_j^{\mathrm{T}}} \tag{11.78}$$

In this case, there are only two items. This approach can be used to obtain the convolution result as follows: multiply the rows and columns of all images by u_1 and u_2, and weigh the sum of the two resulting images appropriately to obtain an error-free result output. Based on this principle, an example of remote sensing image SVD restoration based on a MATLAB simulation is presented, as shown in Fig. 11.15. Observing the original image from the visual viewpoint and restoring the image shows that the recovery effect is high. In terms of quantitation, the mean square error is only 0.0056, the energy loss of the restored image is only 5.74%, and this effect is obtained only when the largest singular value of the diagonal matrix is retained. Therefore, if more singular values are involved in image recovery, the effect will be superior, although the amount of computation will increase. In summary, SVD decomposition can decrease the number of operations under the premise of the maximum restoration of the image.

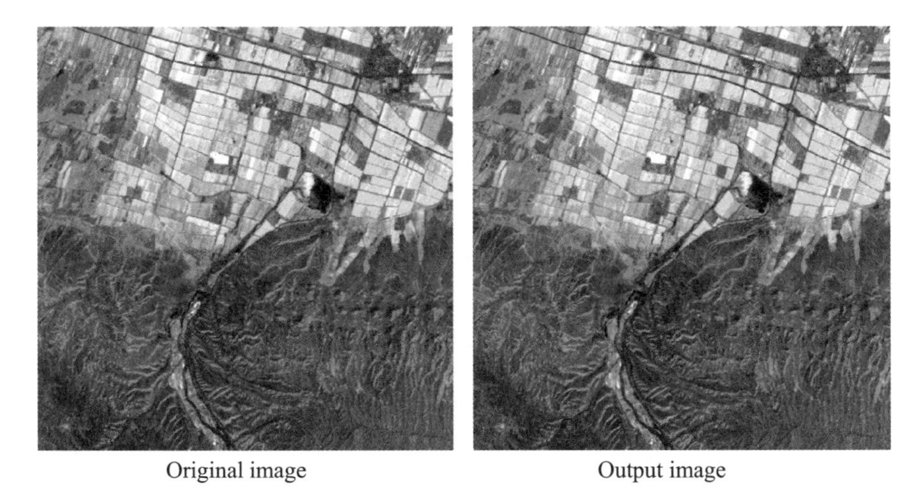

Fig. 11.15 Example of SVD decomposition of remote sensing images

In summary, SVD decomposition can restore the image to the maximum extent while decreasing the number of operations.

2. SGK Decomposition

Combining SVD with small-scale nuclear techniques allows the decomposition of any $M \times M$ core into a set of smaller cores, which can then be convolved sequentially. As shown in the previous section, SVD and each separable matrix in the formula are the outer products of two M-dimensional columns. These vectors can be decomposed through SGK decomposition into a set of 3 × 1 nuclei sequential convolutions. Using this method, $M \times M$ matrix convolution with a rank R can be obtained using only $R(M-1)$ iterations with a 3 × 1 kernel convolution. As shown in Fig. 11.16, 5 × 5 cores are decomposed into two 3 × 3 nuclei.

According to the convolution of the law

$$\boldsymbol{y}=\boldsymbol{h}*\boldsymbol{x}=\boldsymbol{f}*[\boldsymbol{g}*\boldsymbol{x}] \tag{11.79}$$

$$\boldsymbol{h}=\boldsymbol{f}*\boldsymbol{g} \tag{11.80}$$

In the frequency domain,

$$H(\boldsymbol{s})=F(\boldsymbol{s})G(\boldsymbol{s}) \tag{11.81}$$

In this case, the problem to be solved is to decompose the transfer function $H(u, v)$ into two transfer functions that are suitable for obtaining a 3 × 1 convolution. According to the definition of a one-dimensional discrete Fourier transform,

$$H(\boldsymbol{s})=\sum_{i=0}^{M-1} h_i \mathrm{e}^{-\mathrm{j}2\pi s \frac{i}{M}} \tag{11.82}$$

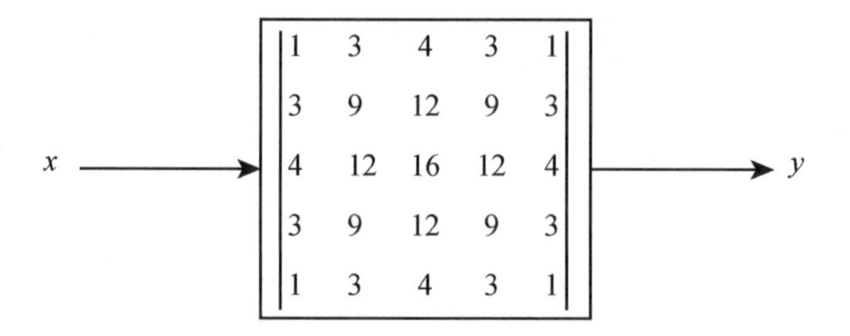

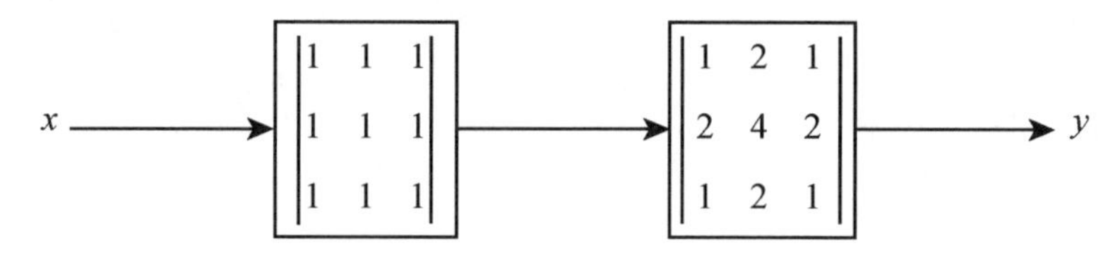

Fig. 11.16 Small generation nuclear decomposition

Through a conversion to the corresponding z transform,

$$z = \mathrm{e}^{-\mathrm{j}\,2\pi s \frac{i}{M}} \tag{11.83}$$

(11.82) can be rewritten as

$$H(z) = h_0 z^{-4}\left[z^4 + \frac{h_1}{h_0} z_3 + \frac{h2}{h0} z^2 + \frac{h_3}{h_0} z + \frac{h_4}{h_0}\right] \tag{11.84}$$

A polynomial of z is present in the brackets. For $\boldsymbol{f}$ and $\boldsymbol{g}$ using the same method, Eq. (11.84) becomes

$$\begin{aligned} H(z) &= h_0 z^{-4}\left(z^4 + \frac{h_1}{h_0} z_3 + \frac{h2}{h0} z^2 + \frac{h_3}{h_0} z + \frac{h_4}{h_0}\right) \\ &= F(z)G(z) = f_0 z^{-2}\left(z^2 + \frac{f_1}{f_0} z + \frac{f_2}{f_0}\right) g_0 z^{-2}\left(z^2 + \frac{g_1}{g_0} z + \frac{g_2}{g_0}\right) \end{aligned} \tag{11.85}$$

For factor z^{-4} on both sides of the equation, substituting the value of an in the example shown in Fig. 11.17,

$$z^4 + 3z^3 + 4z^2 + 3z + 1 = f_0 g_0\left(z^2 + \frac{f_1}{f_0} z + \frac{f_2}{f_0}\right)\left(z^2 + \frac{g_1}{g_0} z + \frac{g_2}{g_0}\right) \tag{11.86}$$

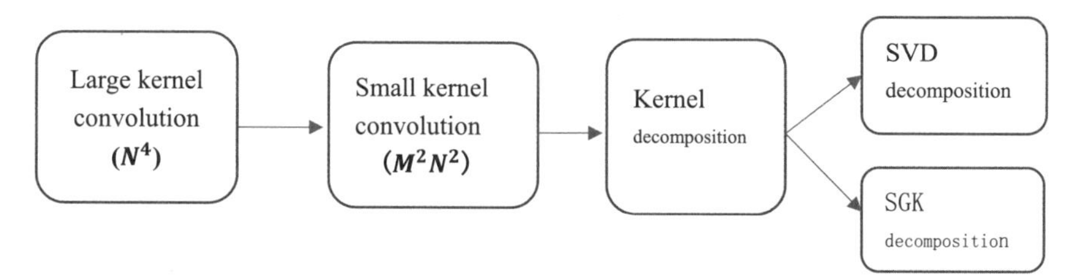

Fig. 11.17 Software hardware process and corresponding calculation

Polynomials can be decomposed into the left end in the form of four ($z - r_i$) product terms, and each r_i is one of the four roots for this polynomial (possibly complex roots). If a root exists, they appear in conjugate pairs, which are grouped by conjugation and multiplied to produce a real quadratic term. Formula (11.86) is decomposed into a quadratic polynomial to yield

$$\left(z^2+z+1\right)\left(z^2+2z+1\right)=f_0g_0\left(z^2+\frac{f_1}{f_0}z+\frac{f_2}{f_0}\right)\left(z^2+\frac{g_1}{g_0}z+\frac{g_2}{g_0}\right) \quad (11.87)$$

Thus, for a solution of $\boldsymbol{f} = (1, 1, 1)^{\mathrm{T}}$ and $\boldsymbol{g} = (1, 2, 1)^{\mathrm{T}}$ Note that, here must make a selection of any of: the constraint $f_0 g_0 = 1$. lower, selected $f_0 = g_0 = 1$. Two 3 × 3 nuclei are $\boldsymbol{ff}^{\mathrm{T}}$ and $\boldsymbol{gg}^{\mathrm{T}}$:

Therefore, the result of SGK decomposition is

$$\begin{bmatrix} 1 & 3 & 4 & 3 & 1 \\ 3 & 9 & 12 & 9 & 3 \\ 4 & 12 & 16 & 12 & 4 \\ 3 & 9 & 12 & 9 & 3 \\ 1 & 3 & 4 & 3 & 1 \end{bmatrix} = \begin{bmatrix} 1 & 1 & 1 \\ 1 & 1 & 1 \\ 1 & 1 & 1 \end{bmatrix} * \begin{bmatrix} 1 & 2 & 1 \\ 2 & 4 & 2 \\ 1 & 2 & 1 \end{bmatrix} \quad (11.88)$$

In general, SGK decomposition cannot reconstruct the original nucleus without error. This example is a unique case. However, errors are introduced during the SVD process, and items with smaller singular values in the summation are ignored. The error is not generated in the factorization of the z-transformation, and thus, the degree of approximation can be controlled by determining the number of singular values to be retained. The symmetry of the nuclear matrix concentrates the amplitude on one or more singular values. Because the mean square error is equal to the sum of the singular values that are discarded, a balance can be ensured between the accuracy and efficiency.

11.6.5 Image Kernel Decomposition Convolution Calculation System and Software/Hardware

When the operation of a computer can be applied directly to a special hardware unit, the operation speed is considerably enhanced. The main steps of the image restoration calculation process involve decomposition until the hardware unit can be utilized as a computer to enhance the processing efficiency. The above-mentioned method described the hardware to gradually realize the process, and the corresponding calculation is shown in Fig. 11.17.

Convolution is a large core unmodified calculation method, and two $N \times N$ matrix convolution is defined as in Formula (11.89).

$$\boldsymbol{G}(i, j) = \sum_{k=1}^{N} \sum_{l=1}^{N} \boldsymbol{A}(k, l) \times \boldsymbol{B}(i - k + 1, j - l + 1) \tag{11.89}$$

For any point $\boldsymbol{G}(i, j)$, the number of computations is N^2, for the entire image with N^2 points, the total number of computations is N^4.

After subsampling or a truncated nuclear generating $M \times M$ after the nuclear matrix, is convolved with a small core matrix and the original image, for any point, to reduce the amount of calculation, the total calculation amount is reduced;

By nuclear decomposition, may be decreasing the size of the convolution kernel. SVD decomposition is generated rank R of the $M \times M$ matrix, is decomposed into R a rank 1 base matrix and weighting, which base matrix can original outer product of the eigenvector matrix to obtain, using the base matrix are convoluted with the image matrix, as shown in Fig. 11.18, calculation amount of each base matrix is convolved with the image, then the total amount of computation.

SGK decomposition is based on the decomposition of SVD and decomposition of small nuclei, and the M-order matrix continues to decompose until the smallest 9 $\times$ 9 or 3 $\times$ 3 size to directly use the computer hardware unit. An $M \times M$ small core is subjected to SVD decomposition, thereby obtaining a rank 1-based core matrix, the decomposition of which generates two matrices until decomposition to a 3 $\times$ 3 base

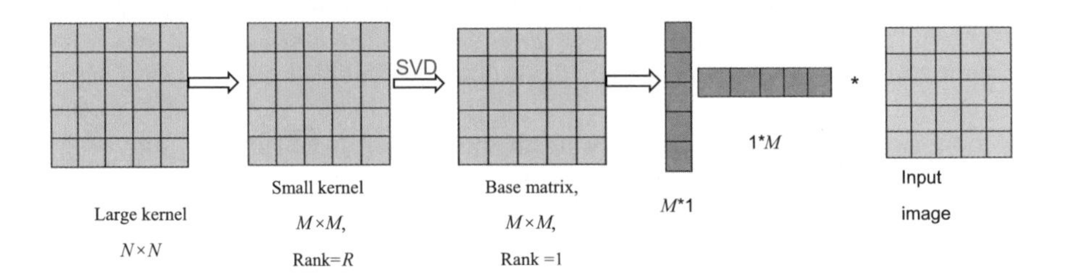

Fig. 11.18 Kernel decomposition

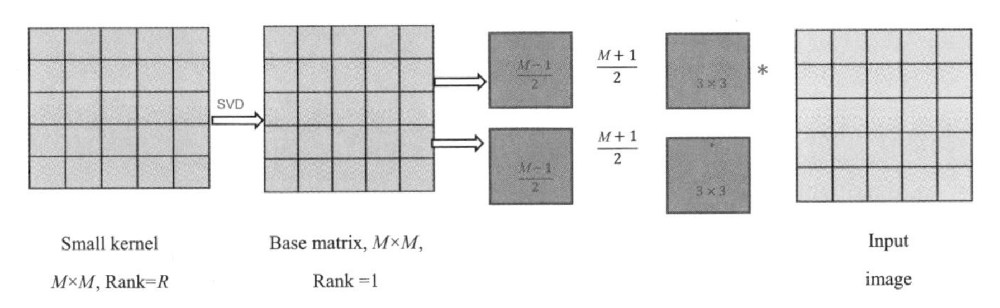

Fig. 11.19 SGK decomposition

matrix, and original image convolution, convolution single amount, a total of R a basic nuclear matrix, the size of 3×3 basis of the total number of moments, the total amount of computation. The decomposition process is illustrated in Fig. 11.19.

11.7 Summary

Figure 11.20 shows a summary of the chapter contents. M, N, R are consistent with those mentioned in this chapter: N is an image (two-dimensional matrix) in any one dimension of length, M is the length of the small core, R is the rank of the matrix, and N^4, M^2N^2 $(N<M)$, $2RM\,N^2$, $R(M-1)N^2$ denote the number of operations of convolution for a large nucleus, small nuclear convolution, SVD decomposition and SGK decomposition, respectively. Formulas (1–9) following the chapter (classic restoration in Sect. 11.2) represent the core mathematical formulas of the chapter, as shown in Table 11.1.

Many of the processing of digital image processing is dedicated to image restoration, including the study of algorithms and the preparation of image processing procedures for specific problems. Fourier transform, content filtering, the optical system, and the linear system are all the basis of this chapter. In addition, this chapter is also the basis of the following pattern recognition, image segmentation, etc. After the remote sensing image is restored, all kinds of useful information can be used for subsequent applications.

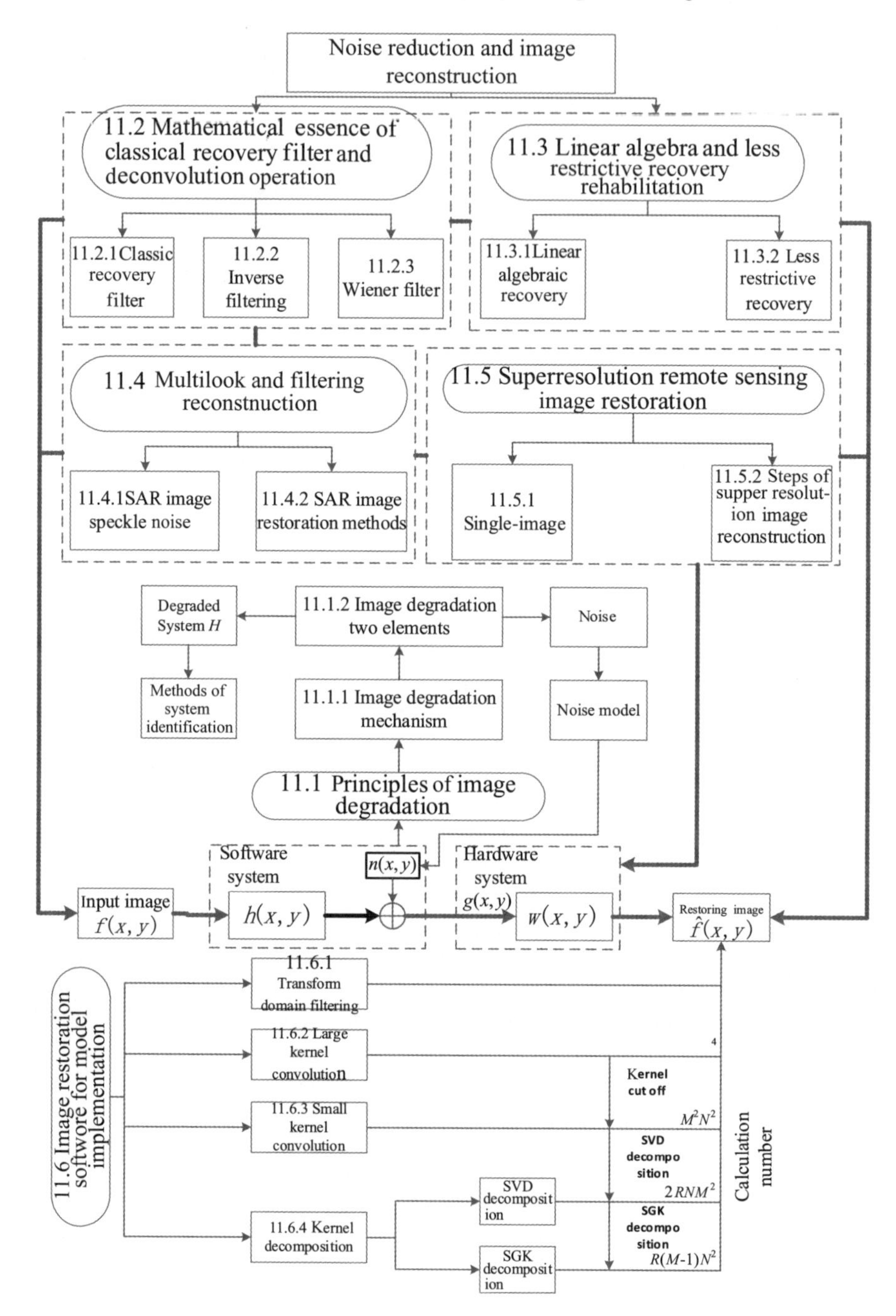

Fig. 11.20 Summary of the contents of this chapter

Table 11.1 The core mathematical formulas of the chapter

formula (1)	$f(x,y) = \Im^{-1}[F(u,v)] = \Im^{-1}\left[\frac{G(u,v)}{H(u,v)}\right]$ (11.20)	11.2.2
formula (2)	$\boldsymbol{g} = \boldsymbol{Hf} + \boldsymbol{n}$ (11.32)	11.3.1
formula (3)	$p(D) = \frac{n^n}{K(n-1)!\langle I\rangle^n}\exp\left(\frac{nD}{K} - \frac{n\exp(D/K)}{\langle I\rangle}\right)$ (11.57) $p(D) = \frac{\exp(\frac{D}{K} - \frac{\exp(D/K)}{\langle I\rangle})}{K\langle I\rangle}$ (11.54)	11.4.1
formula (4)	$H = f_1 \cdot f_2 \cdot A \cdot G$ (11.5)	11.1.2
formula (5)	$\overline{n} = E[n(x,y)]$ (11.8)	11.1.2
formula (6)	$g(x,y) = h(x,y) * f(x,y) + n(x,y)$ (11.1)	11.1.1
formula (7)	$F(s) = \left[\text{III}(2s^{\mathrm{T}})F(s)\right] * \frac{2\pi\sin(2\pi s^{\mathrm{T}})}{2\pi s^{\mathrm{T}}}$ (11.73)	11.5.1
formula (8)	$\boldsymbol{M} = \boldsymbol{USV}^{\mathrm{T}}$ (Fig. 11.14)	11.6.4
formula (9)	$H(s) = F(s)G(s)$ (11.81)	11.6.4

References

1. Jensen JR. Introductory digital image processing: a remote sensing perspective. Columbus: University of South Carolina; 1986. (In Chinese).
2. Peleg S, Keren D, Schweitzer L. Improving image resolution using subpixel motion. Recogn Lett Pattern. 1987;5(3). (In Chinese).
3. Irani M, Peleg S. Improve resolution by image registration. CVGIP Graph Models Image Process. 1991;53(91). (In Chinese).
4. Patti AJ, Sezan MI, Tekalp AM. High-resolution image reconstruction from a low-resolution image sequence in the presence of time-varying motion blur. In: Proceedings of ICIP-94 IEEE international conference on image processing 1994; 1994, no 1. IEEE (In Chinese).
5. Hardie RC, Barnard KJ, Bognar JG, et al. High-resolution image reconstruction from a sequence of rotated and translated frames and its application to an infrared imaging system. Opt Eng. 1998;37(1). (In Chinese).
6. Merino MT, Nunez J. Super-resolution of remotely sensed images with variable-pixel linear reconstruction. IEEE Trans Geosci Remote Sens. 2007;45(5):1446–57 (In Chinese).
7. Chen BY, Guo Q, Chen GL, et al. Amplificatory noise raised by super resolution image reconstruction and filter. J Infrared Millimeter Waves. 2011;30(1):15–20 (In Chinese).

Chapter 12
Technology II of Remote Sensing Digital Image Processing: Digital Image Compression

Chapter guidance To eliminate redundant data and highlight useful information, it is generally necessary to compress the image. The typical compression process can be divided into two categories: lossless compression, in which redundancy is eliminated based on the information entropy theory, and lossy compression, which is based on the theory of the rate distortion function, with the minimum amount of data stored to retain the maximum possible amount of useful information while minimizing engineering errors.

The basic theory of image compression was derived from the Shannon information theory proposed in the late 1940s, and the early source coding methods were established. After several decades of research, several mature technologies have been established in the field of image compression. In addition, the International Standards Organization has developed a series of international standards for image compression, including static image compression and motion image compression standards. Thus, advanced compression technologies have been derived from the theoretical bases. This chapter focuses on the Fourier transform, discrete image transform and wavelet transform, which are mathematical analysis tools for image compression. The logical framework of this chapter is shown in Fig. 12.1.

12.1 Definition of Compression and Common Compression Methods

12.1.1 Meaning of Compression

With the increase in time, space and spectral resolutions of remote sensing images, the corresponding data scale geometrically increases. A higher image resolution corresponds to a higher correlation of the adjacent sampling points and greater

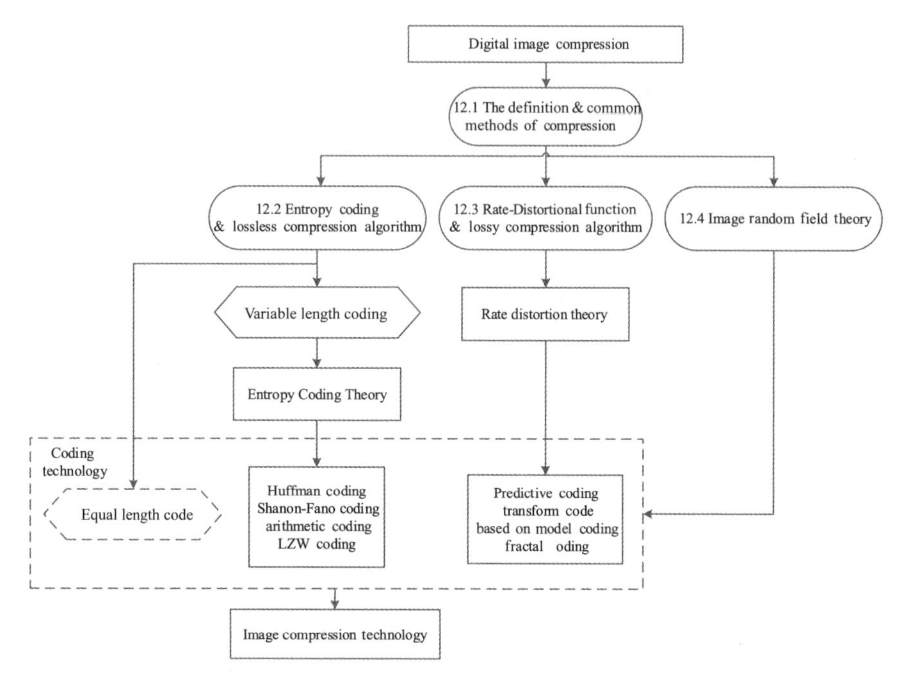

Fig. 12.1 Logical diagram of this chapter

data redundancy. These aspects render the transmission and storage of data challenging. Data compression represents an effective way to solve this problem and has gained increasing attention in the field of remote sensing. The compression of remote sensing data can help decrease the communication channel resources and increase the transmission rate of information. In addition, the overall reliability of the system is increased after data compression.

1. Quantification of Data Redundancy

Data redundancy, which is a major challenge to data compression, can be mathematically quantified. If n_1 and n_2 represent the number of information elements in the dataset representing the same information, the relative data redundancy R_D of the first dataset (n_1) can be defined as

$$R_D = 1 - \frac{1}{C_R} \tag{12.1}$$

where C_R is known as the compression rate and defined as

$$C_R = \frac{n_1}{n_2} \tag{12.2}$$

When $n_2 = n_1$, $C_R = 1$, and $R_D = 0$, the first expression of information does not contain redundant data relative to the second dataset.

When $n_2 \ll n_1$, $C_R \to \infty$, and $R_D \to 1$, significant compression can be realized, and large amounts of redundant data exist.

When $n_2 \gg n_1$, $C_R \to 0$, and $R_D \to -\infty$, the second set of data contained in the data than the original expression of the amount of data. Normally, C_R and R_D take values in open intervals $(0, \infty)$ and $(-\infty, 1)$, respectively.

2. Factors Affecting Data Redundancy

The redundancy in an image mainly pertains to coding redundancy, pixel redundancy and visual redundancy.

(a) **Coding Redundancy**

According to information theory, it is necessary to allocate the corresponding bit number according to the size of the information entropy for a pixel that represents the image data. However, each pixel in the actual image is generally represented by the same number of bits, and thus, there exists a certain redundancy.

(b) **Pixel Redundancy**

The inherent statistical properties of the image information show that there exists a strong correlation between adjacent pixels, adjacent rows or adjacent frames. Specifically, the value of a pixel can be inferred from its neighboring pixel values, indicating that pixel redundancy occurs.

(c) **Visual Redundancy**

In most cases, the final recipient of an image is the human eye. However, due to the limited resolution of the human eye, certain loss in image information cannot be distinguished by the human eye. Therefore, in the process of image compression, a certain distortion is allowed as long as the distortion is difficult to detect by the human eye.

12.1.2 Remote Sensing Image Compression

Usually, based on the distortion of the restored image after compression, image compression coding can be divided into two parts: lossless compression and lossy compression. In daily life, an image is usually evaluated based on the quality of the visual effect. Therefore, the lossy compression algorithm has been widely used to achieve visual effects with an acceptable quality. However, applications involving remote sensing data often require higher accuracy, especially when high-resolution remote sensing images are used. Lossy compression is implemented at the expense of high-frequency information in the image, and thus, the lossy compression algorithm is usually not applied for remote sensing images to be used for research and analysis.

(a) Original image (b) JPEG lossy compression image

Fig. 12.2 Examples of lossy compression details (Dalian Xinghai Square; compression ratio is approximately 10:1)

Figure 12.2b shows the JPEG high compression ratio of the restoration effect diagram. For decades, most of the studies on remote sensing data compression focused on lossless compression. Because the entropy is the lower bound of compression efficiency, the compression ratio is usually 4:1. In recent years, with the growth of remote sensing information, the compression ratio is required to be higher. However, an effective compression technology for remote sensing remains to be established.

The commonly used methods for remote sensing image compression are pulse code modulation (PCM), predictive coding, transform coding (principal component transform, KL transform, or discrete cosine transform MT), interpolation and extrapolation methods (spatial subsampling, subsampling, and adaptive), statistical coding (Huffman coding, arithmetic coding, Shannon–Fano coding and stroke coding), vector quantization and subband coding. These encoding methods are based on the frequency of pixel appearance. The frequency of frequent pixels is assigned to fewer bits. Therefore, few pixels can express the number of bits pertaining to the gray level to achieve data compression. A new generation of data compression methods, such as model-based compression methods and fractal compression and wavelet transform methods, have been developed to practical levels. The radar image compression method is based on the rasterization of the center of gravity compression method, taking into account the appropriate characteristics based on the quadtree tree division, single-trigger unit, discrete cosine transform, wavelet analysis and other compression methods. These approaches are based on the true three-dimensional TIN compression algorithm.

Image coding refers to the expression of information in an image with fewer bits while satisfying a certain quality (signal-to-noise requirement or subjective evaluation score). The image coding system is composed of two parts (Fig. 12.3).

The role of the source encoder is to alleviate or eliminate the coding redundancy in the input image, interpixel redundancy and mental vision redundancy.

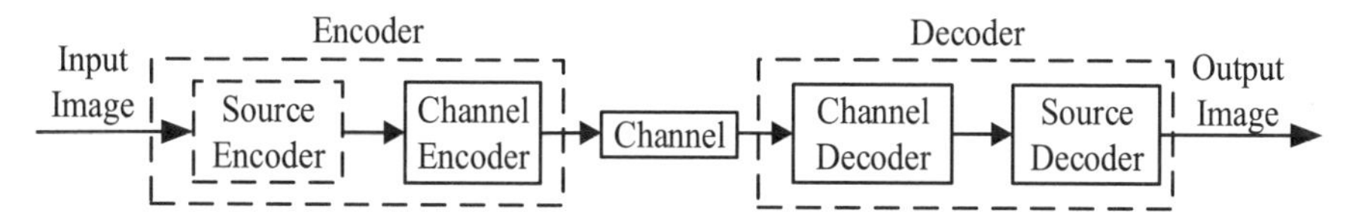

Fig. 12.3 Generic system coding model

12.2 Entropy Coding and Lossless Compression Algorithm

Lossless compression is aimed at eliminating unwanted redundant information, which theoretically leads to superior results. The information entropy specifies the removal efficiency.

12.2.1 Relationship Between Image Coding Compression and Entropy

The concept of information entropy (*H*) was proposed by an American mathematician Shannon [1] in the theory of information established in 1948 to measure the amount of information contained in an entity (that is, the mean or mathematical expectation of the amount of self-information $I(s_i) = \log_2 \frac{1}{p_i}$, where the logarithm of the base number is the number of encoded numbers). The base number in this book is 2:

$$H(S) = \sum_i p_i \log_2 \frac{1}{p_i} \tag{12.3}$$

where H is the information entropy (in bits), S is the source, and p_i is the probability that symbol S_i appears in S.

For example, in a 256-level grayscale image, if the probability of each gray pixel $p_i = 1/256$, $I = \log_2 \frac{1}{p_i} = \log_2 256 = 8$, and $H = \sum_{i=0}^{256} p_i \log_2 \frac{1}{p_i} = \sum_{i=0}^{256} \frac{1}{256} \log_2 256 = 8(\text{bit})$.

Each pixel requires 8 bits (I), and an average of 8 bits (H) exist per pixel.

The amount of redundancy in the information after a certain coding method is applied is the difference in the average code length of a code and entropy of the information source, that is,

$$R = E\{L_w(a_k)\} - H \tag{12.4}$$

Here, $\{L_w(a_k)\}$ represents the code length in terms of a_k (in bits for binary coding). If an encoding method produces an average word length equal to the entropy of the information source, it must remove all redundant information. This result can be achieved as long as a code can be designed. Therefore, the character a_k encoding the word length is

$$L_w(a_k) = -\log_2[P(a_k)] \tag{12.5}$$

This formula indicates the lower limit of the average word length. The average code length specifies the criterion for judging the lossless compression efficiency, while the entropy defines the efficiency limit of lossless compression coding.

Truncation error leads to a gap between the coding algorithm and entropy. Since the measurement of the frequency of occurrence of the information is continuous, and the coding bits are discrete when a code is stored, the frequency of occurrence of the information is discretized when the data are stored in a discrete coding manner, which decreases the efficiency of the information coding compression.

12.2.2 *Entropy Coding and Lossless Compression*

Entropy coding is a class of nondestructive data streams that are compressed using statistical data. The entropy of the information source is the limit of lossless coding compression. Commonly used entropy coding methods are LZW encoding, Shannon coding, Huffman coding and arithmetic coding.

1. Shannon–Fano Coding

The Shannon–Fano coding method is a variable code length encoding, performed first in accordance with the probability of symbolic sort and later from top to bottom using the recursive method. The symbol group is divided into two parts, such that each part has an approximation of the same frequency, marked on both sides as 0 and 1. Each symbol from top to bottom of the 0/1 sequence refers to the binary code.

2. Huffman Coding

The Huffman code for compressed text files was created by Huffman in 1952. The algorithm determines the number of bits of a symbol based on the probability distribution of that symbol in the message.

Huffman encoding process: ① Arrange all symbols in the order of increasing probability from left to right as leaf nodes; ② Connect two minimum-level top-level nodes to form a parent node, and mark 0 and 1 on the two lines to the left and right subnodes, respectively; ③ Repeat step ② until the root node is obtained to form a binary tree; ④ The entities from the root node to each symbol of the leaf node 0/1 string form the symbol of the binary code. Since the symbols are arranged from left to right, the marking of branches as 0 or 1 is insignificant, and the final result may not be unique. However, this framework is only the assigned code, and the average length of the code is the same.

For example, consider a 40-pixel gray image having a total of five gray levels, with symbols A, B, C, D and E. The number of pixels in all grayscale levels is shown in Table 12.1.

Table 12.1 Number of occurrences of symbols in the image

Symbol	A	B	C	D	E
Number of occurrences	15	7	7	6	5

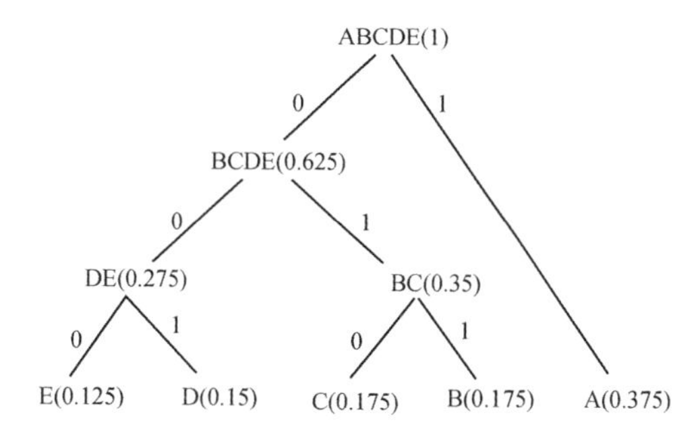

Fig. 12.4 Huffman coding process

We consider this image to illustrate the process of Huffman coding, as indicated in Fig. 12.4 and Table 12.2.

The average code length is defined as

$$L(S) = \sum_i p_i \cdot l_i \tag{12.6}$$

The average code length is 90/40 = 2.25. The compression ratio is 120/90 = 4/3 $\approx$ 1.333:1.

According to Eq. (12.3), the information entropy is 2.1966 in binary coding, less than the average code length for the Huffman coding. This result objectively confirms the previous entropy coding theory.

3. LZW Coding

LZW coding refers to a dictionary-based coding method [2]. A variable length input symbol string is mapped to a fixed-length code forming a phrase dictionary index (string table). This method utilizes the frequency redundancy of the character and

Table 12.2 Example of Huffman algorithm

Symbol	Frequency (p_i)	$\text{Log}_2(1/p_i)$	Code	Bits
A	15(0.375)	1.4150	1	15
B	7(0.175)	2.5145	011	21
C	7(0.175)	2.5145	010	21
D	6(0.150)	2.7369	001	18
E	5(0.125)	3.0000	000	15
Total	40(1)	12.1809		90

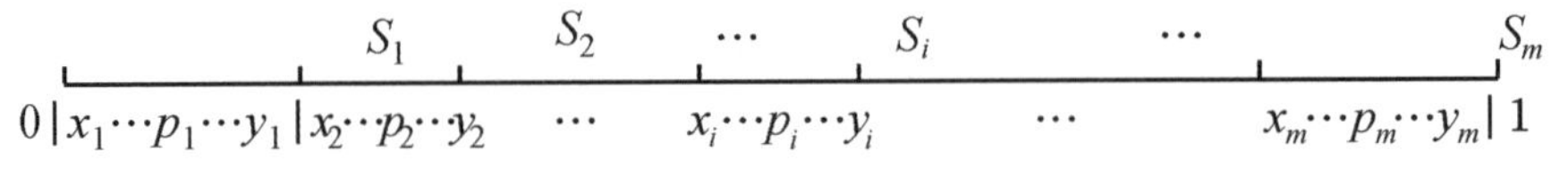

S_1 S_2 ... S_i ... S_m

$0 | x_1 \cdots p_1 \cdots y_1 | x_2 \cdots p_2 \cdots y_2$ ... $x_i \cdots p_i \cdots y_i$ ... $x_m \cdots p_m \cdots y_m | 1$

Fig. 12.5 Probability interval of the arithmetically encoded characters

high redundancy of the string mode to achieve compression. The algorithm only needs to scan once and can be adapted without saving and transmitting the string table.

4. Arithmetic Coding

Arithmetic coding is a lossless compression algorithm. From the information theory perspective, this algorithm is similar to Huffman coding with the optimal variable length of the entropy coding. The main advantage is that the limitation of Huffman coding being an integer bit is overcome, as this limitation may result in a large difference between the actual code length and real number.

The encoding method arranges the symbol table used in the string in increasing order of the original encoding (such as ASCII encoding of characters or binary encoding of characters) and calculates the probability p_i of each symbol S_i in the table. Subsequently, the method determines the corresponding intercell range $[x_i, y_i]$ in the period [0, 1] based on the magnitude of the symbol probabilities p_i:

$$x_i = \sum_{j=0}^{i-1} p_i y_i = x_i + p_i \quad i = 1, \ldots, m \tag{12.7}$$

where $p_0 = 0$. The width of the cell corresponding to symbol Si is its probability p_i (Fig. 12.5).

Subsequently, the input symbol string is encoded: Let the j-th symbol c_j in the string be the i-th symbol S_i in the symbol table. The coding interval $I_j = \left[l_j, r_j\right)$ is calculated from the upper and lower limits (x_i and y_i) of the corresponding interval in the symbol table, that is,

$$l_j = l_{j-1} + d_{j-1} \cdot x_i, \quad r_j = l_{j-1} + d_{j-1} \cdot y_i, \quad j = 1, \ldots, n \tag{12.8}$$

where $d_j = r_j - l_j$ is the width of interval I_j, $l_0 = 0$, $r_0 = 1$, and $d_0 = 1$. l_j increases when d_j and r_j decrease. The lower bound l_n of the interval corresponding to the last symbol of the string is the arithmetic coding value of the symbol string.

12.3 Rate Distortion Function and Lossy Compression Algorithm

In addition to the redundant information removed by lossless compression, a large amount of useless information can be removed by lossy compression methods with a certain distortion. For a distortion source, the rate distortion function measures the compression performance to preserve as much useful information as possible with a minimal amount of stored data.

12.3.1 Rate Distortion Function

1. Definition of the Rate Distortion Function

The distortion D is the root mean square error between the image $g(x, y)$ reconstructed from the compressed image and original image $f(x, y)$, defined as

$$D = E\{[f(x, y) - g(x, y)]^2\} \tag{12.9}$$

Shannon defined the information rate distortion function *R(D)* and discussed the basic theorem regarding this function: the information transfer rate of the source output can be decreased to the *R(D)* value when a certain degree of distortion *D* is allowed. This aspect theoretically yields the relationship between the information transmission rate and allowable distortion and establishes the basis of the theory of information rate distortion. When the source is known and the distortion function has been defined, the information transmission rate *R* of the source to be transmitted to the recipient must be as small as possible when the distortion function is satisfied. This lower limit is related to *D*. From the receiver side, the minimum average amount of information that must be obtained is identified to reproduce the source message under the fidelity criterion. The average amount of information obtained at the receiving end can be expressed by the average mutual information *I (U; V)*. This value is the minimum value for finding the average mutual information *I (U; V)*, that is,

$$R(D) = \min_{P(v_j/u_i) \in B_D} \{I(U; V)\} \tag{12.10}$$

The information rate distortion function *R(D)*, which is known as the rate distortion function and has units of Knight/source symbol or bit/source symbol, specifies the relation between the minimum entropy rate and distortion that the entropy compression coding may cause. The inverse function is known as the distortion function, which represents the minimum average distortion that can be achieved at a certain information rate.

2. Characteristics of the Rate Distortion Function

The definition field for $R(D)$ is $0 \leq D_{\min} \leq D \leq D_{\max}$.

Where $D_{\min} = \sum_x p(x) \min_y d(x, y)$ and $D_{\max} = \min_y \sum_x p(x)d(x, y)$. The lower limit of the allowable distortion D can be zero, which means no distortion is allowed.

R(D) is the lower convex function of the average distortion. Let D_1andD_2 be any two mean distortions, and $0 \leq a \leq 1$. In this case,

$$R(aD_1 + (1 - a)D_2) \leq aR(D_1) + (1 - a)R(D_2) \tag{12.11}$$

R(D) is a continuous and strictly monotonically decreasing function over $(D_{\min}, D_{\max})$. According to the convexity of the information rate distortion function, *R(D)* is continuous on $(D_{\min}, D_{\max})$.*R(D)*is a strictly monotonically decreasing function on the interval $(D_{\min}, D_{\max})$, and *R(D)* is nonadditive and nonconstant. When the algorithm involves binary coding, the units of the rate distortion function as bit/source symbols. Figure 12.6 shows the continuous and discrete graphs of the rate distortion function, and Fig. 12.7 shows the general shape of the rate distortion function.

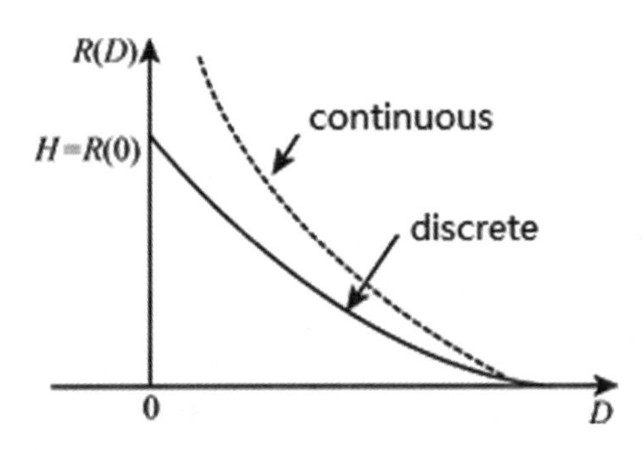

Fig. 12.6 Continuous and discrete graphs of the rate distortion function

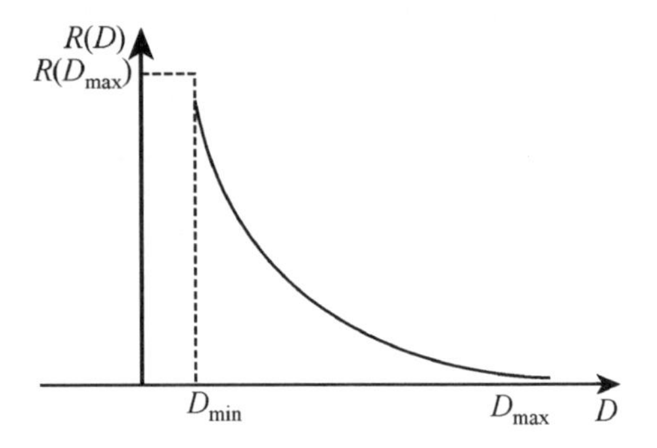

Fig. 12.7 General shape of the rate distortion function

Let *R(D)* represent the information rate distortion function for a discrete memoryless source with a finite distortion measure. For any arbitrary $D \geq 0$, $\varepsilon > 0$, and any long code length *k*, there must be a source code *C* with a number of code words $M \geq 2^{k[R(D)+\varepsilon]}$. Therefore, the average distortion of the encoded code $\bar{D} \leq D$.

When the code length *k* is adequately large, a source code can be found such that the resulting information transfer rate is slightly greater (until infinite approximation) than the rate distortion function *R(D)*, and the average distortion of the code is not greater than the given permissible distortion, that is, $\bar{D} \leq D$. *R(D)* is the lower limit that the source code may reach under a given *D*. Thus, Shannon's third theorem indicates that the optimal source code that achieves this lower limit exists.

3. Relationship Between the Rate Distortion Function and Entropy

In lossless compression coding, the amount of information output from the source is measured in terms of the entropy. In the distortion source, the distortion rate *D* is expressed by the information rate distortion function *R(D)*. This discussion highlights that *R(D)* is the minimum information rate that the source must transmit under the condition of maximum definite distortion, which is theoretically achievable. $R(0)$ is the minimum information transfer rate without distortion and approximates entropy H of the information source (Fig. 12.8). When $D = 0$, no distortion is allowed. In the study of different compression methods, the rate distortion function is as steep as the rate distortion function. Specifically, the slope of the rate distortion function is as small as possible, and the entropy is approximated when no distortion exists.

However, Shannon's third theorem is still only an existential theorem regarding the identification of the best coding method; the theorem is not given and needs to

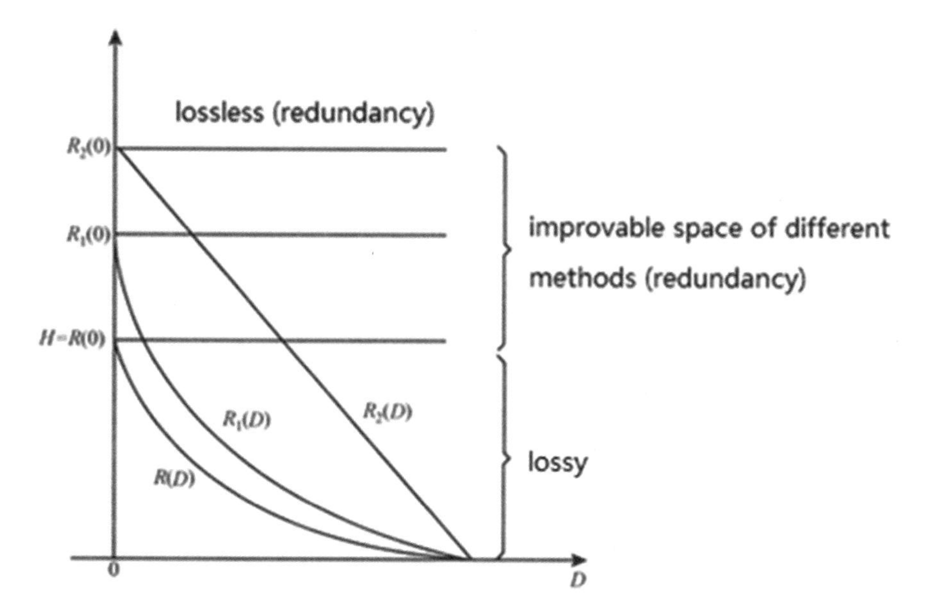

Fig. 12.8 Relationship between the rate distortion function and entropy

be established experimentally. This section discusses the methods to calculate the information rate distortion function *R(D)* according to the actual source and identify the best coding method to achieve the information compression limit *R(D)*. In practical engineering applications, for different coding methods, there exist different actual information transmission rates $R_1(D)$ and $R_2(D)$ due to technical reasons, and these values are consistent with the optimal theoretical value (i.e., entropy H) difference, which pertains to the advantages and limitations of the different source codes. Because $D = 0$ in the absence of distortion, the average bit rate of the information is greater than the entropy of the information source, that is, $R_1(0) > H$.

When no distortion exists, the distance between $R_1(0)$, $R_2(0)$ and H is the space that can be enhanced when the algorithm is lossless. The purpose of algorithm research is to more effectively eliminate this part of data redundancy. When the rate R is less than H, the algorithm is detrimental.

12.3.2 Lossy Compression Algorithm

1. Predictive Coding

Predictive coding is aimed at predicting the next signal based on one or more of the preceding signals according to the correlation characteristics between the discrete signals. Subsequently, the algorithm encodes the difference between the actual and predicted values (prediction error). If the prediction is accurate, the error is extremely small. Under the same precision requirements, fewer bits can be used for the encoding to realize data compression.

In the predictive coding for the symbolic sequence with M values, the entropy of the L-th symbol satisfies

$$\begin{aligned}\log_2 M &\geq H(x_L) \geq H(x_L|x_{L-1}) \geq H(x_L|x_{L-1}, x_{L-2}) \\ &\geq \cdots \geq H(x_L|x_{L-1}, x_{L-2}, \ldots, x_1) > H_\infty \end{aligned} \tag{12.12}$$

The compression capacity of predictive coding is limited. For example, the DPCM can only be compressed to samples having 2–4 bits.

2. Transform Coding

Transform coding refers to a certain function of the signal transformation from one signal (space) to another (space). Next, the signal is encoded, for instance, by transforming the time domain signal into the frequency domain. The transform does not perform data compression and only maps the signal to another domain, making the signal easy to compress in the transform domain. Moreover, the transformed samples are more independent and orderly.

The transform coding method involves the discrete cosine transform (DCT) and wavelet transform. When the one-dimensional and two-dimensional DCT transform

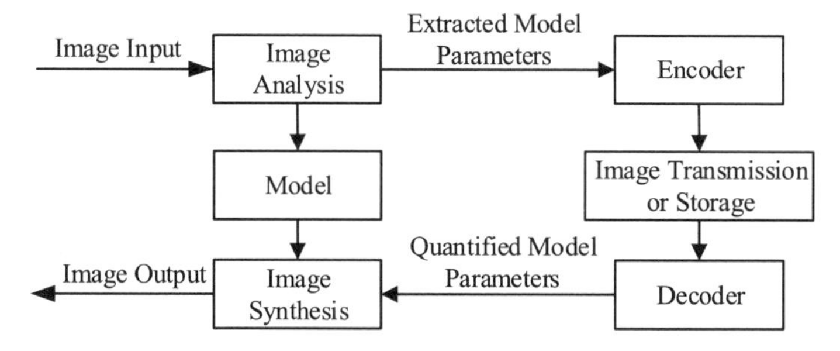

Fig. 12.9 Basic block diagram of model-based image coding

is performed to transform the DCT transform coefficients, the amplitude becomes the frequency:

$$F(u, v) = \frac{4C(u)C(v)}{n^2} \sum_{j=0}^{n-1} \sum_{k=0}^{n-1} f(j, k) \cos\left[\frac{(2j+1)u\pi}{2n}\right] \cos\left[\frac{(2k+1)v\pi}{2n}\right] \tag{12.13}$$

3. Model-Based Coding

The basic concept of model coding is that at the sending end, the image analysis module is used to extract the compact and necessary information to describe the input image, and several model parameters with a small data volume are obtained. At the receiving end, the original image is reconstructed using the image synthesis module. The basic principle of the image information synthesis is shown in Fig. 12.9.

4. Fractal Coding

Fractal compression is achieved using the self-similarity principle in fractal geometry. First, the image is segmented, and the similarity between the blocks is determined by affine transformation. After the affine transformation of each piece, the coefficient of the affine transformation is preserved, and the image is considerably compressed because the amount of data per block is considerably larger than the affine transformation coefficient.

In fractal coding, after the image block for each domain, the geometric transformation, isomorphic transformation and grayscale transformation are defined, and a large number of domain pool definitions can be obtained. The fractal coding of the range block R_i is aimed at identifying the best ϕ_i, τ_i, and G_i and the best definition domain block D_j in the domain pool.

$$E^2(R, D) = \sum_{i,j}^{N} [(r_{i,j} - (s \cdot d_{i,j} + o))]^2 \tag{12.14}$$

where $r_{i,j}$ and $d_{i,j}$ are values of the range block R and pixel value of the defined domain block $D''_j = \tau_i(D'_j) = \tau_i \cdot \phi_i(D_j)$ after the first two transformations, respectively.

12.3.3 JPEG Static Coding Compression Standards and Related Methods

The joint photographic experts group (JPEG) is a static image compression standard developed under the leadership of the International Organization for Standardization (ISO), which combines quantitative and lossless compression coding to eliminate redundant information from the image and corresponding data. The process flow of the JPEG compression encoder algorithm is shown in Fig. 12.10. The decoding (decompression) and compression coding processes are opposite in nature.

JPEG compression coding involves three main steps: ① The forward discrete cosine transform (FDCT) is implemented to transform the image in the spatial domain into that in the frequency domain; ② the best effect for the human visual system is achieved using the quantization function to quantify the DCT coefficients; and ③ the Huffman variable word length encoder is used to encode the quantization coefficients. The process of decoding (decompressing) is opposite to that of compression coding.

The JPEG compression coding algorithm involves the following steps.

1. Forward Discrete Cosine Transform

JPEG encoding is performed separately for each individual color image component. Before performing the FDCT, it is necessary to divide the component image into 8 × 8-pixel image blocks. Insufficient parts can be filled by repeating the last row/column of the image. These image blocks are used as inputs to the 2D FDCT (Fig. 12.11).

2. Quantization

Quantization is performed for the FDCT after the conversion of the frequency coefficient to a smaller range of values. The purpose of quantification is to decrease the magnitude of the nonzero coefficients, thereby decreasing the number of required bits and increasing the number of "0" coefficients. Quantification is the main reason for the decline in image quality.

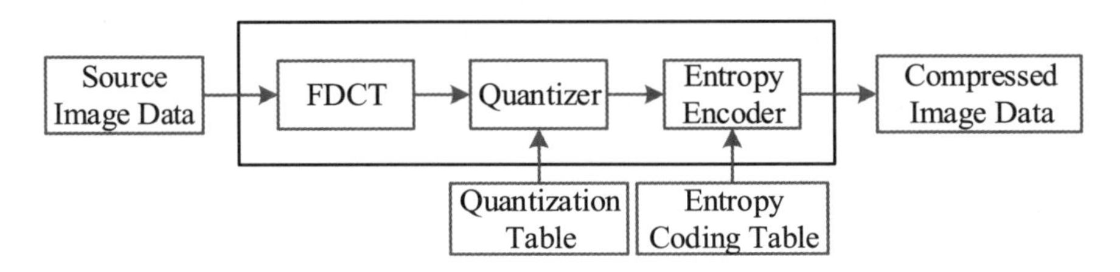

Fig. 12.10 Block diagram of JPEG compression encoding

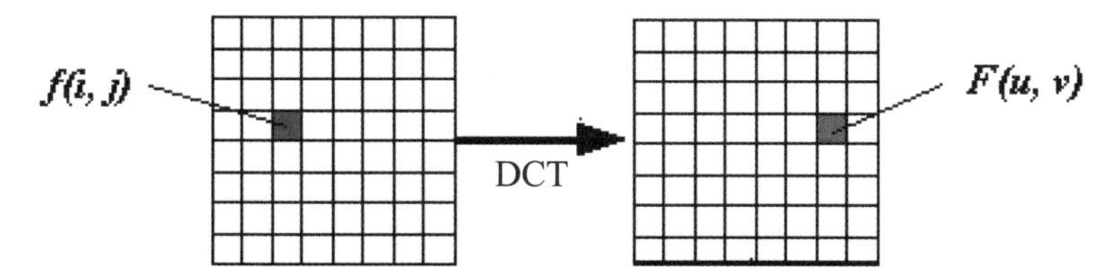

Fig. 12.11 Discrete cosine transform

In the lossy compression algorithm of JPEG images, the uniform scalar quantizer is used, as shown in Fig. 12.12, and the quantization is based on the position of the coefficient and hue value of each color component.

3. Z-Shaped Arrangement

The quantized two-dimensional coefficients are rearranged and converted to one-dimensional coefficients. To increase the number of consecutive "0" coefficients, which is the run length of "0", the zigzag pattern is used in JPEG encoding, as shown in Fig. 12.13.

The serial number of the DCT coefficients is shown in Table 12.3. Such an 8×8 matrix is transformed to a 1×64 vector. The lower frequency coefficients are placed on the head of the vector.

4. DC Coefficient Coding

The 8×8 image block after DCT conversion yields the DC coefficient. The DC coefficient has two characteristics: the coefficient is large, and the DC coefficient values of the adjacent 8×8 image blocks do not change considerably. Considering these characteristics, the JPEG algorithm uses a differential pulse code modulation (DPCM) technique to encode the difference Δ of the DC coefficients between adjacent image blocks:

$$\Delta = \mathrm{DC}(0,\ 0)_k - \mathrm{DC}(0,\ 0)_{k-1} \tag{12.15}$$

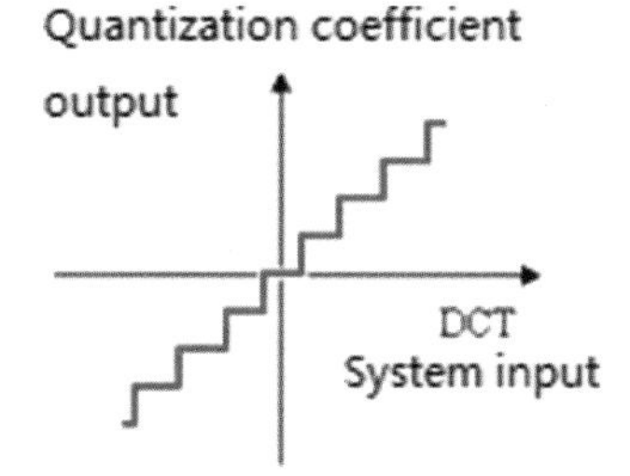

Fig. 12.12 Uniform quantizer

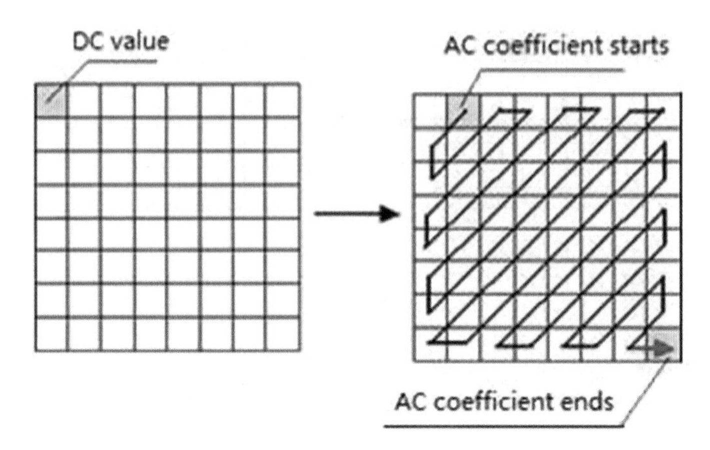

Fig. 12.13 Quantification of the arrangement of DCT coefficients

Table 12.3 Number of quantized DCT coefficients for the zigzag arrangement

0	1	5	6	14	15	27	28
2	4	7	13	16	26	29	42
3	8	12	17	25	30	41	43
9	11	18	24	31	40	44	53
10	19	23	32	39	45	52	54
20	22	33	38	46	51	55	60

5. AC Coefficient Coding

The quantization of the AC coefficient is characterized by the fact that the 1×63 vector contains many "0" coefficients that are continuous. Therefore, the coefficients are encoded using the simple and intuitive run-length coding (RLE).

JPEG uses the upper 4 bits of 1 byte to represent the number of consecutive "0s" and the lower 4 bits to represent the number of bits required to encode the next nonzero coefficient. The value of the quantized AC coefficient is obtained.

6. Entropy Coding

Entropy coding can be performed to compress the DC coefficients obtained by the DPCM and AC coefficients after RLE coding. In JPEG lossy compression algorithms, Huffman or arithmetic coding can be performed to decrease the entropy. The Huffman encoder is often used because an extremely simple lookup table method can be used to accelerate coding.

7. Composition Bit Data Stream

The final step in JPEG encoding is to combine the various tag codes and encoded image data into one frame. The objective is to facilitate the transmission, storage and decoding aspects of the decoder. The data thus organized are referred to as a JPEG bitstream.

For remote sensing images, people may often be interested in only one of the regions. In such scenarios, the JPEG algorithm can be combined with fractal coding.

12.4 Space Object Lossy Compression Theory of the Image Random Field

A remote sensing image is rich in texture that is random. The random field model is a stochastic process that can use the connection between the model parameters and texture to simulate the random texture. The image texture can be described by two elements: the gray level or color of the primitives that constitute the image texture, and the interaction or interdependence between primitives. This dependency can be random, functional, or structured, and this stochastic dependency can be represented by a conditional probability model.

12.4.1 Markov Process

Because of the strong similarity and correlation between the pixels in the local area of an image, the image can be regarded as a Markov random field in the analysis based on probability statistics. The theory of the Markov process is described in the following text.

A static sequence is known as a first-order Markov sequence, and the conditional probability of each element in the sequence depends only on its previous element. The covariance matrix of an $N \times 1$ Markov sequence has the following form:

$$\boldsymbol{C}=\begin{bmatrix} 1 & \rho & \rho^2 \ldots & \rho^{N-1} \\ \rho & 1 & \rho \ldots & \rho^{N-2} \\ \rho^2 & \rho & 1 \ddots & \vdots \\ \vdots & \vdots & \vdots \ddots & \rho \\ \rho^{N-1} & \rho^{N-2} & \ldots \rho & 1 \end{bmatrix} \tag{12.16}$$

where $0 \leq \rho \leq 1$. The eigenvalues of this covariance matrix are

$$\lambda_k = \frac{1-\rho^2}{1-2\rho\cos(\omega_k)+\rho^2} \tag{12.17}$$

The characteristic vector is

$$v_{m,k} = \sqrt{\frac{2}{N + \lambda_k}} \sin\left[\omega_k\left(m - \frac{N-1}{2}\right) + (k+1)\frac{2}{\pi}\right] \tag{12.18}$$

where $0 \leq m, k \leq N - 1$. ω_k is the root of the following transcendental equation:

$$\tan(N\omega) = -\frac{\left(1 - \rho^2\right)\sin\omega}{\cos\omega - 2\rho + \rho^2\cos\omega} \tag{12.19}$$

Therefore, the base vector of the K–L transform can be calculated by setting the value of ρ.

The Markov model aims to estimate the spatial correlation of the image. The Markov property of the image is that the gray value of pixel x_n is related only to the neighboring pixels, regardless of the other pixels, that is,

$$p(x_n|x_l, l \neq n) = p(x_n|x_l, l \in N_n) \tag{12.20}$$

where N_n represents the neighboring pixels of pixel x_n. $p(x_n|x_l, l \in N_n)$ is characterized by the relationship between pixel x_n at a certain position in the image space and its neighboring pixels, that is, the local statistical properties of the image.

The following section describes the Markov model defined by the state transition probability.

When decoding the k-th bit-plane X^k, the previously decoded bit-plane $X^8X^7\ldots X^{k+1}$ and low-density parity-check code (LDPC) decoder in this iteration process are reassembled as symbols determine the current bit-plane X^k. The results are reassembled into symbols $\left\{\widehat{x}_n = x^8x^7\ldots x^{k+1}v_n\right\}$ to the judgment result of the current bit-plane X^k. The judgment result is regarded as a grayscale image $\widehat{x}$ having a pixel accuracy of (8−k) bits. The Markov model is applied to the image $\widehat{x}$, the state transition probability of the Markov model is estimated, and the forward–backward process is used to estimate the conditional probability of each pixel.

In principle, the statistical properties of a statistically even image can be represented by a Markov process. However, almost all the actual image signals are not statistically uniform, and the statistical properties are usually nonstationary. To exploit the spatial statistical properties of the image, the method divides the image into nonoverlapping blocks. Assume that the statistical properties of each block are uniform and can be described using a Markov process. If the size of block $B_h \times B_v$ is reasonably set, the statistical uniformity of each block can be ensured, and there exist adequate pixels for each block to estimate the Markov model.

1. 1-D Markov Model

The pixels of each block are scanned as 1-D data, as shown in Fig. 12.14, i.e., odd and even-numbered lines are scanned from left to right and from right to left, respectively. In this case, the two pixels of the generated 1-D data adjacent to each other are adjacent in the spatial position of the image and are strongly correlated. The first-order Markov chain is used to model the 1-D data of each block. Thus, the

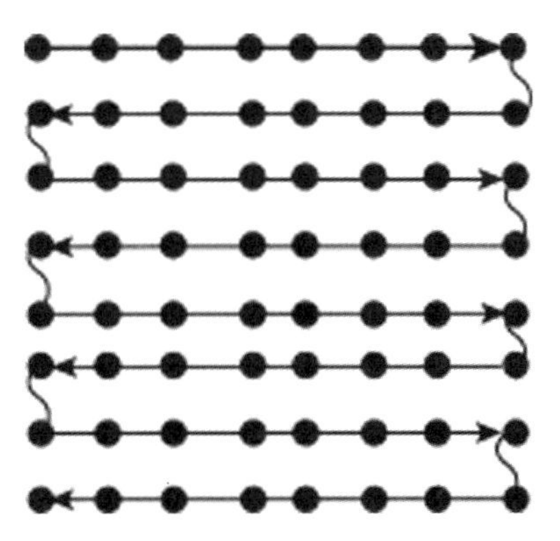

Fig. 12.14 Scanning order of pixels within a 1-D Markov model block

gray value of each pixel is related only to the two adjacent pixels, that is,

$$p(x_n|x_l, l \neq n) = p(x_n|x_{n-1}, x_{n+1}, x_n) \tag{12.21}$$

According to the information g_n of the current bit-plane transmitted from the LDPC decoder and the previously decoded bit-plane $X^8X^7 \ldots X^{k+1}$, the probability of each pixel can be obtained:

$$u_n(j) = \begin{cases} g_n(b), j = x_n^8 x_n^7 \ldots x_n^{k\ 1} b, b \in \{0, 1\} \\ 0, \text{ else} \end{cases} \tag{12.22}$$

The first order Markov chain is defined as

$$\lambda = \{\boldsymbol{A}, \pi\} \tag{12.23}$$

where π is the initial state probability, and $\boldsymbol{A} = (a_{ij})$ is the state transition probability matrix, which represents the one-step transition probability from state i to state j.

We define

$$\alpha_n(j) = p(x_n = j|\lambda, x_1x_2 \ldots x_{n-1}) \tag{12.24}$$

$$\beta_n(j) = p(x_n = j|\lambda, x_{n+1}x_{n+2} \ldots x_N) \tag{12.25}$$

where $N = B_h \times B_v$ is the number of pixels per block. $\alpha_n(j)$ indicates the effect of the forward data $x_1x_2 \ldots x_{n-1}$ on pixel x_n, corresponding to the forward process; $\beta_n(j)$ represents the effect of the backward data on $x_{n+1}x_{n+2} \ldots x_N$, corresponding to the backward process. According to the first-order memory of the Markov chain defined by Eq. (12.20), the state is modulated by the LDPC decoding information, and the forward process can be expressed as a recursive process:

$$\alpha_1(j) = \pi_1(j)u_1(j) \tag{12.26}$$

$$\alpha_n(j) = \sum_i \alpha_{n-1}(i) a_{ij} u_n(j) \tag{12.27}$$

Assuming that the 1-D data of each block represent a reversible smooth Markov process, the backward process can be expressed as a recursive process:

$$\beta_1(j) = \pi_2(j) u_N(j) \tag{12.28}$$

$$\beta_n(j) = \sum_i \beta_{n+1}(i) a_{ij} u_n(j) \tag{12.29}$$

where π_1 and π_2 are the initial state probabilities of the forward and backward processes, respectively.

2. 2-D Markov Model

In considering each block as a two-dimensional first-order Markov process, the gray value of each pixel is only related to its 4-neighborhood pixel, i.e.,

$$p\left(x_{h,v} | x_{h'v'}, \left(h', v'\right) \neq (h, v)\right) = p\left(x_{h,v} | x_{h-1,v}, x_{h,v-1}, x_{h+1,v}, x_{h,v+1}\right) \tag{12.30}$$

The 2-D first-order Markov model can also be defined by formula (12.23). However, the state transition probability matrix $A = \left(a_{ijk}\right)$ is a three-dimensional matrix, where $a_{ijk} = p\left(x_{h,v} = k | x_{h-1,v} = i, x_{h,v-1} = j\right)$ represents the one-step transition probability from the state in which the upper pixel value is i and left pixel value is j to the current pixel value k.

The forward process is defined as

$$\alpha_{h,v}(j) = p\left(x_{h,v} = j | \lambda, X_F\right) \tag{12.31}$$

The backward process can be defined as

$$\beta_{h,v}(j) = p\left(x_{h,v} = j | \lambda, X_B\right) \tag{12.32}$$

with

$$X_F = \left\{x_{h',v'} | 1 \leq h' \leq h, 1 \leq v' \leq v, \left(h', v'\right) \neq (h, v)\right\} \tag{12.33}$$

$$X_B = \left\{x_{h',v'} | h \leq h' \leq H, v \leq v' \leq W, \left(h', v'\right) \neq (h, v)\right\} \tag{12.34}$$

where H and W are the height and width of the image, respectively. X_F and X_B denote the upper left pixel set of pixel $x_{h,v}$ and lower right pixel set of pixel $x_{h,v}$, respectively, as shown in Fig. 12.15. Similar to the 1-D Markov chain, the forward process of a 2-D Markov model can be expressed as a recursive process:

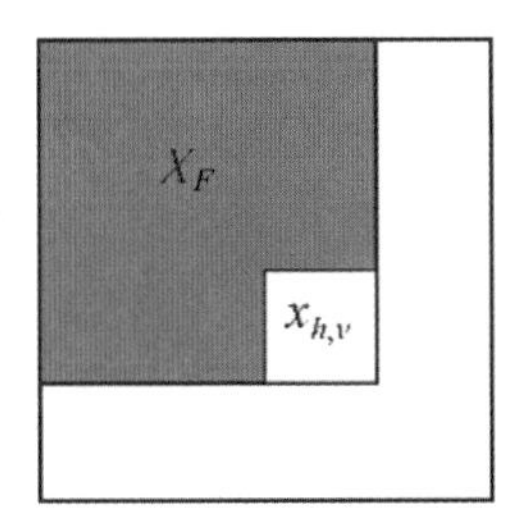

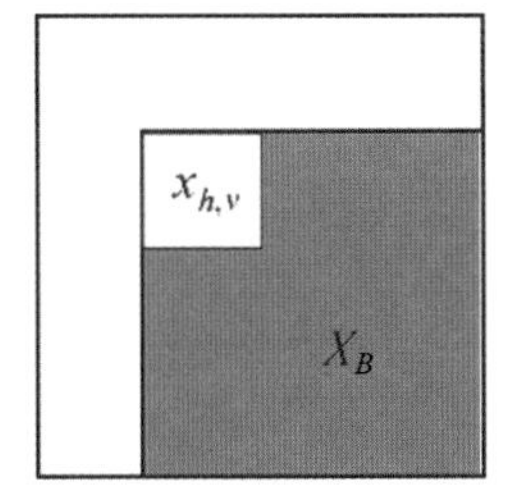

Fig. 12.15 Forward–backward process of the 2-D Markov model

$$\alpha_{1,1}(k) = \pi_1(k)u_{1,1}(k) \tag{12.35}$$

$$\alpha_{h,k}(k) = \sum_{i,j} \alpha_{h-1,v}(i)\alpha_{h,v-1}(j)a_{ijk}u_{h,v}(k) \tag{12.36}$$

Similarly, the backward process can be expressed as

$$\beta_{H,W}(k) = \pi_2(k)u_{H,w}(k) \tag{12.37}$$

$$\beta_{h,k}(k) = \sum_{i,j} \beta_{h+1,v}(i)\alpha_{h,v+1}(j)a_{ijk}u_{h,v}(k) \tag{12.38}$$

3. Estimation of Local Statistical Characteristics of Images

For the 1-D and 2-D Markov models, the forward and backward processes represent the probability. Therefore, the values must be normalized:

$$\alpha_n(j) = \frac{\alpha_n(j)}{\sum_j \alpha_n(j)} \tag{12.39}$$

$$\beta_n(j) = \frac{\beta_n(j)}{\sum_j \beta_n(j)} \tag{12.40}$$

For the 2-D Markov model, each block is converted from that at the top in the order from left to right, and if $x_{h,v}$ is the n pixel, $\alpha_n = \alpha_{h,v}$, $\beta_n = \beta_{h,v}$.

The local statistical characteristics of pixel x_n are

$$p(x_n = j|x_l, l \in N_n) = \frac{\alpha_n(j)\beta_n(j)}{\sum_i \alpha_n(j)\beta_n(j)} \tag{12.41}$$

The Markov node passes this information to the variable node.

4. Estimation of the State Transition Probabilities

The state transition probability reflects the memory characteristics of the Markov model and characterizes the spatial correlation of the image. The corresponding

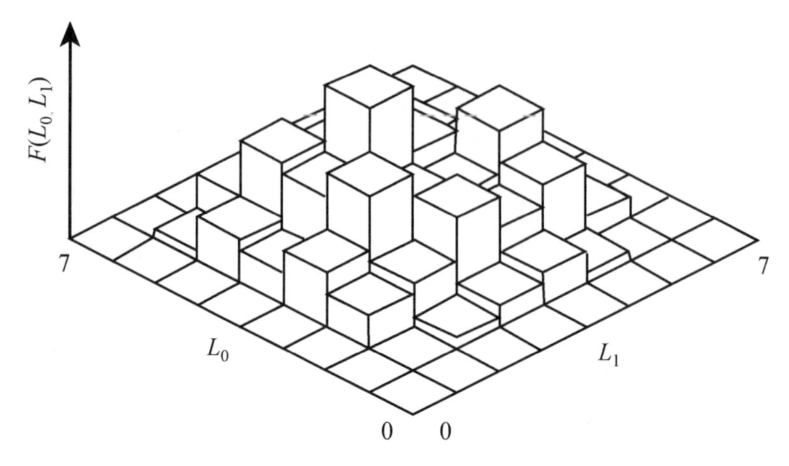

Fig. 12.16 2-D histogram of the first order Markov chain

Markov model is established for each block of image *X*, and the corresponding state transition probability is estimated by evaluating the Markov model high-dimensional histogram. Figure 12.16 illustrates the estimation algorithm of the state transition probability with the first-order Markov chain as an example.

For first-order Markov chains, two-dimensional histograms are required. L_0 is the gray value of pixel x_n, L_1 is the gray value $(1 \leq n < N)$ of pixel x_{n+1}, and $F(i, j)$ is the frequency of $(L_0 = i, L_1 = j)$. $(F(i, j))$ is termed the two-dimensional histogram of the image, as shown in Fig. 12.16, which specifies the probability of state transition of the first-order Markov chain:

$$a_{ij} = \frac{F(i, j)}{\sum_i F(i, j)} \tag{12.42}$$

Similarly, a 3-D histogram is required for the 2-D first-order Markov model.

When decoding k bit-plane X^k, the pixels $\hat{x}_n = x^8 x^7 \ldots x^{k+1} v_n$ of X are composed of (8-*k*) bits. Since v_n is the LDPC, which decodes the decision result in an iteration process before the correct decoding $\{v_n\} \neq \{x_n^k\}$, the value of X based on the histogram is the noisy form of the thermal signal histogram $\overline{X} = \{\hat{x}_n = x^8 x^7 \ldots x^{k+1} v_n\}$. The number of pixels per block is limited, resulting in sparse spurs of the histogram in the spatial points. If there exists a combination $(L_0 = i, L_1 = j)$ in the real signal X and no such combination in the noisy $\hat{x}$, the forward–backward process will be adversely influenced. The estimated state transition probability is $a_{ij} = 0$. This aspect affects the estimation accuracy of the local statistical characteristics of the image. To alleviate the effect of the sparse histogram on the estimated state transition probability, the scheme uses the Parzen window to smooth the histogram and extend the value of the effective point to the surrounding zero.

Let $Z_S = (L_0 = i, L_1 = j)$ be a nonzero value in the 2-D histogram and $z = \left(L_0 = i^{'}, L_1 = j^{'}\right)$ be the zero point in the histogram. The value expanded from Z_S *to Z* by the Parzen window is.

Let $Z_S = (L_0 = i, L_1 = j)$ be a nonzero value in the 2-D histogram and $z = \left(L_0 = i^{'}, L_1 = j^{'}\right)$ be the zero point in the histogram. The value expanded from Z_S to Z by the Parzen window is

$$F(z) = \frac{1}{w^d} K\left(\frac{1}{w}(z - Z_S)\right) \tag{12.43}$$

where w is the window size, d is the dimension of the histogram, and $K\,()$ is the kernel function. The method uses the standard multidimensional Gaussian density function as the kernel function, that is,

$$K(z) = \frac{1}{(2\pi)^{d/2}} \exp\left(-\frac{1}{2} z^{\mathrm{T}} z\right) \tag{12.44}$$

The optimized window size is

$$w = \sigma \left[\frac{4}{L(2d+1)}\right]^{l/(d+4)} \tag{12.45}$$

where L is the number of all points of the histogram, and σ^2 is the variance of the histogram.

With the continuous correction of the LDPC decoding information, the Markov model can obtain an increasingly accurate state transition probability. Therefore, the estimated conditional probability of each pixel approaches the spatial correlation of the original image, which accelerates the convergence of the LDPC decoding.

12.4.2 Random Field Model and Entropy

The image can be considered a source of discrete symbols in information theory. To express the mean uncertainty, the concept of entropy is introduced. This value represents the amount of information in the symbol sequence of the source output.

In the random variable sequence, the joint entropy of the first N random variables is averaged, that is,

$$H_N(X) = \frac{1}{N} H(X_1 X_2 \ldots X_N) \tag{12.46}$$

$H_N(X)$ is termed the average symbol entropy. If N is infinite, the upper limit exists, and $\lim_{N\to\infty} H_N(X)$ is defined as the entropy rate or limit entropy:

$$H_\infty = \lim_{N\to\infty} H_N(X) \tag{12.47}$$

When the source is in state E_i, the average amount carried by a source symbol is determined. Specifically, the conditional entropy for issuing a symbol in the state is

$$H(X|S=E_i) = -\sum_{k=1}^{q} P(X_k|E_i)\log P(X_k|E_i) \tag{12.48}$$

The entropy of the m-order Markov source is defined as in Eq.(12.47):

$$H_\infty = H(X_{m+1}|X_1X_2...X_m) \tag{12.49}$$

This expression indicates that the entropy limit of the m-order Markov source is equal to the m-order entropy.

Assuming that the m-order Markov source symbol sets have q symbols, the source has q^m different states. The entropy of the Markov source is defined as the limit probability of each state. The entropy of the Markov source is

$$H_\infty = \sum_{i=1}^{J} Q(E_i)H(X|E_i) = \sum_{i=1}^{J}\sum_{k=1}^{q} Q(E_i)P(X_k|E_i)\log P(X_k|E_i) \tag{12.50}$$

12.4.3 Image Random Field Compression Theory Based on the K-L Transform

The K-L transform coding can be enhanced using the Markov stochastic process. The distribution of most of the image gray values can be simulated using the first-order Markov process. If the correlation coefficient is known, the covariance matrix of the original data can be determined, and its eigenvalues and eigenvectors can be calculated. In this case, the optimal orthogonal transform of the mean square error of the covariance matrix is the K-L transform. The corresponding eigenvector is the base vector of the K-L transform, and the corresponding transform coding can be performed [3].

The K-L transform is based on a statistical feature-based transformation, which is proposed for a wide range of random images. This tool pertains to the most widely used linear transformation in remote sensing image enhancement and information extraction. The K-L transform can be expressed as

$$\boldsymbol{Y} = \boldsymbol{A} - \boldsymbol{X}(\boldsymbol{m}_x) \tag{12.51}$$

The transformation can be understood as follows: The transformed image vector $\boldsymbol{Y}$ is obtained by multiplying the transformed image vector $\boldsymbol{X}-\boldsymbol{m}_x$ with the transformation matrix $\boldsymbol{A}$. The composition of the image vector $\boldsymbol{Y}$. $\boldsymbol{Y}$ is the same as that of vector $\boldsymbol{X}$. After the K-L transformation is applied to the image, the resulting restored image is statistically the best approximation of the original image. The advantage of the K-L transform is that it has a high correlation and is the best transformation in terms of the mean square error (MSE). This tool is valuable in data compression technology.

However, it is not easy to solve eigenvalues, especially in the case of high dimensions. Therefore, we can use a DCT transform with a correlation coefficient of 1, with the K-L transform having a guiding role.

12.5 Summary

12.5.1 Framework of This Chapter

Figure 12.17 summarizes the structure of this chapter, and the main formulas are listed in Table 12.4.

Information theory is the theoretical basis of image compression. Information theory is valuable in image compression to quantify the image information and prove that the average code length of the codewords assigned to each source symbol can be arbitrarily close to the entropy of the source under the premise that distortion does not occur by a reasonable and effective coding algorithm without generating distortion source entropy. In the case of distortion, the rate distortion function represents the minimum average distortion that may be achieved at a certain information rate. Under this theoretical framework, a variety of image compression methods have been developed [4].

12.5.2 Relationship Between Distortion-Free Coding and Distortion Coding

The encoder is treated as a channel, and the source coding model is shown in Fig. 12.18. The distortion-free code corresponds to the lossless determination channel, and the distortion source coding corresponds to the noisy channel. In the absence of a source code, the input symbol of the channel exhibits a one-to-one correspondence with the output symbol.

In the general case of communication, the amount of information acquired by the recipient is equal to the amount of elimination (reduction) of uncertainty before and after communication.

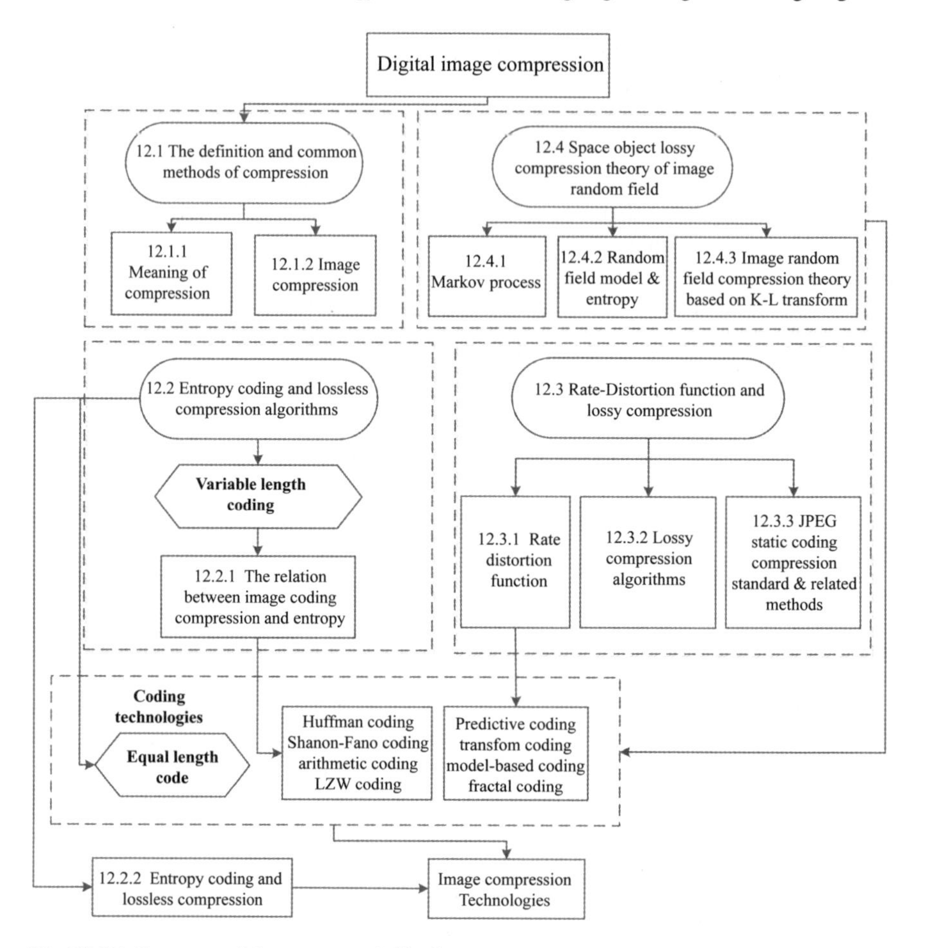

Fig. 12.17 Summary of the contents of this chapter

Table 12.4 Summary of mathematical formulas in this chapter

formula (1)	$I(S_i) = \log_2 \frac{1}{p_i}$	12.2.1
formula (2)	$H(S) = \sum_i p_i \log_2 \frac{1}{p_i}$	12.2.1
formula (3)	$L(S) = \sum_i p_i \cdot l_i$	12.2.2
formula (4)	$R = E\{L_w(a_k)\} - H$	12.2.1
formula (5)	$R(D) = \min_{P(v_j/u_i)\in B_D}\{I(U;V)\}$	12.3.1
formula (6)	$p\left(x_{h,v}\vert x_{h'v'}, \left(h', v'\right) \neq (h, v)\right) = p\left(x_{h,v}\vert x_{h-1,v}, x_{h,v-1}, x_{h+1,v}, x_{h,v+1}\right)$	12.4.1

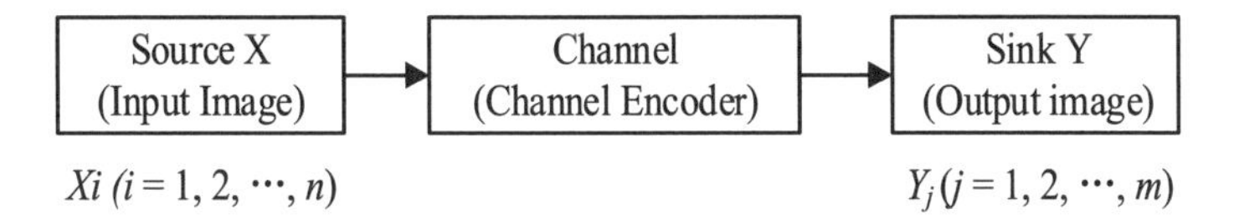

Fig. 12.18 Source encoder diagram

The mutual information $I(x_i; y_j)$ indicates that the recipient has received the information amount of x_i from y_j after receiving y_j. $I(x_i; y_j)$ reflects the elimination of the uncertainty of x_i after y_j has been received. This value is equal to the difference when y_j is received before x_i (existence of uncertainty) and y_j is received after x_i:

$$I(x_i; y_j) = \log \frac{p(x_i/y_j)}{p(x_i)} = \log \frac{p(x_i y_j)}{p(x_i)p(y_j)} = \log \frac{p(y_j/x_i)}{p(y_j)} \tag{12.52}$$

The mutual information $I(x_i; y_j)$ represents the information specified by an event y_j regarding event x_i, which changes with changes in y_j and x_i.

To represent the amount of information regarding another random variable X specified by a random variable Y, the statistical mean $I(X; Y)$ of the mutual information $I(x_i; y_j)$ is defined as the average mutual information between random variables X and Y in the joint probability space of X–Y:

$$\begin{aligned} I(X; Y) &= \sum_{i=1}^{n}\sum_{j=1}^{m} p(x_i y_j) I(x_i; y_j) \\ &= \sum_{i=1}^{n}\sum_{j=1}^{m} p(x_i y_j) \log \frac{p(x_i/y_j)}{p(x_i)} \\ &= \sum_{i=1}^{n}\sum_{j=1}^{m} p(x_i y_j) \log \frac{1}{p(x_i)} - \sum_{i=1}^{n}\sum_{j=1}^{m} p(x_i y_j) \log \frac{1}{p(x_i/y_j)} \\ &= H(X) - H(X/Y) \end{aligned} \tag{12.53}$$

that is

$$I(X; Y) = H(X) - H(X/Y) \tag{12.54}$$

Since $H(X)$ is the entropy or uncertainty of X, and $H(X/Y)$ is the uncertainty of X when Y is known, the information of X obtained when Y is known is the average mutual information $I(X; Y)$. The average mutual information $I(X; Y)$ may also be regarded as the average amount of information transmitted over the perturbed discrete channel. In the lossy coding rate distortion function, the minimum value of the average mutual information $I(X;Y)$ is identified under the condition that the fidelity criterion is satisfied.

For lossless coding, Y is a one-to-one function determined by X. When Y is known, the conditional probability of X is zero. Since X and Y exhibit a one-to-one relationship, if X and Y satisfy the determination function, the conditional probability must be 1. If X and Y do not satisfy the deterministic function, the conditional probability ratio is zero. Specifically, $I(X;Y) = H(X)$. Thus, the information obtained is the X uncertainty or entropy, and the framework can be regarded as a lossless channel encoder [5].

References

1. Shannon CE. A mathematical theory of communication. Bell Syst Tech J. 1948;27(379–423):623–56.
2. Welch TA. High speed data compression and decompression apparatus and method, united states patent No. 4, 558, 302, Dec. 19, 1985 (assigned to Unisys Corporation, PO Box 500, Blue, PA 19424–0001).
3. Zhang L, Guodong FJ. Image coding and wavelet compression techniques: principles, algorithms and standards. Beijing: Tsinghua University Press; 2004. (In Chinese).
4. Qiang L. Image compression coding method. Xi'an: Xi'an University of Electronic Science and Technology Press; 2013. (In Chinese)
5. Jianwei W, et al. Practical hyperspectral remote sensing image compression. Beijing: National Defense Industry Press; 2012. (In Chinese)

Chapter 13
Technology III of Remote Sensing Digital Image Processing: Pattern Recognition (Image Segmentation)

Chapter Guidance The recognition of information in images is the most important purpose of remote sensing digital image processing. To extract useful information in the image, it is necessary to subject the image to pattern recognition processing. Pattern recognition involves three steps: segmentation, extraction and classification, that is, image segmentation, feature extraction and classification, respectively. This section describes image segmentation, which is a key part of remote sensing applications. Feature extraction and classification are described in the subsequent chapter.

The central task of pattern recognition is to determine the essential attributes of a "class", that is, to divide the pattern to be identified into the appropriate schema class based on certain measures and observations.

For a given digital image involving multiple object conditions, the pattern recognition process consists of three stages, as shown in Fig. 13.1.

The first stage pertains to image segmentation or object separation. The second stage is pertains to the feature extraction stage in which the n-dimensional feature vector $X_i = (x_{i1}, x_{i2}, \ldots, x_{id})^T (i = 1, 2, \ldots, N)$ is extracted. The third stage refers to classification.

This chapter focuses on the first phase of pattern recognition: image segmentation or object separation. The objects are detected, and their images are separated from the remaining scene. The contents of this chapter are as follows: Sect. 13.1 describes the logical structure of image segmentation; Sect. 13.2 describes the identification of a suitable threshold image pixel assigned to the appropriate category; Sect. 13.3 describes the detection of an edge using a boundary and segmented nature; Sect. 13.4 uses the object heterogeneity to split or combine the iterative division; Sect. 13.5 pertains to the binary image optimization processing; Sect. 13.6 describes the data structures of the images.

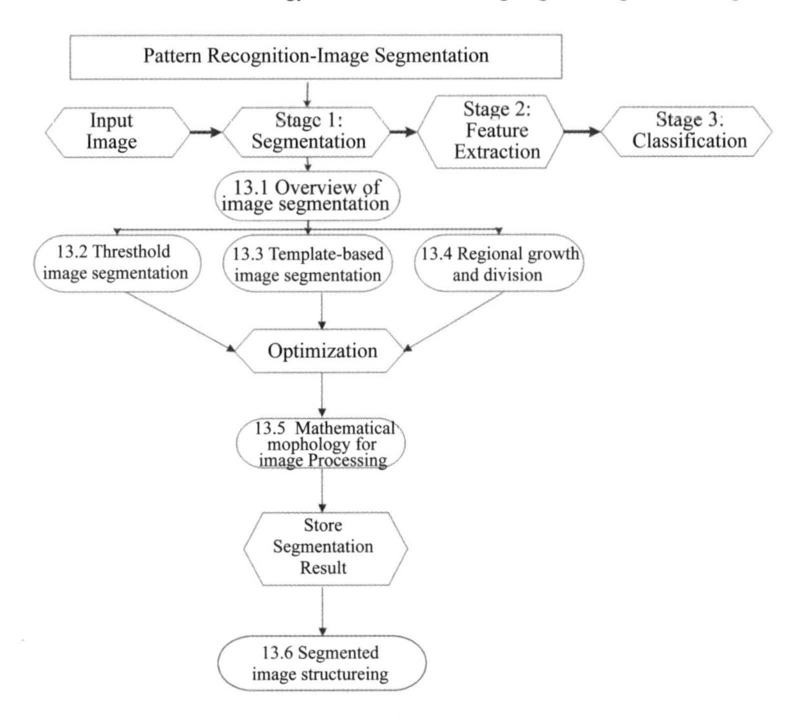

Fig. 13.1 Pattern recognition (Stage 1) and logical framework of this chapter

13.1 Overview of Image Segmentation

Figure 13.2 shows the relationship of the sections in this chapter pertaining to image segmentation. Image segmentation refers to the division of the image into specific regions with unique properties and processing of the object of interest. Image processing is a key step for image analysis. Conventional image segmentation methods are divided into the following categories: threshold-based segmentation methods, region-based methods, edge-based segmentation methods and other segmentation methods based on particular theories.

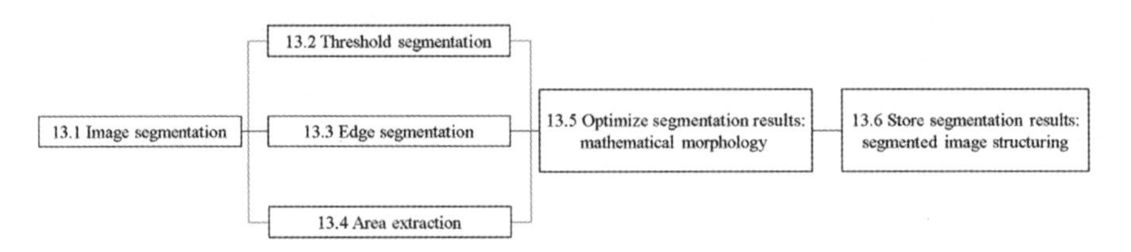

Fig. 13.2 Relationship between the sections of this chapter

Since 1998, researchers have continued to improve the existing image segmentation and proposed new methods for image segmentation and mathematical morphologies. The objects extracted after image segmentation can be used in the fields of image semantic recognition and image search. This chapter discusses the segmentation results in terms of structured storage.

Image segmentation is the process of dividing a digital image into nonintersecting and nonoverlapping regions [1]. The basic task of image segmentation is to separate different elements based on similarity parameters or differences between adjacent regions. R represents the entire image area. The image with region R that is nonempty is divided into n regions, R_1 , R_2 ,..., R_n, which satisfy the following conditions:

(1) $\bigcup_{i=1}^{n} R_i = R$;
(2) R_i is a communicated region; $i = 1, 2, 3,\ldots,n$;
(3) $R_i \cap R_j = \emptyset$ for all i and j, and $i \neq j$;
(4) $P(R_i) =$ TRUE for $i = 1, 2, 3,\ldots,n$;
(5) $P(R_i \cup R_j) =$ FALSE for $i \neq j$;

$\emptyset$ donates an empty set. $P(R_i)$ indicates the uniformity of test criteria in the set point of R_i.

Condition (1) indicates that the division must be thorough. Condition (2) indicates that points in the same region must communicate with one another. Condition (3) does not hold for mutually overlapping regions. Condition (4) must be satisfied for the same region of the segmented properties of pixels; in terms of the test criteria for P, regions R_i and R_j are different.

13.2 Threshold Image Segmentation

The threshold method (gray level threshold technique) is a regional segmentation technique that is particularly useful for the segmentation of objects that exhibit a strong contrast against the background. This method is simple, practical and not computationally intensive, and closed boundaries can be used to define the nonoverlapping areas.

13.2.1 Basic Idea

The basic idea is to use a threshold to select one (or more) optimum threshold gray gradation characteristics of the image, and the image gradation value is compared with a threshold value for each pixel. The pixel corresponding to the last comparison result is assigned to the appropriate category. If Image X is an $M \times N$-dimensional image with the pixel gray value at (i, j) on the image being $f(i, j)$, t is a gray-level threshold image. The threshold value t for the target and background is defined as follows:

Target section: $O = \{f(i, j) < t|(i, j) \in X\}$

Background section: $B = \{f(i, j) > t|(i, j) \in X\}$

The threshold is divided into the global threshold (the entire image uses a threshold value) and local threshold (different regions of the image use a different threshold).

1. Global Threshold Method

If a background grayscale value is considered constant throughout the image, and all objects and backgrounds have similar grayscale values, a fixed global threshold can be used to achieve satisfactory results.

2. Local Threshold Method

In many cases, the gray value of the background is not constant, and the contrast of the object and background changes in the image. In this case, it is appropriate to set the grayscale threshold to a function that gradually changes with the position in the image.

13.2.2 Analysis of Point Objects

In many important cases, it is necessary to find objects that are generally round. The following methods are limited to circular objects. The optimal threshold selection that cannot be obtained in other cases can be obtained under conditions that limit the circular object.

1. Definition

Assume that a point in the image has the largest gray value. In the case of polar coordinates, the image can be represented as contour shown in Fig. 13.3, and the corresponding value is

$$B_p(r_1, \theta) \geq B_p(r_2, \theta), if \ \ r_2 > r_1 \tag{13.1}$$

If the formula does not have an equality, a monotonous point is considered. Along any direction of the straight line, the gray value is strictly decreasing. For monotonic points, a flat peak is not allowed to exist, and the peak is unique.

A notable special case is that the outer edges of all monotonous points are circles centered on (x_0, y_0). We term this special case the concentric point (concentric circular spot, CCS). This special case approximates the noise-free astronomical images acquired using a telescope, certain cells in the microscopic image, and other important types of images.

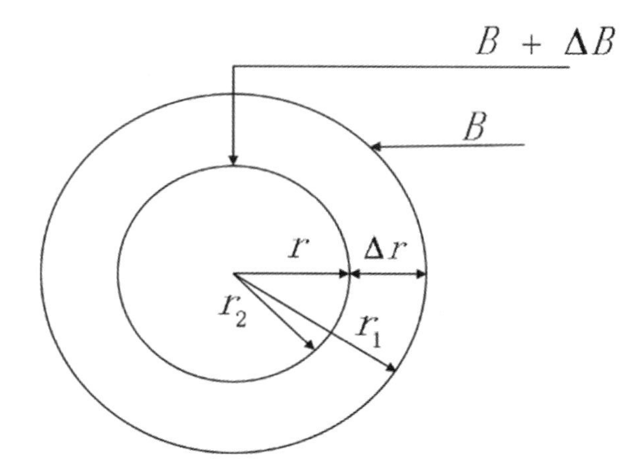

Fig. 13.3 Thresholding for the concentric circle point

The function of the CCS is independent of the value of θ, known as the contour point (spot profile function). This curve is useful for selecting thresholds. For example, we can determine the turning point and select the grayscale threshold such that the boundary is defined as a point with the largest slope value, which is similar to the boundary defined by the human eye when observing a smooth edge image. This value is smooth and exhibits considerable noise stability.

If we grade the threshold value T to a binary monotonic point, we can define an object having a certain area and perimeter. When the change in the gray threshold T is within the range, we define the threshold function and Zhou Chang area functions. For any point object, the two functions are unique. The monotonic points are contiguous, and any of the conditions can fully define a CCS. By definition, if two points have the same perimeter function, the two points are p-equivalent, and if they have the same histogram, they are H-equivalent The H-equivalent points have the same function of the area threshold value.

2. Histograms and Contours

The CCS framework assumes that an image is a profile function. We identify an expression that describes the histogram through a contour function. Suppose we use the grayscale thresholds B and $B + \Delta B$ to threshold $B(x + y)$. These expressions indicate two radii and circular contours, as shown in Fig. 13.3. The contours of the two circles are

$$\Delta A = \pi r^2 - \pi (r + \Delta r)^2 \approx -2\pi r \Delta r \tag{13.2}$$

In this case, Δr is small, and Δr^2 is negligible. Therefore,

$$\frac{\Delta A}{\Delta r} \approx -2\pi r \tag{13.3}$$

In this case, the histogram of the image is defined as

$$H_B(D) \approx \lim_{\Delta D \to 0} \frac{\Delta A}{\Delta D} \tag{13.4}$$

In addition, we can use Δr to divide the numerator and denominator, and formula (13.3) can be discretized to obtain

$$H_B(D) \approx \lim_{\Delta D \to 0} \frac{\Delta A}{\Delta D} = \frac{-2\pi r}{\mathrm{d}/\,\mathrm{d}r\, B_p(r)} \tag{13.5}$$

Note that the equality in 13.4 is based on Δr and ΔD. The derivative of the denominator is set to zero to obtain the profile function. Since $B(x, y)$ is a monotonic point image, $B(x + y)$ is a monotonic function of r. The inverse function is

$$r(D) = B_p^{-1}(D) \tag{13.6}$$

In this way, we can ensure that the histogram a function of the desired grayscale value. Because the profile function $B_p(r)$ monotonically decreases with r, the molecule can be removed in the negative direction, and thus, the histogram achieves the target value.

3. The Contour Function of the Area Function Holds at All Points

We determine a histogram to describe the expression profile of the CCS. The radius of a circular object obtained by thresholding a CCS with a gray threshold T is

$$R(T) = \left[\frac{1}{\pi} A(T)\right]^{\frac{1}{2}} = \left[\frac{1}{\pi} \int_T^{\infty} H_B(D)\mathrm{d}D\right]^{\frac{1}{2}} \tag{13.7}$$

For a monotonic point, the gray level maximum and minimum of histogram $H_B(D)$ are nonzero, which means that the functions $A(T)$ and $R(T)$ increase monotonically. The inverse of this equation exists, which is the contour function.

4. The contour function of the perimeter function holds at all points

Thresholding the CCS with grayscale T generates a circular object with a radius

$$R(T) = \frac{1}{2\pi} P(T) \tag{13.8}$$

where P(T) is the circumference function. According to the abovementioned method, the contour function is the inverse of this equation.

13.2.3 Average Boundary Gradient

For round point-like objects, a gradation value cannot be used for *H*- or *P*-equivalent CCS entities. For an arbitrarily shaped object, the average gradient around the edge can be used as a threshold grayscale function for defining the boundary. If a circular nonmonotonic point of the object in the grayscale is vertical with respect to a boundary, it is located in the direction of the gradient vector of point a. The magnitude of the gradient vector on the outer boundary is

$$|\nabla B| = \lim_{\Delta D \to 0} \frac{\Delta D}{\Delta r} \tag{13.9}$$

Since only the average gradient around the boundary is of interest, the average only along the outer boundary can be considered. If Δr is extremely small compared with the circumference, the area of the two boundaries is approximately

$$\Delta A = p(D)\overline{\Delta r} \tag{13.10}$$

where $\overline{\Delta r}$ is the average vertical distance from the outer boundary to the inner boundary, and $p(D)$ is the perimeter function. To obtain the average gradient around the boundary, the value can be substituted into Eq. 13.5 to yield

$$\overline{|\nabla B|} = \lim_{\Delta D \to 0} \frac{\Delta D}{\Delta A} p(D) = \frac{p(D)}{H_B(D)} \tag{13.11}$$

This expression indicates that the average boundary gradient is equal to the ratio of the circumference function and histogram.

13.2.4 Best Threshold Selection Method

Unless the object in the image has a steep edge, the value of the grayscale threshold considerably influences the positioning of the extracted object boundary and overall size. Therefore, we need an optimal or at least a heterogeneous method to determine the threshold.

1. Threshold Selection Based on Gray Histogram

(a) General histogram technique

It is generally assumed that each peak of the grayscale histogram represents a target region, and the valley is the transition point from one target region to another. The histogram threshold segmentation is aimed at segmenting the target region represented by these peaks (Fig. 13.4).

The object area using the gray threshold T is defined as

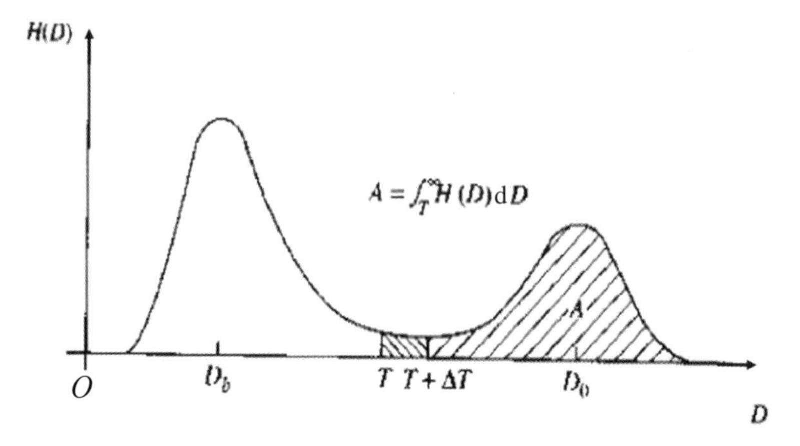

Fig. 13.4 Bimodal histogram

$$A = \int_{T}^{\infty} H(D)\mathrm{d}D \tag{13.12}$$

If the threshold corresponds to the valley of the histogram, the threshold increases from T to $T + \Delta T$, with only a slight decrease in the area. Therefore, setting the threshold at the valley of the histogram minimizes the effect of small errors in the threshold selection on area measurements.

(b) Watershed algorithm

The watershed algorithm is associated with adaptive binarization. Figure 13.5 illustrates the working mechanism of this method. Assume that the gray value of the object on the way is lower and the background gray value is higher. The figure shows the grayscale distribution along a scanning line that passes through two extremely close objects.

The image is originally binarized on a low grayscale value. The grayscale value divides the image into the correct number of objects, but the boundaries are biased toward the inside of the object. The threshold is incremented. When the boundaries are in contact with one another, the objects are not merged; this value refers to the threshold.

2. Threshold Selection Based on Image Difference

An ideal segmentation method increases the gap between the target and background, corresponding to a higher contrast between the target and background. Therefore, the thresholds can be selected based on the gap measure of the image. The Otsu method, also known as maximum category variance, is a commonly used approach. We assume that the threshold value t for image L has gradations X divided into two categories:$C_0 \in [0, t]$, $C_1 \in [t + 1, L - 1]$. The optimal threshold value t^* corresponds to classes (C_0, C_1) for which the variance σ_B^2 is maximized:

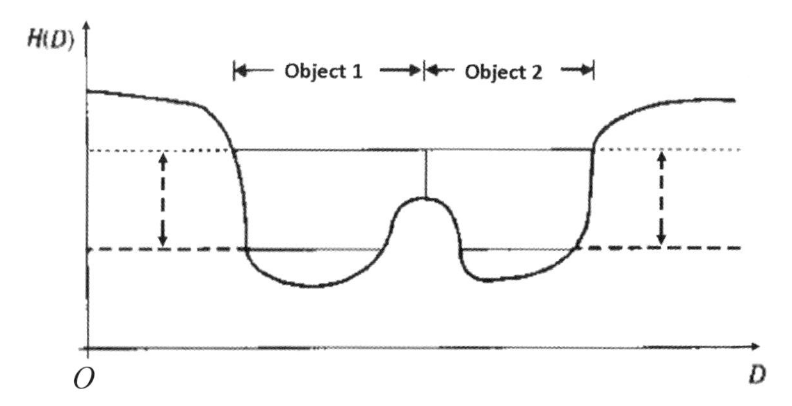

Fig. 13.5 Watershed algorithm

$$t^* = \arg \max_{0 \le t \le L-1} \left(\sigma_B^2\right) \tag{13.13}$$

The Otsu method determines the threshold to maximize the difference between the segmented target and background. The optimal threshold value can be obtained in this following manner: the gap between the target and divided original image is $d_{\mathrm{OA}}(t)$, and the difference between the original image and background is $d_{\mathrm{BA}}(t)$. Therefore, the maximum sum of the two thresholds t^* is the optimal threshold:

$$t^* = \arg \max_{0 \le t \le L-1} [d_{OA}(t) + d_{BA}(t)] \tag{13.14}$$

Thus, the product of the maximum threshold t^* is the optimal threshold:

$$t^* = \arg \max_{0 \le t \le L-1} [d_{OA}(t) \cdot d_{BA}(t)] \tag{13.15}$$

3. Threshold Selection Based on Image Fuzzy Measures and Its Improved Mlgorithm

The fuzzy threshold method selects the segmentation threshold by calculating several fuzzy measures of the image. According to the fuzzy set theory, one $M \times N$ two-dimensional image L with X gradations can be expressed as

$$X = \bigcup_{i=1}^{M} \bigcup_{j=1}^{N} \frac{P_{ij}}{X_{ij}} \tag{13.16}$$

where $\frac{P_{ij}}{X_{ij}}\left(0 \le P_{ij} \le 1\right)$ represents a pixel (i, j) with property P. Property P converts an image from the spatial domain to a mapping function of a fuzzy domain (i.e., fuzzy membership function).

The fuzzy threshold method commonly uses standard G as a function of the mapping function, defined as

$$p_{ij} = G_{x_{ij}} \begin{cases} 0, & 0 \le x_{ij} \le t - \Delta t \\ 2\left[\left(x_{ij} - t + \Delta t\right)/2\Delta t\right]^2, & t - \Delta t \le x_{ij} \le t \\ 1 - 2\left[\left(x_{ij} - t + \Delta t\right)/2\Delta t\right]^2, & t \le x_{ij} \le t + \Delta t \\ 1, & t + \Delta t \le x_{ij} \le L - 1 \end{cases} . \quad (13.17)$$

After determining the mapping function and performing the image blur matrix mapping, the fuzzy blur measure and spatial image blur are determined by calculating the ratio of X or fuzzy entropy. The fuzzy entropy and rate are defined as

(a) Fuzzy rate

$$\Upsilon(x) = \frac{2}{MN} \sum_{i=0}^{M} \sum_{j=1}^{N} \min\left\{p_{ij}, 1 - p_{ij}\right\} \quad (13.18)$$

(b) Fuzzy entropy

$$E(x) = \frac{1}{MN \ln 2} \sum_{i=1}^{M} \sum_{j=1}^{N} S_n\left(p_{ij}\right) \quad (13.19)$$

The Shannon function is

$$S_n\left(p_{ij}\right) = -p_{ij} \ln p_{ij} - \left(1 - p_{ij}\right) \ln\left(1 - p_{ij}\right) \quad (13.20)$$

For appropriate segmentation of the target and background image, a smaller blur or fuzzy entropy rate is desirable (calculated using the haze ratio $\gamma(\mathrm{x})$ to select the threshold value).

The mapping function is determined considering the window width $c = 2\,\Delta\,t$ and parameter t. After the window width c is set, the blur rate $\gamma(x)$ is only related to parameter t. When parameter t changes, $\gamma(x)$ changes. We set $\gamma(x)$such that t_0 as the optimal threshold for the segment image is minimized.

$$t_0 = \arg \min_{0 \le t \le L-1} \Upsilon_t(X) \quad (13.21)$$

13.3 Template-Based Image Segmentation

In the abovementioned approach, threshold segmentation was realized by dividing the image into internal and external point sets. In contrast, the boundary method uses the nature of the boundary to detect the edge and connection to achieve segmentation.

13.3.1 Edge Detection

The edge is the most significant part for the local brightness change of an image. The edge mainly exists between two targets, a target and the background, and two regions (with different colors).

Based on the nature of the edge, as shown in Fig. 13.6, the edge portion of the gradation becomes steep at the edge points. The first derivative is the maximum or minimum value of the neighborhood, and the second derivative value is zero. Thus, the edge detection method can be divided into two categories: based on the search for a class (first derivative) and based on zero passes through a class (second derivative) [2].

1. Gradient

Given a scalar function $f(x, y)$ and a unit vector in the x-axis direction, i, the gradient vector function for the y-axis direction of the unit vector in the j coordinate system is

$$\nabla f(x, y) = \boldsymbol{i}\frac{\partial f(x, y)}{\partial x} + \boldsymbol{j}\frac{\partial f(x, y)}{\partial y} \left(\boldsymbol{G}(x, y) = [G_X G_Y]^{\mathrm{T}} = \begin{bmatrix} \frac{\partial f}{\partial x} \frac{\partial f}{\partial y} \end{bmatrix}^{\mathrm{T}} \right) \tag{13.22}$$

∇ is the vector gradient operator. Vector $\nabla f(x, y)$ points to the maximum slope of the direction of the amplitude (length) equal to the size of the slope, and the scalar function can be defined by the following gradient:

$$|\nabla f(x, y)| = \sqrt{\left(\frac{\partial f}{\partial x}\right)^2 + \left(\frac{\partial f}{\partial y}\right)^2} \tag{13.23}$$

Since the calculation of the square root is time-consuming, formula 13.12 can be approximated as follows:

$$|\nabla f(x, y)| \approx \max[|f(x, y) - f(x+1, y)|, |f(x, y) - f(x, y+1)|] \tag{13.24}$$

Usually, a 2×2 matrix is used to determine the first-order differential template for the partial derivatives of x and y:

$$\boldsymbol{G}_x = \begin{bmatrix} 1 & -1 \\ 1 & -1 \end{bmatrix}, \boldsymbol{G}_y = \begin{bmatrix} -1 & -1 \\ 1 & 1 \end{bmatrix} \tag{13.25}$$

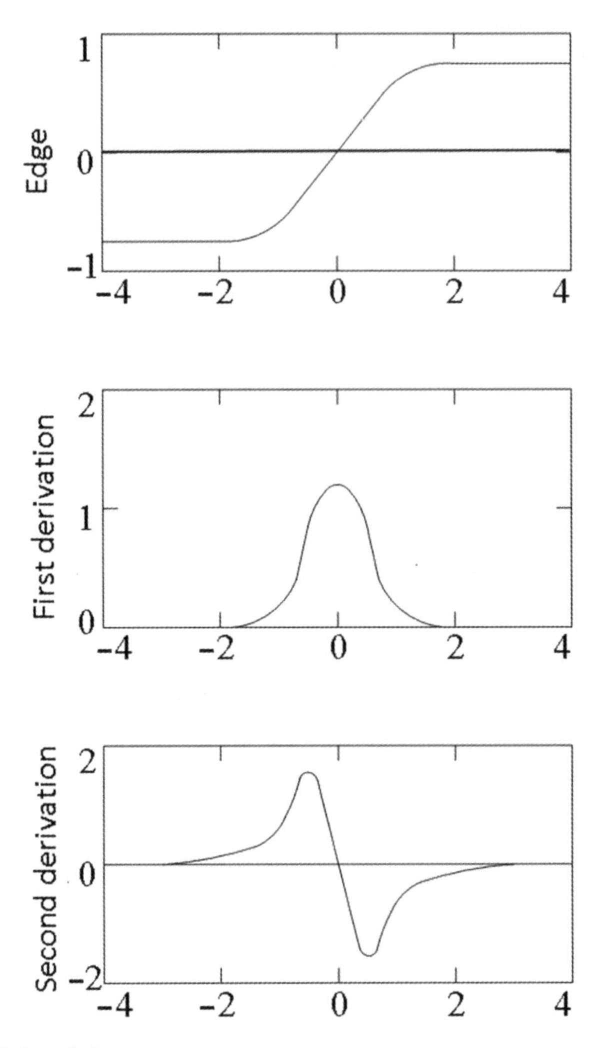

Fig. 13.6 Gradation of edges

The gradient is extremely large at the edge and small in the uniform area.

2. Boundary Tracking

For boundary tracking, the gradient amplitude image is processed from the highest point of the gray level (i.e., the point at which the gradient is the highest in the original image) as the starting point. If there exist several points with the highest gray level, we can choose one value.

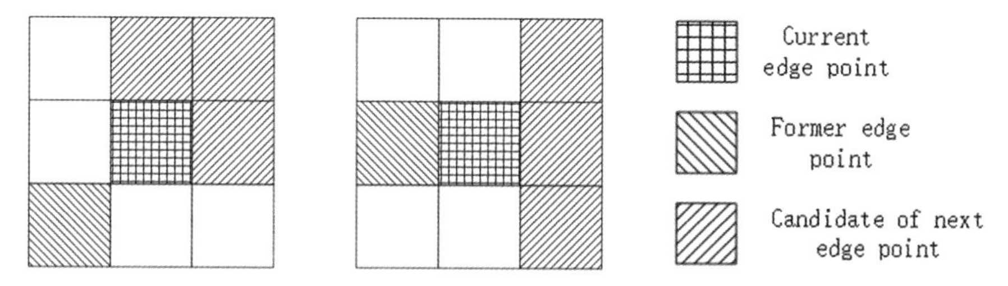

Fig. 13.7 Boundary tracking

In this case, the search starting point to the boundary as the center of the 3 × 3 neighborhood (Fig. 13.7) finds the maximum gray level having a field point as the second boundary point. If there are two domain points with the same maximum gray level, any one point is selected. Continuous iterations are performed until the start point is found.

3. Gradient Image Binarization

If a gradient image is binarized using a moderate threshold, the hit point inside the background object is below a threshold, above which the majority of the edge points lie (Fig. 13.8). The Kirsch segmentation method exploits this phenomenon. This aspects pertains to the application of the watershed algorithm in gradient images.

4. Edge Detection Operator

The edge detection operator checks the neighborhood of each pixel to quantify the rate of gray change, usually including the direction of determination. Several methods

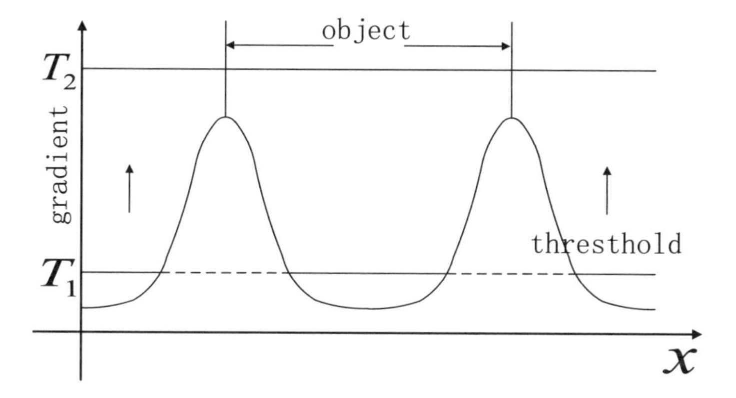

Fig. 13.8 Kirsch segmentation algorithm

can be used, most of which are based on the direction derivative mask for convolution [3].

(a) Roberts edge operator

The Roberts edge detection operator is a difference operator utilizing a local operator to find the edge:

$$g(x, y) = \left\{ \left[\sqrt{f(x, y)} - \sqrt{f(x+1, y+1)} \right]^2 + \left[\sqrt{f(x+1, y)} - \sqrt{f(x, y+1)} \right]^2 \right\}^{1/2} \tag{13.26}$$

where $f(x, y)$ is an input image pixel with integer coordinates. To facilitate the calculation, we set

$$g(x, y) \approx |f(x, y) - f(x+1, y+1)| + |f(x+1, y) - f(x, y+1)| \tag{13.27}$$

where $G(x, y) = |\boldsymbol{G}_x| + |\boldsymbol{G}_y|$, and $\boldsymbol{G}_x$ and $\boldsymbol{G}_y$ are calculated using the following template:

$$\boldsymbol{G}_x = \begin{bmatrix} 0 & -1 \\ 1 & 0 \end{bmatrix}, \boldsymbol{G}_y = \begin{bmatrix} 1 & 0 \\ 0 & -1 \end{bmatrix} \tag{13.28}$$

(b) Sobel edge operator

The Sobel operator is used to calculate the gradient considering the gradient of a pixel value adjacent to the region in accordance with the computed set of distances, which is a constant weight. For example, around the pixels, the vertically adjacent pixel weight value is 2, and the oblique direction is 1:

$$G(x, y) = \sqrt{\mathbf{G}_x^2 + \mathbf{G}_y^2} \tag{13.29}$$

$$\begin{aligned} g(x) &\approx |f(x+1, y+1) - f(x-1, y+1)| + 2|f(x+1, y) - f(x-1, y)| \\ &\quad + |f(x+1, y-1) - f(x-1, y-1)| \\ g(y) &\approx |f(x-1, y-1) - f(x-1, y+1)| + 2|f(x, y-1) - f(x, y+1)| \\ &\quad + |f(x+1, y-1) - f(x+1, y+1)| \end{aligned} \tag{13.30}$$

$$g(x, y) \approx \max[g(x), g(y)] \tag{13.31}$$

The Sobel operator is a 3×3 operator template, and convolution kernels form the $\boldsymbol{G}_x$ and $\boldsymbol{G}_y$ and Sobel operators, according to Formulas 13.21. The maximum value of two convolutions is the output value of this point.

$$G_x = \begin{bmatrix} -1 & 0 & 1 \\ -2 & 0 & 2 \\ -1 & 0 & 1 \end{bmatrix},\ G_y = \begin{bmatrix} -1 & -2 & -1 \\ 0 & 0 & 0 \\ 1 & 2 & 1 \end{bmatrix} \tag{13.32}$$

(c) Prewitt operator edge

The Prewitt operator and Sobel operator can be used. However, when the gradient values for the neighboring region are considered to calculate a gradient of a pixel, the weight is set differently. Prewitt operator to set the right periphery of the pixel neighborhood space 8 a weight of one.

Specifically,

$$G(x, y) = \sqrt{G_x^2 + G_y^2} \tag{13.33}$$

$$\begin{aligned} g(x) &\approx |f(x-1, y+1) - f(x+1, y+1)| + |f(x-1, y) - f(x+1, y)| \\ &\quad + |f(x-1, y-1) - f(x+1, y-1)| \\ g(y) &\approx |f(x-1, y-1) - f(x-1, y+1)| + |f(x, y-1) - f(x, y+1)| \\ &\quad + |f(x+1, y-1) - f(x+1, y+1)| \end{aligned} \tag{13.34}$$

$$g(x, y) \approx \max[g(x), g(y)] \tag{13.35}$$

For the 3 × 3 Prewitt operator template, Eq. (13.25) indicates the two convolution kernels that form G_x, G_y and the Prewitt operator. The maximum value of the two convolutions is the output value of this point.

$$G_x = \begin{bmatrix} 1 & 0 & -1 \\ 1 & 0 & -1 \\ 1 & 0 & -1 \end{bmatrix},\ G_y = \begin{bmatrix} -1 & -1 & -1 \\ 0 & 0 & 0 \\ 1 & 1 & 1 \end{bmatrix} \tag{13.36}$$

(d) Krisch edge operator

The following 8 convolution kernels form the Kirsch edge operator. Each point in the image is convolved with 8 masks, and each mask yields the maximum response to a particular edge direction. The maximum value for all 8 directions is output as the edge-magnitude image.

$$G_1 = \begin{bmatrix} 5 & 5 & 5 \\ -3 & 0 & -3 \\ -3 & -3 & -3 \end{bmatrix},\ G_2 = \begin{bmatrix} -3 & 5 & 5 \\ -3 & 0 & 5 \\ -3 & -3 & -3 \end{bmatrix},\ G_3 = \begin{bmatrix} -3 & -3 & 5 \\ -3 & 0 & 5 \\ -3 & -3 & 5 \end{bmatrix},\ G_4 = \begin{bmatrix} -3 & -3 & -3 \\ -3 & 0 & 5 \\ -3 & 5 & 5 \end{bmatrix},$$

$$G_5 = \begin{bmatrix} -3 & -3 & -3 \\ -3 & 0 & -3 \\ 5 & 5 & 5 \end{bmatrix},\ G_6 = \begin{bmatrix} -3 & -3 & -3 \\ 5 & 0 & -3 \\ 5 & 5 & -3 \end{bmatrix},\ G_7 = \begin{bmatrix} 5 & -3 & -3 \\ 5 & 0 & -3 \\ 5 & -3 & -3 \end{bmatrix},\ G_8 = \begin{bmatrix} 5 & 5 & -3 \\ 5 & 0 & -3 \\ -3 & -3 & -3 \end{bmatrix},$$

(e) Laplacian edge detection

The Laplacian operator is defined as

$$\nabla^2 f(x, y) = \frac{\partial^2 f(x, y)}{\partial x^2} + \frac{\partial^2 f(x, y)}{\partial y^2} \tag{13.37}$$

For digital image processing, the expression may be simplified as

$$\nabla^2 f(x, y) = |4f(x, y) - f(x+1, y) - f(x-1, y) - f(x, y+1) - f(x, y-1)|$$

or

$$\begin{aligned}\nabla^2 f(x, y) = |&8f(x, y) - f(x+1, y) - f(x-1, y) - f(x, y+1) - f(x, y-1) \\ &- f(x-1, y-1) - f(x-1, y+1) - f(x+1, y-1) - f(x+1, y+1)|\end{aligned} \tag{13.38}$$

If only 4 connected pixels of the target pixel (i.e., up, down, left, and right) are considered, convolution kernel $\boldsymbol{G}_x$ is formed. If the 8-connected pixel of the target pixel is considered, convolution kernel $\boldsymbol{G}_y$ is formed.

$$\boldsymbol{G}_x = \begin{bmatrix} 0 & -1 & 0 \\ -1 & 4 & -1 \\ 0 & -1 & 0 \end{bmatrix}, \boldsymbol{G}_y = \begin{bmatrix} -1 & -1 & -1 \\ -1 & 8 & -1 \\ -1 & -1 & -1 \end{bmatrix} \tag{13.39}$$

The Laplacian filtered image has an average grayscale of zero.

If a noise-free image has steep edges, the edges can be found using the Laplace operator. However, due to the presence of noise, low-pass filtering must be performed before using the Laplacian operator.

A Gaussian low-pass filter is usually used, and the associative convoluted Laplacian and Gaussian impulse response can be combined into a single Gaussian Laplacian core LOG:

$$f(x, y) = \frac{1}{2\pi\sigma^2} e^{-\frac{x^2+y^2}{2\sigma^2}} \tag{13.40}$$

If $r^2 = x^2 + y^2$,

$$\frac{\partial^2}{\partial x^2} f(x, y) = \frac{1}{2\pi\sigma^2} e^{-\frac{r^2}{2\sigma^2}} \left(-\frac{1}{\sigma^2} + \frac{x^2}{\sigma^4}\right) = -\frac{1}{\pi\sigma^4} e^{-\frac{r^2}{2\sigma^2}} \left(\frac{1}{2} - \frac{x^2}{2\sigma^2}\right) \tag{13.41}$$

Similarly,

$$\frac{\partial^2}{\partial y^2} f(x, y) = -\frac{1}{\pi\sigma^4} e^{-\frac{r^2}{2\sigma^2}} \left(\frac{1}{2} - \frac{y^2}{2\sigma^2}\right) \tag{13.42}$$

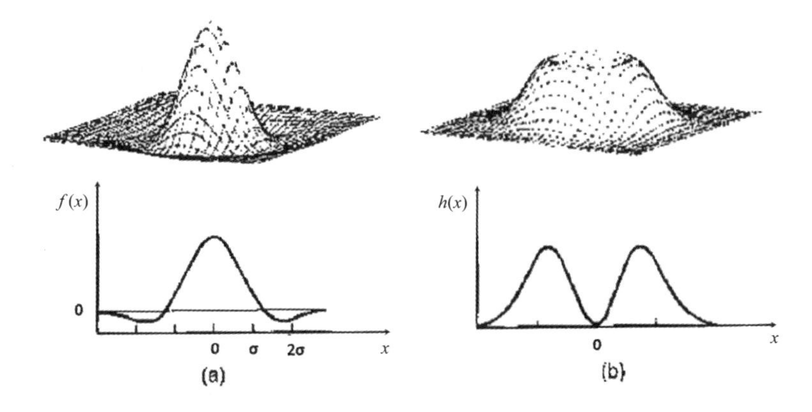

Fig. 13.9 Laplacian of Gaussian convolution: (**a**) an impulse response; (**b**) the transfer function

and

$$\nabla^2 f(x,y) = \nabla^2\left[\frac{1}{2\pi\sigma^2}\mathrm{e}^{-\frac{x^2+y^2}{2\sigma^2}}\right] = -\frac{1}{\pi\sigma^4}\left(1-\frac{x^2+y^2}{2\sigma^2}\right)\mathrm{e}^{-\frac{x^2+y^2}{2\sigma^2}} \tag{13.43}$$

which indicates that

$$-\nabla^2 f(x,y) = \frac{1}{\pi\sigma^4}\left(1-\frac{x^2+y^2}{2\sigma^2}\right)\mathrm{e}^{-\frac{x^2+y^2}{2\sigma^2}} \tag{13.44}$$

This impulse response is separable from x and y and can therefore be effectively implemented, as shown in Fig. 13.9. The parameter σ controls the width of the center peak and thus the degree of smoothness.

Assume that an image is expressed as $\boldsymbol{g}(x, y)$, and the transformed image is $\boldsymbol{h}(x, y)$. The two-dimensional discrete convolution expressions are $\boldsymbol{h}(i, j) = \boldsymbol{f}[i, j] * \boldsymbol{g}[i, j] = \sum_{k=0}^{n-1}\sum_{l=0}^{m-1} f[k, l]g[i-k, j-l]$ and $\boldsymbol{h}(x, y) = \{(x, y, \sigma)|\nabla^2[\boldsymbol{g}(x, y)] * \boldsymbol{f}(x, y) = 0\}$ based on $\nabla^2 g(x, y) = \frac{1}{2\pi}(\frac{x^2+y^2-2\sigma^2}{\sigma^6})e^{-\frac{x^2+y^2}{2\sigma^2}}$, when $\sigma = 1$:

$$\boldsymbol{h} = \begin{bmatrix} -1 & -2 & -2 & -2 & -1 \\ -2 & 0 & 4 & 0 & -2 \\ -2 & 4 & 12 & 4 & -2 \\ -2 & 0 & 4 & 0 & -2 \\ -1 & -2 & -2 & -2 & -1 \end{bmatrix} \tag{13.45}$$

5. Principle and Physical Essence of the Template

The process of convolution is equivalent to decomposing the signal into an infinite number of pulse signals and then superimposing the impulse response. If there exist only one pulse, the image shape does not change, and the value is the product of this value and the original image. The passage of the linear system can be viewed as a superposition of the result after a series of pulses have passed through the time delay sequence.

If an image is always convoluted, how do the image size and value change?

After a number of images have been convolved, the image size increases. However, because the image size is usually considerably larger than the template size, the change is insignificant. The gray value of the image pixel depends on the sum of the values in the convolution template. If the sum of the values in the template is greater than 1, the number of images gradually increases in the form of a Gaussian distribution after multiple convolutions of an image. When the values in the template are less than 1, the value of the image will. If the sum of the gray values of the image remains constant after the number of convolutions, the image size increases. If the sum of the values in the template is 1, the average gray value slightly decreases, although the change is negligible when the template radius is considerably smaller than the image size. Although the average gray value is almost constant, multiple convolutions may change the distribution of the grayscale. After multiple convolutions, the image center area increases the grayscale, and the relative edge of the region decreases the grayscale. The gray value is concentrated toward the center in the form of a Gaussian distribution.

When the sum of the values in the template is 1, the convolution of the image is convoluted several times with the template, and thus, the overall energy remains unchanged. This phenomenon occurs because the edge of the template extraction can ensure that the total energy remains unchanged under the premise of the edge of information to enhance the effect of the nonedge information suppression.

Under normal circumstances, because the number of convolutions is limited, the image range and energy do not change significantly compared with those of the original image. Therefore, under certain conditions, the template before the coefficient can be ignored. However, if the number of convolutions is large, the template of the relationship between the values cannot be ignored.

13.3.2 *Edge Connection*

An edge map usually outlines the outline of each object with edge points, but it rarely forms the closed and connected boundary needed for image segmentation. Edge point connection connects adjacent edge points to create a closed communication boundary. This process fills the gap created by noise and shadows [4].

1. Heuristic Search

To address the gap, the edge quality function calculated on an arbitrary path connecting two end points (A and B) assumes an upper edge of the image as a striped boundary.

First, we evaluate the neighborhood points of A and identify the point that is a candidate for the first step toward B.Then,we can choose the point that maximizes the edge quality function from point A to the point, and this point is used in the next iteration as the starting point. When connected to B, the edge quality function of the newly created path is compared to a threshold. If the newly created edge does not satisfy the threshold condition, it is discarded.

2. Curve Fitting

If the edge points are sparse, it may be necessary to fit these points with a piecewise linear or higher-order spline curve to form a boundary to which the object is extracted. We introduce a piecewise linear method known as iterative endpoint fitting.

Suppose there exists a set of edge points that are interspersed between two edge points A and B. It is desirable to select a subset of these points as a set of nodes on a piecewise linear path from A to B. First, we draw a straight line from A to B. Next, we calculate the vertical distance from each other edge point to the line. The farthest point becomes the other node on the desired path, and thus, this path has two branches. This process is repeated for each new branch on the path until the distance between the remaining edge point and its nearest branch is not greater than a certain value. Performing this process on all pairs of points (A, B) around the object produces a polygonal approximation of the boundary.

3. Hough Transform

The straight line $y = mx + b$ can be expressed as follows in polar coordinates:

$$\rho = x\cos(\theta) + y\sin(\theta) \tag{13.46}$$

Here, $(\rho,\ \theta)$ defines a vector from the origin of the nearest point, as shown in Fig. 13.10. This vector is perpendicular to the line.

Consider a two-dimensional space defined with a parameter ρ, θ. Any line in the x, y plane corresponds to a point in the space. Therefore, the Hough transform of any straight line in the x, y plane is a point in the ρ, θ space.

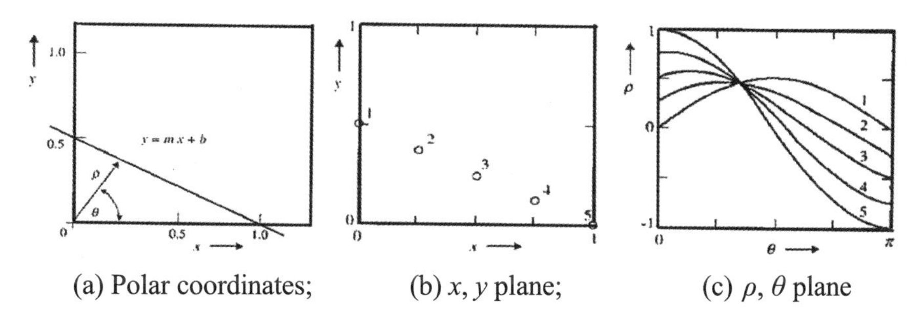

(a) Polar coordinates; (b) x, y plane; (c) ρ, θ plane

Fig. 13.10 Hough transform

Next, we consider a specific point (x_1, y_1) in the x, y plane. A straight line can pass through many points, each of which corresponds to a point in space ρ,θ. These points must satisfy Formula 13.42 when x_1 and y_1 are constant. Therefore, in a parameter space with x, the y space of all the straight lines corresponding to the locus of points is a sinusoidal curve, and any point on the x , y plane corresponds to a sine curve in the ρ, θ space.

If there exists a set of edge points on a line determined by parameters ρ_0 and θ_0, each edge point corresponds to a sine-type curve in the ρ, θ space. All these curves intersect at point (ρ_0, θ_0), as this is the argument with which they share a straight line.

To determine the straight lines formed by these points, a spatial two-dimensional histogram can be built in the ρ, θ space. Each edge point (x_i, y_i) for all points corresponds to the Hough transform ρ, θ. The space of the histogram is incrementally increased. After all the edge points have been considered, the (ρ_0, θ_0) squares exhibit a local maximum. We identify ρ, θ between the histogram for the local maximum search to obtain the parameters of the boundary line segment.

4. Boundary Curvature Analysis

The curvature at a point on a curve is defined as the change in the tangent angle along the curve. The curvature of the boundary of an object is positive value at the bump and negative in the depression.

For example, the curvature in Fig. 13.11 at the boundary corresponds to the two recesses at the negative peak. If the object must be convex, this phenomenon indicates a split error. Between a and b, a cut line separating the two objects can be introduced. Thus, the boundary curvature function can facilitate automatic detection and correction of the segmentation errors.

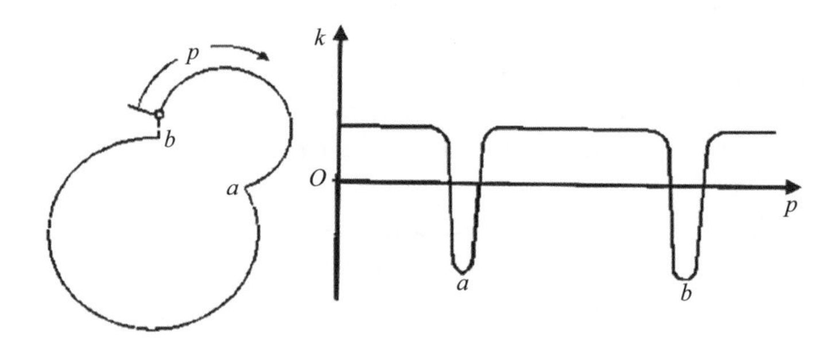

Fig. 13.11 Boundary curvature function

13.4 Regional Growth and Division

13.4.1 Splitting Method

The regional growth and split merger method are two typical serial area techniques, and the subsequent steps of the segmentation process are determined based on the results of the previous steps.

The basic idea of regional growth is to combine pixels with similar characteristics to form regions. First, a seed pixel is determined as the starting point of the growth for each area that needs to be divided, and the pixels for which the surrounding characteristics are the same or similar are integrated in the area in which the seed pixels are located according to a certain growth criterion. These new pixels continue to grow as seeds until pixels that do not meet the conditions can be included. At this time, growth is terminated, and a region is formed.

The basic idea of the split merger method is to obtain the various regions from the whole image through continuous splitting.

13.4.2 Regional Growth Merge Based on the Heterogeneity Minimum Criterion

The basic idea of regional growth merge is to combine pixels with similar properties to form the region, and the regional growth merging algorithm based on the heterogeneity minimum criterion fully considers the spectrum and shape [5].

To minimize the heterogeneity, two factors are determined before segmentation: the spectral factor and shape factor. The heterogeneity of any image object can be expressed by w_{color} (spectral factor weight), w_{shape} (shape factor weight), h_{color} (spectral heterogeneity), h_{shape} (shape heterogeneity). The shape factor consists of two parts: compactness h_{compact} and smoothness h_{smooth}, depending on the number of pixels that constitute the object, object perimeter and smallest rectangle that contains the object.

In this section, the qualitative spectral divergence metrics and shape heterogeneity metrics proposed by Baatz and Schape are used as regional heterogeneity metrics for regional consolidation. The heterogeneity *f* is calculated as [6]

$$f = w_{\text{color}} \cdot h_{\text{color}} + w_{\text{shape}} \cdot h_{\text{shape}} \tag{13.47}$$

where $w_{\text{color}} \in [0, 1]$, $w_{\text{shape}} \in [0, 1]$, $w_{\text{color}} + w_{\text{shape}} = 1$.

The spectral anisotropy value of the object is calculated using Eq. 13.37:

$$h_{color} = \sum_{c} w_c \left(n_{merge} \cdot \sigma_c^{merge} - \left(n_{obj1} \cdot \sigma_c^{obj1} + n_{obj2} \cdot \sigma_c^{obj2}\right)\right) \tag{13.48}$$

n_{merge} indicates the number of pixels contained in the merged object, n_{obj1} and n_{obj2} denote the number of pixels in premerger objects 1 and 2, respectively; σ_c represents the standard deviation for the objects in the *c-th* layer data; and obj1, obj2 refer to objects 1 and 2, respectively; w_c represents the share of the weight layer data, ranging between 0 and 1.

The shape heterogeneity value h_{shape}, which indicates the smoothness and compactness of the composition of the object, reflects the change in the shape of the object after the merger, as represented in Formula 13.38:

$$h_{shape} = w_{compact} \cdot h_{compact} + w_{smooth} \cdot h_{smooth} \tag{13.49}$$

where $h_{compact}$ and h_{smooth} represent the object smoothness and compactness, respectively, defined as in Eqs. 13.39 and 13.40; $w_{compact}$ and w_{smooth} are the weights of the compactness and smoothness, respectively, which range from 0 to 1, and $w_{compact} + w_{smooth} = 1$.

$$h_{compact} = n_{merge} \cdot \frac{l_{merge}}{\sqrt{n_{merge}}} - \left(n_{obj1} \cdot \frac{l_{obj1}}{\sqrt{n_{obj1}}} + n_{obj2} \cdot \frac{l_{obj2}}{\sqrt{n_{obj2}}}\right) \tag{13.50}$$

$$h_{smooth} = n_{merge} \cdot \frac{l_{merge}}{b_{merge}} - \left(n_{obj1} \cdot \frac{l_{obj1}}{b_{obj1}} + n_{obj2} \cdot \frac{l_{obj2}}{b_{obj2}}\right) \tag{13.51}$$

where l represents the perimeter of the object, n represents the number of pixels contained in the object, and b represents the perimeter of the smallest rectangle that contains the object. The smooth heterogeneity of the object is the ratio of the perimeter of the object to the perimeter of the smallest rectangle containing the object. The compactness heterogeneity of the object is the ratio of the perimeter of the object to the square root of the pixels contained in the object.

13.4.3 Multiscale Division

Multiscale segmentation is a type of image segmentation. According to the requirements different scales can be selected for segmentation. The steps for multiscale segmentation are shown in Fig. 13.12.

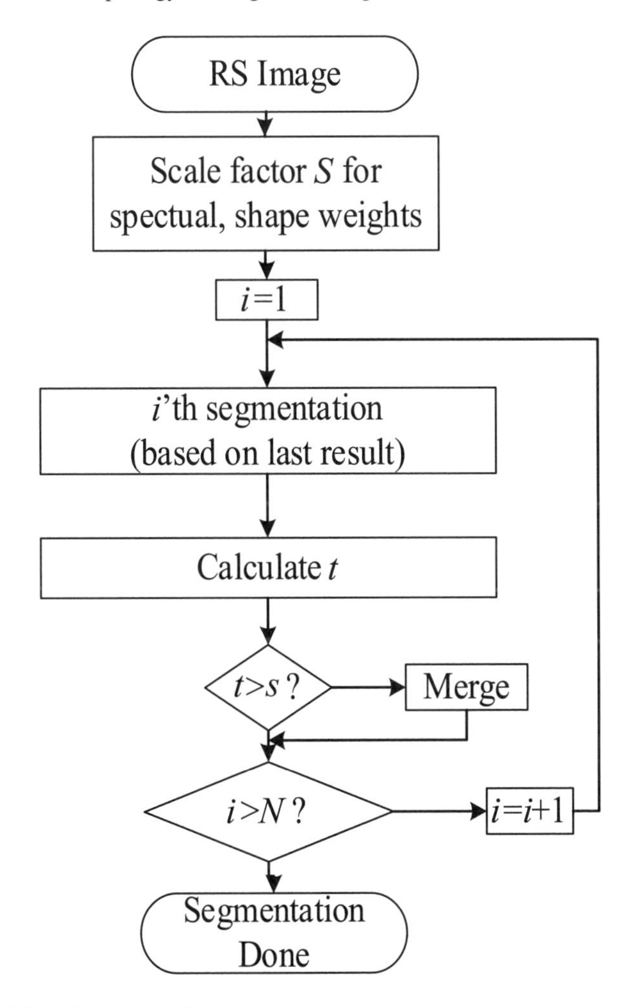

Fig. 13.12 Multiscale segmentation

13.5 Mathematical Morphology for Image Processing

13.5.1 Basic Concept

The idea of mathematical morphology can be used to process the segmented image (binary image) to enhance the segmentation effect. The basic idea of mathematical morphology is to treat the image as a collection, and the various objects to be extracted on the image constitute a subset of the image. Next, a structural element (Fig. 13.13) with a certain shape and size is used as a probe to detect the image. The structure and topological relationship of objects are identified [7].

Fig. 13.13 Common structural elements

Mathematical morphological operations involve a set of morphological algebraic operators. There exist four basic operations: expansion, erosion, and opening and closing operations.

13.5.2 Erosion and Dilation

Mathematical morphological operations are based on dilation and erosion. By definition, the boundary point is inside the object, but there is at least one pixel whose neighbor is outside the object.

1. Erosion

The set of corroded structural elements X by B is denoted as $\varepsilon_B(X)$. In other words, when B is contained in X, B is the set of the origin point x, as shown in Fig. 13.14. Specifically,

$$\varepsilon_B(X)=\{x|B_x \subseteq X\} \tag{13.52}$$

2. Dilation

X is an expanded set of structural unit B and can be expressed as $\delta_B(X)$. When B hits X, the set of points x in which the origin of B is located is shown in Fig. 13.15. Specifically,

$$\delta_B(X) = \{x|B_x \cap X \neq \emptyset\} \tag{13.53}$$

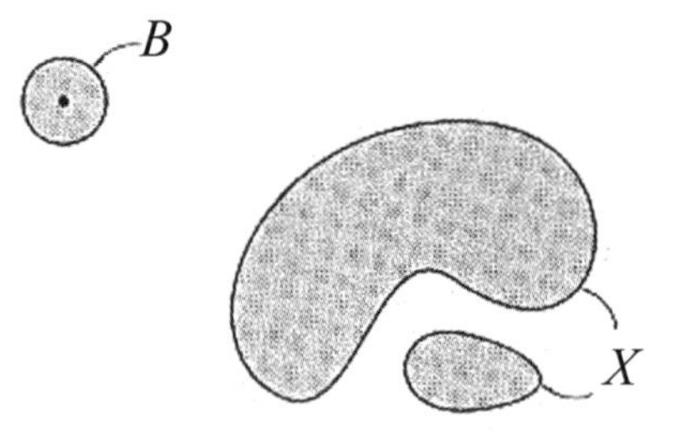

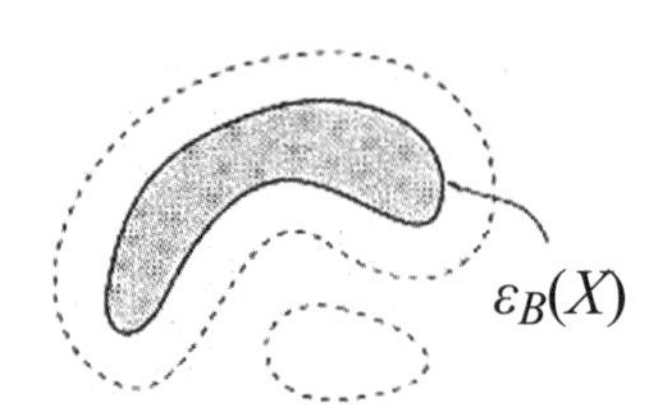

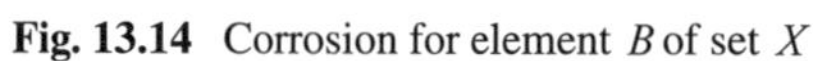

Fig. 13.14 Corrosion for element B of set X

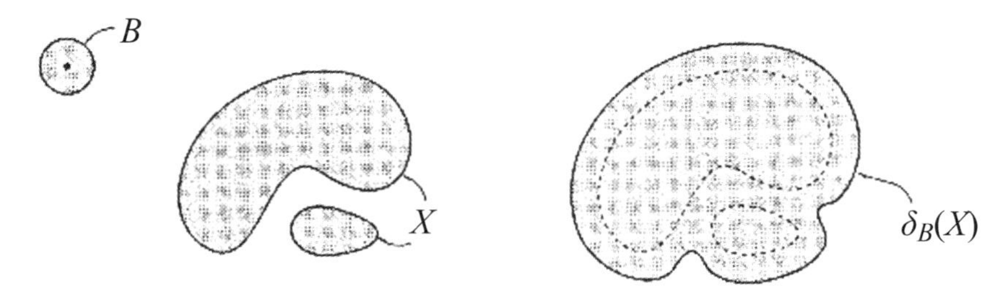

Fig. 13.15 Expansion of the structure element B for set X

Simple expansion refers to the process of incorporating all the background points that are in contact with an object into the object.

13.5.3 Open and Closed Operations

1. Open Operation

The opening operation $\Upsilon_B(f)$ is defined as the first operation of etching the image with structural element B. Subsequently, the dilation operation is performed on the etched image with symmetrical structural element B (Fig. 13.16):

$$\gamma_B(f) = \delta_{\check{B}}[\varepsilon_B(f)] \tag{13.54}$$

When structural element B is contained by set X, all of B is retained (in contrast to the corrosion case).

2. Closed Operation

In the closed operation of gray value image $\phi_B(f)$, the image is defined in terms of the dilation structural elements B, and the etching operation (Fig. 13.17) is performed on the image obtained by the expansion -B symmetrical structural elements:

$$\phi_B(f) = \varepsilon_{\check{B}}[\delta_B(f)] \tag{13.55}$$

Fig. 13.16 Structural element B for opening operation of set X

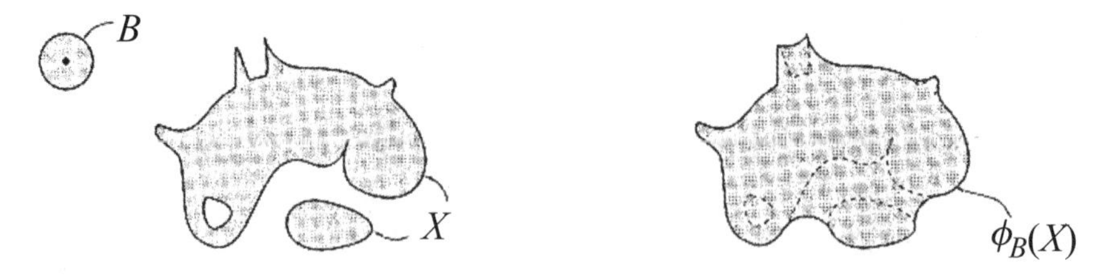

Fig. 13.17 Structuring element B for closing operation of set X

If structural element B is the complement included in X, B belongs to the complement of the closing calculation result.

Usually, when a noisy image is binarized with the threshold, the resulting boundary is extremely smooth, the object area exhibits certain misplacement holes, and the background area has several small noise objects. Continuous open and closed operations can significantly improve this situation. In certain cases, corrosion occurs after successive iterations, and expansion for the same number of times must be performed before the desired effect is achieved.

13.6 Segmented Image Structuring

In general, each object must be marked with a serial number when it is detected. This object number can be used to identify and track objects in the scene. In this section, three methods for structuring divided images (i.e., storing the divided images) are discussed.

13.6.1 Object Affiliation Diagram

Segmentation information can be stored by additionally generating an image of the same size as the original image. In this image, the object affiliation is encoded pixel by pixel. For example, all pixels belonging to the 27th object in the image will have the 27th grayscale value in the membership graph [8].

13.6.2 Boundary Chain Code

A more compact form of storing image segmentation information is the border chain code. Only the boundaries are stored, and no internal points exist.

The chain code begins with the (x, y) coordinate of a starting point arbitrarily chosen on the object boundary. This starting point has 8 adjoining points, at least one

3	2	1
4		0
5	6	7

Fig. 13.18 Boundary direction code

of which is the boundary point. The boundary chain code specifies the direction that must be set for the step from the current boundary point to the next boundary point.

Because there are 8 possible directions, they can be numbered from 0 to 7. Figure 13.18 shows one of the eight possible encoding schemes. The boundary chain code therefore contains the coordinates of the starting point and coding sequence used to determine the path around the boundary path [9].

13.6.3 Line Segment Coding

Line segment coding is a line-by-line processing technique for storing extracted objects. Consider the example shown in Fig. 13.19. We aim to segment an image with a gray level threshold T [10].

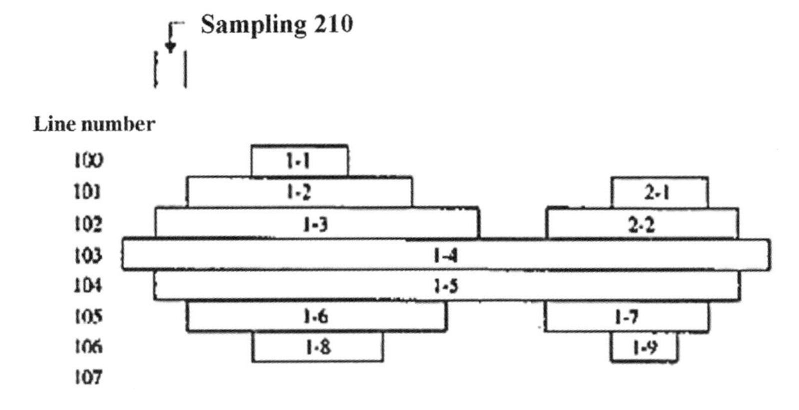

Fig. 13.19 Line segment of an object

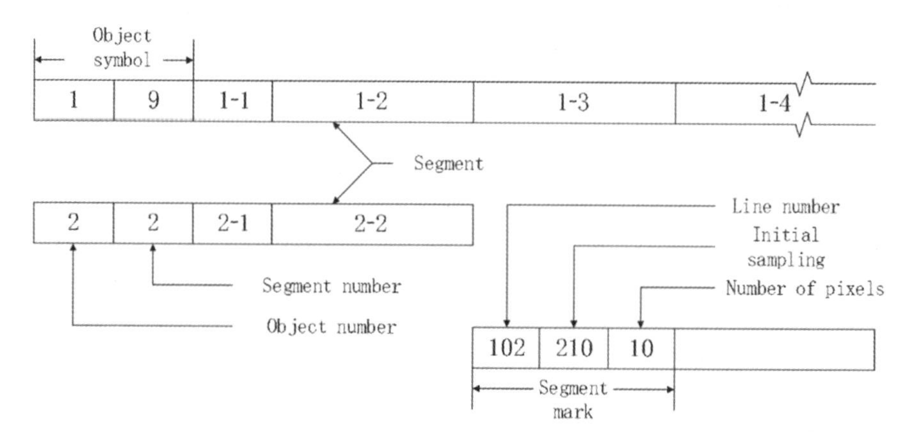

Fig. 13.20 Object segment file

Figure 13.20 shows a way to organize the object segment information on the disk. Each time a new object is located, the program generates a new object file. This file starts with a flag that contains the object number and number of segments in the object. The latter entry must continue to be modified until the segmentation of the object is complete.

13.7 Summary

See Fig. 13.21 and Table 13.1.

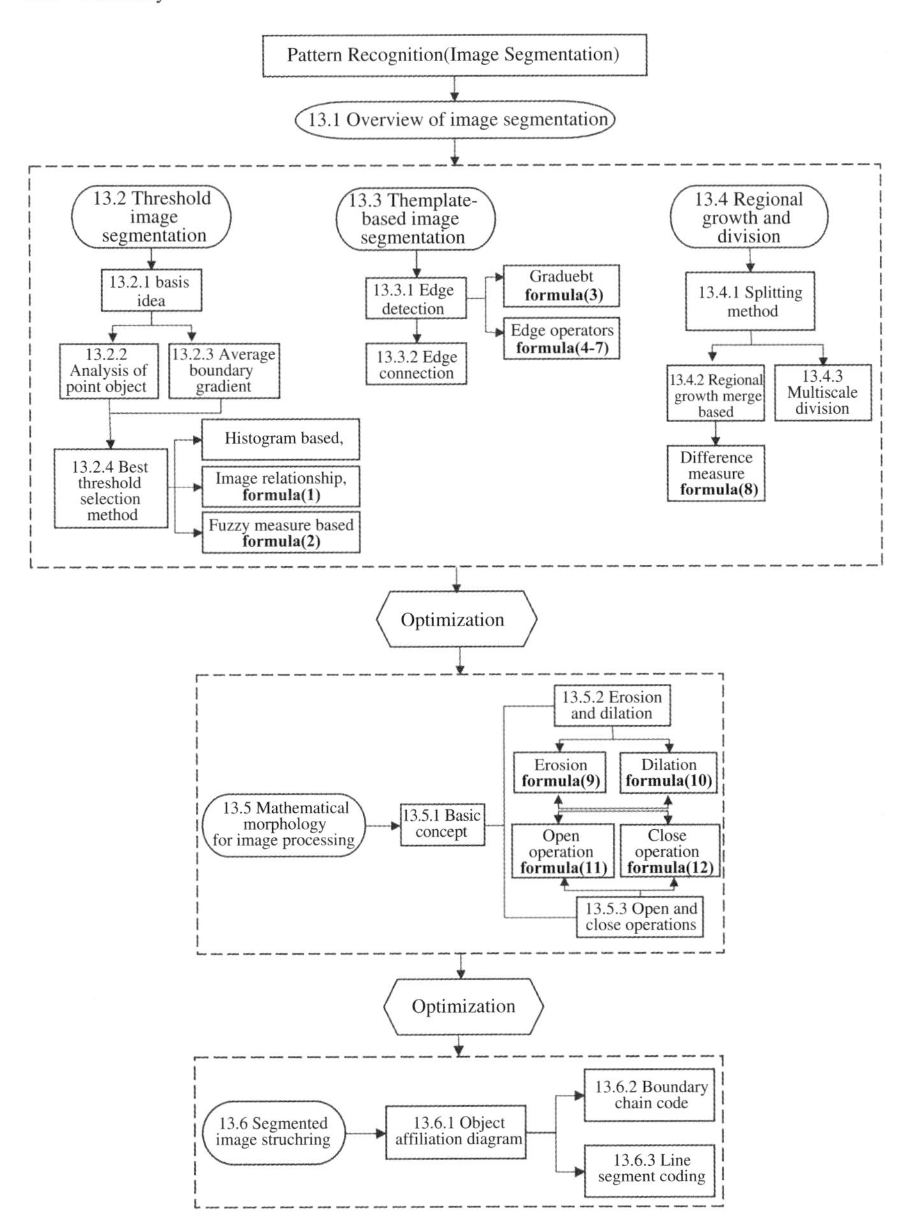

Fig. 13.21 Summary of this chapter

Table 13.1 Main formulas

formula (1)	$t^* = \arg \max_{0 \le t \le L-1} (\sigma_B^2)$	13.2.4
formula (2)	$t_0 = \arg \min_{0 \le t \le L-1} \Upsilon_t(X)$	13.2.4
formula (3)	$\nabla f(x, y) = \boldsymbol{i}\frac{\partial f(x,y)}{\partial x} + \boldsymbol{j}\frac{\partial f(x,y)}{\partial y}$	13.3.1
formula (4)	$G_x = \begin{bmatrix} 0 & -1 \\ 1 & 0 \end{bmatrix}, G_y = \begin{bmatrix} 1 & 0 \\ 0 & -1 \end{bmatrix}$	13.3.1
formula (5)	$G_x = \begin{bmatrix} -1 & 0 & 1 \\ -2 & 0 & 2 \\ -1 & 0 & 1 \end{bmatrix}, G_y = \begin{bmatrix} -1 & -2 & -1 \\ 0 & 0 & 0 \\ 1 & 2 & 1 \end{bmatrix}$	13.3.1
formula (6)	$G_x = \begin{bmatrix} 1 & 0 & -1 \\ 1 & 0 & -1 \\ 1 & 0 & -1 \end{bmatrix}, G_y = \begin{bmatrix} -1 & -1 & -1 \\ 0 & 0 & 0 \\ 1 & 1 & 1 \end{bmatrix}$	13.3.1
formula (7)	$G_x = \begin{bmatrix} 0 & -1 & 0 \\ -1 & 4 & -1 \\ 0 & -1 & 0 \end{bmatrix}, G_y = \begin{bmatrix} -1 & -1 & -1 \\ -1 & 8 & -1 \\ -1 & -1 & -1 \end{bmatrix}$	13.3.1
formula (8)	$f = w_{\text{color}} \cdot h_{\text{color}} + w_{\text{shape}} \cdot h_{\text{shape}}$	13.4.2
formula (9)	$\varepsilon_B(X)=\{x \mid B_x \subseteq X\}$	13.5.2
formula (10)	$\delta_B(X) = \{x \mid B_x \cap X \neq \emptyset\}$	13.5.2
formula (11)	$\gamma_B(f) = \delta_{\overset{\vee}{B}}[\varepsilon_B(f)]$	13.5.3
formula (12)	$\phi_B(f) = \varepsilon_{\overset{\vee}{B}}[\delta_B(f)]$	13.5.3

References

1. Rosenfield A. Connectivity in digital pictures. J ACM. 1970(17).
2. Marr D, Hildreth E. Theory of edge detection . Proc R Soc London, Series B. 1980(207).
3. Panda DP, Rosenfeld A. Image segmentation by pixel classification in (Gray level, Edge value) Space. Comput, IEEE Trans. 1978(C-27).
4. Freeman H. Boundary encoding and processing. In: Lipkin B, Rosenfeld A, Picture processing and psychopictorics. New York: Academic Press; 1970.
5. Serra J. Image Analysis and Mathematical Morphology. New York: Academic Press; 1982.
6. Weszka J. A survey of threshold selection techniques. Comp Graph Image Proc. 1978.
7. Dacis LS. A survey of edge detection techniques. CGIP; 1975(4).
8. Wall RJ, Klinger A, Cattleman KR. Analysis of Image Histograms. In: Joint conference on pattern recognition (IEEE Pub. 74CH-0885–4C), Copenhagen, p. 341–44.
9. Necatia R. Locating object boundary in textured environments. Comput IEEE Trans. 1976(C-25).
10. ZuckerS. Region growing: childhood and adolescence. Comput Graph Image Proc.1976(5).

Chapter 14
Technology III of Remote Sensing Digital Image Processing: Pattern Recognition (Feature Extraction and Classification)

Chapter Guidance Image information extraction is a key purpose of remote sensing digital image processing. To extract useful information in the image, it is necessary to perform pattern recognition processing on the image. Pattern recognition is generally divided into three major steps: the 'search' of features, 'extraction' and 'classification', i.e., image segmentation, feature extraction and classification, respectively. Image segmentation as a means of remote sensing application has been described in Chap. 13. Feature extraction and classification are described in this chapter, which are key purposes of remote sensing applications and the final element of classification performance.

The last chapter focused on image segmentation or object separation in pattern recognition. In this chapter, we discuss how to extract features from objects in images and classify them using statistical decision theory. This chapter presents an overview of feature extraction and classification (Sect. 14.1) and describes four methods of pattern recognition. Features of remote sensing images (Sect. 14.2): This section describes the use of artificial methods to quantitatively evaluate the features of targets in remote sensing images. Curve and surface fitting (Sect. 14.3): This section describes the fitting of target features by computer methods based on feature measurements. Classification of the remote sensing image object (Sect. 14.4): This section describes the division of the feature space to achieve target classification and identification. Estimation of the error rate (Sect. 14.5): This section describes the evaluation of the credibility of the target classification results. Thus, the fundamental goal of realizing remote sensing applications is target object identification. Figure 14.1 shows a logical block diagram of the sections in this chapter.

The first stage of pattern recognition is known as the image segmentation or object separation phase, as described in Chap. 13. The second stage is the feature extraction stage, in which the object is measured. The third stage pertains to the classification decision-making and estimation of the correctness of the decision. The decision is aimed at determining the category to which an object belongs.

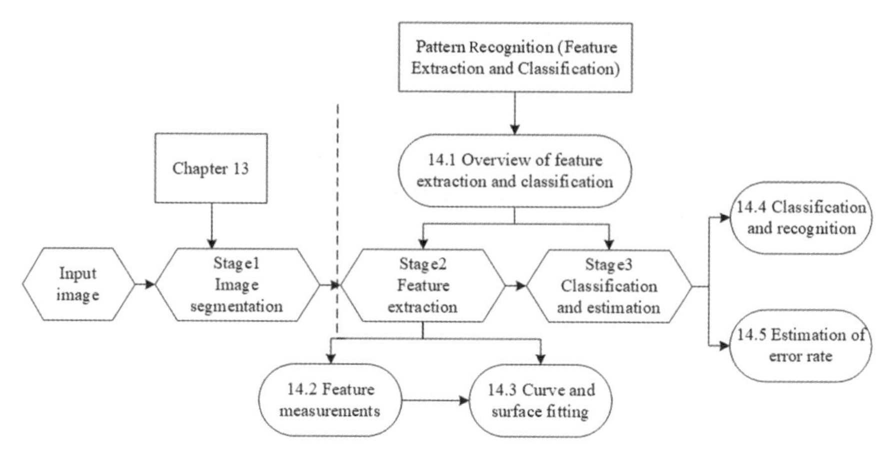

Fig. 14.1 Logical block diagram for this chapter

14.1 Overview of Feature Extraction and Classification

Four basic pattern recognition methods are available: statistical, structural, fuzzy and neural network-based pattern recognition methods [1]. This section describes the four pattern recognition methods. The focus is on statistical pattern recognition.

14.1.1 Pattern Recognition Methods

1. Statistical Pattern Recognition

The basic principle of statistical pattern recognition is that the samples with similarity are close to each other in the pattern space and form a 'group', i.e., 'things are clustered' [2]. The analysis is based on the pattern of the measured vector $\boldsymbol{x}_i = (x_{i1}, x_{i2},\ldots, x_{id})^{\mathrm{T}}$, where $i=1,2,\ldots,N$, T is the transpose, N is the number of samples, and d is the number of sample features. A given pattern is classified into c categories $\omega_1, \omega_2,\ldots, \omega_c$, and the classification is performed according to the distance function between the modes.

2. Structural Pattern Recognition

In certain image recognition problems, the purpose of recognition is not only to assign an image to specific categories (classification) but also to describe the shape of the image. The language structure method can be used to identify the image.

3. Fuzzy Pattern Recognition

In 1965, Zadeh proposed his famous fuzzy set theory, from which a new discipline, fuzzy mathematics, was created. Fuzzy set theory is a generalization of traditional set theory. In fuzzy sets, each element belongs to a certain set to a certain degree, and it can simultaneously belong to several sets at different levels. This principle can be applied in the field of pattern recognition to perform fuzzy pattern recognition.

4. Neural Network-Based Pattern Recognition

A key feature of the neural network-based pattern recognition method is that it can effectively solve many nonlinear problems. However, there exist many important problems in neural networks that have not been solved theoretically. Therefore, there remain many factors that need to be determined empirically, such as the number of network nodes, initial weight, and learning steps. In addition, local minima problems and overlearning and underlearning problems also occur in many neural network methods. Statistical learning theory provides a superior theoretical framework for studying pattern recognition and neural network problems.

14.1.2 Feature Extraction and Classification Estimation of Remote Sensing Digital Images

Feature extraction can be performed using two methods: by selecting the measurement that can effectively represent the object features (Sect. 14.2), and by curve or surface fitting to directly simulate the object (Sect. 14.3).

In terms of classification estimation (Sect. 14.4), we describe the Bayesian estimation method and maximum likelihood method as statistical decision methods. Other statistical decision methods are similar to these two methods and are thus not repeated. Proportion estimation is described in Sect. 14.5.

14.2 Feature Measurements of Remote Sensing Images

Features of the target in the remote sensing image are quantitatively measured using an artificial method, and the problem of object measurements is emphasized. Through these measurements, we specify the eigenvector of the object to be identified, decompose the object into numerous small triangles, and lay the foundation for curve and surface fitting. This section also introduces the area, perimeter, shape analysis, and texture features commonly used to reflect the size of the object.

14.2.1 Area and Perimeter of the Image Target

The area and perimeter can be easily calculated in the process of extracting objects from the segmented image.

1. Boundary Definition

Before specifying the algorithm for calculating the area and perimeter of an object, the definition of the boundary of the object must be specified to confirm that the perimeter and area of the same polygon are measured. The question that must be addressed is whether the boundary pixels are all or partially contained in the object. In other words, it must be determined whether the actual boundaries of the objects pass through the center of the boundary pixels or are located around the outer edges.

2. Pixel Count Area

The simplest (uncalibrated) area calculation method is to count the number of pixels inside the boundary (including the boundary). Corresponding to this definition, the perimeter is the length of the outer boundary around all the pixels. Typically, this distance involves many 90° turns, which exaggerates the perimeter value.

3. Perimeter of the Polygon

A more effective technique to measure the perimeter of an object is to define the object boundary as a polygon with vertices at the center of each boundary pixel. Thus, the corresponding perimeter is the sum of the intervals of the series of vertical and horizontal directions ($\Delta p = 1$) and diagonal directions ($\Delta p = 2$). This value can be summed during the process of tracing the boundary when the object is accumulated or programmed with a trip code. The perimeter of an object can be expressed as

$$p = N_{\text{even}} + \sqrt{2}N_{\text{odd}} \tag{14.1}$$

where N_{even} and N_{odd} are the even- and odd-number steps in the boundary chain code, respectively. The perimeter can also be simply obtained from the object block file by calculating the sum of the center distances of adjacent pixels on the boundary. The corresponding perimeter is equal to the sum of the sides of the polygon. If all the boundary points of the polygon are used as vertices, the perimeter is the sum of all the measured values in the horizontal and diagonal directions.

4. Area of the Polygon

Determined with reference to the center of the pixel, the area of the polygon is equal to the difference in the number of all pixels and half of the number of pixels in the boundary plus 1, i.e.,

$$A = N_O - \left[N_B/2 + 1\right] \tag{14.2}$$

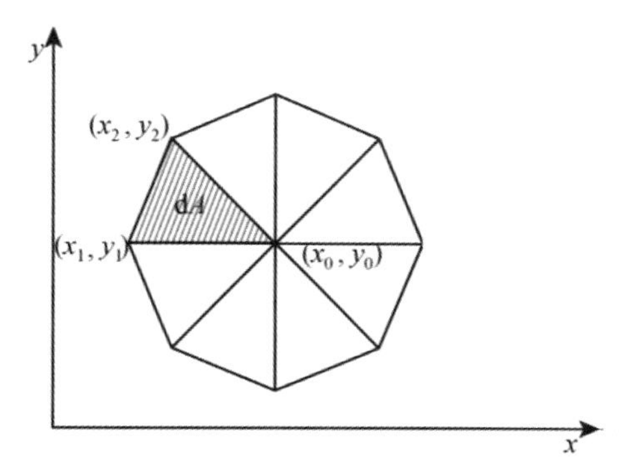

Fig. 14.2 Area of a polygon

where N_O and N_B denote the number of pixels of the object (including the number of boundary pixels) and on the boundary, respectively. In general, half of the boundary pixels are inside the object, and the other half are outside the object. Since the object is generally convex around a closed curve, the additional area corresponding to half the pixels lies outside the object. In other words, the area derived from the number of pixels can be approximatively corrected by subtracting half of the perimeter.

A simplified method can be used to calculate the area and perimeter of a polygon in a traversal, i.e., the area of a polygon is equal to the sum of areas of all triangles formed by the connection of each fixed point and any internal point, as shown in Fig. 14.2. For the sake of generality, this point can be set as the origin of the image coordinate system.

As shown in Fig. 14.3, the area of the triangle is

$$\mathrm{d}A = x_2 y_1 - \frac{1}{2}x_1 y_1 - \frac{1}{2}x_2 y_2 - \frac{1}{2}(x_2 - x_1)(y_1 - y_2) \tag{14.3}$$

which can be written as

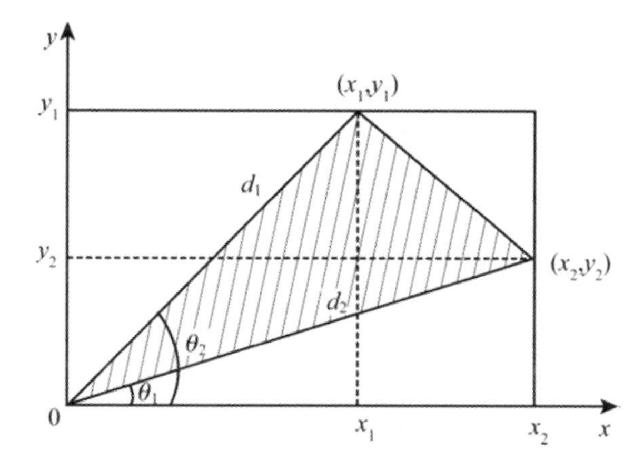

Fig. 14.3 Area of a triangle

$$\mathrm{d}A = \frac{1}{2}(x_2 y_1 - x_1 y_2) \tag{14.4}$$

The area of the polygon is

$$A = \frac{1}{2}\sum_{i=1}^{N}(x_{i+1}y_i - x_i y_{i+1})$$

$$\begin{aligned} \mathrm{d}A &= \frac{1}{2}\sin(\theta_1 - \theta_2)d_1 d_2 \\ &= \frac{1}{2}(\sin\theta_2\cos\theta_1 - \cos\theta_2\sin\theta_1)d_1 d_2 \\ &= \frac{1}{2}\left(\frac{y_1}{d_1}\frac{x_2}{d_2} - \frac{x_1}{d_1}\frac{y_2}{d_2}\right)d_1 d_2 \\ &= \frac{1}{2}(y_1 x_2 - y_2 x_1) = \frac{1}{2}\sum_{i=1}^{N}(x_{i+1}y_i - x_i y_{i+1}) \end{aligned} \tag{14.5}$$

where N is the number of boundary points.

If the origin is outside the object, all triangles occupy a certain area not within the polygon. Moreover, the area of a triangle can be positive or negative, determined by the direction of boundary traversal. When the boundary had been completely traversed, the area falling outside the object was considered to have been subtracted. Using Green's theorem, we can construct a simpler method with the same result. According to the integral calculation, the area enclosed by a closed surface in the *xy* plane can be determined using the contour integral.

$$A = \frac{1}{2}\oint(x\mathrm{d}y - y\mathrm{d}x) \tag{14.6}$$

where the integral is along the closed curve, consistent with the expression of the Green integral. To discretize this integral, formula (14.6) can be expressed as

$$A = \frac{1}{2}\sum_{i=1}^{N}\left[x_i(y_{i+1} - y_i) - y_i(x_{i+1} - x_i)\right] \tag{14.7}$$

14.2.2 Shape Analysis of the Image Target

Generally, a class of objects can be distinguished from other objects by the shape of the objects. Shape features can be used independently or in combination with size measurements. This section describes several common shape parameters.

1. Rectangularity

The rectangular fitting factor reflects the rectangularity of an object:

$$R = A_O / A_R \tag{14.8}$$

where A_O is the area of the object, and A_R is the area of the minimum enclosing rectangle (MER). R reflects the extent to which an object is filled with its MER. For a rectangular object, the maximum R is 1.0; for a round object, R is $\pi/4$; for slender or curved objects, R takes a small value. The value of the rectangular fitting factor ranges from 0 to 1.

2. Ratio of the Length to Width

The ratio of the length to width equals the ratio of the MER width W to MER length L, i.e.,

$$A = W / L \tag{14.9}$$

This feature distinguishes finer objects and square or round objects.

3. Roundness

The circularity index is a set of shape features that take the minimum value for a circular shape calculation. The magnitude reflects the complexity of the boundaries being measured. The most commonly used roundness indicator is

$$C = P^2 / A \tag{14.10}$$

that is, the ratio of the square of perimeter P to area A. This feature takes a minimum value of 4π for a circular shape. A more complex shape corresponds to greater value. The circularity index C is related to the border complexity.

Another roundness indicator is the boundary energy. Assume that the perimeter of an object is P, and the variable p represents the distance from a certain point on the boundary. At any point, the boundary has an instantaneous radius of curvature $r(p)$, which is radius of the tangent to the point (Fig. 14.4). The curvature function at this point is

$$K(p) = 1 / r(p) \tag{14.11}$$

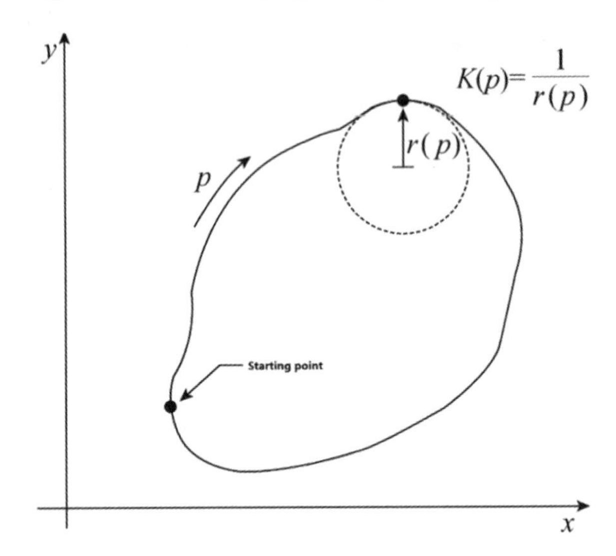

Fig. 14.4 Radius of curvature

Function $K(p)$ is a periodic function of period P. The average energy per unit boundary length can be calculated as

$$E = \frac{1}{P}\int_0^P |K(p)|^2 \mathrm{d}p \tag{14.12}$$

For a fixed area value, the circle has the minimum boundary energy:

$$E_0 = \left(\frac{2\pi}{P}\right)^2 = \left(\frac{1}{R}\right)^2 \tag{14.13}$$

where R is the radius of the circle. The curvature can be easily calculated from the chain code, and thus, the boundary energy can be easily calculated. The boundary energy is more consistent with the perceived complexity of the boundary than the circularity index defined in Eq. (14.10).

The third roundness indicator uses the average distance from the point on the boundary to the point within the object. This distance is

$$\overline{d} = \frac{1}{N}\sum_{i=1}^{N} x_i \tag{14.14}$$

where x_i is the distance from the i-th point in an object with N points to the nearest boundary point. The corresponding shape measure is

$$g = \frac{A}{\overline{d}^2} = \frac{N^3}{(\sum_{i=1}^{N} x_i)} \tag{14.15}$$

The sum of the denominations of (14.15) is the integrated optical density (IOD) of the transformed image. The gray level of a pixel in the transformed image reflects the distance of the pixel from its nearest boundary. Figure 14.5 shows the results of a binary image and its distance transformation.

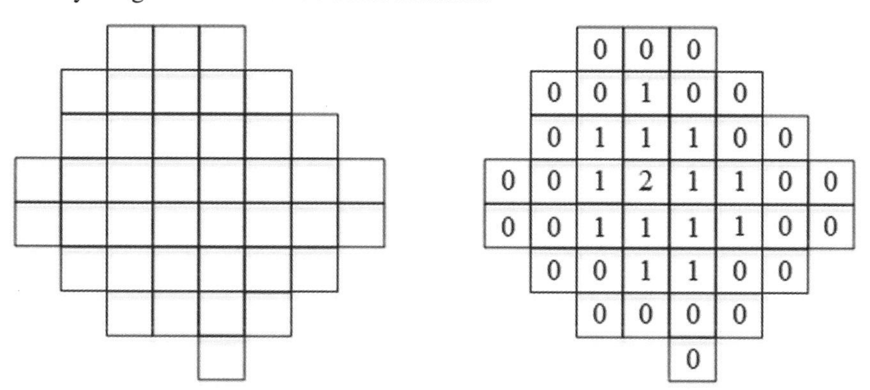

Fig. 14.5 Distance transformation

For round and regular polygons, the corresponding values are specified by Eqs. (14.10) and (14.15). However, for more complex shapes, Eq. (14.15) yields a higher resolution.

4. Invariant Moments

① Background

Moments of functions are often used in probability theory. Several expectation values derived from moments can be used in the shape analysis.

We define the moment set of the bounded function $f(x, y)$ with two variables:

$$M_{jk} = \int_{-\infty}^{\infty} \int_{-\infty}^{\infty} x^i y^k f(x, y) \mathrm{d}x \, \mathrm{d}y \tag{14.16}$$

where j and k are nonnegative integers. The probability density function (PDF) is widely used in probability theory.

Since j and k are nonnegative integers, an infinite set of moments can be obtained. This set can also yield function $f(x, y)$. In other words, the set M_{jk} is unique for function $f(x, y)$, and only $f(x, y)$ corresponds to the particular set of moments.

To describe the shape, we assume that $f(x, y)$ is 1 in the object and 0 in the other positions. This silhouette function reflects only the shape of the object and ignores the internal gray level details. Each shape has a specific contour and specific set of moments.

The parameter $j + k$ is known as the order of moments. There exists only one zero order moment:

$$M_{00} = \int_{-\infty}^{\infty}\int_{-\infty}^{\infty} f(x, y)\mathrm{d}x\,\mathrm{d}y \tag{14.17}$$

This value is the area of the object. There are two first-order moments and more than two higher-order moments. M_{00} is divided by all the 1st-order moments and higher-order moments, and thus, the values are independent of the size of the object.

② Center distance

The coordinates of the center of gravity of an object are

$$\overline{x} = \frac{M_{10}}{M_{00}}, \quad \overline{y} = \frac{M_{01}}{M_{00}} \tag{14.18}$$

The center moment is calculated with the center of gravity as the origin

$$u_{jk} = \int_{-\infty}^{\infty}\int_{-\infty}^{\infty} (x - \overline{x})^j (y - \overline{y})^k f(x, y)\mathrm{d}x\,\mathrm{d}y \tag{14.19}$$

Therefore, the center moment is independent of the position.

③ Spindle

The second-order central moment is defined as $\mu_2(x)$. In this case, the k-order central moment is $\mu_k(x)$, and the rotation angle θ for which $\mu_k(x)$ is minimized is

$$\tan 2\theta = \frac{2\mu_{11}}{\mu_{20} - \mu_{02}} \tag{14.20}$$

The x and y-axes are rotated by angle θ to obtain the axis (x', y'), which is the main axis of the object. The uncertainty in θ when θ is 90° can be specified as

$$\mu_{20} < \mu_{02}, \mu_{30} > 0 \tag{14.21}$$

If the object is rotated by angle θ before calculating the moment or the moments relative to the x', y' axes are calculated, the moment is considered to be rotation invariant.

The center moment calculated with respect to the principal and normalized by the area is maintained constant when the object is enlarged, translated, and rotated. Only the third or higher order of the moment after such standardization is not invariant.

The magnitude of these moments reflects the shape of the object and can be used for pattern recognition. Invariant moments and their combinations have been used in the identification and chromosome analysis of printed characters.

14.2.3 Texture of the Image Target

In remote sensing, texture is defined as 'an attribute that reflects the spatial distribution of pixel gray levels in a region'.

The textural feature is a value calculated from the image of the object, which quantifies the features of the grayscale changes within the object [3]. In general, textural features are related to the position, orientation, size, and shape of the object but are independent of the average gray level (brightness).

Simple grayscale statistics include the standard deviation, variance, inclination and kurtosis. These values can be calculated as the moments of the grayscale histogram of the object as the computational module feature:

$$I=\sum_{i=1}^{N}\frac{H_i-M/N}{\sqrt{\frac{H_i(1-H_i/M)+M(1-1/N)}{N}}} \tag{14.22}$$

where M is the number of pixels in the object, N is the number of grayscale levels, I is the moment of the grayscale histogram, and H_i is the number of pixels in the i-th grayscale.

I is based on statistics; in addition, for a given image, a two-dimensional Fourier transform can obtain all of the textural information. Thus, in addition to the textural features derived from the object, the values can be derived from the spectrum. This spectrum-based textural feature is the spectral textural feature.

14.3 Curve and Surface Fitting

The target feature is fitted using a computer method, based on feature measurements. In image analysis, one-dimensional function fitting is often performed on a set of data points by using a polynomial or a Gaussian function, and two-dimensional surface fitting is performed on an image by using a two-dimensional polynomial or a Gaussian function. Fitting can be performed to eliminate noise and determine various parameter values (such as the position, size, shape, and amplitude) of the object, which can be used as measurement functions [4]. This section describes five fitting methods.

14.3.1 Minimum Mean Square Error Fitting

Given a subset (x_i, y_i), the commonly used fitting technique is to find the function $f(x)$ to minimize the mean square error:

$$\mathrm{MSE}=\frac{1}{N}\sum_{i=1}^{N}[y_i-f(x_i)]^2 \tag{14.23}$$

for N data points (x_i, y_i).

For example, a parabolic $f(x)$ can be expressed as

$$f(x)=c_0+c_1x+c_2x^2 \tag{14.24}$$

Curve fitting is performed to determine the optimum values for coefficients c_0, c_1, and c_2. It is desirable to determine the values of these coefficients to minimize the mean square error at the given points.

14.3.2 One-Dimensional Parabolic Fitting

Matrix algebra can be used to solve the problem. First, we construct matrices $\boldsymbol{B}$ containing x values, $\boldsymbol{Y}$ containing y values, and $\boldsymbol{C}$ containing the undetermined coefficients, i.e.,

$$\boldsymbol{Y}=\begin{bmatrix} y_1 \\ y_2 \\ \vdots \\ y_N \end{bmatrix},\ \boldsymbol{B}=\begin{bmatrix} 1 & x_1 & x_1^2 \\ 1 & x_2 & x_2^2 \\ \vdots & \vdots & \vdots \\ 1 & x_N & x_N^2 \end{bmatrix},\ \boldsymbol{C}=\begin{bmatrix} c_0 \\ c_1 \\ c_2 \end{bmatrix} \tag{14.25}$$

The column vector representing the error of each data point can be written as

$$\boldsymbol{E}=\boldsymbol{Y}-\boldsymbol{BC} \tag{14.26}$$

where matrix product $\boldsymbol{BC}$ is the column vector of the $y=f(x)$ value calculated using Eq. (14.24).

The mean square error in Eq. (14.23) can be defined as

$$\mathrm{MSE}=\frac{1}{N}\boldsymbol{E}^{\mathrm{T}}\boldsymbol{E} \tag{14.27}$$

Substituting Eq. (14.26) into Eq. (14.27), differentiating C and setting the derivative as zero yields (14.28).

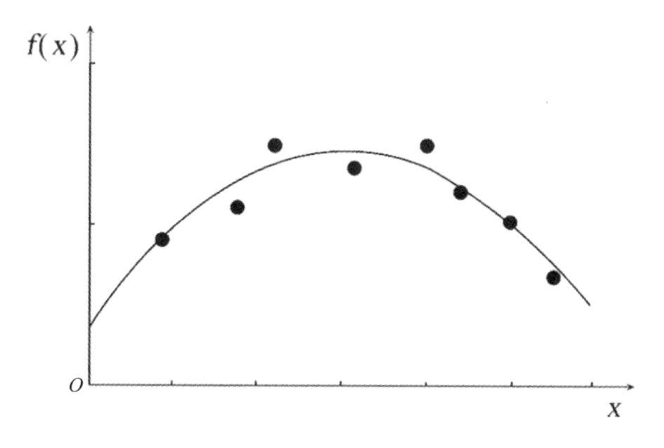

Fig. 14.6 Fitting a parabola with five data points

$$C = [B^{T}B]^{-1}[B^{T}Y] \tag{14.28}$$

This coefficient vector has an extremely small mean square error. The square matrix $[B^{T}B]^{-1}B^{T}$ is termed the pseudoinverse matrix of B, and this method is known as the pseudoinverse method.

If the number of points and number of coefficients are equal, B is a square. If this entity is nonsingular, it can be directly inversed. In this case, Eq. (14.28) can be simplified as

$$C = B^{-1}Y \tag{14.29}$$

In this case, the problem is to solve the linear equations containing multiple unknown quantities.

Next, a parabola is fitted with five data points with the following values:

$$X = \begin{bmatrix} 0.9 \\ 2.2 \\ 3 \\ 4 \\ 5 \end{bmatrix}, \quad Y = \begin{bmatrix} 1.8 \\ 3 \\ 2.5 \\ 3 \\ 2 \end{bmatrix}, \quad B = \begin{bmatrix} 1 & 0.9 & 0.81 \\ 1 & 2.2 & 4.84 \\ 1 & 3 & 9 \\ 1 & 4 & 16 \\ 1 & 5 & 25 \end{bmatrix} \tag{14.30}$$

Figure 14.6 shows the set of given points and the best fitting parabola determined using this method. The calculation process is described in the following text.

$$B^{T}B = \begin{bmatrix} 5 & 15 & 56 \\ 15 & 56 & 227 \\ 56 & 227 & 986 \end{bmatrix}, \quad B^{T}Y = \begin{bmatrix} 12.3 \\ 37.7 \\ 136.5 \end{bmatrix}, \quad C = \begin{bmatrix} 0.747 \\ 1.415 \\ -0.230 \end{bmatrix} \tag{14.31}$$

We compare the calculated and observed values and determine the error vector:

$$
\boldsymbol{Y}=\begin{bmatrix}1.8\\3\\2.5\\3\\2\end{bmatrix},\ \boldsymbol{BC}=\begin{bmatrix}1.83\\2.75\\2.92\\2.73\\2.07\end{bmatrix},\ \boldsymbol{E}=\begin{bmatrix}-0.03\\+0.25\\-0.42\\+0.27\\-0.07\end{bmatrix} \tag{14.32}
$$

For example, for an autofocus application, it is necessary to determine the parabolic vertex position. We set the derivative of Eq. (14.24) as zero:

$$
x_{\max}=\frac{-c_2}{2c_3}=3.076,\ f(x_{\max})=2.923 \tag{14.33}
$$

If these points are grayscale values along the same scan line, x_i is equispaced. In practice, however, the possible point arrangements are diverse, and any discrete set of points may exist. The only limitation is that $f(x)$ is a function of x, and thus, all values of x must be unique. In other words, $f(x)$ cannot be updated to fit the data.

The first factor on the right side of (14.28) is the inverse of a matrix, which may lead to computational burden. Regardless of the number of points used in the fitting, the matrix is 3-sided. Therefore, the computational complexity is not high.

14.3.3 Two-Dimensional Cubic Fitting

The abovementioned techniques can be extended to polynomial fitting techniques higher than the second order and extended to two-dimensional functions.

There exists an effective background correction technique, which can be implemented through the binomial fitting of several background points selected according to the principle of low gray value. The resulting function is subtracted from the image to flatten the background.

This method is illustrated by an example of fitting a two-variable cubic function. This function has 10 terms:

$$
f(x,y)=c_0+c_1x+c_2y+c_3xy+c_4x^2+c_5y^2+c_6x^2y+c_7xy^2+c_8x^3+c_9y^3 \tag{14.34}
$$

Matrix $\boldsymbol{B}$ is expressed as an N-order matrix

$$
\boldsymbol{B}=\begin{bmatrix}1 & x_1 & y_1 & x_1y_1 & x_1^2 & y_1^2 & x_1^2y_1 & x_1y_1^2 & x_1^3 & y_1^3\\ \vdots & \vdots & \vdots & \vdots & \vdots & \vdots & \vdots & \vdots & \vdots & \vdots\end{bmatrix} \tag{14.35}
$$

Thus, it is necessary to perform a 10-dimensional square matrix inverse, as indicated in Eq. (14.30). Figure 14.7 shows an example of background subtraction with two-dimensional cubic fitting.

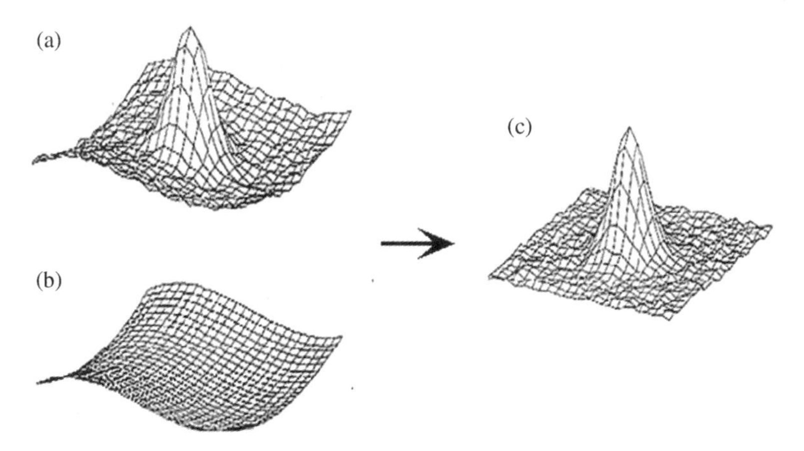

Fig. 14.7 Performing a two-dimensional cubic fitting on the background. (a) Image background, containing noise and shadows. (b) Cubic fitting of the background points. (c) Image subtracted from the background

14.3.4 Two-Dimensional Gaussian Fitting

Circular or elliptic objects in an image can be measured by fitting the image with a two-dimensional Gaussian surface [5]. A two-dimensional Gaussian function can be expressed as

$$z_i = A \exp\left[-\frac{(x_i - x_0)^2}{2\sigma_x^2} - \frac{(y_i - y_0)^2}{2\sigma_y^2}\right] \tag{14.36}$$

where A is the amplitude, (x_0, y_0) is the central position of the two-dimensional Gaussian function, and σ_x and σ_y are the standard deviations in the x- and y-axis directions, respectively.

By taking the logarithm on both sides of the equation and expanding and organizing the square terms, the quadratic terms of x and y can be obtained. We multiply both sides by z_i:

$$z_i \ln(z_i) = \left[\ln(A) - \frac{x_0^2}{2\sigma_x^2} - \frac{y_0^2}{2\sigma_y^2}\right] z_i + \frac{x_0}{\sigma_x^2}[x_i z_i] + \frac{y_0}{\sigma_y^2}[y_i z_i] + \frac{-1}{2\sigma_x^2}[x_i^2 z_i] + \frac{-1}{2\sigma_y^2}[y_i^2 z_i] \tag{14.37}$$

In the matrix form,

$$\boldsymbol{Q} = \boldsymbol{C}\boldsymbol{B} \tag{14.38}$$

where $\boldsymbol{Q}$ is an N-dimensional vector with elements

$$q_i = z_i \ln(z_i) \tag{14.39}$$

C is a five-dimensional vector consisting entirely of Gaussian parameters:

$$\boldsymbol{C}^{\mathrm{T}} = \left[\ln(A) - \frac{x_0^2}{2\sigma_x^2} - \frac{y_0^2}{2\sigma_x^2} \frac{x_0}{\sigma_y^2} \frac{y_0}{\sigma_y^2} \frac{-1}{2\sigma_x^2} \frac{-1}{2\sigma_y^2} \right] \tag{14.40}$$

B is an *N*-order matrix, and its *i*-th row is

$$[b_i] = [z_i, z_i x_i, z_i y_i, z_i x_i^2, z_i y_i^2] \tag{14.41}$$

Matrix ***C*** is calculated using formula (14.28), and the Gaussian parameters can be obtained as

$$\sigma_x^2 = \frac{-1}{2c_4}, \sigma_y^2 = \frac{-1}{2c_5} \tag{14.42}$$

$$x_0 = c_2\sigma_x^2, y_0 = c_3\sigma_y^2 \tag{14.43}$$

$$A = \exp\left[c_1 + \frac{x_0}{2\sigma_x^2} + \frac{y_0}{2\sigma_y^2} \right] \tag{14.44}$$

where only one matrix of five-dimensional squares must be inversed, independent of the number of points *N* used for fitting.

Figure 14.8a shows an example of Gaussian fitting for a noise peak. The original image is a Gaussian surface with random noise. Table 14.1 compares the parameters that produce the image with the fitted parameters.

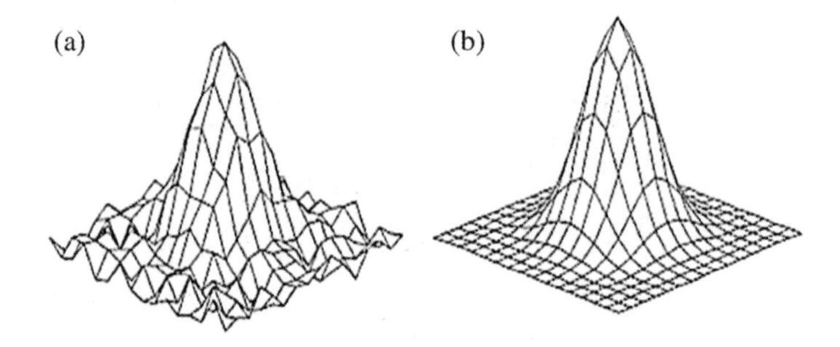

Fig. 14.8 Two-dimensional Gaussian fitting performed on a noise peak. (a) Original image. (b) Gaussian fitting

Table 14.1 Actual and Gaussian fitting parameters

	A	x_0	y_0	σ_x	σ_y
Actual	10	4	4	2	2
Fitting	10.17	4.04	4.06	2.00	2.06

Due to the presence of noise, the original Gaussian image is not precisely reconstructed. However, the estimated values of these parameters are reasonable estimations of the original parameters. Figure 14.8b shows a highly accurate copy of the noisy form, and the root mean square (RMS) or standard error are 6% of the peak.

14.3.5 Elliptical Fitting

In many images, the objects of interest are round or elliptical. Therefore, it is of significance to fit an ellipse of an arbitrary size, shape and orientation according to a set of boundary points.

The general equation of a quadratic curve is

$$ax^2 + bxy + cy^2 + dx + ey + f = 0 \tag{14.45}$$

This expression can represent an ellipse if it satisfies

$$b^2 - 4ac < 0 \tag{14.46}$$

An ellipse is determined by five parameters: the center of the x-axis and y-axis coordinates; length of the long and short half axes; and angle between the spindle and horizontal axis. The ellipse can be fitted by substituting the coordinates of the five points into Eq. (14.45) to identify the solutions of the five equations. By identifying a series of ellipses passing through five points, the average (or median) value of the parameters is considered to obtain an optimal fit.

We can set $a = 1$ to normalize Eq. (14.45) and derive the mean square error.

$$\varepsilon^2 = \sum_i (x_i^2 + bx_iy_i + cy_i^2 + dx_i + ey_i + f)^2 \tag{14.47}$$

If we obtain the partial derivatives of coefficients b, c, d, e, and f in Eq. (14.47) and set them be zero, we can obtain five equations for the square terms of x_i and y_i and their multiplication and determine these coefficients. This process can be implemented through the inverse of the five-dimensional square matrix described earlier.

14.4 Classification and Recognition of Remote Sensing Image Targets

This section describes the division of the feature space to realize object classification and identification, which is a key goal of remote sensing imaging applications. Image classification refers to the selection of feature parameters by analyzing the spectral features of all kinds of objects, division of the feature space into nonoverlapping subspaces, and classification by dividing the pixels within the image into subspaces [6]. This section describes two classification methods: the statistical decision method and syntactic structure method. The former method is described in depth, and the latter method is only introduced.

14.4.1 Statistical Decision Methods

We consider two statistical decision methods: the Bayesian (Bayes) estimation method, and maximum likelihood method, which is commonly used in remote sensing classification.

1. Bayesian Estimation Method

$P(h)$ denotes the initial probability of h before the training data are acquired, defined as the prior probability of h. The prior probabilities reflect the possibility that hypothesis h holds. Without this prior knowledge, each candidate hypothesis can simply be assigned the same prior probabilities.

Similarly, $P(D)$ represents the prior probability of training data D, and $P(D|h)$ represents the probability of D when hypothesis h is tenable. In machine learning, we focus on $P(h|D)$, known as the posterior probability of h.

Suppose that each object has n measurements. Since each object is not a single eigenvalue, it is composed of a feature vector $\boldsymbol{F} = [x_1,x_2,\ldots,x_n]^{\mathrm{T}}$ and corresponds to a point in the n-dimensional feature space. In addition, we assume that the object category is not two but m. In these conditions, the posterior probability of the i-th class membership, based on the Bayesian theorem is

$$p(C_i|x_1, x_2, x_3, \ldots, x_n) = \frac{p(x_1, x_2, x_3, \ldots, x_n|C_i)p(C_i)}{\sum_{i=1}^{m} p(x_1, x_2, x_3, \ldots, x_n|C_i)p(C_i)} \tag{14.48}$$

where the condition $\boldsymbol{F} = [x_1,x_2\ldots,x_n]^{\mathrm{T}}$ corresponds to an n-dimensional vector.

Assuming that there exists a sample set D and the prior probabilities can be calculated and that the sample categories are independent of one another, the Bayesian formula is

$$P(\tilde{\omega}_i|x, D) = \frac{p(x|\tilde{\omega}_i, D)P(\tilde{\omega}_j|D)}{\sum_{j=1}^{c} p(x|\tilde{\omega}_j, D)P(\tilde{\omega}_j|D)} \tag{14.49}$$

Integrating the joint probability density $p(x, \theta|D)$ yields

$$p(x|D) = \int p(x, \theta|D)\mathrm{d}\theta \tag{14.50}$$

This formula transforms the two-dimensional distribution into two independent one-dimensional distributions, and the selection of sample set D and test sample x are independent. Thus, this formula is the most central formula in Bayesian estimation.

The Bayesian decision is that the unknown state is estimated by the subjective probability under incomplete information, and the probability of occurrence is corrected using the Bayesian formula. Subsequently, the expected value and corrected probability are used to determine the optimal value.

The Bayesian estimation method is implemented as follows: When the sample follows the Gaussian normal distribution as a univariate distribution, the only unknown value is the mean μ. If the distribution is $p(\mu) \sim N(\mu_0, {\sigma_0}^2)$, μ_0 represents the best priori estimation for μ, and ${\sigma_0}^2$ represents the uncertainty of this estimation. Prior knowledge regarding the mean μ is contained in the prior probability density function $p(\mu)$. After selecting μ, the Bayesian formula is

$$p(\mu|D) = \frac{p(D|\mu)p(\mu)}{\int p(D|\mu)p(\mu)\mathrm{d}\mu} = a\prod_{k=1}^{n} p(x_k|\mu)p(u) \tag{14.51}$$

where a is a normalized coefficient that depends only on the sample set D. We can obtain

$$\mu_n = \left(\frac{n\sigma_0^2}{n\sigma_0^2 + \sigma_0^2}\right)\hat{\mu}_n + \frac{\sigma^2}{n\sigma_0^2 + \sigma^2}\mu_0 \tag{14.52}$$

$$\sigma_n^2 = \frac{\sigma_0^2\sigma^2}{n\sigma_0^2 + \sigma^2} \tag{14.53}$$

where μ_n is the best estimation of the true value of μ after observing n samples, ${\sigma_n}^2$ reflects the uncertainty of the estimation, and $\hat{\mu}_n = \frac{1}{n}\sum_{k=1}^{n} x$ is the mean of the samples.

${\sigma_n}^2$ is a monotonically decreasing function of n, i.e., for each additional sample, we decrease the uncertainty of μ. As n increases, the waveform of $p(\mu|D)$ becomes increasingly sharp and approaches the Dirac function when n approaches infinity [7]. This phenomenon is also known as the Bayesian learning process, as shown in Fig. 14.9.

2. Maximum Likelihood Method

The most commonly used method for the classification of remote sensing image data is the maximum likelihood method.

In general, if the totality X has a probability density $p(x; \theta_1, \theta_2, \ldots, \theta_k)$, where $\theta_1, \theta_2, \ldots, \theta_k$ are unknown parameters, and $(x_1, x_2, \ldots, x_n)$ is a set of observations of

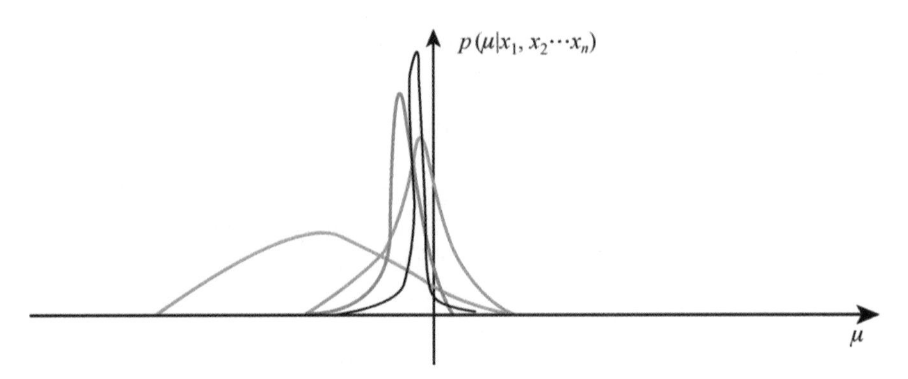

Fig. 14.9 Bayesian learning process

the sample, the probability that the samples (X_1, X_2, ..., X_n) lie in the neighborhood of (x_1, x_2, ..., x_n) is $\prod_{i=1}^{n} p(x_i; \theta_1, \theta_2, ..., \theta_k)\mathrm{d}x_i$, which is a function of $\theta_1, \theta_2, ..., \theta_k$. The intuitive concept of the maximum likelihood estimation is that since the observed value is obtained experimentally, the event must have the greatest likelihood that the samples lie in the neighborhood of the observed values (x_1, x_2, ..., x_n). Therefore, we must select the parameter value that maximizes this probability as the estimation of the true value of the parameter.

$$L(x_1, x_2, \ldots, x_n; \theta_1, \theta_2, \ldots, \theta_k) \hat{=} L(x; \theta) = \prod_{i=1}^{n} p(x_i; \theta_1, \theta_2, \ldots, \theta_k) \tag{14.54}$$

is the likelihood function.

For fixed (x_1, x_2, ..., x_n), if θ= (θ_1,θ_2,...,θ_k), select $\hat{\theta}$= ($\hat{\theta}_1$,$\hat{\theta}_2$,...,$\hat{\theta}_k$) to ensure that $L(x; \hat{\theta}) = \max L(x; \theta)$. In this case, $\hat{\theta}$ is the maximum likelihood estimation of θ.

If X is discrete, the likelihood function is

$$L(x; \theta) = \prod_{i=1}^{n} P(X = x_i; \theta_1, \theta_2, \ldots, \theta_k) \tag{14.55}$$

where $P(X = x_i; \theta_1,\theta_2,...,\theta_k) i = (1, 2, ...)$ is the probability distribution of X.

The maximum likelihood estimation of θ pertains to the calculation of the maximum point of the likelihood function $L(x; \theta)$. If the partial derivative of $L(x; \theta)$ to θ_i (i = 1, 2, ..., k) exists, according to the calculus theory, the maximum likelihood estimation $\hat{\theta}$ must satisfy

$$\frac{\partial L}{\partial \theta_i} = 0,\ i = 1, 2, \ldots, k \tag{14.56}$$

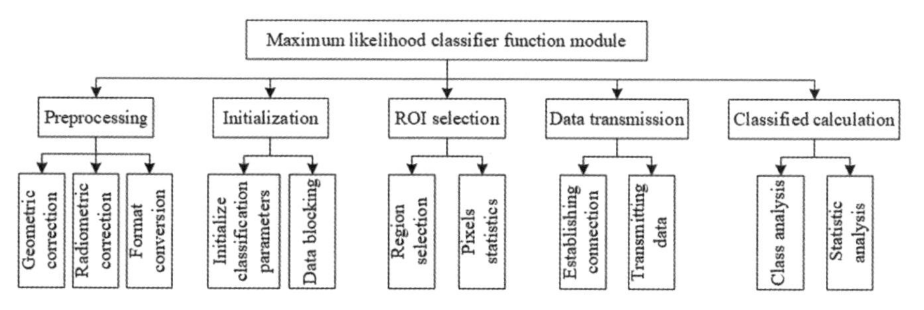

Fig. 14.10 Maximum likelihood classification

Equation (14.56) is known as a likelihood equation group. In many cases, the maximum point of $\ln L(x;\theta)$ is relatively simple, and $\ln x$ is a strictly increasing function of x. Thus, when the partial derivative of $\ln L(x;\theta)$ of $\theta_i (i = 1, 2, ..., k)$ exists, $\hat{\theta}$ can be calculated using (14.57):

$$\frac{\partial \ln L}{\partial \theta_i} = 0,\ i = 1, 2, \ldots, k \tag{14.57}$$

Equation (14.57) is known as a logarithmic likelihood equation. If the stagnation point of $\ln L(x;\theta)$ is unique and can be proven to be a maximal point, it must be the maximum point of $\ln L(x;\theta)$, i.e., the maximum likelihood estimation. However, if the stagnation point is not unique, it is necessary to determine the maximum point. If the partial derivative of $L(x;\theta)$ to $\theta_i (i = 1, 2, ..., k)$ does not exist, we cannot obtain the set of equations, and the maximum point of $L(x;\theta)$ must be obtained according to the definition of the maximum likelihood estimation.

Figure 14.10 shows the process of maximum likelihood classification in remote sensing.

14.4.2 Syntactic Structure Methods

Traditional remote sensing image classification methods, which are based on the statistical relationship between the statistical characteristics of the remote sensing image data and training sample data, involve limitations owing to the complexity of the distribution pattern of the feature type. In recent years, syntactic structure methods have been used for image classification to achieve satisfactory results. These methods involve neural networks, fuzzy mathematics, particle swarm optimization, decision trees, support vector machines, and artificial intelligence methods.

14.5 Estimation of the Error Rate

The error rate is estimated to identify the confidence level of the target classification results. In many applications, in addition to classifying and counting the number of objects per class, it is necessary to estimate one or more proportions, i.e., the proportion of each class of objects in the total set of objects to validate whether the classification is reasonable.

Suppose there exist K classes of objects. We can obtain a vector P, each element of which is $p_{i,j} = P\{\text{randomly selected object belonging to class } i\}$, where $i=1,\ldots, K$. The classifier error rate can be expressed by a confusion matrix C, with components $c_{i,j} = P\ \{i\text{-th class objects are divided in the } j\text{-th class}\}$, where $j=1,\ldots, k$. The confusion matrix is a classifier probability array.

We define q as the classifier probability vector for the classifier, with components $q_j = P\ \{\text{randomly selected object belonging to class } j\}$. The value can be calculated from the following formula:

$$q_j = \sum_{i=1}^{k} p_i C_{ij} \text{ or } q = C^{\mathrm{T}} p \tag{14.58}$$

If the classifier classifies N objects, where n_j is divided in the j-th class, the maximum likelihood estimation yields the estimated value of q as $\widehat{q}$, the component of which is

$$\hat{q} = \frac{n_j}{N} \tag{14.59}$$

14.6 Summary

Figure 14.11 summarizes the contents of this chapter. The main formulas are listed in Table 14.2.

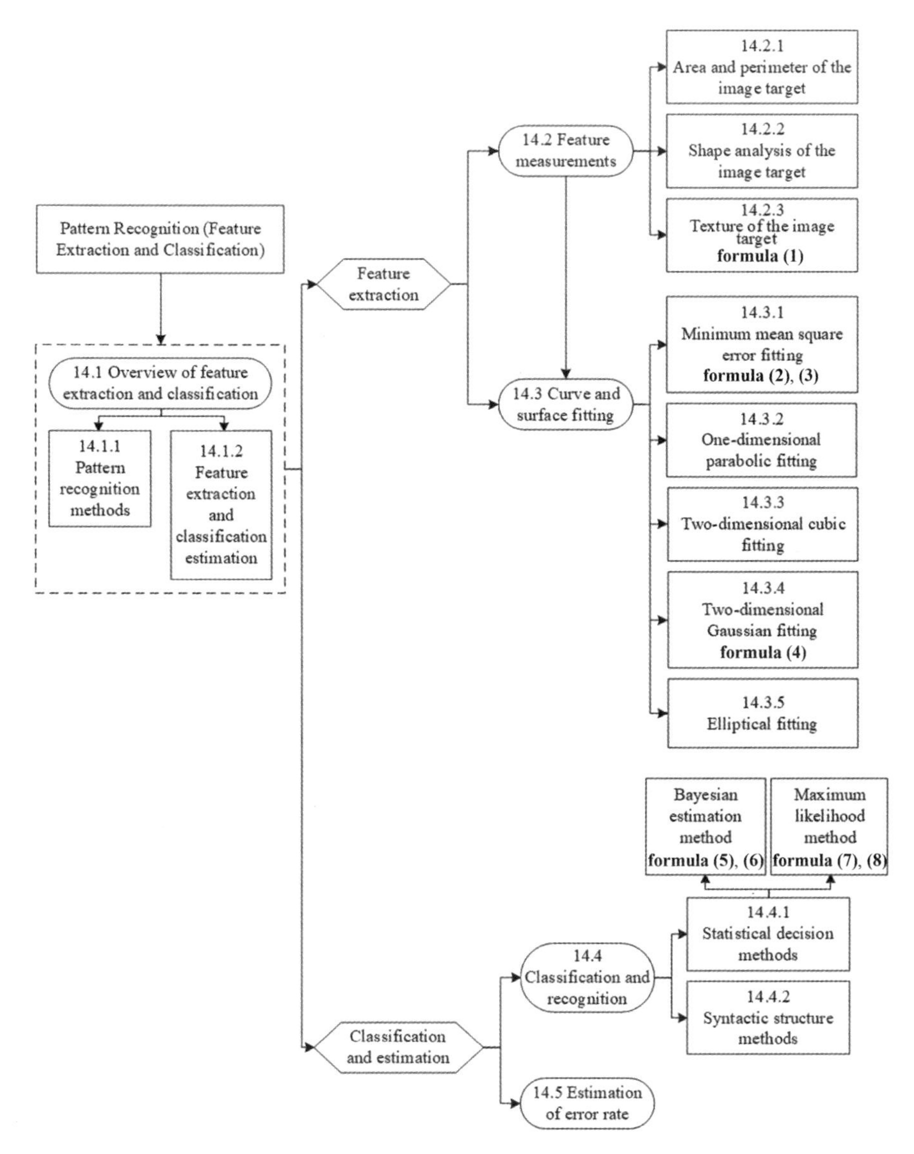

Fig. 14.11 Summary of the contents of this chapter

Table 14.2 Summary of mathematical formulas in this chapter

formula (1)	$I = \sum_{i=1}^{N} \frac{H_i - M/N}{\sqrt{\frac{H_i(1-H_i/M)+M(1-1/N)}{N}}}$	14.2.3
formula (2)	$\mathrm{MSE} = \frac{1}{N}\sum_{i=1}^{N}[y_i - f(x_i)]^2$	14.3.1
formula (3)	$f(x) = c_0 + c_1 x + c_2 x^2$	14.3.1
formula (4)	$z_i = A\exp[-\frac{(x_i-x_0)^2}{2\sigma_x^2} - \frac{(y_i-y_0)^2}{2\sigma_y^2}]$	14.3.4
formula (5)	$p(\mu\|D) = \frac{p(D\|\mu)p(\mu)}{\int p(D\|\mu)p(\mu)\mathrm{d}\mu} = a\prod_{k=1}^{n} p(x_k\|\mu)p(u)$	14.4.1
formula (6)	$\mu_n = (\frac{n\sigma_0^2}{n\sigma_0^2+\sigma_0^2})\hat{\mu}_n + \frac{\sigma^2}{n\sigma_0^2+\sigma^2}\mu_0$	14.4.1
formula (7)	$L(x_1, x_2, \ldots, x_n; \theta_1, \theta_2, \ldots, \theta_k) \hat{=} L(x;\theta) = \prod_{i=1}^{n} p(x_i; \theta_1, \theta_2, \ldots, \theta_k)$	14.4.1
formula (8)	$L\left(x;\hat{\theta}\right) = \max_{\theta} L(x;\theta)$	14.4.1

References

1. Alt FL. Digital pattern Recognition by Moments. JACM 1962;(9).
2. Levi G, Montanari U. A. Gray-weighted skeleton. *Information and Control*, 1970;(17).
3. Li, H. Spectral analysis of signals [Book Review]. Sig Proc Magaz IEEE. 2007;24(1).
4. Richards JA. Remote sensing digital image analysis. Remote sensing digital image analysis: Springer-Verlag; 1986.
5. Sa JPMD. Pattern recognition: concepts, methods and applications. Springer; 2001.
6. Castleman, KR, White BS. The tradeoff of cell classifier error rates. Cytometry, 1980;(1).
7. Theodoridis S, Koutroumbas K. Pattern recognition. In: Pattern recognition, 4th edn. Academic Press; 2008.

Chapter 15
Technology IV of Remote Sensing Digital Image Processing: Color Transformation and 3D Reconstruction

Chapter Guidance After appropriately processing remote sensing images and extracting useful information, it is necessary to properly express the information to ensure that the image is suitable for human eyes and mimics the real world. To this end, proper color transformation and 3D reconstruction of the image processing results must be performed.

The contents of this chapter can be summarized as follows:

Section 15.1: Color transformation and vector space expression. Introduces the color synthesis and color fusion display of remote sensing images to ensure consistency with human vision filtering characteristics;

Section 15.2: Data foundation of three-dimensional image reconstruction from two-dimensional images (sequence). Describes the optical transfer function and related techniques for the application of the convolution method in small-scale, near-distance optical slice processing and 3D reconstruction;

Section 15.3: Imaging fundamentals for 3D image reconstruction pertaining to tomographic scan technology. Introduces the inverse Fourier transform at the mesoscale, distance 3D tomography and reconstruction application method;

Section 15.4: 3D image reconstruction parameter basis pertaining to the 3D measurement technology. Laser scanning is used as an example to describe the spatial transformation and linear equations pertaining to large-scale, long-distance 3D imaging stereo images and parametric characteristics of the alignment. The system display technology of colors is described, along with the three-dimensional reproduction of remote sensing images.

The logical relationship between the sections of this chapter is shown in Fig. 15.1.

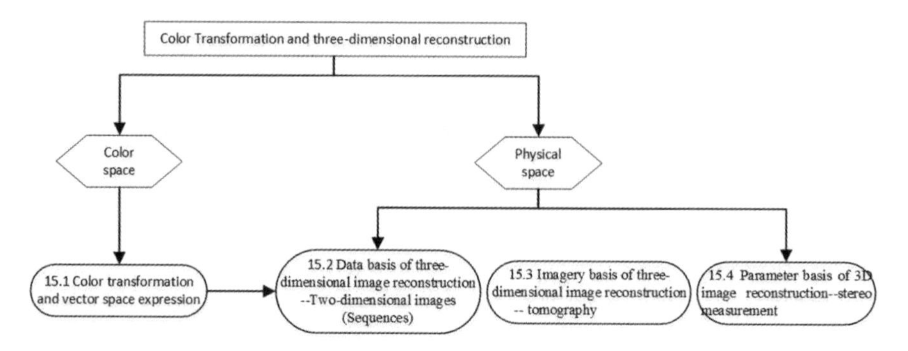

Fig. 15.1 Logical framework of this chapter

15.1 Color Transformation and Vector Space Expression

The human visual system is sensitive to the red, green, and blue spectral bands, resulting in color image vision. The process of generating human eye vision is equivalent to the mathematical model process of image degradation and restoration. The criteria for obtaining the optimal image by this mathematical model are based on the principle of the least mean squared error filter. The process of color image transformation processing is similar to that of human eye vision formation, and its purpose is to restore the degraded color image to the image with the best visual effect for adults.

15.1.1 Color Vision

The human eye can convert light signals into different sensitized chemical characteristics of nerve impulses. When the light conditions are satisfactory, the cones on the retina divide the visible spectrum of the electromagnetic spectrum into three bands according to the obtained photochemical characteristics: red, green, blue. The three types of cone cells in the human visual system are sensitive to the three colors red, green and blue. Figure 15.2 shows the sensitivity curve of human photoreceptor cells.

Three primary colors, red, green and blue, can be mixed in equal ratios to produce white. Any color can be obtained from the mixture of the three primary colors (red, green and blue). In 1802, T. Young proposed the three-color principle, which states that $C = aC_1 * bC_2 * cC_3$, where C_1, C_2, and C_3 are the three primary colors, a, b, and c are the weights of the three primary colors (ratio of three primary colors or concentration), and C is the synthesized color (which can be any color). This theory is generally termed the principle of three primary colors.

Color is a subjective feeling produced by external light acting on human visual organs. Any color vision can be measured according to three characteristics: hue,

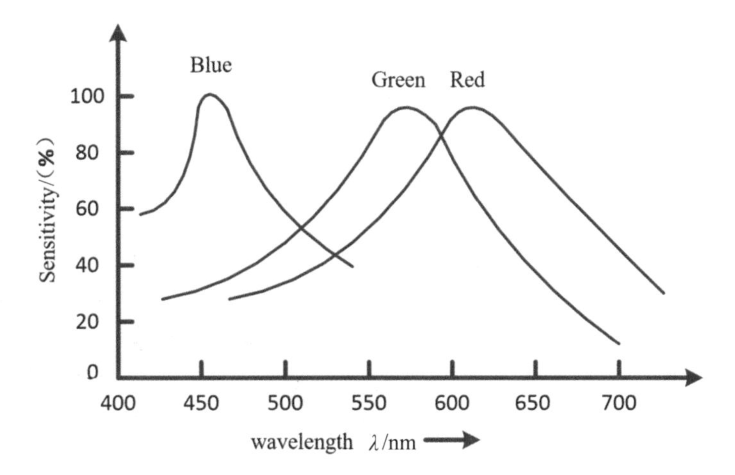

Fig. 15.2 Sensitivity of human photoreceptor cells

brightness, and saturation. Currently, three types of color space models are commonly used: RGB, HIS, and HLS.

1. RGB Color Model

The RGB model is a color cube in the 3D Cartesian color system. R (red), G (green), and B (blue) are the three coordinate axes of the color space. Each coordinate axis is quantized between 0 and 255, with 0 corresponding to the darkest (black) and 255 corresponding to the brightest. The origin is the starting point of the three coordinates of the three axes corresponding to the three primary colors: red (255, 0, 0), green (0, 255, 0), blue (0, 0, 255). Black (0, 0, 0) is the origin of the coordinate system, and white (255, 255, 255) is the opposite corner of the origin of the coordinate system. The three primary colors that lie on the diagonal line are of equal brightness and form a gray line.

2. HIS Color Model

The HIS color model is based on the visual characteristics of the human eye and represents a color in terms of the hue, intensity, and saturation (Fig. 15.3). The HIS model is based on a cylindrical coordinate system whose center axis is a straight line in RGB space ($R = G = B = 1$). The color that can be displayed in the HIS system is included in the RGB pyramid. The range of the saturation values of the light intensity (close to white) and light intensity (close to black) is small. The hue around the vertical axis H is represented in terms of the angle, $0 \leq H \leq 360$, with 0°, 120°, and 240° corresponding to red, green, and blue, respectively.

3. HLS Color Model

The color space represented by the HLS model is a double-hexagonal pyramid. The color is represented in terms of the hue, lightness, and saturation, as shown

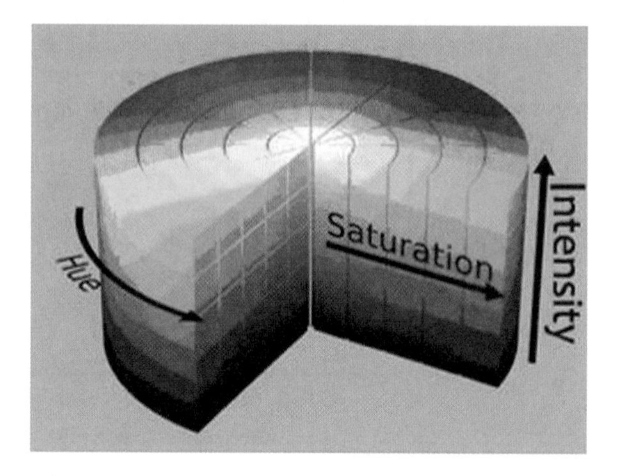

Fig. 15.3 HIS color model

in Fig. 15.4. Hue H surrounds the circumference of the vertical axis. The hue is represented by the angle of red as the starting point, with the trend occurring in the counter clockwise direction. The colors appear in the following order: red, yellow, green, cyan, blue and magenta. The vertical axis represents brightness L, with black and white representing 0 and 1, respectively. The radius of divergence along the horizontal plane from the vertical axis represents the saturation S. The axis is 0, and the maximum saturation is 1.

HLS model features:

(1) **HLS** is a canonical transformation of the HIS model
(2) **HLS** with $S = 1$ and $L = 0.5$ corresponds to the HIS-defined base plane
(3) The three-dimensional vector space satisfies the fork multiplication right-handed spiral definition

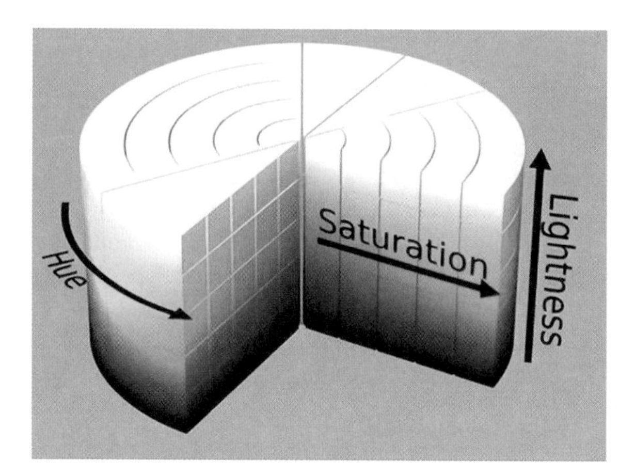

Fig. 15.4 HLS color model

$$\mathbf{HLS} = k_1 \boldsymbol{H} \times \boldsymbol{L} \times \boldsymbol{S} = k_2 \boldsymbol{S} \times \boldsymbol{H} \times \boldsymbol{L} = k_3 \boldsymbol{L} \times \boldsymbol{S} \times \boldsymbol{H}$$

(4) Different ***L*** slices represent the maximum and reasonable configuration of ***H*** and ***S*** under the brightness slice.

15.1.2 Color Space Transformation

Usually, image processing is performed in the RGB space. The display result of the image is often in a single color, and the display effect is inferior. Adjusting any one of the three components of RGB affects other components, and thus, a mutually independent color space is needed. In the HIS space, the correlations of brightness, chroma, and saturation are extremely weak, and the entities can be considered independent of one another. Therefore, we can independently process the three variables in the HIS space. The separation of the three components enables the attainment of saturation. The dynamic range is directly expanded, which decreases the correlation between the brightness of the various bands and fundamentally enhances the image quality. In addition, the brightness, chroma, and saturation can be simultaneously changed to different degrees for specific purposes. To exploit the display and quantitative calculations of the RGB and HIS systems, it is necessary to establish a conversion relationship between the systems.

1. RGB to HIS Conversion

We establish the rectangular coordinate system (x, y, z), rotate the RGB cube, ensure that the diagonal line coincides with the z-axis, and set the R axis in the xz plane. Next, we convert the polar coordinates defined in the xy plane into a cylindrical coordinate system. Model conversion is performed from the RGB space to HIS space, and this process is known as HIS transformation (Fig. 15.5).

The rotation matrix is

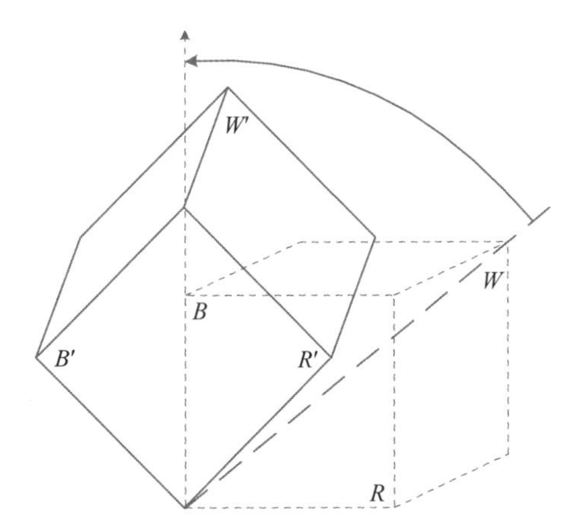

Fig. 15.5 Rotation of the RGB cube

$$\begin{bmatrix} x \\ y \\ z \end{bmatrix} = \begin{bmatrix} \frac{2}{\sqrt{6}} & -\frac{1}{\sqrt{6}} & -\frac{1}{\sqrt{6}} \\ 0 & \frac{1}{\sqrt{2}} & -\frac{1}{\sqrt{2}} \\ \frac{1}{\sqrt{3}} & \frac{1}{\sqrt{3}} & \frac{1}{\sqrt{3}} \end{bmatrix} \begin{bmatrix} R \\ G \\ B \end{bmatrix} \tag{15.1}$$

Moreover, it is necessary to define the cylindrical coordinate system, which is converted from polar coordinates (H, S) to HIS in the rectangular coordinate xy plane.

$$S = \sqrt{x^2 + y^2}, \begin{bmatrix} x \\ y \end{bmatrix} = \begin{bmatrix} \cos H \\ \sin H \end{bmatrix} S, \ H = \arctan\left(\frac{y}{x}\right), \ H \in (0, 2\pi), \ I = z \tag{15.2}$$

The definition of saturation S involves two limitations: (1) Saturation S and intensity I are not independent, which is inconsistent with the requirements of the HIS model; (2) the fully saturated color lies on a hexagon on the xy plane and not on the circle. Saturation S can be normalized by dividing by the maximum value of saturation S corresponding to different hues H. Saturation S can thus be defined as

$$S = \frac{s}{s_{\max}} = 1 - \frac{\sqrt{3}}{I} \min(R, G, B) \tag{15.3}$$

This completely saturated color falls on a unit circle in the xy plane.

$$H = \begin{cases} \arctan\left(\frac{y}{x}\right), & G \geq B \\ 2\pi - \arctan\left(\frac{y}{x}\right), & G \leq B \end{cases} \tag{15.4}$$

2. HIS to RGB

The model conversion from the HIS space to the RGB space is termed HIS transformation. The inverse transformation is performed analogously to the RGB transformation. The rotation matrix is defined as follows:

$$\begin{bmatrix} R \\ G \\ B \end{bmatrix} = \begin{bmatrix} \frac{1}{\sqrt{3}} & 0 & \frac{2}{\sqrt{6}} \\ \frac{1}{\sqrt{3}} & \frac{1}{\sqrt{2}} & -\frac{1}{\sqrt{6}} \\ \frac{1}{\sqrt{3}} & -\frac{1}{\sqrt{2}} & -\frac{1}{\sqrt{6}} \end{bmatrix} \begin{bmatrix} I \\ S \cos H \\ S \sin H \end{bmatrix} \tag{15.5}$$

The HIS transform depends on the sector of the color circle that the point of conversion falls on. There exist several variants of HIS conversion. From the perspective of color image processing, as long as hue H is an angle and the saturation and gradation are independent, the chosen form does not influence the processing result. A sample form is described in the following expressions:

When $0° \leq H < 120°$,

$$R = \frac{1}{\sqrt{3}}\left[1 + \frac{S\cos H}{\cos(60^\circ - H)}\right], B = \frac{I}{\sqrt{3}}(1 - S), G = \sqrt{3I - R - B} \tag{15.6}$$

When $120^\circ \leq H < 240^\circ$,

$$G = \frac{I}{\sqrt{3}}\left[1 + \frac{S\cos(H - 120^\circ)}{\cos(180^\circ - H)}\right], R = \frac{I}{\sqrt{3}}(1 - S), B = \sqrt{3I - R - G} \tag{15.7}$$

When $240^\circ \leq H < 360^\circ$,

$$\mathrm{B} = \frac{I}{\sqrt{3}}\left[1 + \frac{S\cos(H - 240^\circ)}{\cos(300^\circ - H)}\right], G = \frac{I}{\sqrt{3}}(1 - S), R = \sqrt{3}I - G - B \tag{15.8}$$

15.1.3 Color Features and Physical Essence of Remote Sensing Multispectral Images

All colors of a multispectral image result from the selective reflection of wavelengths of one band and absorption of other wavelengths. Multispectral color image synthesis can be realized through four methods: true color synthesis, pseudocolor synthesis, false color synthesis, and simulated true color synthesis. Pseudocolor synthesis refers to the conversion of a single-band grayscale image into a color image. True color and false color syntheses are color synthesis methods. Simulated true color synthesis is a color synthesis method that generates pseudocolors by simulation. These syntheses are also known as color enhancements.

In true color synthesis, the wavelength of the band selected in the color synthesis is the same as or similar to the wavelengths of red, green, and blue, and the color of the obtained image is similar to the true color.

In false color synthesis (or false color enhancement), for any multiband remote sensing image, any three bands are selected, and the three primary colors (red, green, and blue) are assigned. The wavelengths of the selected band in the color synthesis are different from those of red, green, and blue.

Pseudocolor density segmentation converts different gray levels in a single-band grayscale image to a certain color according to a specific functional relationship through the density division method. Subsequently, different colors are assigned to each segment of the range of luminance values for segmentation. Theoretically, a single-band image gray can be divided into 256 levels, and 256 colors can be assigned. The number of levels can be determined according to the application requirements.

15.1.4 Application of Color Transformation in Remote Sensing

1. Color Balance

Color balance is implemented through a color check. Specifically, we check whether all gray objects are displayed in gray. Second, we check whether the color of the high saturation S has a normal hue H. If the image has a distinct black or white background, significant peaks occur in the histogram. If the peaks in each histogram are at different gray levels of the three primary colors, the color is imbalanced.

The color balance method employs a linear grayscale transform for each individual R, G, B image. Usually, only two of the component images need to be transformed to match the third image. The following methods are commonly used for designing grayscale transformation functions: (1) selecting relatively uniform light gray and dark gray areas in the image; (2) calculating the average gray values of all three component images in the two areas; and (3) using linear contrast scaling for the two component images to match it with the third. If the two regions have the same gray level in all three component images, the color balance is established. Figures 15.6 and 15.7 show the comparison charts before and after color balance.

2. Saturation and Color Enhancement

When the image saturation is insufficient, the image is not vivid, and it is challenging to distinguish the details in the image. Therefore, saturation S enhancement and hue H conversion must be performed.

Fig. 15.6 Original image pertaining to standard pseudocolor synthesis

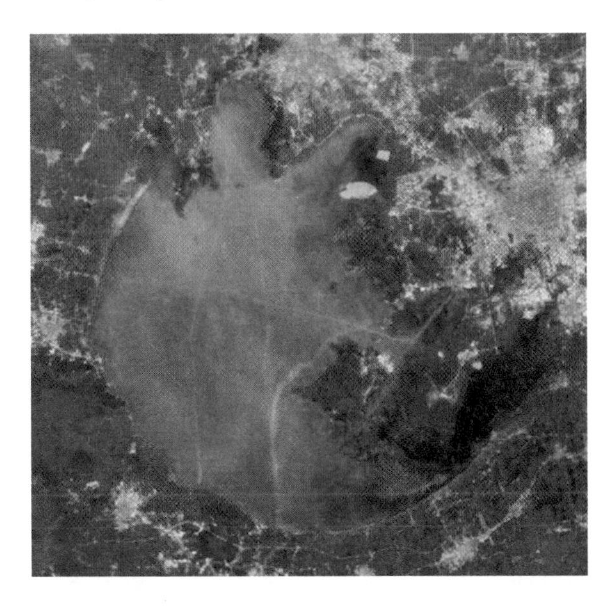

Fig. 15.7 Color balanced image (Equalization transformation in RGB channels)

Saturation *S* enhancement: First, the remote sensing image is transformed from the RGB color space to the HIS color space, and the saturation *S* component is stretched and converted to the RGB color space display to increase the image saturation.

Hue *H* Transform: In the HIS color space, the hue is expressed as an angle. Two methods are available:

(a) Each pixel in the *H* component of the remote sensing image is added or subtracted by the same degree. If the degree is small, the color of the image will become cooler or warmer. If the degree is large, the color of the image drastically changes.
(b) The hue *H* component in the remote sensing image is hue-stretched, and each pixel in the hue *H* component is multiplied by a number greater than 1. The difference in color in the corresponding spectral range on the remote sensing image increases.

Figures 15.8, 15.9, 15.10 and 15.11 show the corresponding details.

3. Pseudocolor

The pseudocolor method is used to match each gray level (located on the center axis of the color cylinder) to a point in the color space by a density division method and map the image into a color image. First, the gray histogram of the image is statistically obtained. The gray level of the image is divided according to the application, and a suitable color is selected to assign different gray levels. The original images shown in Figs. 15.12 and 15.13 are thermal infrared images of the Taihu Lake area in October 2009.

Fig. 15.8 Raw image(RGB)

Fig. 15.9 HIS color composite image after HIS conversion

4. Image Fusion Using HIS Transform

The images acquired by different sensors have different characteristics. Through color conversion (HIS transform and RGB transform) for image fusion, the advantages of different data can be exploited. In this manner, the rich spectral information of the image can be preserved while enhancing the spatial or temporal resolution of the image and practicality of the image. The method of image fusion based on color transformation (HIS transform and RGB transform) can be divided into four

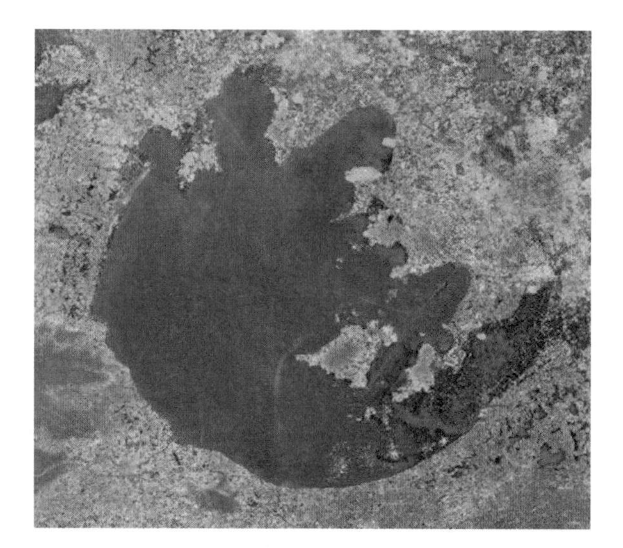

Fig. 15.10 Saturation enhanced image

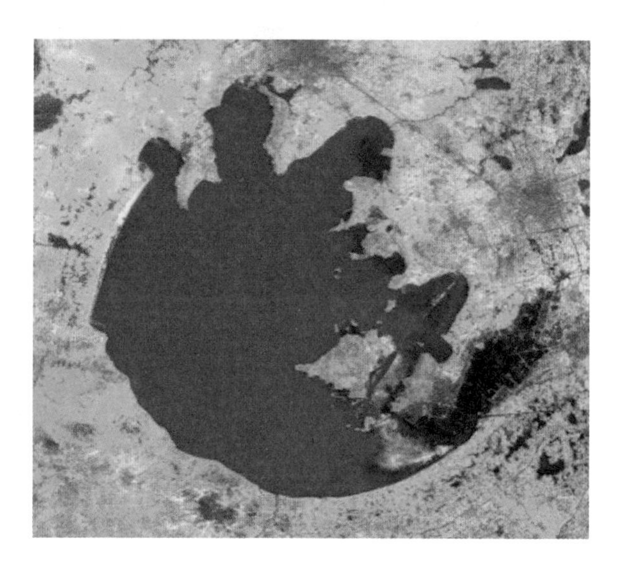

Fig. 15.11 Image after hue conversion

steps: (1) registration; (2) HIS transform; (3) replacement of I luminance component; (4) RGB transform. The original images shown in Figs. 15.14, 15.15 and 15.16 are examples of thermal infrared images of the Taihu Lake area in October 2009, the comparison of which shows the changes before and after fusion.

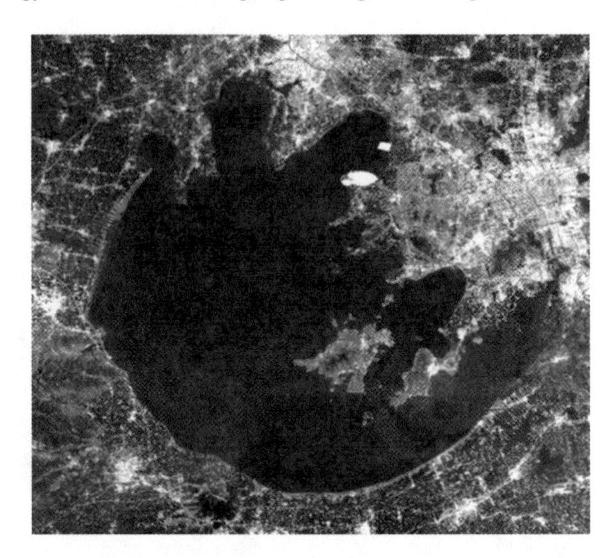

Fig. 15.12 Single-band thermal infrared grayscale image

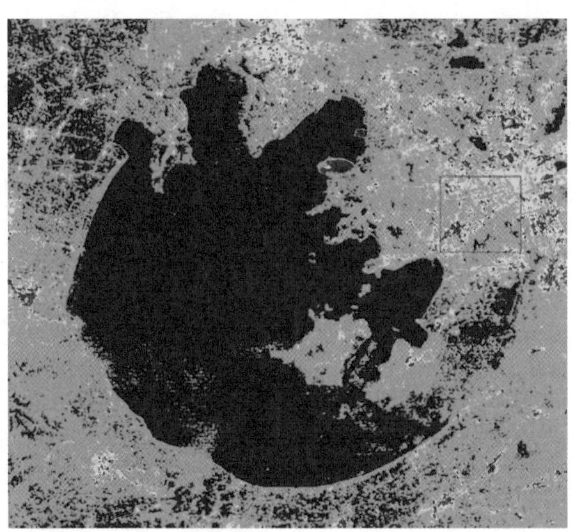

Fig. 15.13 Pseudocolor density-segmented image

15.2 Date Basis of 3D Image Reconstruction: Two-Dimensional Images (Sequences)

A two-dimensional digital image can be considered an image consisting of gray values determined by a bivariate function, which can be directly generalized to three dimensions, allowing us to process images in which the gray value is a function of three spatial variables. However, three dimensions impose higher requirements for data processing. In this section, the three-dimensional model of the optical slice is dissected from a biological viewpoint. Considering this aspect, thick sample imaging is examined to realize reconstruction from a sequence of two-dimensional images to

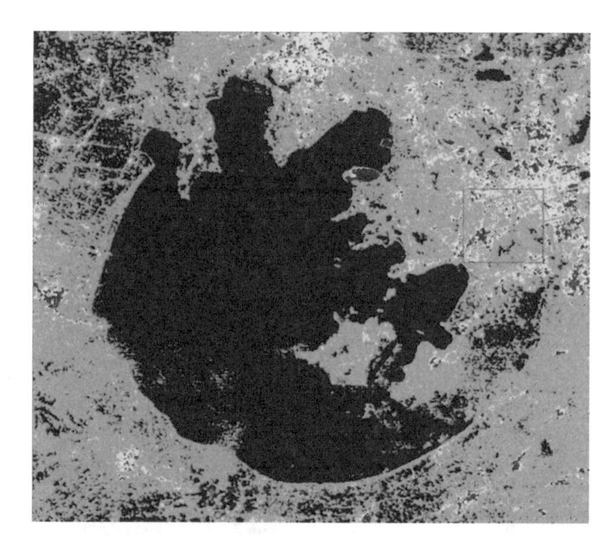

Fig. 15.14 Pseudocolor thermal infrared image

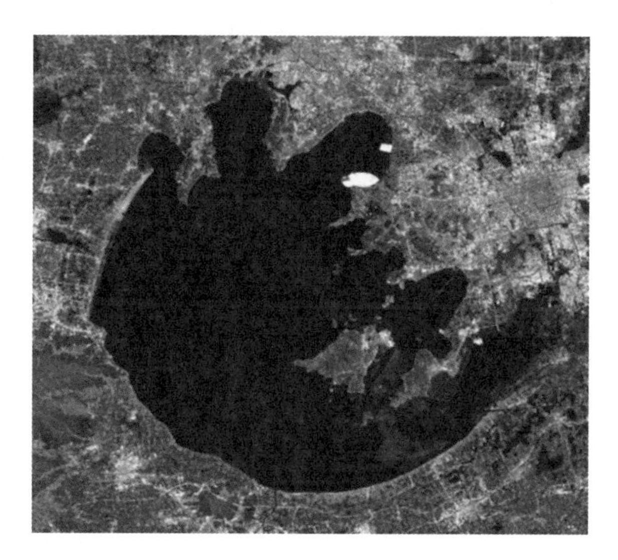

Fig. 15.15 High-resolution panchromatic image

three-dimensional images. Finally, we describe the application of three-dimensional reconstruction based on aerial remote sensing images. The logic diagram for this section is presented.

15.2.1 Optical Slices

Optical microscopy is a common tool for histology and microanatomy. The physiological sections are three-dimensional in microscopic dimensions. However, this

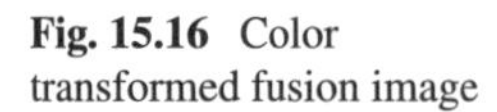
Fig. 15.16 Color transformed fusion image

aspect poses a problem in analyses based on conventional optical microscopes. First, only the structures at or near the focal plane are visible. Second, although the structure outside the focal plane is visible, it may be obscure. The structures far from the focal plane are not visible, but they influence the recorded images.

This three-dimensional effect can be overcome by using multiple cut edges. Specifically, the sample is cut into a series of slices, and each part is separately examined to understand the three-dimensional structure of the sample. The sample can be digitized by setting the focal plane at different positions on the optical axis and processing each layer of the image to weaken or remove the defocus information generated by the structure on the adjacent plane.

15.2.2 Thick Specimen Imagery

Figure 15.17 shows the optical system of a microscope imaging a sample of thickness T. The origin of the three-dimensional coordinate system of the system is located at the bottom of the sample, and the z-axis coincides with the optical axis of the microscope. The image distance d_i is fixed, and the focal plane is located at $z = z'$, that is, at position d_f below the objective lens. The image plane has its own coordinate system (x', y') whose origin lies on the z-axis.

According to the imaging equation of the lens, the focal length of the objective lens determines the distance from the lens to the focal plane d_f:

$$\frac{1}{d_i} + \frac{1}{d_f} = \frac{1}{f} \tag{15.9}$$

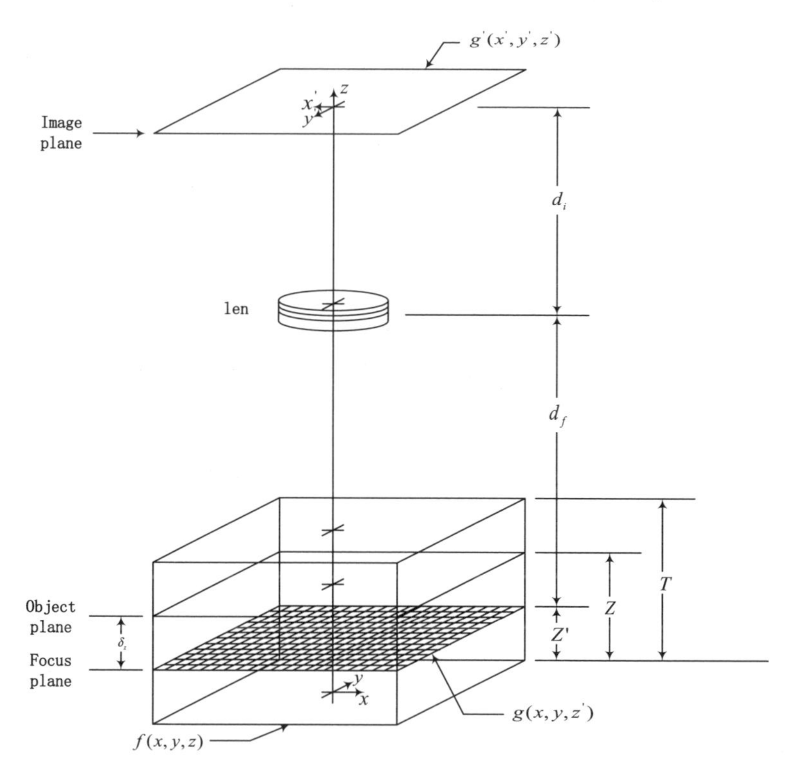

Fig. 15.17 Thick specimen imagery

This value determines the magnification of the objective lens:

$$M = \frac{d_i}{d_f} \tag{15.10}$$

For further analysis, the intensity (brightness or optical density) distribution can be described by function $f(x, y, z)$. The imaging result when the focal plane is located at z' is represented by $g(x', y', z')$. First, we define an ideal projection that projects the image plane back to the focal plane. Projection from $g(x', y', z')$ to $g(x, y, z')$ counteracts the amplitude variation introduced by the image system and the 180° rotation of the projection image. Therefore, a point within the sample entity (x, y, z') maps to a point (x, y, z') on the focal plane. Finally, it is desirable to establish a link between image (x, y, z') and sample function $f(x, y, z)$, in which the sample has an intensity of zero everywhere except at the object plane at $z = z_1$:

$$f(x, y, z) = f_1(x, y)\delta(z - z_1) \tag{15.11}$$

This expression corresponds to the two-dimensional imaging process of the object at z-z' beyond the focal plane. Because the defocused lens is a linear system, the following convolution relationship exists:

$$g_1(x, y, z') = f(x, y, z_1) * h(x, y, z_1 - z') \tag{15.12}$$

where h is the impulse response function PSF of the optical system, and the defocus amount is z_1-z'.

A three-dimensional sample can be represented by a model of a stack of object planes with a slight separation Δz along the z-axis:

$$\sum_{i=1}^{N} f(x, y, i\Delta z)\Delta z, N = \frac{T}{\Delta z} \tag{15.13}$$

The image of the plane of this stack at focal plane z' is the sum of the individual plane images, i.e.,

$$g(x, y, z') = \sum_{i=1}^{N} f(x, y, i\Delta z) * h(x, y, z' - i\Delta z)\Delta z \tag{15.14}$$

If $z = i\Delta z$, and Δz approaches 0, the summation is transformed to an integral:

$$g(x, y, z') = \int_{0}^{T} f(x, y, z) * h(x, y, z' - z)\mathrm{d}z \tag{15.15}$$

If we set the value of $f(x, y, z)$ outside of the field of view excluding $0 \ll z \ll T$ as 0 and expand the 2D convolution, the following expression can be obtained:

$$g(x, y, z') = \int_{-\infty}^{\infty}\int_{-\infty}^{\infty}\int_{-\infty}^{\infty} f(x', y', z')h(x - x', y - y', z' - z)\mathrm{d}x'\mathrm{d}y'\mathrm{d}z \tag{15.16}$$

Therefore, the imaging of a thick sample with a microscope includes a three-dimensional convolution of a sample function and a point spread function.

15.3 Imagery Basis of 3D Image Reconstruction: Tomography

This section describes the application of the inverse Fourier transform in three-dimensional tomography and reconstruction of mesoscale and intermediate distances. We focus on the acquisition and reconstruction of images from tomography and introduce the mathematical nature and applications of remote sensing imaging.

15.3.1 Principles

The steps of three-dimensional reconstruction for the tomography technique are shown in Fig. 15.18. Our goal is to reconstruct a series of two-dimensional cross-sectional images into three-dimensional images through relevant steps [1]. This section describes the acquisition and surface reconstruction of tomographic images.

The principle of image acquisition in tomography is shown in Fig. 15.19a. A plane-arranged X-ray passes through the object, and the passing ray line is evaluated using a linear X-ray detector. In this manner, the transmission intensity function shown in Fig. 15.19b [2] is generated.

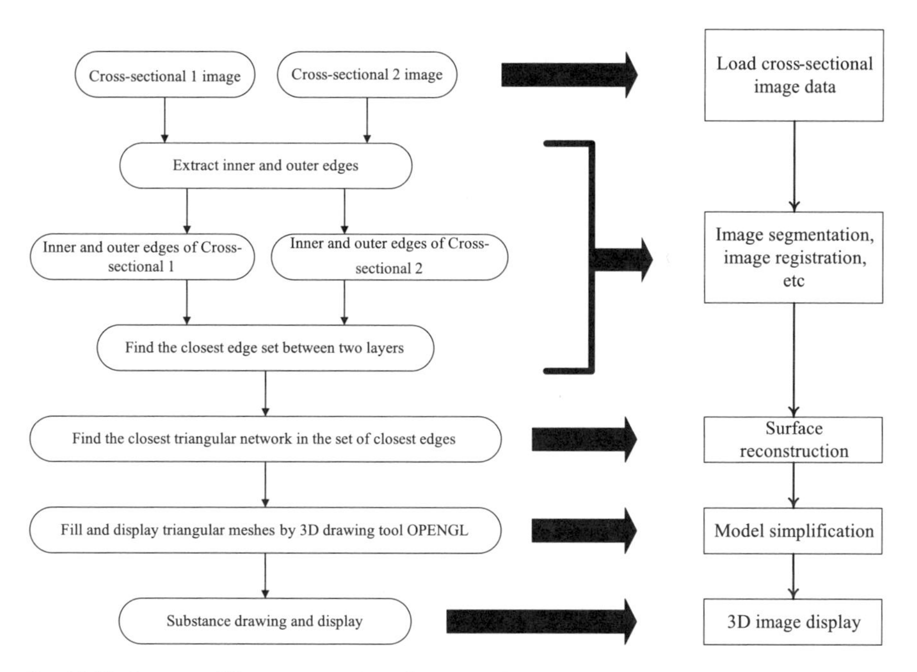

Fig. 15.18 Process of 3D reconstruction of tomography

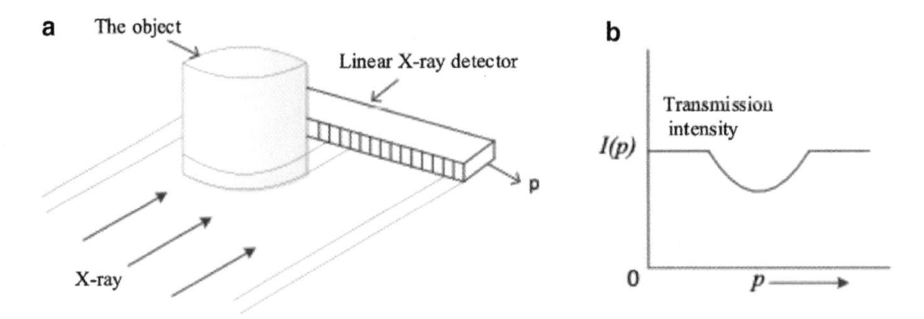

Fig. 15.19 Tomography

The obtained one-dimensional intensity function set can be used to determine the two-dimensional image of the object interface in the ray irradiation plane. A series of cross-sectional images are generated after multiple repetitions, and the three-dimensional image of the object is formed by stacking these images [3].

This process can be analytically described using the Radon transform:

$$d_r(p,\theta) = \int_{-\infty}^{\infty}\int_{-\infty}^{\infty} d(x,y)\delta[x\cos(\theta) + y\sin(\theta) - s]\mathrm{d}x\,\mathrm{d}y \tag{15.17}$$

where $d(x, y)$ is the density distribution of the object in the plane with an elevation of z, and the ray direction forms an angle θ with the y-axis (Fig. 15.19). For any p and θ, the value of $d_r(p,\theta)$ is the total amount of density on the ray at a distance p from the origin in direction θ relative to the y-axis. $d_r(p,\theta)$ is the one-dimensional projection of $d(x, y)$ to a line with an angle θ to the x-axis.

15.3.2 Image Reconstruction

The resolution along the ray direction is lost during projection, and the reconstruction uses multiple projections to restore the two-dimensional resolution of the image. The Radon transform yields an entity that can be reconstructed by projection. Considerable effort has been expended to obtain an optimal mathematical formula. The classic image reconstruction algorithms are the Fourier transform reconstruction algorithm and filtered back projection method. The following section describes the Fourier transform reconstruction algorithm associated with the classic parallel beam CT image reconstruction.

1. Central Slice Theorem

The theoretical basis of tomographic imaging is the central slice theorem, also known as the projection slice theorem or Fourier center slice theorem. The central slice theorem is based on the principle that the Fourier transform of a two-dimensional (one-dimensional) projection of a three-dimensional (two-dimensional) object is equal to the central section (centerline) of the Fourier transform of the object. In this case, when the projection is rotated, the central section (centerline) of its Fourier transform rotates with it. Therefore, the process of image reconstruction can be described as follows: First, the projection transformations at different angles are combined to form a complete Fourier transform of the object, and the object is reconstructed through the inverse Fourier transform.

When the detector is revolved 180° around the object, the central fragment of the object's two-dimensional Fourier transform corresponding to the direction of the detector can cover the entire Fourier space. In other words, the detector can be revolved 180° around the object. In this manner, we can obtain the full Fourier transform function. When these lines cover the entire full space, the original image can be obtained by a two-dimensional inverse Fourier transform.

2. Mathematical Derivation of the Fourier Transform Image Reconstruction

The 3D image reconstruction technology inputs a series of projection images, and the output is a reconstruction image. Let $f(x, y)$ be a two-dimensional image whose Fourier transform is

$$F(u,v)=\int_{-\infty}^{+\infty}\int_{-\infty}^{+\infty}f(x,y)\mathrm{e}^{-\mathrm{j}2\pi(ux+vy)}\mathrm{d}x\,\mathrm{d}y \tag{15.18}$$

The projection of $f(x, y)$ on the x-axis is

$$g_y(x)=\int_{-\infty}^{+\infty}f(x,y)\mathrm{d}y \tag{15.19}$$

The transform is implemented to obtain

$$G_y(u)=\int_{-\infty}^{+\infty}g_y(x)\mathrm{e}^{-\mathrm{j}2\pi ux}\mathrm{d}x=\int_{-\infty}^{+\infty}\int_{-\infty}^{+\infty}f(x,y)\mathrm{e}^{-\mathrm{j}2\pi ux}\mathrm{d}x\,\mathrm{d}y=F(u,0) \tag{15.20}$$

According to Eq. 15.21, the Fourier transform of the two-dimensional image $f(x, y)$ projected on the x-axis is equal to the two-dimensional Fourier transform $F(u, v)$ of $f(x, y)$ when the straight line $v = 0$. $F(u, v)$ is defined as

$$g_y(x)G_y(u)=F(u,0)|_{v=0}=G(R,\theta)|_{\theta=0} \tag{15.21}$$

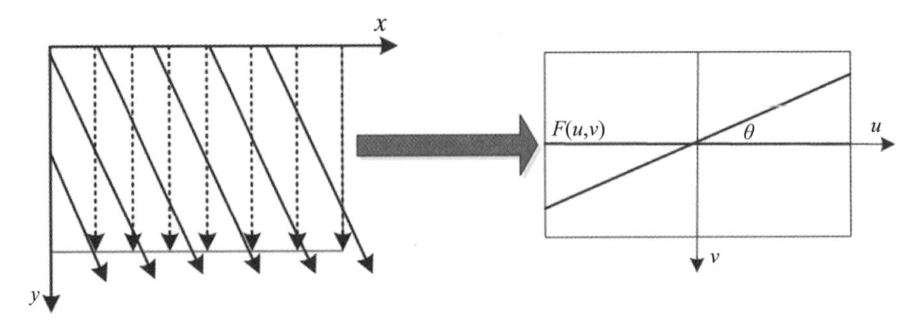

Fig. 15.20 New perspective projection

As shown in Fig. 15.20, the image projection value in the other direction can be obtained by changing the angle θ. At this time, the projection is in the s direction with angle θ to the x-axis. The relationship between the new projection axis coordinate system and original coordinate system is

$$\begin{bmatrix} s \\ t \end{bmatrix} = \begin{bmatrix} \cos\theta & \sin\theta \\ -\sin\theta & \cos\theta \end{bmatrix} \begin{bmatrix} x \\ y \end{bmatrix} \tag{15.22}$$

Projecting the image $f(x, y)$ to the s-axis yields

$$g(s, \theta) = \int\limits_{(s,\theta)} f(x, y)\mathrm{d}t \tag{15.23}$$

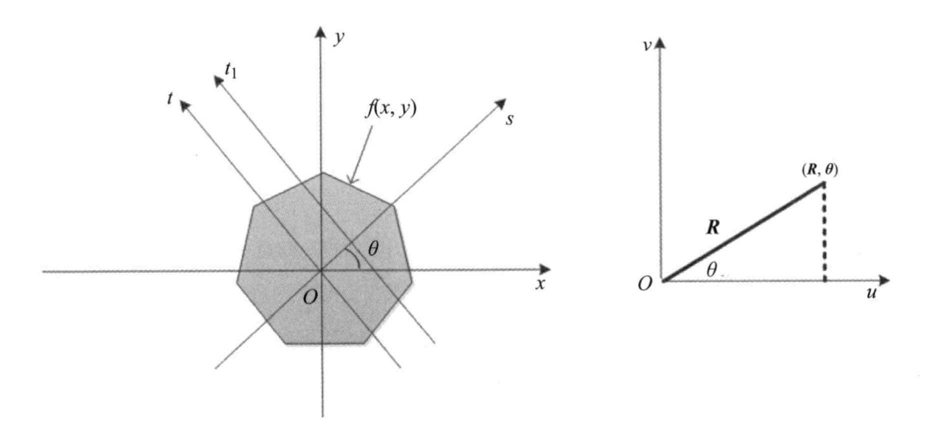

Fig. 15.21 Projection integral path

The integration path (Fig. 15.21) is a straight line t_1 parallel to t, i.e.,

$$s_1 = x\cos\theta + y\sin\theta \tag{15.24}$$

The one-dimensional Fourier transform of this projection is

$$G(R,\theta) = \int_{-\infty}^{+\infty} g(s,\theta)\mathrm{e}^{-\mathrm{j}2\pi Rs}\mathrm{d}s = \int_{-\infty}^{+\infty}\int_{-\infty}^{+\infty} f(x,y)\mathrm{e}^{-\mathrm{j}2\pi R(x\cos\theta + y\sin\theta)}\mathrm{d}x\mathrm{d}y$$

$$\begin{cases} u = R\cos\theta \\ v = R\sin\theta \end{cases} \tag{15.25}$$

Moreover,

$$G(R,\theta) = F(u,v) \tag{15.26}$$

The Fourier transform of the image in a certain direction is equal to the 2D Fourier transform of the central section cut by the angle corresponding to the image.

According to this discussion, to reconstruct an image, we only need to apply an inverse transform to the projected Fourier transform $G(R,\theta)$:

$$f(x,y) = \int_{-\infty}^{+\infty}\int_{-\infty}^{+\infty} F(u,v)\mathrm{e}^{\mathrm{j}2\pi(ux+vy)}\mathrm{d}u\mathrm{d}v \tag{15.27}$$

To accurately reconstruct the original image, adequate lines of the radiation must be projected. A set of images is stacked to form a three-dimensional image.

These aspects form the traditional theoretical basis for reconstructing three-dimensional images by tomography. According to this series of formula derivations, we can determine the three-dimensional information $d(x, y, z)$ of each point of the object to construct a three-dimensional model. Actual three-dimensional digital image reconstruction is based on the following methods: image preprocessing, image interpolation, image segmentation, image registration, surface reconstruction, model simplification, and image display. References [4–6] focused on image preprocessing and image registration.

The mathematical basis of image acquisition and image reconstruction from tomography—Radon transform and filtered back projection—pertain to the Fourier transform and inverse Fourier transform. Therefore, the tomography technique represents the practical application of the Fourier transform technique introduced earlier in this book.

15.4 Parameter Basis of 3D Image Reconstruction: Stereo Measurement

This section describes the parameter basis of three-dimensional image reconstruction, which is the mathematical essence of three-dimensional measurement, and its application in remote sensing. Laser scanning is considered as the example to highlight the parameter characteristics of space transformation and linear equations in stereoscopic pairs of large-scale and long-distance three-dimensional images.

15.4.1 Principle of Stereo Measurement

The existing methods for stereoscopic measurements include traditional stereoscopic projection and new 3D reconstruction based on laser scanning. Two viewpoints (sites) are used to obtain data, and data registration and integration are realized to describe the entire object.

Stereo projection directly simulates the way in which a human eye observes a scene and observes the same scene from two viewpoints. That is, one or two cameras in different positions are moved or rotated to shoot the same scene, and the spatial point is calculated by the principle of triangulation. The parallax between the two image pixels restores the depth information of the target object information. Finally, the surface shape of the object is recovered by the depth information [7].

In particular, 3D reconstruction based on laser scanning data includes four parts [8]: data acquisition, data registration, model construction and texture mapping. Effective preprocessing must be performed after data acquisition, and effective data fusion must be performed after data registration. To obtain a complete 3D data model, we need to scan the object from different directions, and thus, data registration and fusion are performed.

In general, the reconstruction process of stereoscopic projection (traditional three-dimensional reconstruction) is more complicated than reconstruction based on laser radar. Calibration, matching, and depth calculation processes must be implemented in stereoscopic projection.

15.4.2 Observation Equation

1. Observation Equation

If $P(X_0, Y_0, Z_0)$ is placed in front of the camera and imaged on two camera planes, similar triangles are used in the planes of zx and yz. Figure 15.22 shows that the line passing through the center of the lens from point P intersects the (image) plane at

$$X_l = -X_0 \frac{f}{Z_0}, Y_l = -Y_0 \frac{f}{Z_0} \tag{15.28}$$

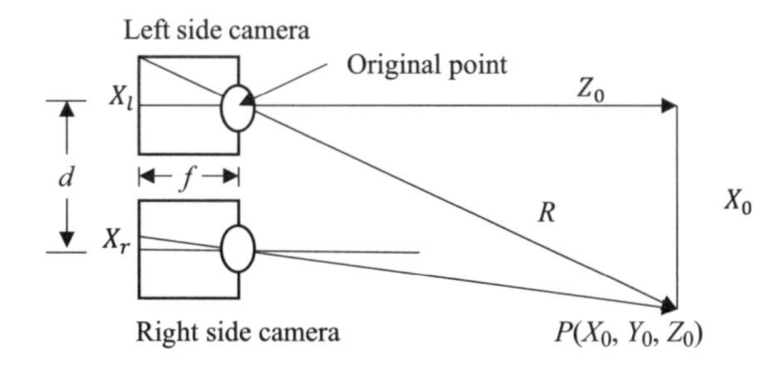

Fig. 15.22 Stereoscopic imaging

Similarly, the line passing through the center of the right camera from point P intersects the image plane at

$$X_r = -(X_0 + d)\frac{f}{Z_0} - d, Y_r = -Y_0 \frac{f}{Z_0} \tag{15.29}$$

Therefore, we set a two-dimensional coordinate system on each imaging plane. For convenience, the two coordinate system positions are rotated by 180° to counteract the inherent rotation in the imaging process. In this case,

$$x_l = -X_l, y_l = -Y_l, x_r = -X_r - d, y_r = -Y_r \tag{15.30}$$

The coordinates of the point in the image are

$$x_l = X_0 \frac{f}{Z_0}, y_l = Y_0 \frac{f}{Z_0} \tag{15.31}$$

and

$$x_r = (X_0 + d)\frac{f}{Z_0}, y_r = Y_0 \frac{f}{Z_0} \tag{15.32}$$

The y coordinates of the two images are the same.

Refining Eqs. (15.31) and (15.32) yields

$$X_0 = x_l \frac{Z_0}{f} = x_r \frac{Z_0}{f} - d \tag{15.33}$$

We thus obtain Z_0 and the normal depth equation

$$Z_0 = \frac{fd}{x_r - x_l} \tag{15.34}$$

Additionally, in the three-dimensional space, using similar triangles, the following expression can be established:

$$\frac{R}{Z_0} = \frac{\sqrt{f^2 + x_l^2 + y_l^2}}{f} \tag{15.35}$$

Rearranging and replacing the terms in Eq. 15.35 yields

$$R = \frac{d\sqrt{f^2 + x_l^2 + y_l^2}}{x_r - x_l} \tag{15.36}$$

This equation defines the actual depth, which specifies the total length from the origin to point P. For a narrow field (telescope) optical system, $X_{0,} Y_0 \leq Z_0$, and x_i and y_i are small relative to f. Thus, Eq. (15.36) can be approximated by Eq. (15.34).

2. Depth Calculation

The depth of the stereoscopic projection can be calculated as follows. First, for each pixel in the left image, we determine which pixel in the right image corresponds to the same point on the object. For the parallel alignment system of an image, the calculation can be performed in a row-by-row manner, since any point on the object is mapped to the same vertical position of the image and thus to the same scan line. Subsequently, $x_r - x_i$, is calculated to generate an offset map in which the grayscale represents the pixel offsets in proper proportions. Equation (15.34) is used to calculate each pixel by offsetting the image to generate a normal distance map. Finally, we calculate the x, y coordinates of each point.

$$X_0 = x_l \frac{Z_0}{f}, Y_0 = y_l \frac{Z_0}{f} \tag{15.37}$$

This procedure allows us to compute the x, y, z coordinate values for each point on the object, each of which corresponds to a pixel in the camera. We use Eq. (15.36) to calculate R as a function of x and y, and an actual depth map is generated. In both cases, we successfully map the visible surface of the three-dimensional object.

According to this theory, it is possible to calculate the 3D coordinates of the studied object to achieve 3D reconstruction. However, at present, because the main body of digital images is a discrete grid pattern, the three-dimensional reconstruction of the image involves the spatial coordinate system of the object, and a more practical reconstruction procedure must be applied when using a computer to simply perform the calculation.

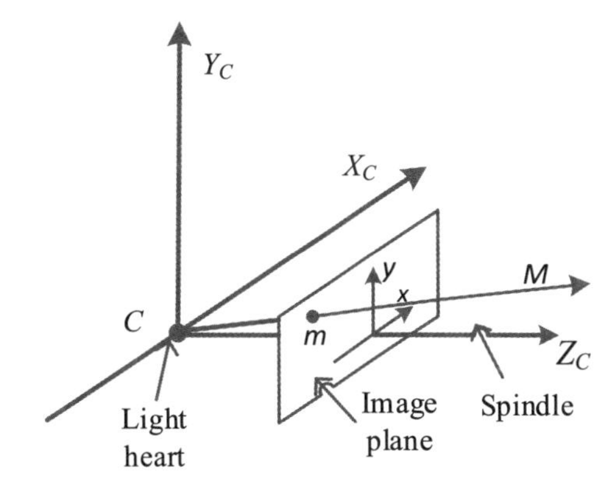

Fig. 15.23 Camera internal model

15.4.3 Three-Dimensional Reconstruction

1. Three-Dimensional Reconstruction of Stereoscopic Projection

(a) Principles

Stereo projection involves the following five steps: image acquisition, camera calibration, image preprocessing, stereo matching, and space reconstruction. We introduce camera calibration, stereo matching and postprocessing separately.

The first step involves image acquisition and camera calibration. To determine the camera parameters, camera model and coordinate system, camera calibration must be performed [9]. Figure 15.23 shows the principle and process of camera calibration.

① The relationship between the plane coordinates (x, y) (photo coordinates) and pixel coordinates (u, v) is

$$\begin{bmatrix} u \\ v \\ 1 \end{bmatrix} = \begin{bmatrix} \frac{1}{\mathrm{d}x} & r' & u_0 \\ 0 & \frac{1}{\mathrm{d}y} & v_0 \\ 0 & 0 & 1 \end{bmatrix} \begin{bmatrix} x \\ y \\ 1 \end{bmatrix} \tag{15.38}$$

where (u_0, v_0) is the azimuth element in the image, and r' is the tilt factor.

② The relationship between the image coordinates (X_c, Y_c, Z_c) and image plane coordinates (x, y) is

$$Z_c \begin{bmatrix} x \\ y \\ 1 \end{bmatrix} = \begin{bmatrix} f & 0 & 0 & 0 \\ 0 & f & 0 & 0 \\ 0 & 0 & 1 & 0 \end{bmatrix} \begin{bmatrix} X_c \\ Y_c \\ Z_c \\ 1 \end{bmatrix} \tag{15.39}$$

where f is the camera focal length.

③ The relationship between the photographic coordinates (X_c, Y_c, Z_c) and geodetic coordinates (X_w, Y_w, Z_w) is

$$\begin{bmatrix} X_c \\ Y_c \\ Z_c \\ 1 \end{bmatrix} = \begin{bmatrix} R & t \\ 0 & 1 \end{bmatrix} \begin{bmatrix} X_w \\ Y_w \\ Z_w \\ 1 \end{bmatrix} \tag{15.40}$$

where R is the amount of rotation, and t is the amount of translation.

④ Through this derivation, we obtain the photo coordinates (x, y):

$$Z_c \begin{bmatrix} u \\ v \\ 1 \end{bmatrix} = \begin{bmatrix} \frac{1}{dx} & r' & u_0 \\ 0 & \frac{1}{dy} & v_0 \\ 0 & 0 & 1 \end{bmatrix} \begin{bmatrix} f & 0 & 0 & 0 \\ 0 & f & 0 & 0 \\ 0 & 0 & 1 & 0 \end{bmatrix} \begin{bmatrix} R & t \\ 0 & 1 \end{bmatrix} \begin{bmatrix} X_w \\ Y_w \\ Z_w \\ 1 \end{bmatrix}$$

$$= \begin{bmatrix} f_u & r & u_0 \\ 0 & f_v & v_0 \\ 0 & 0 & 1 \end{bmatrix} [R\ t] \begin{bmatrix} X_w \\ Y_w \\ Z_w \\ 1 \end{bmatrix} = \boldsymbol{K}[R\ t]\boldsymbol{M} \tag{15.41}$$

where $f_u = \frac{f}{dx}$, $f_v = \frac{f}{dy}$, $r = r'f$, $\boldsymbol{K}$ is the camera parameter matrix, $\boldsymbol{P} = \boldsymbol{K}[R\ t]$ is the camera projection matrix, and $\boldsymbol{M}$ is the geodetic coordinate. Finally, the image coordinates $(u, v, 1)$ and geodetic coordinates (X_w, Y_w, Z_w) are obtained [10, 11].

The subsequent image preprocessing can be divided into three steps, as shown in Fig. 15.24.

The essence of stereo matching is to identify a point in an image and find the corresponding point in another image such that the two points are projections of the same object point in space [12]. Stereo matching involves three steps. First, an image feature corresponding to the actual physical structure is obtained from an image in the stereoscopic image pair, such as the left graph. Second, as shown in the right graph, we determine the same physical structure of the corresponding image features in the other image. Third, we determine the relative position between the two features. The second step is the key to realize matching.

At present, the limitations in the domestic and international research of feature point matching pertain to the scale invariant feature transform (SIFT). The matching ability of two images pertain to translation and rotation. Nevertheless, the affine transformation of the matching problem at any angle of the image, to a certain extent, exhibits more stable feature matching capabilities [13].

Finally, we introduce the algorithm of space point reconstruction. If image points p_1 and p_2 on the two cameras C_1 and C_2 can be detected separately from the two images, p_1 and p_2 are the corresponding points of the same point P. Assuming that C_1 and C_2 have been calibrated, their projection matrices are M_1 and M_2, respectively.

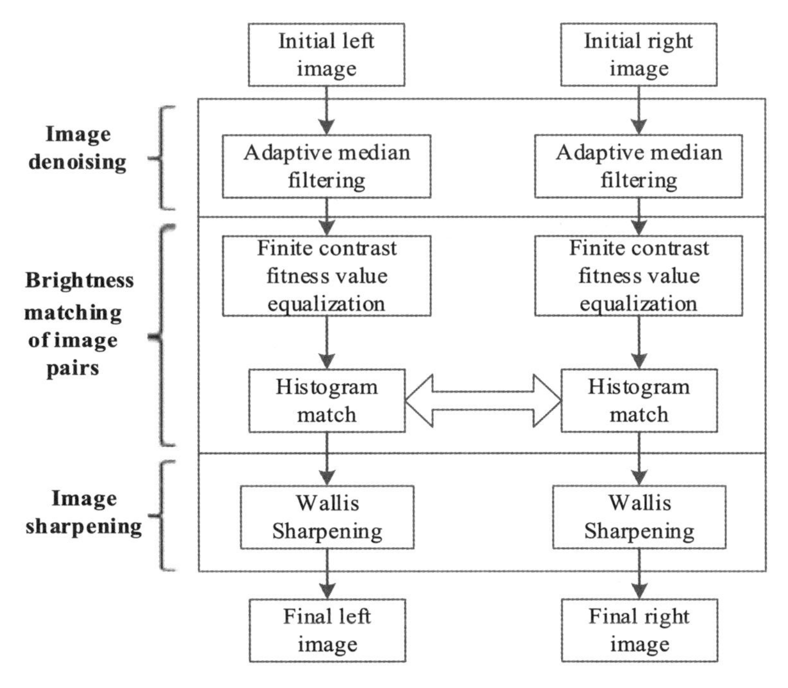

Fig. 15.24 Stereoscopic image preprocessing steps

The following expression can be obtained:

$$Z_{C1}\begin{bmatrix} u_1 \\ v_1 \\ 1 \end{bmatrix} = \boldsymbol{M}_1 \begin{bmatrix} X \\ Y \\ Z \\ 1 \end{bmatrix} = \begin{bmatrix} m_{11}^1 & m_{12}^1 & m_{13}^1 & m_{14}^1 \\ m_{21}^1 & m_{22}^1 & m_{23}^1 & m_{24}^1 \\ m_{31}^1 & m_{32}^1 & m_{33}^1 & m_{34}^1 \end{bmatrix} \begin{bmatrix} X \\ Y \\ Z \\ 1 \end{bmatrix} \tag{15.42}$$

$$Z_{C2}\begin{bmatrix} u_2 \\ v_2 \\ 1 \end{bmatrix} = \boldsymbol{M}_2 \begin{bmatrix} X \\ Y \\ Z \\ 1 \end{bmatrix} = \begin{bmatrix} m_{11}^2 & m_{12}^2 & m_{13}^2 & m_{14}^2 \\ m_{21}^2 & m_{22}^2 & m_{23}^2 & m_{24}^2 \\ m_{31}^2 & m_{32}^2 & m_{33}^2 & m_{34}^2 \end{bmatrix} \begin{bmatrix} X \\ Y \\ Z \\ 1 \end{bmatrix} \tag{15.43}$$

where $(u_1, v_1, 1)$ and $(u_2, v_2, 1)$ are the image homogeneous coordinates of p_1 and p_2 in their images, respectively; $(X, Y, Z, 1)$ is the homogeneous coordinate in the world coordinate system; and $m_{ij}^k (k = 1, 2; i = 1, \ldots, 3; j = 1, \ldots, 4)$ is the element of line i, column j of M_k. According to the camera's linear model formula, we can eliminate Z_{C1} and Z_{C2} and obtain the four linear equations pertaining to X, Y, Z:

$$\begin{cases} \left(u_1 m_{31}^1 - m_{11}^1\right)X + \left(u_1 m_{32}^1 - m_{12}^1\right)Y + \left(u_1 m_{33}^1 - m_{13}^1\right)Z = m_{14}^1 - u_1 m_{34}^1 \\ \left(v_1 m_{31}^1 - m_{21}^1\right)X + \left(v_1 m_{32}^1 - m_{22}^1\right)Y + \left(v_1 m_{33}^1 - m_{23}^1\right)Z = m_{24}^1 - v_1 m_{34}^1 \end{cases} \tag{15.44}$$

Fig. 15.25 Relative images of the experiment

$$\begin{cases} \left(u_2 m_{31}^2 - m_{11}^2\right)X + \left(u_2 m_{32}^2 - m_{12}^2\right)Y + \left(u_2 m_{33}^2 - m_{13}^2\right)Z = m_{14}^2 - u_2 m_{34}^2 \\ \left(v_2 m_{31}^2 - m_{21}^2\right)X + \left(v_2 m_{32}^2 - m_{22}^2\right)Y + \left(v_2 m_{33}^2 - m_{23}^2\right)Z = m_{24}^2 - v_2 m_{34}^2 \end{cases} \quad (15.45)$$

Expressions (15.44) and (15.45) indicate a straight line passing $O_1 p_1$ and $O_2 p_2$. Since the spatial point $P(X, Y, Z)$ is the intersection of $O_1 p_1$ and $O_2 p_2$, it must fit the two equations. Therefore, the above two equations can be used to obtain coordinates (X, Y, Z) of P[14, 15]. In practical applications, the data are always noisy. Therefore, we usually use the least squares method to obtain the spatial coordinates of the three points. These points can be surface fitted to obtain the spatial three-dimensional maps.

(2) Case Analysis

First, we consider the image acquisition (Fig. 15.25) in which image pairs are experimentally collected.

Second, camera calibration and image preprocessing is performed, including the correction and filtering of the camera lens distortion for subsequent processing to eliminate lens error noise interference [16]. The preprocessed images are shown in Fig. 15.26. Next, SIFT match is performed, the result of which is shown in Fig. 15.27.

After matching, we obtain the corresponding points of the left and right images, which exhibit a certain correspondence with the points on the actual objects. As described in the previous chapter, after preprocessing and matching are completed to obtain the three-dimensional coordinates of the spatial points, the data can be used to reconstruct the object in three dimensions. We present an example of the three-dimensional reconstruction of the Kunning Palace after SIFT matching (Fig. 15.29) using the Kunning Palace image dataset (Fig. 15.28).

2. Laser Scanning Technology for 3D Reconstruction

Two types of equipment are commonly used in three-dimensional scanning: three-dimensional laser scanners, and fault scanning equipment such as industrial CT.

Fig. 15.26 Preprocessed image pairs

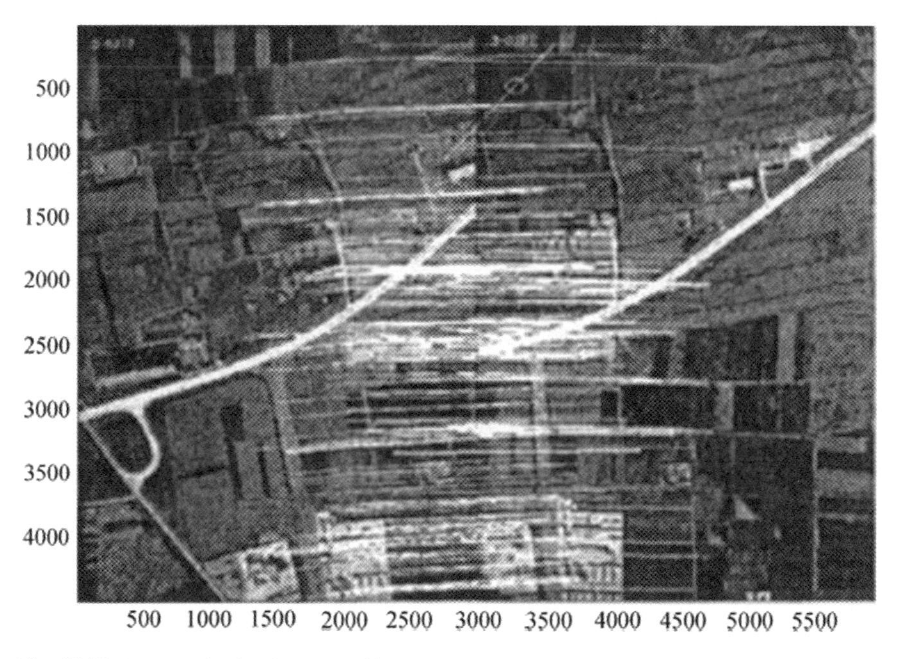

Fig. 15.27 Image pair after SIFT matching

Generally, the laser scanning method is used for a wide range of scenes, and fault scanning is performed for smaller objects, such as cultural relics and fossils.

(1) Principle

Generally, the scanner has a laser launcher and laser detection device, and two stepper motors control laser beams in the horizontal and vertical directions. The scanner emits a laser beam that reflects when the beam encounters an obstacle, and the laser detector in the scanner detects the reflected light to measure the scanning point (Fig. 15.30). By measuring the time of flight (TOF) between the laser transmission and reception,

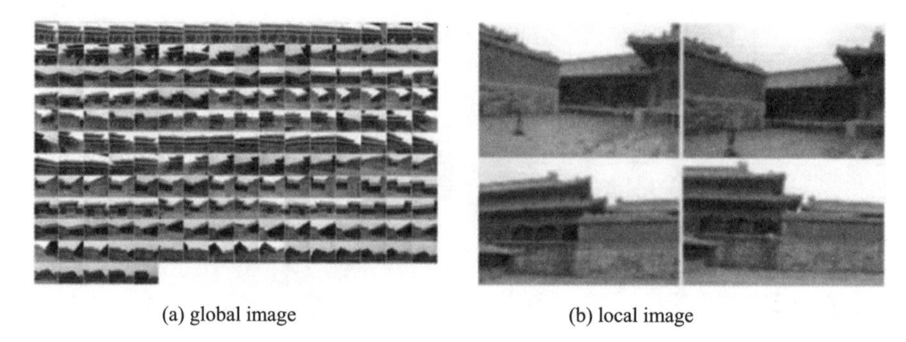

(a) global image (b) local image

Fig. 15.28 Kunning palace image dataset

(a) image 1 (b) image 2

Fig. 15.29 Three-dimensional reconstruction map of the Kunning Palace

the laser propagation distance, that is, the spherical coordinate $r(i, j)$, is calculated. Subsequently, after determining the angle of the laser emission, the angle α_i between the laser beam and X axis and angle β_j between the laser beam and Z axis can be calculated. The spherical coordinate system $(r(i,j),\alpha_i,\beta_j)$ can be transformed to a three-dimensional Cartesian coordinate system $(x(i, j), y(i, j), z(i, j))$ as [17]

$$\begin{cases} x(i, j) = r(i, j)\cos\beta_j\cos\alpha_i \\ y(i, j) = r(i, j)\cos\beta_j\sin\alpha_i \\ z(i, j) = r(i, j)\sin\beta_j \end{cases} \tag{15.46}$$

By repeating this process, the three-dimensional coordinates of all the object-surface points in the current viewpoint can be calculated.

As shown in Fig. 15.30, the raw laser scan data are in the form of a regular matrix, which is stored in “bitmap style”, indexed by rows and columns [18]. The laser scanning data only store the geometric information of the current scanning point, i.e., the local coordinate value (X, Y, Z) and reflectivity value R at that point. Because coordinate information is available at each point, these data are also referred to as the depth data or distance images.

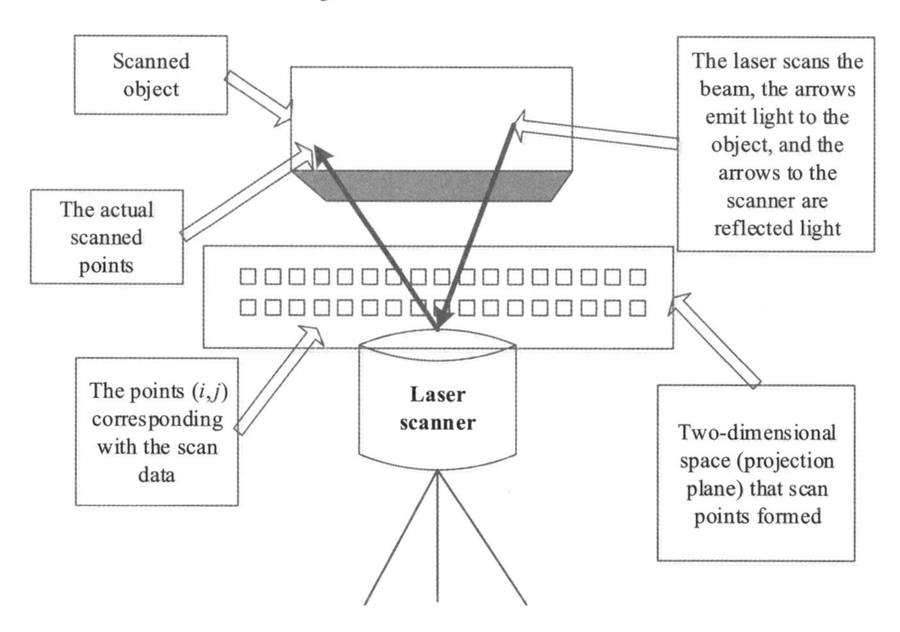

Fig. 15.30 Scanning process of laser scanner

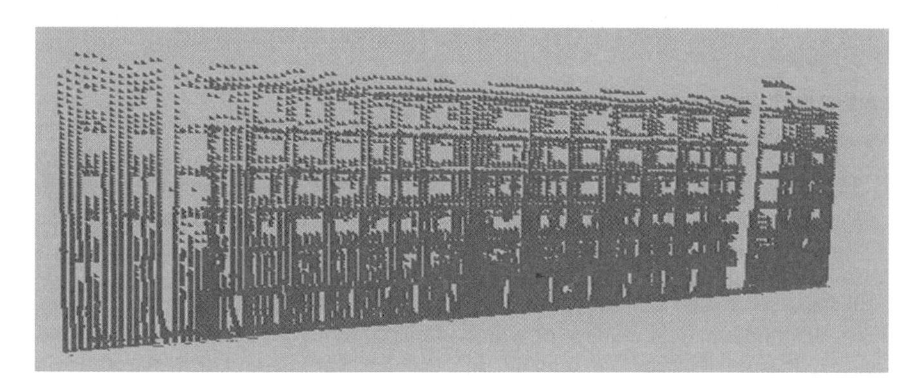

Fig. 15.31 Loading and displaying the cloud data

(2) Case Analysis

The experiment involves the three-dimensional texture reconstruction of the front of a building.

① Load the laser scan of the point cloud data, as shown in Fig. 15.31.
② After preprocessing and coordinate transformation, we create a positive triangulated irregular network (TIN) plot, as shown in Fig. 15.32.

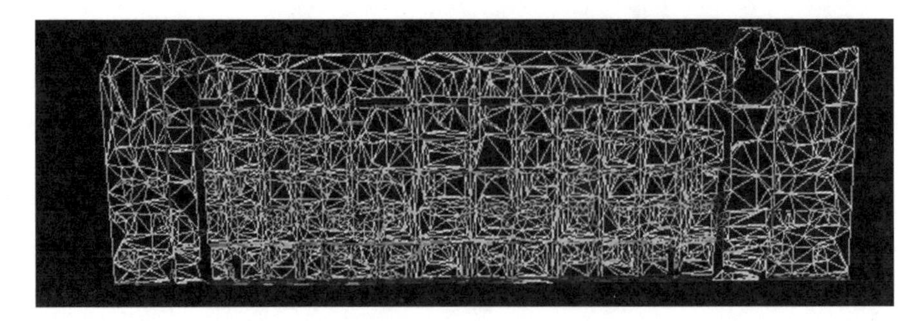

Fig. 15.32 Front TIN image

Fig. 15.33 Three-dimensional map after mapping

③ For visual effects, we can use a metallic glossy rendering surface, and the camera can be used to capture the real scene map for textural mapping. The final three-dimensional reconstruction is obtained, as shown in Fig. 15.33.

In the case of laser scanning, the core of stereo measurement lies in depth acquisition, the mathematical essence of which is spatial transformation and linear equations. To ensure that remote sensing images mimic the real world, Sects. 15.2–15.4 introduce different three-dimensional image processing methods and mathematical essences pertaining to the three dimensions of physical space. In summary, this chapter introduces techniques for color conversion and the three-dimensional rendering of remote sensing images.

15.5 Summary

This section describes the principle of color transformation and three-dimensional reconstruction of remote sensing images. The first part focuses on the color space of the image, and the second part relates to the physical space of the image. To describe the color space, the key task is to master the transformation of color space to satisfy the needs of practical applications. The physical space involves three-dimensional image reconstruction, which is divided into angle tomography and planar tomography, both

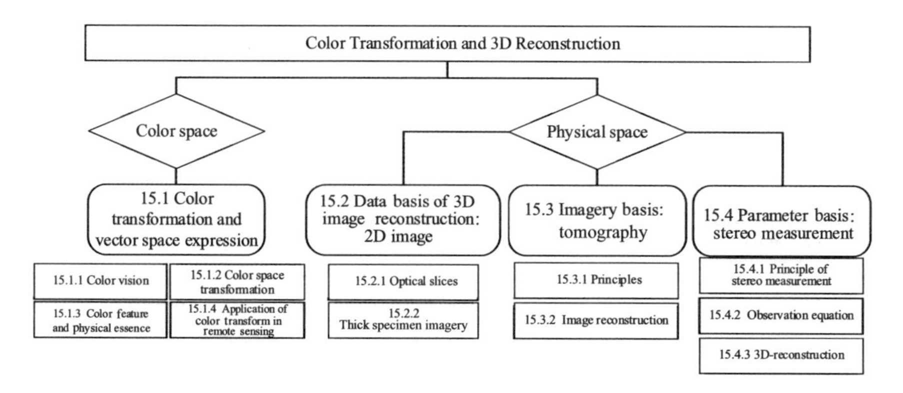

Fig. 15.34 Summary of this chapter

of which have inherent advantages and disadvantages. Recent laser three-dimensional scanning technologies represent a special case of angle chromatography. In recent years, 3D reconstruction technology has emerged as a popular research topic. Considerable research has been performed to solve bottleneck-related issues. Figure 15.34 shows the contents of this chapter. Table 15.1 summarizes the mathematical formula in this chapter.

Table 15.1 Formulas in this chapter

formula (1)	$\begin{bmatrix} x \\ y \\ z \end{bmatrix} = \begin{bmatrix} \frac{2}{\sqrt{6}} & -\frac{1}{\sqrt{6}} & -\frac{1}{\sqrt{6}} \\ 0 & \frac{1}{\sqrt{2}} & -\frac{1}{\sqrt{2}} \\ \frac{1}{\sqrt{3}} & \frac{1}{\sqrt{3}} & \frac{1}{\sqrt{3}} \end{bmatrix} \begin{bmatrix} R \\ G \\ B \end{bmatrix}$	15.1.2
formula (2)	$\begin{bmatrix} R \\ G \\ B \end{bmatrix} = \begin{bmatrix} \frac{1}{\sqrt{3}} & 0 & \frac{2}{\sqrt{6}} \\ \frac{1}{\sqrt{3}} & \frac{1}{\sqrt{2}} & -\frac{1}{\sqrt{6}} \\ \frac{1}{\sqrt{3}} & -\frac{1}{\sqrt{2}} & -\frac{1}{\sqrt{6}} \end{bmatrix} \begin{bmatrix} I \\ S\sin H \\ S\cos H \end{bmatrix}$	15.1.2
formula (3)	$g(x, y, z') = \int_{-\infty}^{\infty}\int_{-\infty}^{\infty}\int_{-\infty}^{\infty} f(x', y', z')* h(x - x', y - y', z - z')\mathrm{d}x'\mathrm{d}y'\mathrm{d}z$	15.2.2
formula (4)	$d_r(p, \theta) = \int_{-\infty}^{\infty}\int_{-\infty}^{\infty} d(x, y)\delta[x\cos\theta + y\sin\theta - s]\mathrm{d}x\mathrm{d}y$	15.3.1
formula (5)	$R = \frac{d\sqrt{f^2 + x_l^2 + y_l^2}}{x_r - x_l}$	15.4.2
formula (6)	$\begin{cases} x(i, j) = r(i, j)\cos\beta_j \cos\alpha_i \\ y(i, j) = r(i, j)\cos\beta_j \sin\alpha_i \\ z(i, j) = r(i, j)\sin\beta_j \end{cases}$	15.4.3

References

1. Su Y, Li Y, Han. Research on 3D reconstruction and visualization of CT data. Mech Des Manuf. 2009;(1).
2. Gonzalez, (Geng Qiuqi, Yan Yuzhi Translated). Digital image processing, 2nd ed. Beijing: Electronic Industry Press; 2007.
3. Jain AK. Fundamentals of digital image processing. PrenticeHall, Englewood Cliffs, NJ; 1989.
4. Fei Z, Junzhou J, Shijiang C. Implementation of three-dimensional crack reconstruction of rock CT fault layer sequences. Metal Mining. 2009(4).
5. Zheng Z, Fang-hua D. Three-dimensional Reconstruction of CT Tomography Images Using MATLAB. CT Theory Appl. 2004(13).
6. Tang Z. 3D data field visualization. Beijing: Tsinghua University Press; 1999.
7. Dang L. The study of three-dimensional reconstruction method based on binocular stereoscopic vision. Xi'an: Chang'an University; 2005 (In Chinese).
8. He W. Depth image registration in large-scale 3D reconstruction. Beijing: Peking University; 2004 (In Chinese).
9. Xiaosong Z, Hongwei Z, Guoxiong Z, et al. Research on camera calibration technology. J Mech Eng. 2002(38) (In Chinese).
10. Abdel-Aziz YI, Karara HM. Direct linear transformation from comparator coordinates into object space coordinates in close-range photogrammetry. In: Proceedings symposium on close-range photogrammetry, 1971. p. 1–18.
11. Zhang Z. A Flexible new technique for camera calibration. Pattern Anal Mach Intell, IEEE Trans. 2000(22).
12. Yi X, Jun Z, Yuanhua Z. Stereo vision matching technology. Comput Eng Appl. 2003(39) (In Chinese).
13. Zhao, H. Image registration algorithm based on point feature. Shandong: Shandong University (In Chinese).
14. Koenderink. The structure of images. Biological Cybernetics, 1984.
15. Lindeberg. Scale-Space for discrete Signals. IEEE Trans. PAMI, 1980;20:7–18.
16. Jiangang P. Key technology of 3D reconstruction based on laser scanning data. Beijing: Capital Normal University; 2005 (In Chinese).
17. Pan J. The key technology research of three-dimensional reconstruction based on laser scanning data. Beijing: Capital Normal University (In Chinese).
18. Han JW. Data mining concepts and technology. Beijing: Machinery Industry Press; 2001 (In Chinese).

Chapter 16
Applications of Remote Sensing Digital Image Processing

Chapter Guidance Based on the systematic theories and main methods of remote sensing digital image processing, this chapter introduces four typical applications to strengthen and deepen our understanding of remote sensing digital image processing.

This chapter describes the representation of remote sensing data in the base vector form in Case 1 (Sect. 16.1) to highlight that we can process remote sensing images even at the pixel level. The geometric correction of remote sensing images is described in Case 2 (Sect. 16.2), focusing on the gray level histogram, point operation, algebraic computing and geometric operation threshold segmentation of remote sensing images. This correction pertains to Case 3 (Sect. 16.3), which indicates how the gray histogram theory and fuzzy measure theory are applied to process remote sensing digital images. The minimum noise fraction (MNF) rotation of remote sensing images is described in Case 4 (Sect. 16.4), which demonstrates the integration of the image denoising theory and principal component analysis (PCA).

These four applications and their order have been selected considering the fundamentals of image processing. Using the basic theories of digital images, including base functions, base images and base vectors, we can mathematically analyze images by applying the gray histogram theory, fuzzy measurement theory and orthogonal transformation theory. The logical relationship of the sections in this chapter is shown in Fig. 16.1. The theoretical basis of each application is described in Table 16.1.

16.1 Case 1: Vector Representation of Remote Sensing Data

The base function, base image and base vector of the image are the core theories of image processing, and the representation of the image information is the core of image processing technology. Description of the image information depends largely on the choice of the base function. In the representation of image information based on the generation model, the image generation process can be simulated by the linear superposition of a set of base functions. By transforming the original image

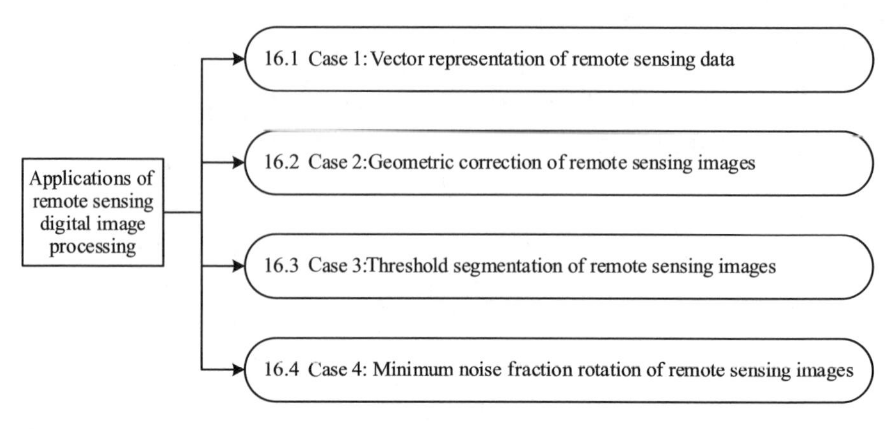

Fig. 16.1 Logical framework of this chapter

Table 16.1 Theoretical basis of the four considered cases

Application case	Example	Related contents	Mathematical essence
Base vector representation	Haar transformation	Discrete image transformation	Base function
Geometric correction	Geometric correction algorithms	Geometric computing	Image histogram
Threshold segmentation	Image mosaicing	Pattern recognition	Histogram–fuzzy measurement
MNF rotation	Denoising of Hyperion data	Image restoration	Orthogonal transformation

into the projection coefficient representation of the base function space, the inner structure of the image can be easily obtained, thereby enhancing the effectiveness of processing, including image recognition, noise reduction and compression. Therefore, the vector representation of remote sensing data can help understand the base functions, base images and base vector theory and their significance. The visualization of a vehicle monitoring system is a typical example. Moreover, online publishing of spatiotemporal data, online operation service of the map, feature query and map calculation have been performed (Fig. 16.2), and real-time tracking and monitoring of the dynamic target, polling display, path preset or track playback and other functions can be realized.

The theoretical method of image display and storage is the basis of the image representation mechanism. Normal images are stored in a continuous form of ordinary storage. However, for high-frequency-update dynamic target data, the amount of data is expected to be extremely large. Through the discrete image representation of the document (Fig. 16.3), an image can be encoded in a more compact data format

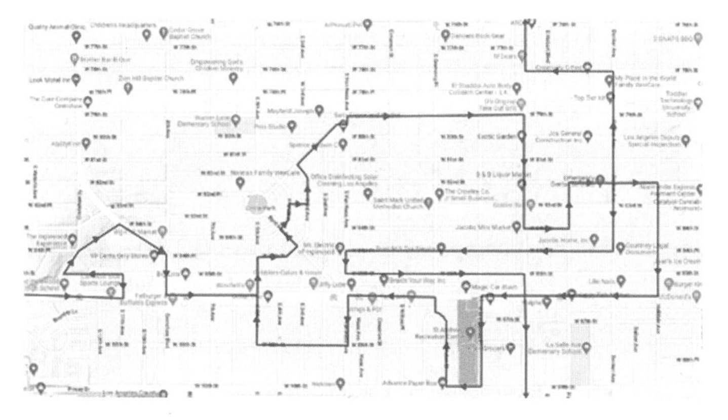

Fig. 16.2 Visualization of a dynamic target in map service

Fig. 16.3 Principle of discrete representation of images

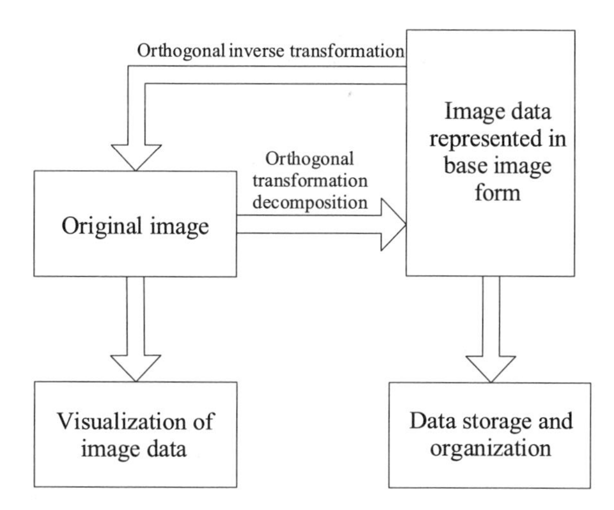

without losing any or losing only a small portion of the information. This "representation" of an image is a specific expression (or data format) that defines the image, in the form of a matrix or vector.

We define the linear transformation from a matrix $\boldsymbol{F}$ sized $N \times N$ to another matrix $\boldsymbol{G}$ sized $N \times N$ as

$$G_{m,n} = \sum_{i=0}^{N-1} \sum_{k=0}^{N-1} F_{i,k} \Im(i, k, m, n) \tag{16.1}$$

In the following matrix, $\Im(i, k, m, n)$ is the kernel function of the transformation. The matrix is a block matrix sized $N^2 \times N^2$, corresponding to N rows of N blocks, which represent a matrix sized $N \times N$. The blocks are indexed as m, n, while elements from child blocks are indexed as i, k.

$$
\begin{array}{c} \\ m=1 \\ m=2 \\ \\ m=N \end{array}
\begin{array}{c} n=1 \; n=2 \; n=N \\ \begin{bmatrix} [] & [] & \cdots & [] \\ [] & [] & \cdots & [] \\ \vdots & \vdots & \ddots & [] \\ [] & [] & \cdots & [] \end{bmatrix} \end{array}
$$

If $\Im(i, k, m, n)$ can be decomposed to the product of the row and column vector functions, i.e., if

$$\Im(i, k, m, n) = T_r(i, m) T_c(k, n) \tag{16.2}$$

the transformation can be considered separable.

The two-dimensional inverse transform reconstructs the original image by summing a set of appropriately weighted base images. Each element in the transformation matrix $\boldsymbol{G}$ is the coefficient of the corresponding base image used for multiplication in the summing process.

A base image can be generated by inversely transforming a coefficient matrix containing only one nonzero element (the value is 1). There exist N^2 such matrices, producing N^2 base images. We set one of the sparse matrices as

$$G^{p,q} = \left\{\delta_{i-p,j-q}\right\} \tag{16.3}$$

where i and j represent rows and columns, respectively, and p and q represent the positions of nonzero elements. Thus, the inverse transformation of Eq. (16.1) is

$$F_{m,n} = \sum_{i=0}^{N-1} T(i, m) \left[\sum_{k=0}^{N-1} \delta_{i-p,k-q} T(k, n) \right] = T(p, m) T(q, n) \tag{16.4}$$

Thus, for a separable unitary transformation, each base image is the outer product (vector product) of two rows of the transformation matrix. Similar to a one-dimensional signal, the base image set can be considered the unit set component of the original image, which is also the basic structural element that composes the original image. The transformation is implemented by determining the coefficients to be decomposed, while the inverse transformation is implemented by weighting the images to be reconstructed.

In the Haar transform, the Haar function is used to perform the symmetric and separable unitary transformation of the base function. The Haar function is defined as

$$h_0(x) = \frac{1}{\sqrt{N}}, h_k(x) = \frac{1}{\sqrt{N}} \begin{cases} 2^{p/2}, & \frac{q-1}{2^p} < x \leq \frac{q-\frac{1}{2}}{2^p} \\ -2^{p/2}, & \frac{q-\frac{1}{2}}{2^p} < x \leq \frac{q}{2^p} \\ 0, & \text{others} \end{cases} \tag{16.5}$$

where $0 \leq k \leq N-1$, $K = 2^p + q - 1$, $(p, q \in Z, q - 1 < 2^p)$, through which the base vectors of Haar transformation, i.e., each row of the kernel matrix, can be obtained, as shown in Eq. (16.6).

$$H_r = \frac{1}{\sqrt{8}} \begin{bmatrix} 1 & 1 & 1 & 1 & 1 & 1 & 1 & 1 \\ 1 & 1 & 1 & 1 & -1 & -1 & -1 & -1 \\ \sqrt{2} & \sqrt{2} & -\sqrt{2} & -\sqrt{2} & 0 & 0 & 0 & 0 \\ 0 & 0 & 0 & 0 & \sqrt{2} & \sqrt{2} & -\sqrt{2} & -\sqrt{2} \\ 2 & -2 & 0 & 0 & 0 & 0 & 0 & 0 \\ 0 & 0 & 2 & -2 & 0 & 0 & 0 & 0 \\ 0 & 0 & 0 & 0 & 2 & -2 & 0 & 0 \\ 0 & 0 & 0 & 0 & 0 & 0 & 2 & -2 \end{bmatrix} \tag{16.6}$$

Furthermore, we can obtain all the base images of the Haar transform ($N = 8$), as shown in Fig. 16.4. The base image is the row vector of the kernel matrix, i.e., the external product of the base functions. Thus, when representing image data, it is necessary to determine the base vector corresponding to the determination of such data. The minimal simple representation of the discrete image transformation based on the base function, base vector and base image can help realize lossless reduction along with superior data compression and storage. This approach provides a reasonable theoretical foundation for the compressed storage management of dynamic target space–time data and efficient visualization of data.

16.2 Case 2: Geometric Correction of Remote Sensing Images

Geometric correction of remote sensing images is a kind of overall image processing based on the grayscale histogram theory, point operation, algebraic operations, geometric computing and other related theories. Geometric correction is an important part of remote sensing information processing, which is directly related to the accuracy and practicality of the extracted information. From a physical viewpoint, image distortion refers to the misplacement of pixels, i.e., the pixel value belongs to a point that is located elsewhere. Therefore, through the image grayscale histogram calculation for geometric correction, image distortion can be corrected, and the image availability can be enhanced.

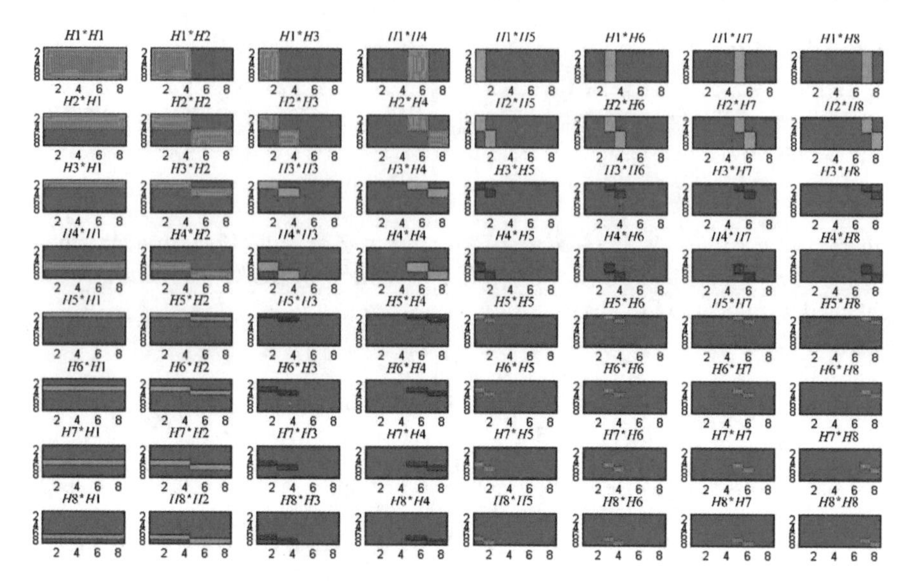

Fig. 16.4 Base images of Haar transformation ($N = 8$)

16.2.1 Causes of Geometric Distortion

The deformation error of remote sensing images can generally be divided into two kinds: internal error and external error. The internal error is mainly caused by the sensor performance, structure and other factors. The external error is caused by factors other than the sensor characteristics, such as the Earth's curvature and rotation and topographic relief.

1. Influence of the Imaging Geometry of Sensors

Figure 16.5 shows the deformation of the ground grid after imaging. This distortion is also known as panoramic distortion. f is the focal length, H is the camera height, and S is the lens. The imaging deformation of the slanting projection is shown in Fig. 16.6. The side-looking radar yields a slanting projection.

2. Influence of Distortion of Exterior Orientation Elements

Changes in the elements of the exterior orientation of the sensor correspond to the remote sensing platform attitude in the six degrees of freedom: three-axis directions (X, Y, Z) and three attitude angles (ϕ, ω, κ). Figure 16.7 shows the representation of the distortion caused by the six degrees of freedom on a ground-based grid image, obtained using a scanned image.

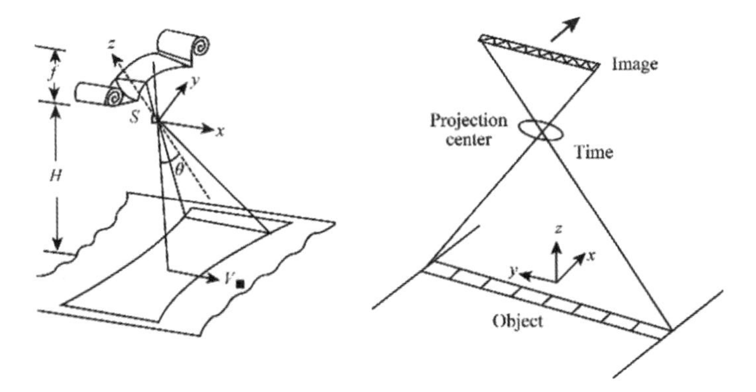

Fig. 16.5 Panoramic distortion

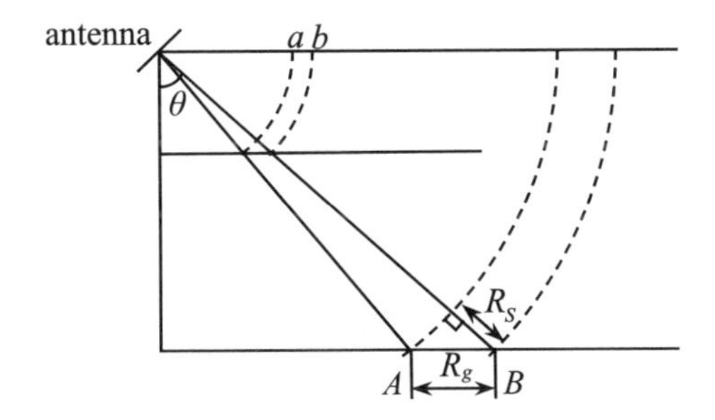

Fig. 16.6 Deformation of side-looking radar

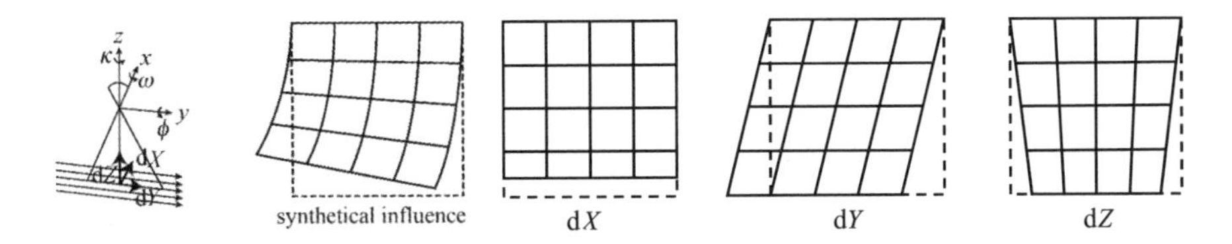

Fig. 16.7 Distortion caused by elements of exterior orientation

3. Influence of the Earth's Rotation

As shown in Fig. 16.8, the Earth's rotation does not influence the instantaneous optical imaging remote sensing. However, for the scanned image, the distortion is parallel to the direction, i.e.,

$$\Delta V_e = t_e v_\phi \tag{16.7}$$

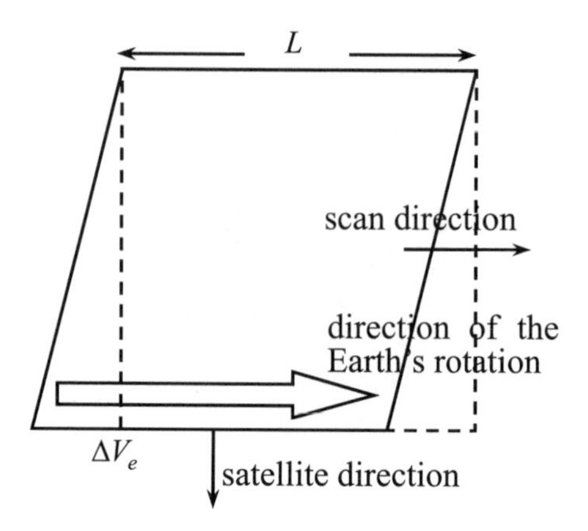

Fig. 16.8 Influence of the Earth's rotation

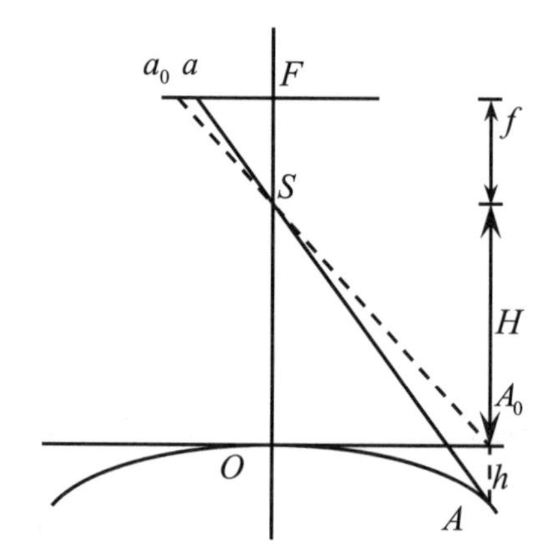

Fig. 16.9 Influence of the Earth's curvature

4. Influence of the Earth's Curvature

The effect of the Earth's curvature is illustrated in Fig. 16.9. The principle of error caused by the terrain relief is the same as that of the aerial images.

16.2.2 Related Theory of Geometric Correction

1. Convex Equation of Center Projection

The space coordinate relationship between image point a and object point A can be presented as collinear equations for central projection imaging.

$$
\begin{aligned}
x - x_0 &= -f\frac{a_1(X-X_s)+b_1(Y-Y_s)+c_1(Z-Z_s)}{a_3(X-X_s)+b_3(Y-Y_s)+c_3(Z-Z_s)} \\
y - y_0 &= -f\frac{a_2(X-X_s)+b_2(Y-Y_s)+c_2(Z-Z_s)}{a_3(X-X_s)+b_3(Y-Y_s)+c_3(Z-Z_s)}
\end{aligned}
\tag{16.8}
$$

In Eq. (16.9), x, y represent the coordinates in the coordinate system with the origin at the principal point, and x_0, y_0 is the deviation of the point coordinates from true values. f is the main focal length. X, Y, Z represent the coordinates of the corresponding point in the real world, while X_S, Y_S, Z_S represent the coordinates of the camera center. $a_i, b_i, c_i (i = 1, 2, 3)$ are direction cos values of the three angles pertaining to the exterior orientation ϕ, ω, κ.

$$
\begin{cases}
a_1 = \cos\phi\cos\kappa - \sin\phi\sin\omega\sin\kappa,\ b_1 = \cos\omega\sin\kappa \\
c_1 = \sin\phi\cos\kappa + \cos\phi\sin\omega\sin\kappa, \\
a_2 = -\cos\phi\sin\kappa - \sin\phi\sin\omega\cos\kappa,\ b_2 = \cos\omega\cos\kappa \\
c_2 = -\sin\phi\sin\kappa + \cos\phi\sin\omega\cos\kappa \\
a_3 = \sin\phi\cos\omega,\ b_3 = -\sin\omega,\ c_3 = \cos\phi\cos\omega
\end{cases}
\tag{16.9}
$$

2. Process of Geometric Correction

Two basic methods can be used to realize the geometric correction of remote sensing images. The conventional method of geometric correction, or analog geometric correction, is based on the principle of optical simulation and optical mechanical equipment. The digital geometry correction method, or digital differential correction, is based on the principle of digital processing and electronic computers. The latter method is a type of point element correction. The basic unit of digital correction is the pixel, represented as $g(x, y)$ in the image.

The process of digital correction is as follows:

(1) Establish the digital image of the photo, i.e., obtain the gray value of each pixel in an aerial photograph;
(2) Resolve the coordinate relationship between object points and corresponding map points;
(3) Photogrammetrically interpolate the gray value, i.e., resample the image gray value;
(4) Output the digital orthophoto map (DOM). Figure 16.10 shows the digital correction, with pixels of the original image $g(i, j)$ (i.e., each small cell in the left grid) considered the basic correction unit. An uncorrected remote sensing image is selected and geometrically corrected using the polynomial correction method, as shown in Fig. 16.11. In the process of geometric correction, the changes can be identified in terms of the the overall operation of the image or rearrangement of pixels according to certain geometric rules.

The histograms for the case before and after geometric correction are shown in Figs. 16.12 and 16.13, respectively.

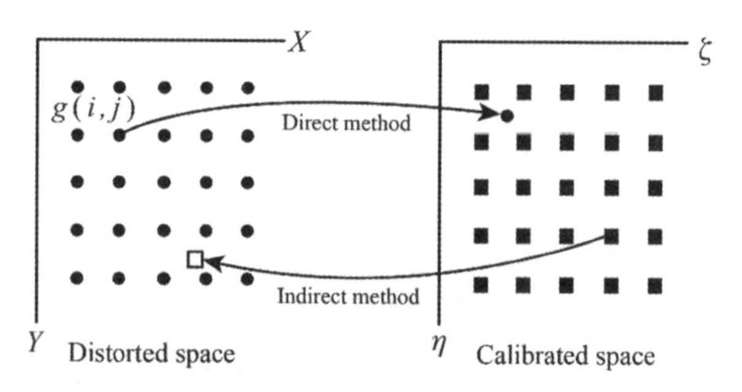

Fig. 16.10 Schematic of digital correction

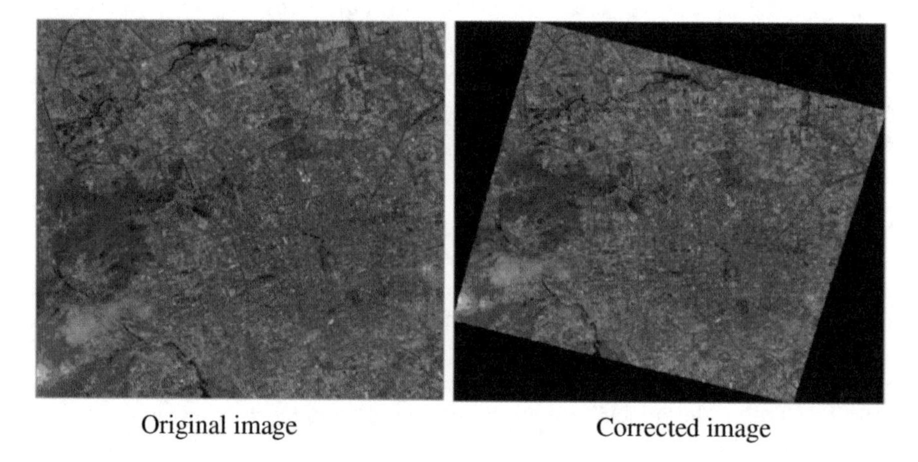

Fig. 16.11 Satellite image before and after geometric correction

Fig. 16.12 Histogram before geometric correction

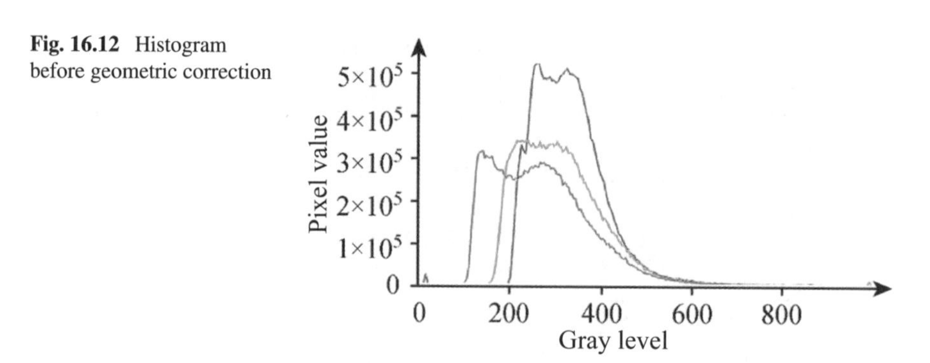

The change in the image histogram before and after correction is small, indicating consistency. Geometric correction generally does not change the pixel value of the original image and only rearranges the pixel values in accordance with certain conversion rules. In the process of resampling, the bilinear interpolation method or cubic convolution method can be used to equalize the pixel values to a certain extent,

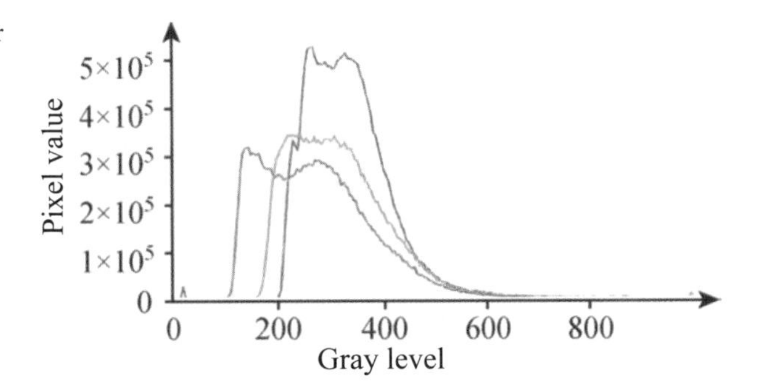

Fig. 16.13 Histogram after geometric correction (zero values excluded)

which makes the image smoother, leading to a small change in the corrected image histogram (as the histogram becomes smoother).

16.2.3 Key Algorithms for Geometric Correction

Due to the confidentiality of data parameters and other reasons, ground control points (GCPs) are usually used in practical applications for the geometric correction of remote sensing images. The commonly used algorithms are the general polynomial algorithms, collinear equation models, rational function models, direct linear transformation models, self-checking direct linear transformation models, extensible direct linear transformation models, and strict projection affine transformation models. The first two algorithms are introduced in Sect. 3.4.5, and the following text lists the core formulas of the geometric correction algorithms.

1. Rational Function Model

The rational function model is a rational function approximation of a 2-dimensional image plane and 3-dimensional spatial correspondence. The positive solution of the rational function is expressed as

$$\begin{cases} r_n = \dfrac{p_1(X_n, Y_n, Z_n)}{p_2(X_n, Y_n, Z_n)} \\ c_n = \dfrac{p_3(X_n, Y_n, Z_n)}{p_4(X_n, Y_n, Z_n)} \end{cases} \tag{16.10}$$

2. Linear Transformation Model

$$\begin{cases} x = \dfrac{L_1 X + L_2 Y + L_3 Z + L_4}{L_9 X + L_{10} Y + L_{11} Z + 1} \\ y = \dfrac{L_5 X + L_6 Y + L_7 Z + L_8}{L_9 X + L_{10} Y + L_{11} Z + 1} \end{cases} \tag{16.11}$$

3. Self-Checking Direct Linear Transformation Model

$$\begin{cases} x = \dfrac{L_1 X + L_2 Y + L_3 Z + L_4}{L_9 X + L_{10} Y + L_{11} Z + 1} \\ y = \dfrac{L_5 X + L_6 Y + L_7 Z + L_8}{L_9 X + L_{10} Y + L_{11} Z + 1} + L_{12}xy \end{cases} \tag{16.12}$$

4. Extensible Direct Linear Transformation Model

$$\begin{cases} x = \dfrac{L_1 X + L_2 Y + L_3 Z + L_4}{L_9 X + L_{10} Y + L_{11} Z + 1} + L_{12}x^2 \\ y = \dfrac{L_5 X + L_6 Y + L_7 Z + L_8}{L_9 X + L_{10} Y + L_{11} Z + 1} + L_{13}xy \end{cases} \tag{16.13}$$

5. Strict Projection Affine Transformation Model

$$\begin{cases} \dfrac{f - \frac{Z}{m \cos \alpha}}{f - (x - x_0) \tan \alpha}(x - x_0) = a_0 + a_1 X + a_2 Y + z_3 Z \\ y - y_0 = b_0 + b_1 X + b_2 Y + b_3 Z \end{cases} \tag{16.14}$$

16.2.4 Comparative Analysis of Geometric Correction Algorithms

The algorithms that can be used for geometric correction are summarized in Table 16.2.

16.3 Case 3: Threshold Segmentation of Remote Sensing Images

The threshold segmentation technique of remote sensing images refers to the application of gray histogram theory and fuzzy measurement theory in the general processing of remote sensing images. Threshold segmentation considers the difference between the objects to be extracted in the image and its background in terms of the grayscale characteristics. The image is considered a combination of two regions (target and background) with different gray levels. By selecting an appropriate threshold, each pixel can be classified to the target area or background area, resulting in the corresponding binary image.

Table 16.2 Performance comparison of geometric correction algorithms

Algorithm	Availability	Requirements	Complexity	Number of GCPs	Performance
Quadratic polynomial	Generic	\	10 × addition + 16 × multiplication	6	Simple; applicable for plain images
Cubic polynomial	Generic	\	18 × addition + 40 × multiplication	10	Relatively accurate but complex; applicable for plain images
Quadratic rational function	Generic	\	30 × addition + 53 × multiplication	16	Accurate and complex. Instability caused by the large number of parameters. Easily affected by the GCP distribution
Cubic rational function	Generic	\	57 × addition + 134 × multiplication	30	Accurate and highly complex. Instability caused by the large number of parameters. Easily affected by the GCP distribution
Collinear equation model	Space-borne linear array CCD	Flying height and focal length	20 × addition + 14 × multiplication	6	Relatively accurate. Can consider the optical geometry. Instability caused by the correlation of parameters
Direct linear transformation	Space-borne linear array CCD	\	9 × addition + 11 × multiplication	6	Relatively accurate, simple, and applicable for sensors with no specific information
Self-checking direct linear transformation	Space-borne linear array CCD	\	10 × addition + 11 × multiplication	6	Accurate, simple and stable. Changes in the exterior orientation are considered. Applicable for fine correction
Extensible direct linear transformation	Space-borne linear array CCD	\	11 × addition + 13 × multiplication	7	Accurate, simple, and applicable for sensors with no specific information

(continued)

Table 16.2 (continued)

Algorithm	Availability	Requirements	Complexity	Number of GCPs	Performance
Strict projection affine transformation	Space-borne linear array CCD	Flying height and focal length	11 × addition + 20 × multiplication	5	Accurate, relatively simple, and stable. Does not require prior information of sensors. Comprehensively fine algorithm

16.3.1 Objectives of Threshold Segmentation

An image mosaic synthesizes a series of images of the same scene that overlap with one other to create a large wide viewing angle image. The compositive image must be similar to the original image, with minimized distortion and no clear suture. Image mosaics have been widely used in space exploration, submarine surveys, medical surveys, meteorological surveys, geological surveys, military reconnaissance, video compression and transmission, and 3D reconstruction of objects. Many real aerial image have histograms that exhibit significant differences between the background brightness and target brightness, i.e., the histograms have a bimodal characteristic.

As shown in Figs. 16.14 and 16.15, for images with histograms having a significant bimodal characteristic, the threshold segmentation method can be used for binarization to extract the characteristics of the image, based on which, satisfactory results for pattern matching can be obtained. Generally, the threshold segmentation method is not time-consuming and can satisfy the requirements of real-time stitching of UAV images. Therefore, we adopt the adaptive threshold segmentation algorithm to implement the image mosaic.

Fig. 16.14 Demonstrative image

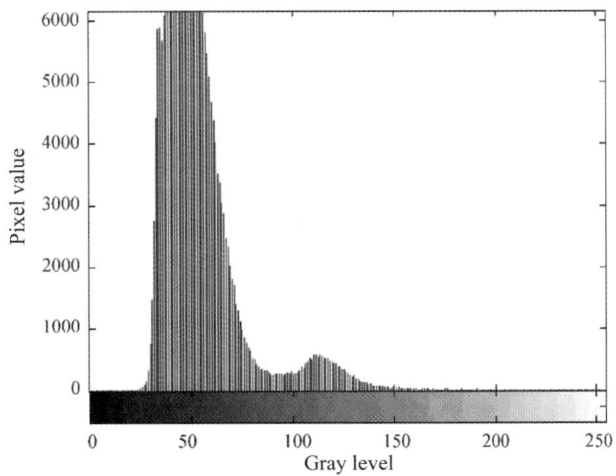

Fig. 16.15 Histogram of the image

16.3.2 Adaptive Threshold Segmentation Algorithm

In 1979, Otsu proposed a commonly used segmentation method, the adaptive threshold segmentation algorithm [1], which is based on the grayscale histogram of an image. The method is described in the following text.

Assume a two-dimensional matrix describing a grayscale image as $F_{P\times Q} = [f(x, y)]_{P\times Q}$, where $P \times Q$ is the image size, $f(x, y)$ is the gray value of pixel (x, y), and $f(x, y) \subset \{0, 1, ..., L-1\}$, L is the gross number of gray levels. n_i is the number of times gray value i occurs. The probability of i is

$$p_i = \frac{n_i}{P \times Q}, p_i \geq 0, \sum_{i=0}^{L-1} p_i = 1 \tag{16.15}$$

Each pixel in the image can be classified into two classes using t as the threshold: class S_1 pertains to the background, containing pixels with gray value $i \leq t$; and class S_2 refers to the foreground, containing pixels with gray value $i > t$. The probabilities of S_1 and S_2 are

$$P_1 = \sum_{i=0}^{t} p_i, P_2 = \sum_{i=t+1}^{L-1} p_i, P_1 + P_2 = 1 \tag{16.16}$$

The clustering centers of S_1 and S_2 can be defined as follows:

$$\omega_1 = \sum_{i=0}^{t} i p_i / P_1 \tag{16.17}$$

$$\omega_2 = \sum_{i=t+1}^{L-1} ip_i / P_2 \quad (16.18)$$

Moreover,

$$P_1\omega_1 + P_2\omega_2 = \omega_0 = \sum_{i=0}^{L-1} ip_i \quad (16.19)$$

Therefore, the variance between the background and foreground classes is

$$\sigma_2 = P_1(\omega_1 - \omega_0)^2 + P_2(\omega_2 - \omega_0)^2 = P_1P_2(\omega_2 - \omega_0)^2 \quad (16.20)$$

To achieve the best image segmentation effect, we must ensure the best classification effect to determine the optimal image segmentation threshold. Otsu used the variance between the two types of classes as a threshold recognition function and noted that the optimal threshold t^* must be the largest gray value that can maximize σ^2. Therefore, there exists an optimal threshold discriminant:

$$\sigma^2(t^*) = \max \sigma^2(t) \quad (16.21)$$

The final segmentation result based on the adaptive threshold segmentation algorithm is shown in Fig. 16.16. The matching obtained by the adaptive threshold segmentation algorithm is not ideal, and there exists a large misplacement phenomenon. The main reason is that the selected threshold discriminant function is relatively simple. Therefore, based on the previous algorithm, the threshold recognition function can be enhanced to achieve a superior segmentation effect [2, 3].

Fig. 16.16 Mosaic result

16.3.3 Improved Adaptive Threshold Segmentation Algorithm

In the adaptive threshold method, it is not adequate to consider only the variance between the foreground and background classes, as this framework ignores the classification information contained in the pixels in each class. To achieve the classification quality required to attain an accurate segmentation result, the two types of distances can be redefined and the concept of dispersion of classes can be introduced to enhance the threshold recognition function.

As mentioned, Eqs. (16.17)–(16.21) define the clustering centers of background class S_1 and foreground class S_2. We define the distance between two classes as

$$D = |\omega_1 - \omega_2| \tag{16.22}$$

To a certain extent, D can reflect the classification effect. A greater D corresponds to a greater distance between two classes and superior separation between S_1 and S_2. In addition, the cohesion of background class S_1 and foreground class S_2 is a direct and important indicator of the classification effectiveness. We define the distance from each pixel to the center of the class as the degree of dispersion, i.e.,

$$\begin{aligned} d_1 &= \sum_{i=0}^{t} |i - \omega_1| \cdot \frac{p_i}{P_1} \\ d_2 &= \sum_{i=t+1}^{L-1} |i - \omega_2| \cdot \frac{p_i}{P_2} \end{aligned} \tag{16.23}$$

A smaller dispersion of each class corresponds to a superior cohesion and classification effect.

Considering these two factors, to achieve the optimal sorting, we must ensure that D is maximized and d_1, d_2 are minimized. In this case, each type of cohesion is optimal, the distance between the two categories is maximized, and the best classification is achieved. The foreground and background in the image can be maximally separated to achieve the best image segmentation effect. The threshold recognition function for defining the background and foreground segmentation can be defined as follows:

$$H(t) = \frac{P_1 \cdot P_2 \cdot D}{P_1 d_1 + P_2 d_2} \tag{16.24}$$

In other words, the best classification is achieved when H is maximized, and the optimal threshold recognition function is

$$H\left(t^*\right) = H_{\max}(t) \tag{16.25}$$

If the gray level t^* of an image can ensure that $H(t^*) = H_{max}(t)$, t^* is the optimal segmentation threshold to classify image $F_{P\times Q} = [f(x, y)]_{P\times Q}$ into background class S_1 and foreground class S_2. In this case,

$$S_1 \cup S_2 = F_{P\times Q} \text{ and } S_1 \cap S_2 = \phi \tag{16.26}$$

According to the improved adaptive threshold selection method, different images have different matrix representations, each of which has a gray value corresponding to a threshold recognition function value H. We attempt to determine H_{max} from H to determine the optimal threshold t^* for segmentation. Figure 16.17 shows the process flow of the adaptive threshold image segmentation method.

The result of the image mosaic obtained using the improved adaptive threshold segmentation algorithm is shown in Fig. 16.18. From the image, we can intuitively infer that the improved adaptive threshold segmentation algorithm yields a satisfactory result.

16.4 Case 4: Minimum Noise Fraction (MNF) Rotation of Remote Sensing Images

The minimum noise separation transform is a representative method of remote sensing image lossy processing, which demonstrates how the principal component analysis (PCA) method can be used in image restoration. PCA is a widely used algorithm that involves a multidimensional orthogonal linear transformation based on statistical features. Through PCA transformation, the useful information in a multiband image can be concentrated into the smallest number of new principal component images. In this case, the principal component images are irrelevant, and the total amount of data is substantially decreased. However, the PCA transform is highly sensitive to noise, i.e., the principal component with a large proportion of information does not necessarily have a high signal-to-noise ratio. When the variance of the noise contained in the principal component pertaining to a large amount of information is larger than that of the signal, the principal component image quality is lower [4].

16.4.1 Fundamental Concepts

MNF rotation divides a remote sensing image into two parts: signal and noise. The spectral feature vector $\boldsymbol{Z}(x)$ is the linear combination of the real signal vector $\boldsymbol{S}(x)$ and noise signal $\boldsymbol{N}(x)$:

$$\boldsymbol{Z}(x) = \boldsymbol{S}(x) + \boldsymbol{N}(x) \tag{16.27}$$

$$\text{Cov}(\boldsymbol{Z}(x)) = \Sigma_S + \Sigma_N \tag{16.28}$$

Fig. 16.17 Process flow of the adaptive threshold image segmentation algorithm

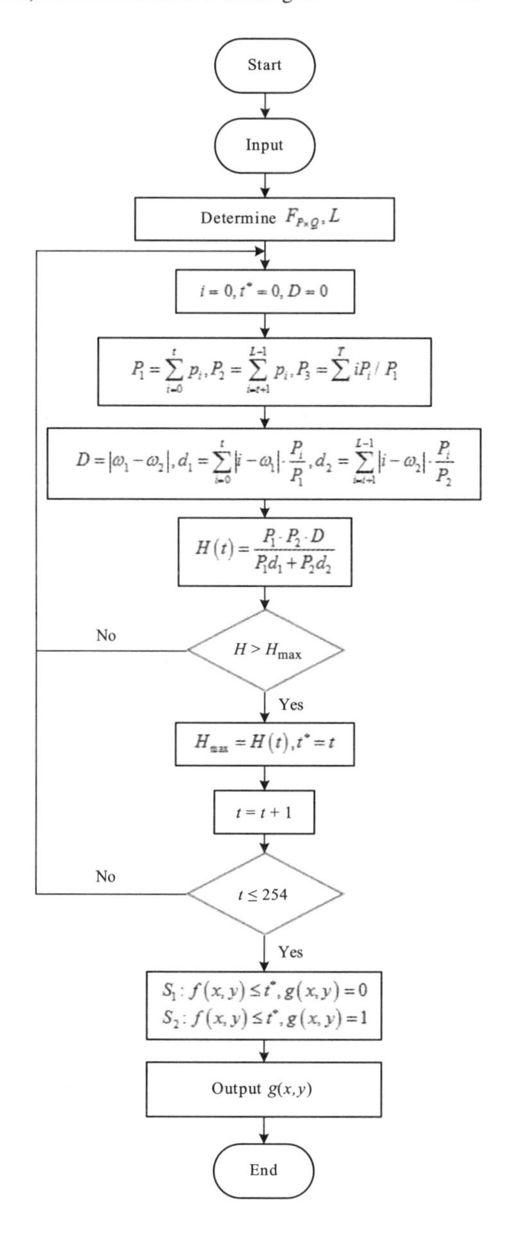

For the signal-to-noise combination in the form of multiplication, the logarithm of both sides of the equation must be taken to transform the combination to the addition form. The MNF transform matrix is $\boldsymbol{Y}(x)=\boldsymbol{A}^{\mathrm{T}}\boldsymbol{Z}(x)$. According to the principle of PCA, it is necessary to identify any two values of the covariances of signal Σ_S, noise Σ_N, and spectral characteristic vector Σ before the transformation. The remaining value can be identified through the linear operation. The central idea of the MNF

Fig. 16.18 Image obtained from mosaicing

method is to use the correlation between the neighborhood pixels in the remote sensing image and the small correlation between the noise in the neighborhood such that the noise can be obtained considering the covariance between the neighborhood pixels.

16.4.2 Noise Estimation by MNF Rotation

In the principal component analysis on which the MNF is based, the selected transformation matrix T exhibits an increasing spatial correlation, i.e., a greater principal component order corresponds to a greater correlation of the transform matrix. For the principal component, the latter component is mainly composed of noise information, which shows that the spatial correlation between the noise is small, and the noise covariance pertaining to the spatial relation is large.

The spatial correlation of the i-th component $\boldsymbol{Y}_i(x) = \boldsymbol{a}_i^{\mathrm{T}} \boldsymbol{Z}(x)$ can be expressed by the correlation of pixels in the neighborhood:

$$\operatorname{Corr}(\boldsymbol{Y}_i(x), \boldsymbol{Y}_i(x+V)) = \operatorname{Corr}\left(\boldsymbol{a}_i^T \boldsymbol{Z}(x), \boldsymbol{a}_i^T \boldsymbol{Z}(x+V)\right) \tag{16.29}$$

This correlation is orthogonal to the transformed component $\boldsymbol{Y}_k(x)(k < i)$ before i, and it is infinitesimal compared to $\boldsymbol{Y}_k$.

The data involved in the process usually pertain to a digital matrix. MAF transforms can usually be expressed by the covariance matrix of the data and difference between neighborhoods. The MAF transform refers to the maximum/minimum autocorrelation factor (MAF) transformation. The spatial characteristics of the remote sensing image are considered. The method estimates the covariance matrix of the image and calculates the difference covariance matrix corresponding to the original and offset data. In the matrix form, a_i is the feature vector of $\Sigma_\Delta \Sigma^{-1}$, where $\Sigma_\Delta = \mathrm{Cov}(\boldsymbol{Z}(x), \boldsymbol{Z}(x+\Delta))$, and Σ is the covariance matrix of the spectral image. Since $\boldsymbol{a}_i$ is also the feature vector of $\Sigma_N \Sigma^{-1}$ in the PCA, $\Sigma_N \Sigma^{-1}$ and $\Sigma_\Delta \Sigma^{-1}$ share the same form of expression. Thus, Σ_Δ is another measure of noise, which does not contain a strong autocorrelation signal [5].

Seitzer et al. introduced a simple proportional covariance model to illustrate the use of Σ_Δ to express the covariance matrix Σ_N. The model assumes that the signal and noise are irrelevant, and

$$\begin{aligned} \mathrm{Cov}\{\boldsymbol{S}(x), \boldsymbol{S}(x+\Delta)\} &= b_\Delta \Sigma_S \\ \mathrm{Cov}\{\boldsymbol{N}(x), \boldsymbol{N}(x+\Delta)\} &= c_\Delta \Sigma_N \end{aligned} \tag{16.30}$$

Considering the correlation in the neighborhood, b_Δ and c_Δ in Eq. (16.30) are constant, and b_Δ is larger than c_Δ. For the simple covariance model, the most notable conclusion is that the correlation coefficient between any band signal (and noise) neighborhood is the same, i.e., $b_\Delta(c_\Delta)$. According to the model,

$$\frac{1}{2}\Sigma_\Delta = (1 - b_\Delta)\Sigma + (b_\Delta - c_\Delta)\Sigma_N \tag{16.31}$$

If signals have a high spatial correlation, $\mathrm{Cov}\{\boldsymbol{S}(x), \boldsymbol{S}(x+\Delta)\}$ is expected to approach Σ_S, i.e., $b_\Delta \approx 1$, and pepper noise $\mathrm{Cov}\{\boldsymbol{N}(x), \boldsymbol{N}(x+\Delta)\}$ tends to 0, i.e., $c_\Delta \approx 1$. In this condition, $\Sigma_N = \Sigma_\Delta/2$. Usually,

(a) The feature vectors of $\Sigma_\Delta \Sigma^{-1}$ and $\Sigma_N \Sigma^{-1}$ are correlated, and the value is not related to b_Δ or c_Δ. Furthermore, the ratio of noise in the transformed principal component is not related to b_Δ or c_Δ.
(b) $\boldsymbol{\lambda}_i$ and $\boldsymbol{\mu}_i$, feature vectors of $\Sigma_\Delta \Sigma^{-1}$ and $\Sigma_N \Sigma^{-1}$, respectively, exhibit the following correlation:

$$\boldsymbol{\mu}_i = \frac{\lambda_i/2 - (1 - b_\Delta)}{b_\Delta - c_\Delta} \tag{16.32}$$

(c) Since $0 \le \lambda_i/2b_\Delta \le 1$,

$$c_\Delta \le 1 - \lambda_i/2 \le b_\Delta \tag{16.33}$$

Equation (16.33) holds for all bands. Therefore, according to the eigenvalues of the neighborhood correlation matrix, we can use the maximum and minimum eigenvalues to determine the upper and lower limits of c_Δ(lower limit is 0) and b_Δ (upper limit is 1), respectively. Therefore, we can use λ_i to estimate the values of b_Δ and c_Δ to determine the eigenvalues and eigenvectors of the noise. After obtaining the eigenvector of the noise covariance matrix and spectral covariance matrix, the eigenvalues can be reordered from large to small, and the transformation of the spectral image to the new principal component can be determined. After transformation, a noise-reduced (enhanced) image can be obtained by performing an inverse transform of the images from the first few bands.

16.4.3 Strip Removal in Hyperion Data

In this experiment, Hyperion data are used. Hyperion data have a large number of bands (70 bands) in the VNIR band and a high spectral resolution (10 nm). Therefore, this platform is suitable for monitoring vegetation characteristics. Moreover, the data have a sufficiently high spectral resolution (3 bands) in the wavelength range of 680–710 nm, and thus, Hyperion data are widely used in quantitative remote sensing.

Stripes exist in the Hyperion image in the direction parallel to the CCD array. As shown in Fig. 16.19, stripes exist on the image, and the stripes must be wiped before Hyperion data can be used. MNF rotation can be used to remove stripes. The principle is that each band of the Hyperion image shares the same variance and covariance, and the bands are strongly correlated. MNF rotation can be effectively used to remove the stripes. The right image in Fig. 16.20 shows part of a band image after the removal of strips. The stripe removal effect is significant. Figure 16.20 shows the contrast effect before and after stripe removal.

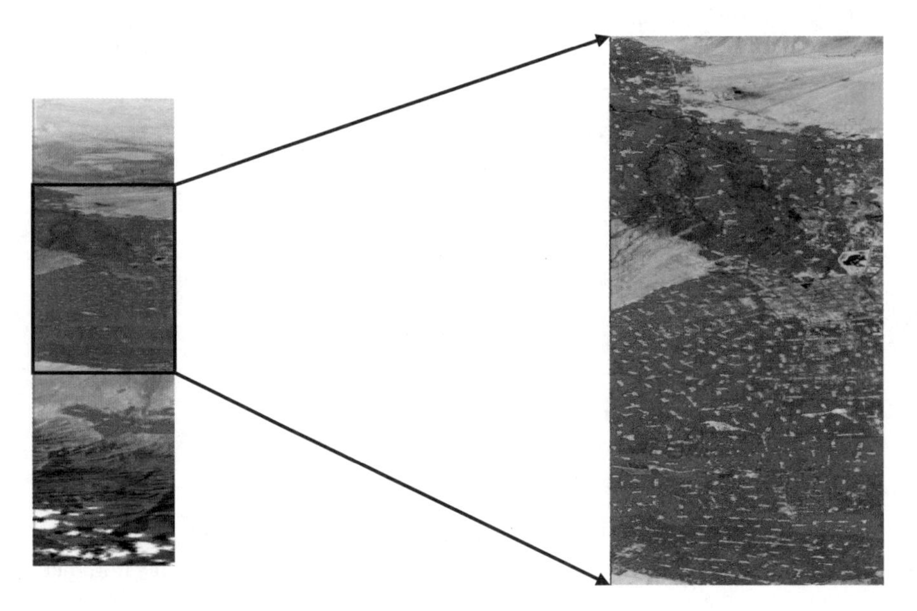

Fig. 16.19 Hyperion image used in the experiment

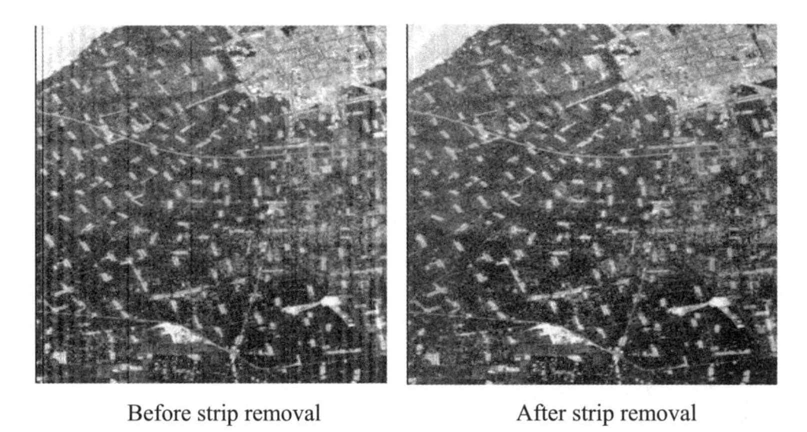

Fig. 16.20 Comparison of images before and after strip removal

16.4.4 Smile Effect

All the Hyperion products exhibit the Smile effect, which refers to the shift in the wavelength of the pixel from the center to both sides in the direction perpendicular to the flight direction. The Smile effect of an image can be detected by MNF rotation. After the image is transformed by MNF, if there exists a significant brightness gradient in the first or second band, the image exhibits a Smile effect. After MNF transforms are performed on the VNIR and SWIR bands, a significant luminance gradient can be observed on the VNIR band but not in the SWIR band, as shown in Fig. 16.21. Figure 16.22 shows the first band after MNF transformation in the VNIR band after stripe processing. No obvious gradient exists on the graph, and the Smile effect is alleviated.

16.5 Summary

Four types of methods for the application of remote sensing digital image processing are introduced: vector representation, geometric correction, threshold segmentation and MNF rotation of remote sensing images. In general, applications of remote sensing images can be divided into preprocessing and application. Geometric correction can be classified as a preprocessing step for remote sensing images, and MNF corresponds to the application of remote sensing images. However, all the abovementioned methods are supported by the basic theory of remote sensing digital image processing. The chapter is summarized in Fig. 16.23, and Table 16.3 lists the formulas in this chapter.

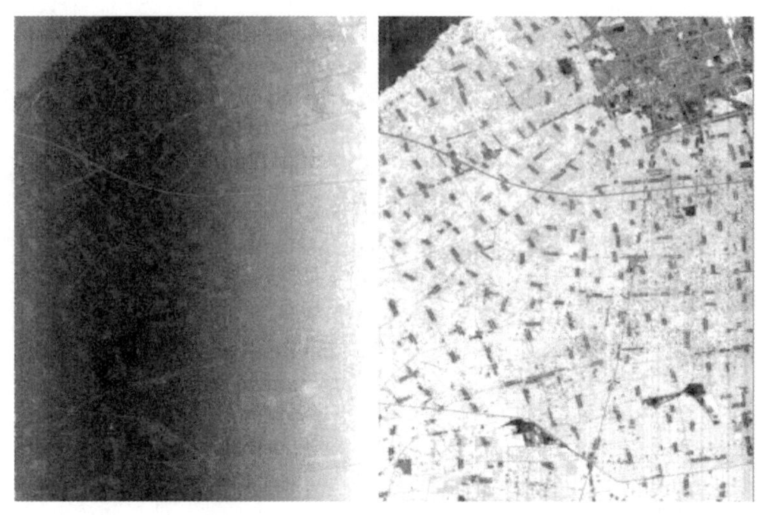

Fig. 16.21 Detection of smile effect

Fig. 16.22 First band of VNIR image with stripes removed by MNF

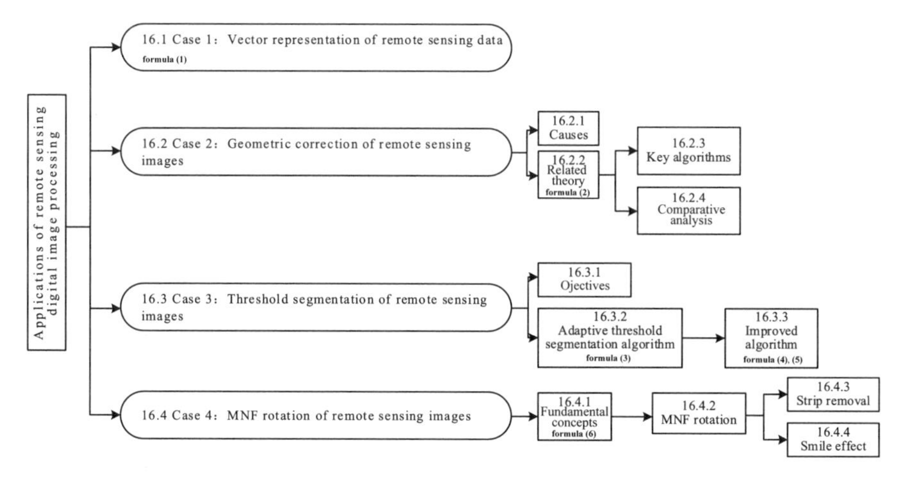

Fig. 16.23 Summary of this chapter

Table 16.3 Formulas in this chapter

formula (1)	$G_{m,n} = \sum_{i=0}^{N-1} \sum_{k=0}^{N-1} F_{i,k} \Im(i, k, m, n)$	16.1
formula (2)	$x - x_0 = -f \frac{a_1(X-X_s)+b_1(Y-Y_s)+c_1(Z-Z_s)}{a_3(X-X_s)+b_3(Y-Y_s)+c_3(Z-Z_s)}$ $y - y_0 = -f \frac{a_2(X-X_s)+b_2(Y-Y_s)+c_2(Z-Z_s)}{a_3(X-X_s)+b_3(Y-Y_s)+c_3(Z-Z_s)}$	16.2.2
formula (3)	$p_i = \frac{n_i}{P \times Q}$	16.3.2
formula (4)	$D = \lvert\omega_1 - \omega_2\rvert$	16.3.3
formula (5)	$d_1 = \sum_{i=0}^{t} \lvert i - \omega_1\rvert \cdot \frac{p_i}{P_1}$ $d_2 = \sum_{i=t+1}^{L-1} \lvert i - \omega_2\rvert \cdot \frac{p_i}{P_2}$	16.3.3
formula (6)	$\mathrm{Cov}(Z(x)) = \sum_S + \sum_N$	16.4.1

References

1. Xiao CY, Zhu WX. Image segmentation algorithm based on Otsu principle and entropy. Comput Eng. 2007;(33). (In Chinese)
2. Ning J, Zhang L, Zhang D, et al. Interactive image segmentation by maximal similarity based region merging. Pattern Recogn. 2010;43(2).
3. Weszka JS. A survey of threshold selection techniques. Comput Graph Image Proc. 1978.
4. Lee JB, Woodyatt AS, Berman M. Enhancement of high spectral resolution remote sensing data by a noise-adjusted principle component transform. IEEE Trans Geosci Remote Sens. 1990;28(3):295–304.
5. Nielsen A. Kernel maximum autocorrelation factor and minimum noise fraction transformations. IEEE Trans Image Proc Publ IEEE Sig Proc Soc. 2011;20(3).

Postscript

Over the past 20 years, the author deeply felt that remote sensing, as the source of spatial information science and technology, was a multidisciplinary and theoretical system that needed to be perfected. Mathematics and physics, with their rigorous logical reasoning interpretation and dimensional analysis in their theoretical systems, have provided the basis for the development of the most basic tools but also have only created the means by which the tools can be used. In 1951, Eric Temple Bell, a member of the American Academy of Sciences, noted in his classic book *Mathematics: Queen and Servant of Science* that mathematics is both a scientific queen and a servant of other disciplines. Furthermore, physics provides the most rigorous and concise description of the essential laws of subjects in different disciplines and promotes the establishment of a sound theoretical system by means of mathematical expressions. Therefore, mathematical expressions and physical analyses have been the only remote sensing spatial information science and technology theoretical systems viewpoints examined, which has also been the conceptual approach used by the author, colleagues and students in teaching and research.

Remote sensing digital image processing plays an important role in spatial information image analysis and application. Most of the spatial information obtained by remote sensing is displayed through remote sensing images, so appropriate methods are used to deal with the acquired remote sensing images to maximize the useful information. Remote sensing spatial information technology workers must master a quantitative method. "Remote Sensing Digital Image Processing' has always been a compulsory course in higher education related to remote sensing spatial information, and it is an effective means to promote the development of the theoretical systems of remote sensing spatial information science by revealing the mathematical physics of the remote sensing image processing methods. The author has been engaged in teaching the course "Advanced Remote Sensing Digital Image Processing" for many years. In his teaching process, he has guided students to understand the nature of remote sensing image processing from the perspective of mathematical abstraction and physical dimensions and has accumulated a significant amount of comprehensive experience and material. Therefore, given the fierce competition, the author was fortunate to be awarded the Beijing Graduate Fine Textbook project funding. The most important features of this book include the explanation of processing methods for remote sensing images, the revealing of mathematical physics in remote sensing image processing methods, the variety of different processing methods, the mathematical correlation process, and physical dimension conversion to promote the development of remote sensing space information theoretical systems.

The compilation of this book has taken 19 years (including 4 postgraduate courses), and it has been revised 23 times. Several graduate students provided their full support, and many of them made valuable contributions to the book. I express to them my sincere thanks!

In this tutorial, the complement of mathematical formulas and the proposals in the mathematical and technical content was mainly composed by the author. Hongying Zhao (the second author) is a postdoctoral teacher at the school, Yi Lin (third author) and Yanbiao Sun (fourth author) are two of the author's doctoral graduate students who have been graduated and who devoted much effort and energy during their PhD study. They made systematic and other related contributions to this book, in which Hongying Zhao mainly wrote the second and sixteenth chapters, Yi Lin mainly wrote the third, seventh and eleventh chapters, and Yanbiao Sun mainly wrote the fourth and fourteenth chapters. The author primarily completed the rest of the chapters.

Limited by the author's knowledge, the book inevitably has some mistakes; therefore, the author hopes the readers will correct them. I hope information experts, scholars, and young students in the field of remote sensing and space will read this reference book and improve it. I am particularly grateful to Peking University Press. I hope this study will fill the gaps in the extant studies of this guiding ideology. Thanks to the editorial board and to science editors Xiaohong Chen and Jianfei Wang for five years of support and for helping me finish this book. Finally, thanks to the Beijing Institute for Collaborative Innovation (BICI), as some of the findings in the book are produced with their full support. Finally, I thank Peking University Press, which provided significant support and editing assistance for the color printing in the book.

Acknowledgments

Here, I would like to express my deepest gratitude to the 400 outstanding students (333 Ph.D. students, the master graduate students recorded in the following list, and the more than 60 students not recorded) of Peking University with whom I have interacted over 19 years. Without you, this book would not have been created. The sum total of your in-depth discussions are contained in the mathematical and physical connotations of this book. I will remember your names and give thanks to you forever. In addition, the 1997–1998 graduate student class, three postgraduate classes and visiting scholars (e.g., Kangnian Zhao, Suoli Guo, and Xiaohong Xiao), and approximately more than 60 postgraduates are some of those not recorded in the list because the time period has been so long, but their dedication will always be remembered in my heart.

1999 session: (14) WenJiang Zhang, Yuanfeng Su, Chengyuan Wang, Guiyun Zhou, Ting Ma, Yuling Liu, Tao Jing, Shengchao Yu, Wenyong Shi, Jing Dou, Lei Mao, Hao Wang, Jiayuan Lin, Xiaoqun Wang.

2000 session: (26) Lei Chang, Yongming Du, Qiang Ge, Yingduan Huang, Haitao Liu, Fenqin Wang, Huan Wu, Hui Xia (dedicator), Dehai Yang, Kaixin Yang, Jianwei Zhao, Wei Zhao, Gaolong Zhu, Lijiang Zhu, Shu Liu, Lingyun Wang, Jinde Zhang, Qun Zhao, Huanjie Zhang, Jingzhu Chen, Fuhao Zhang, Lianxi Wang, Sen Du, Hu Zhao, Daohu Sun, Bin Zhong.

2001 session: (20) Xueping Liu, Sijin Chen, Bin Li, Jizhi Li, Ruihong Liu, Lei Ma, Tingting Mao, Zhiyou Sha, Hongyan Wang (dedicator), Lili Wei, Xiaobo Xiao, Binen Xu, Ming Yan, Xuesong Zhang, Xiaosong Zheng, Longwen Zhu, Jianxun Xia, Yuliang Gan, Zhiyong Fan, Zhiming Gui.

2001 session of on-the-job master: (15) Xiaoying Wu, Tongpao Hou, Lili Jiang, Bin Li, Peng Lin, Yibing Liu, Jinhua Pan, Zhansheng Su, Yusheng Xu, Liqiong Yu, Pubin Zhang, Kefa Zhou (dedicator), Rongjun Zhou, Junke Xu, Hong Yan.

2002 session: (33) Hong Li, Lili Miao, Yinghai Ke, Yunhai Zhang, Quanzhou Bian, Huawei He, Zhe Li, Bin Li, Xiaosong Zheng, Xinpo Li, Daojun Wang, Ning Wu, Zhiyou Sha, Haitao Li, Bo Zhou, Jianxin Wu, Feng Li, Shunyi Wang, Xin Li, Longjiang Du, Bing Hu, Hua Zhang, Yonghong Yi, Zhanli Wang, Qiang

Zhou, Kaicun Wang (dedicator), Jiyong Liang, Dondong Wang, Tingting Mao, Ping Dong, Hui Wen, Shuqiang Lv, Jianghua Zheng.

2003 session: (9) Tao Cheng, Wei Deng, Liang Gao, Hongtao Hu, Xiaofan Li, Yihua Liu, Li Luo, Jianhong Ou, Wenbai Yang (dedicator).

2004 session: (16) Lina He, Yizhen Qu, Xuanyao Long, Lin Si, Xianwei Meng, Wei He, Yangbo Zhu, Ting Zhang, Jun Li, Guangyong Yang, Zhaoyong Zhang (dedicator), Ping Chen, Liang Gao, Yingchun Tao, Wenliang Jiang, Pengqi Gao.

2005 session: (21) Jie Ding, Yi Luo, Dan Yin, Shujie Mao, Weichao Shao, Lei Duan, Ying Li, Chaofeng Han, Xin Jin, Jiancong Guo, Shihu Zhao, Baochang Gong, Hanyu Xiang, Yi Lin, Shilei Hao, Zheng Huang, Defeng Ma (dedicator), Miao Jiang, Liang Yang, Kai Qin, Duo Gao.

2006 session: (19) Shukun Li, Huiran Jin, Xin Tao, Binbin Lu, Zheng Pan, Qingde Yu, Tiantian Shen, Jiayan Hu, Xingwang Ou, Shujing Song, Chenbin Lin, Taixia Wu, Zhiyang Gou, Ran Wei, Yun Xiang, Piyuan Yi, Maodi Su, Xiaogang Han, Jianying Jia (dedicator).

2007 session: (13) Jia Chen (dedicator), Tianxing Chu, Bo Li, Ying Li, Liang Zhao, Xinwen Ning, Yang shen, Anning Tang, Hongzhao Tang, Daihui Wu, Haikuo Yu, Ting Zhang, Guangyong Yang.

2008 session: (19) Dapeng Li, Yaokui Cui, Fang Dong, Songlin Li, Cunren Liang, Yunfei Lu, Benqin Song (dedicator), Quan Sun, Yiting Wang, Rui Wu, Haiqing Xu, Binyan Yan, Zhongyi Yin, Huabo Sun, Yan Ma, Xiai Cui, Meisun Chen, Heshun Wang, Wei Chen.

2009 session: (22) Ding Wu, Shasha Jiang, Peng Cheng, Ying Mai, Yingying Gai, Xiya Zhang, Jinliang Wang, Suihua Liu, Yubin Xu, Xiao Sun, Xiao Zhou, Yini Duan, Xuyang Wang, Jun Li, Jiazhou Geng, Jundong Zhang, Lanlan Huang, Huaihong Su, Xin Jing, Leilei Chai, Jiepeng Zhao (dedicator), Jing Li.

2010 session: (20) Yanbiao Sun, Ye Yun, Sheng Gao, Huiyi Bao, Cai cai, Lianjun Ding, Shiyue Fan, Yang Feng, Long Jiao, Chunhua Liao (dedicator), Xiangfeng Liu, Yuan Liu, Mengjia Luo, Yuzhong Ma, Yuan Tian, Lu Wang, Xueyang Wang, Zemin Yang, Jiahuan Wan, Mingzhi Wang.

2011 session: (17) Yaping Cai, Gaoxing Chen, Qiaohua Deng, Huaiyu Li, Yanran Liao, Jing Liu, Sijie Liu, Yu Liu, Boren Luo, Huiling Wang, Siting Xiong (dedicator), Xiaojie Zhang, Han Zhu, Wenhuan Wu, Hong Xu, Han Xiao, Yulong Guo.

2012 session: (14) Yuzhen Cai (dedicator), Yanfeng Cao, Fulin Chu, Chao Guo, Hanxian He, Junxiao Hu, Xin Li, Huili Liu, Yang Lv, Lu Wang, Xue Wang, Tiantian Xin, Haizhen Zhang, Bin Yang.

2013 session: (19) Shuling Pan (dedicator), Peng Yang, Zhenyu Yang, Jun Du, Junsong Huang, Yan Chen, Ruiyan Fan, Jiannan Jiao, Sa Liu, Tiantian Liu, Jinjie Meng, Xuefeng Peng, Junfeng Rao, Lin Shen, Yunpeng Wei, Hongfeng Yu, Beitong Zhang, Xinlong Zhang, Xiang Tan.

2014 session: (19) Bohao Ban, Yuan Wang (dedicator), Xinyi Wang, Zhongkui Shi, Rui Chen, Haoran Zhou, Baohui Chai, Jiwei Chen, Liqin He, Jianghui Huang,

Lulu Jiang, Jvcai Li, Shaohong Tian, Huanhuan Wang, Xueqi Wang, Weihui Ye, Peng Zhao, Hongyun Zheng, Guixiang Zou.

2015 session: (17) Shuaiyang Zhao, Yi Sun, Xiangshuang Meng, Jie Wan, Zezhong Wang (dedicator), Renqiang Gao, Kaili Han, Xu Jin, Zibo Ke, Dian Liu, Junru Liu, Maolin Liu, Wenting Ma, Dingfang Tian, Yizhen Yan, Mengdan Zhang, Miao Zheng.